The first GaAs/AlGaAs diode laser to demonstrate cw operation at room temperature in June 1970 at AT&T Bell Laboratories, Murray Hill, New Jersey and reported by I. Hayashi, M. B. Panish, P. W. Foy, and S. Sumski. The laser diode, grown by liquid phase epitaxy, is 400 μm ×100 μm in size and is mounted on a heat sink (courtesy of I. Hayashi, Optoelectronics Technology Research Laboratory, Ibaraki, Japan).

A monolithically integrated four-channel transimpedance photoreceiver array made with metal-semiconductor-metal photodiodes and metal-semiconductor FTSs. The GaAs wafer was grown by metalorganic vapor phase epitaxy (courtesy of O. Wada, Fujitsu Limited, Atsugi, Japan).

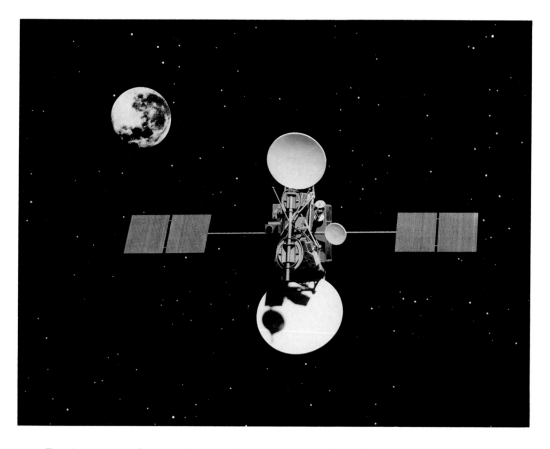

The Advanced Communication Technology satellite with arrays of Si solar cells which can generate almost 2KW of power (courtesy of I. Weingerg, NASA Lewis Research Center, Cleveland, Ohio).

Semiconductor Optoelectronic Devices

Semiconductor Optoelectronic Devices

PALLAB BHATTACHARYA

*Department of Electrical Engineering
and Computer Science
University of Michigan, Ann Arbor*

Prentice-Hall International, Inc.

© 1994 by Prentice-Hall, Inc.
A Paramount Communications Company
Englewood Cliffs, New Jersey 07632

Printed in the United States of America

10   9   8   7   6   5   4   3   2   1

ISBN 0-13-489766-8

PRENTICE-HALL INTERNATIONAL (UK) LIMITED, *London*
PRENTICE-HALL OF AUSTRALIA PTY. LIMITED, *Sydney*
PRENTICE-HALL CANADA, INC., *Toronto*
PRENTICE-HALL HISPANOAMERICANA, S.A., *Mexico*
PRENTICE-HALL OF INDIA PRIVATE LIMITED, *New Delhi*
PRENTICE-HALL OF JAPAN, INC., *Tokyo*
SIMON & SCHUSTER ASIA PTE. LTD., *Singapore*
EDITORA PRENTICE-HALL DO BRASIL, LTDA., *Rio de Janeiro*
PRENTICE-HALL, INC., ENGLEWOOD CLIFFS, *New Jersey*

*Dedicated to the
cherished memory
of my father.*

Preface

SCOPE

Since the demonstration of the first light-emitting diode (LED) and junction laser, optoelectronics has made remarkable progress. The developments in this field have been driven by the needs of lightwave communication systems, alternate energy sources, and optical or optoelectronic counterparts of electronic switching and logic elements. Optical devices and components and optical fibers are selectively replacing electronic devices and circuits, offering unique advantages. In fact, optoelectronic devices and circuits have unobtrusively and efficiently made their way into our daily lives. In the light of this enormous progress in the field, it is hoped that this book will reflect these dramatic changes in the field and serve two main purposes: (1) to formally introduce senior-level undergraduate and graduate students to optoelectronics, thereby helping them to guide their studies and career developments, and (2) to provide, in an accessible textbook format, a good and well-focused reference/tutorial book for practicing engineers and physicists.

PRESENTATION

The text has been developed at two levels, to benefit both the seniors and graduate students. The book is intended to be self-sufficient and extensive reference work should not be necessary. A background of a first course in semiconductors is assumed.

The first four chapters lay the foundations for the optoelectronic devices. The first chapter describes compound semiconductor materials and their epitaxy. Much of the present-day device concepts would not be realized without sophisticated and matured epitaxial techniques. Semiconductor statistics and carrier transport properties are described in Chapter 2. The basic optical processes of absorption and recombination in bulk and quantum well structures are analyzed and described in Chapter 3. Here, detailed quantum mechanical calculations are excluded, since these are found in at least half-a-dozen texts. However, appropriate references are provided as footnotes. Chapter 4 describes junction theory, including metal-semiconductor junctions and heterojunctions. The case of high-level injection, which becomes important for laser operation, is emphasized in this chapter.

The devices themselves are described in Chapters 5–11 in the following order: light-emitting diodes, lasers, photodetectors, solar cells, and light-modulators. Lasers and photodetectors, which are perhaps the more important and common optoelectronic devices, are each described in two chapters.

The principles of the devices are presented with appropriate analyses and derivations. Measurement techniques and recent experimental results are also included to give the reader a feel for real parameter values.

The organization of the device chapters should provide the instructor the flexibility to present material to both undergraduate and graduate students. In discussing the different devices, I have introduced new concepts, within the scope of the text. For example, pseudomorphic materials, quantum wells, distributed-feedback and surface-emitting lasers, modulated barrier photodiodes, coherent and wavelength selective detection, and quantum well modulation devices are all described and analyzed in various levels of detail. Finally, the text is concluded in Chapter 12 with a review of optoelectronic integrated circuits (OEICs), an emerging and important subject.

HOW TO USE THIS BOOK

This book is flexible and can be adapted to your local curriculum and course needs. An example of a one-term senior undergraduate course that could be taught from the book would cover chapters 1, 2, and 4 as review; chapters 3, 5, 6, part of 7, 8, and part of 11 should be treated as essential; and chapter 10 could be treated as optional. The rest of chapters 7 and 11, and chapters 9 and 12 I consider advanced material that you can teach in a graduate level course or material that you can choose from selectively to tailor the course to your desired emphasis and objectives.

READING LIST AND PROBLEMS

Suggested texts for more extensive reading and key articles from journals and periodicals are listed at the end of each chapter and as footnotes. These will help the more inquisitive students to go beyond the confines of the text and course and to enhance their knowledge and understanding. Also included are a set of problems at the end of each chapter, in addition to a few worked-out examples. The purpose of these problems is twofold: (1) to enhance the understanding of the different devices and underlying concepts and (2) to get a feel for practical values of different device and material parameters and their units.

UNITS

The rationalized MKS system of units has been mostly followed, with convenient changes. For example cm is more often used as the unit of length, and the electron volt (eV) is used in place of joule (J) as the unit of energy. The cgs system is sometimes used to keep in line with common use.

ACKNOWLEDGMENTS

It is a pleasure to acknowledge all those who have provided help, inspiration, and encouragement as I proceeded with this arduous task. I would like to thank Professor Ben Streetman for providing invaluable suggestions and warning me of the many pitfalls before I got started. Discussions with Dr. Niloy Dutta are gratefully acknowledged. I am indebted to my colleagues, Professor George Haddad for his encouragement and Professor Jasprit Singh for his many valuable comments and suggestions. The text has greatly benefited from the help and comments of many of my present and former graduate students. Special thanks are due to Larry Davis, Augusto Gutierrez-Aitken, Yeeloy Lam, Sanjay Sethi, and Hsiang-Chi Sun, and Drs. Yaochung Chen, Subrata Goswami, Shantanu Gupta, Weiqi Li, and Doyle Nichols. I would like to thank my professional colleagues who have generously contributed data and photographs. They are J. Goldman (RIBER SA) and Drs. G. A. Antypas (Crystacomm Inc.), S. N. G. Chu (AT&T Bell Laboratories), I. Hayashi (Optoelectronics Technology Research Laboratory), J. L. Jewell (Photonics Research Incorporated), I. Loehr (Wright Laboratories, WPAFB), D. A. B. Miller (AT&T Bell Laboratories), D. Pooladdej (Laser Diode, Inc.), R. Sahai (Rockwell International), J. Singh (University of Michigan), S. Swirhun (Bandgap Technology Corporation), W. T. Tsang (AT&T Bell Laboratories), O. Wada (Fujitsu Limited), I. Weinberg (NASA, Lewis Research Center), and E. Woelk (AIXTRON GmbH). This book would not have seen the light of day without the hard work of three persons: Christina Baydl, who transformed my hieroglyphics into a readable form, and Mary Ann Pruder and Mina Hale, who did all the artwork. The help of Alan Apt and Mona Pompili is greatly appreciated. I will remain indebted to all of you.

Last, and by no means the least, I am indebted to my mother for her support and to my family for their understanding, support, and willingness to sacrifice many evenings, weekends, and holidays.

Pallab Bhattacharya
Ann Arbor, Michigan

Contents

Semiconductor Optoelectronic Devices

1

Elemental and Compound Semiconductors

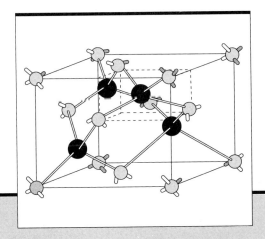

Chapter Contents

1.1 Introduction

1.2 Bonding in Solids

1.3 Crystalline Nature of Solids

1.4 Alloy Semiconductors

1.5 Lattice-Mismatched and Pseudomorphic Materials

1.6 Transmission Media and Choice of Materials

1.7 Crystal Growth

1.8 Device Processing

1.1 INTRODUCTION

Optoelectronics deals with the interaction of electronic processes with light and optical processes. Devices in which such interaction can suitably take place, usually accompanied by an energy conversion process (e.g., from electrical to optical, and vice versa), are called optoelectronic devices. Such devices are conveniently made with semiconductors, and this text is devoted to the understanding of the principles of semiconductor optoelectronic devices and the properties of materials with which they can be made. Research and development of optoelectronic devices and optoelectronic integrated circuits have received a tremendous boost with the development of low-loss optical fibers for long distance communication. These devices and circuits now play an important role in our daily lives.

Although the elemental semiconductors, and in particular Si, have been very useful for the development of microelectronics, they have some important drawbacks. The fundamental bandgap of these semiconductors is indirect. This implies that they emit light very poorly and their absorption coefficients are low. As a solar energy converter Si is technologically good, but because of its small energy gap the conversion efficiency is low. It became clear that Si, considered by many as a universal semiconductor material, cannot perform many important functions. For optoelectronic applications, in particular, it was natural then to turn to other materials. It turned out that compound semiconductor materials offered many of the desired properties and could be synthesized without much difficulty. Compound semiconductors, as the name suggests, are made from elements of different columns of the periodic table. Examples are III–V, II–VI, IV–VI, or IV–IV compounds. Historically, for optoelectronic device applications, the III–V compounds have been the first and most widely used. For the purpose of this text we will be mostly involved with III–V compounds, and in particular the GaAs and InP-based ones. The energy band diagrams of common elemental and compound semiconductors are shown in Fig. 1.1. It may be noted that a compound semiconductor, represented by $A_{III}B_V$ or $C_{II}D_{VI}$, has the same average number of valence electrons per atom as Si. In other words Si has four valence electrons and in III–V or II–VI compounds, the sum of outer electrons is eight.

Indium antimonide (InSb) was the first III–V compound semiconductor to be discovered in 1950. The particular features of this compound that attracted interest were the ease with which it could be synthesized, the electron mobility, and the ionic component in the crystal binding. These properties are still of immense interest, and as advanced epitaxial techniques are being developed, the purity of the crystals continue to improve. Furthermore, because of its low bandgap, $\mathcal{E}_g = 0.17$ eV, InSb has become important for the development of far infrared detector technology. Two important events, the invention of the semiconductor laser and the discovery of the Gunn effect turned the interest to other III–V compounds such as GaAs ($\mathcal{E}_g = 1.43$ eV) and InP ($\mathcal{E}_g = 1.35$ eV). The next in order of importance is GaP, which has its bandgap (2.1 eV) in the visible part of the spectrum and therefore became important for the development of the light-emitting diode (LED). It may be noted that the bandgap of GaP is indirect, but by certain doping techniques, which we shall learn about in

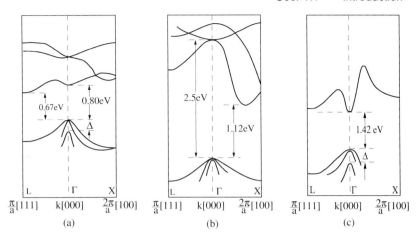

Figure 1.1   Diagrams showing the variations of electron energy with wave number (momentum) in (a) Ge, (b) Si, and (c) GaAs along the [100] and [111] directions in k space. Electrons are located near the minimum of the conduction band, whereas holes are located near the maximum of the valence band. The band structures of Ge and Si are examples of indirect-gap semiconductors, whereas that of GaAs represents a direct bandgap semiconductor. Δ is the spin-orbit splitting (from S. Wang, *Fundamentals of Semiconductor Theory and Device Physics*, Prentice Hall, Englewood Cliffs, NJ, 1989).

Chapter 5, it is possible to improve the radiative efficiency. Compared to Si and Ge, GaAs and InP have high electron mobilities and velocities, properties that are extremely important for the development of high-speed electronic devices. Their direct bandgaps and the consequent high radiative efficiency make them important optoelectronic materials. The bandgaps and lattice constants of common III–V binary compounds are depicted in Fig. 1.2.

An attractive feature of the binary compounds are that they can be combined or *alloyed* to form *ternary* or *quaternary* compounds, or *mixed crystals*. These compounds are made up of three or four group III and group V atoms and are indicated by the tie lines between the binary compounds in Fig. 1.2. Note that by choosing different binary compounds, it is possible to select different bandgaps, and therefore varying emission energies for light sources. However, by alloying it is possible to vary the bandgap *continuously* and monotonically, and together with it the bandstructure, electronic, and optical properties. The formation of ternary and quaternary compounds of varying bandgaps also enables the formation of heterojunctions, which have become essential for the design of high-performance electronic and optoelectronic devices. As an example, the bandgap of the ternary compound $Al_x Ga_{1-x} As$ ($0 \leq x \leq 1$) depends on the mole fraction x of AlAs in the solid solution and changes continuously from 1.43 eV (GaAs, $x = 0$) to 2.1 eV (AlAs, $x = 1$). As we shall see later, the bandstructure, electronic, and optical properties of the mixed crystal also change with change in alloy composition, and these are exploited in the design of electronic and optoelectronic devices. Among the common GaAs and InP-based ternary

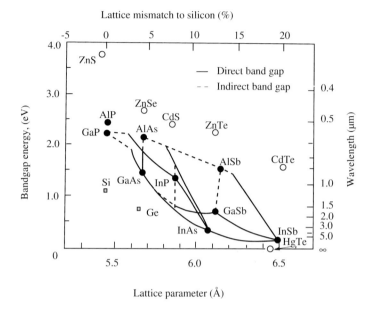

Figure 1.2 Energy bandgap versus lattice constant for common elemental and compound semiconductors. The tie lines joining the binaries represent ternary compositions. The dashed lines represent indirect bandgap material. The vertical dashed line passing through the point representing InP contains the bandgaps for the lattice-matched InGaAlAs and InGaAsP quaternary systems.

and quaternary compounds, the properties of $Al_xGa_{1-x}As$ have been most thoroughly investigated. Other important ternary and quaternary compounds are $GaAs_{1-x}P_x$, $In_{1-x}Ga_xP$, $In_xGa_yAl_{1-x-y}As$, and $In_{1-x}Ga_xAs_yP_{1-y}$. The last two are usually grown on InP substrates. The bandgaps and lattice constants of these compounds can be found from the tie lines in Fig. 1.2. The quaternary compounds mentioned above have emerged as being extremely important for optical communication, since their bandgaps correspond to the spectral window in which silica fibers have their lowest loss and dispersion. Several important observations may be made from Fig. 1.2. First, it may be noted that the lattice constants of GaAs (5.6532 Å) and AlAs (5.6611 Å) are almost identical. This implies that all the mixed crystal compositions of $Al_xGa_{1-x}As$ can be grown *lattice-matched* on GaAs substrates. We shall soon see in Sec. 1.5 that this is very useful because any thickness of the ternary crystal can be grown without having to worry about strain effects or the generation of dislocations. Similarly certain mixed crystal compositions of the quaternary alloys InGaAlAs and InGaAsP are lattice-matched to InP. These range from the end-point compositions $In_{0.53}Ga_{0.47}As$ ($\mathcal{E}_g = 0.74$ eV) to $In_{0.52}Al_{0.48}As$ ($\mathcal{E}_g = 1.45$ eV) and $In_{0.53}Ga_{0.47}As$ to InP (1.35 eV), respectively, for the two quaternary systems. Note that the only ternary $In_xGa_{1-x}As$ composition lattice-matched to InP is with $x = 0.53$. All other compositions from $x = 0$ (GaAs) to $x = 1$ (InAs) are mismatched and the mismatch between these two

end-point binaries is 7%. The second point to note in Fig. 1.2 is that certain compositions of the ternary and quaternary compounds, depending on the nature of the
binary constituents, have indirect bandgaps. These are indicated by dashed tie lines.
For example, $Al_xGa_{1-x}As$ is indirect bandgap for $x \geq 0.43$ and AlAs is, of course,
an indirect bandgap semiconductor. Finally, it may be noted that these mixed crystals
are formed by mixing in the group III sublattice, or the group V sublattice, or both.
This point has an important bearing on the techniques used to grow them.

Many physical parameters of ternary compounds are determined by the parameters
of the constituent binaries and vary roughly linearly with composition. For example,
the lattice constant, a, of $In_xGa_{1-x}As$ is given by Vegard's law as

$$a_{In_xGa_{1-x}As} = xa_{InAs} + (1-x)a_{GaAs} \tag{1.1}$$

Similarly, for a quaternary compound $A_{1-x}B_xC_yD_{1-y}$, a material parameter Q can be
expressed as

$$Q(x, y) = \{\; x(1-x)[(1-y)T_{12}(x) + yT_{43}(x)]$$
$$+ y(1-y)[(1-x)T_{14}(y) + xT_{23}(y)]\}$$
$$[x(1-x) + y(1-y)]^{-1} \tag{1.2}$$

where T_{ij} is the material parameter for the ternary alloy formed by binaries i and j.
The relevant parameters of GaAs and $In_{0.53}Ga_{0.47}As$ together with those for Si, for
the purpose of comparison, are listed in Appendix 1.

1.2 BONDING IN SOLIDS

In the *solid state*, also referred to as *condensed matter*, the atoms forming the solid are
held together by bonding forces. The atoms also maintain a finite, fixed distance from
each other. If the array of atoms have long-range order or *periodicity*, the resulting
solid is crystalline, and we will study the properties of crystals in the next section.
The periodic array of atoms leads the way to the energy band models and conduction
properties. However, before going into all that, we should understand the nature of
the forces that hold the atoms together in their equilibrium positions. There are in
general two types of forces, attractive and repulsive, which are both functions of the
interatomic distance z. At large distances the attractive forces dominate, and therefore
the atoms are drawn nearer to each other. At small interatomic distances the repulsive
forces dominate, and the atoms are pushed further apart. In equilibrium the forces
of attraction $F_A(z)$, and repulsion, $F_R(z)$, must balance to establish the equilibrium
atomic spacing. In other words:

$$F_A(z) = F_R(z) \tag{1.3}$$

The general behavior of the attractive and repulsive forces, as a function of interatomic
distance, is depicted in Fig. 1.3. It is seen that at $z = a_o$ the resultant force is zero.
Since force is a derivative of energy, Eq. 1.3 can be rewritten as

$$\frac{d}{dz}[\mathcal{E}_R(z) - \mathcal{E}_A(z)] = 0 \tag{1.4}$$

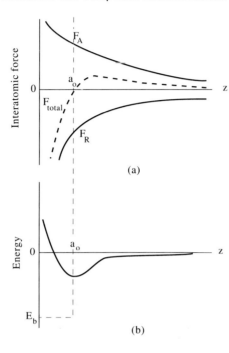

Figure 1.3 Interatomic forces in a solid: (a) attractive, repulsive and total forces and (b) equilibrium lattice constant and binding energy when the resultant force is zero.

and denoting the quantity within the brackets as $\mathcal{E}_{\text{total}}$, we have

$$\frac{d}{dz}\mathcal{E}_{\text{total}} = 0 \tag{1.5}$$

or, at the distance $z = a_o$ where the resultant interatomic force is zero, the total energy of the system also has a minimum value $\mathcal{E}_b$. The energy as a function of interatomic distance is also represented in Fig. 1.3. It is important to remember that although a_o (or simply a) is the equilibrium atomic spacing, which we shall see is the *lattice constant*, in a solid at room temperature the atoms vibrate around their equilibrium position due to thermal energy. Referring to Fig. 1.3, where the reference energy is taken as that of an isolated atom, it is clear that when the atoms in a crystal are spaced at their equilibrium distance a_o, the energy of the system is *lowered* by an amount $\mathcal{E}_b$. This lowering is a result of bonding.

The subject of bonding in solids is a vast one and has been discussed in detail in many texts. In this section we will mainly learn about the aspects that are relevant in the context of compound semiconductors. Some important points should first be reiterated. We have seen that the equilibrium interatomic spacing is a_o, the lattice constant. The energy $\mathcal{E}_b$ is also called the *binding energy*, implying that it is the minimum energy needed to break up the solid. This energy can range from a very small value (0.02 eV in He_2) to a very large value (10 eV in LiF) depending on the type of bonding. Also, it is to be noted that as a_o increases, $\mathcal{E}_b$ decreases until finally atoms are relatively free to move around. Under these conditions we approach the liquid phase and finally the gaseous phase.

Different bonding arrangements have different strengths, and the bonding classification is primarily described by the dominant attractive force. *Van der Waals bonding* is characterized by the formation of dipoles. Atoms and their electrons form dipoles, which interact with other dipoles, giving rise to a small attractive force. Such dipole interactions occur in inert gases. In fact, inert gases have the most stable electron configurations corresponding to complete s and p subshells of the outer shell. As an example, the spectroscopic designation of Ar is $1s^2 2s^2 2p^6 3s^2 3p^6$. In *ionic bonding* one of the atomic constituents yields an outer shell electron to the other atom, producing positive and negative ions, which attract each other by Coulomb-type interactions. For this bonding

$$F_A = \frac{q^2}{4\pi\epsilon_o z^2} \tag{1.6}$$

Equilibrium is reached when the attractive forces are balanced by the repulsive forces such that

$$F_{A(eq)} = \frac{q^2}{4\pi\epsilon_o a_o^2} \tag{1.7}$$

and the binding energy is just the electrostatic potential energy. Ionic bonding exists in alkali halides such as NaCl and KCl. After the electron exchange between the Na and Cl atoms have taken place, N_a^+ and Cl^- ions are formed. The outer orbits of all the atoms are filled and the ions, in fact, have the closed shell configurations of the inert gases Ne and Ar. More importantly, since there are no loosely bound or free electrons, the alkali halides are usually good insulators.

Covalent bonding arises from the tendency of atoms to form closed outer shells. This type of bonding is characterized by the sharing of electrons between neighboring atoms. This type of bonding is found in all organic compounds and nearly all semiconductors. The bonding force is a result of the quantum mechanical interaction between the shared electrons. The simplest example is the H_2 molecule. The H atom with atomic number $Z = 1$ has the structure $1s^1$. So each atom could share its electron to form closed outer shells for both. Similarly the outer (valence) subshells of the elemental semiconductors Si ($Z = 14$) and Ge ($Z = 32$) are characterized by

$$Si : 3s^2 3p^2$$
$$Ge : 4s^2 4p^2 \tag{1.8}$$

As a result, each Si or Ge atom can acquire up to four extra electrons to complete its outer subshell. As we shall see, these elemental semiconductors and compound semiconductors such as GaAs crystallize in the diamond and zincblende structures, respectively. In these structures, each lattice atom has four nearest neighbors. Thus, each Si or Ge atom can get the four deficient electrons in its outer subshells from the nearest neighbors and can share a total of eight valence electrons with its nearest neighbors. Each shared electron pair between neighboring atoms constitutes a covalent bond, illustrated in Fig. 1.4. The two lines between atoms signify the two electrons forming the covalent bond. In order to satisfy Pauli's exclusion principle, the two electrons must have opposite spins, indicated by the arrows. It may be apparent that because of the formation of the closed outer shells by sharing of electrons, these materials should also behave as insulators. The picture we have just described is valid at $0°K$ and at this temperature the semiconductors have very little conduction. However, as the temperature is raised, enough energy may be provided to break a covalent bond (~ 1 eV) and free carriers are available for conduction.

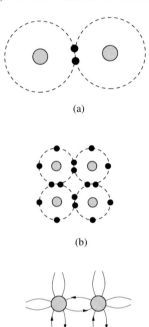

(a)

(b)

(c)

Figure 1.4 Atomic configuration and bonding arrangement in covalently bonded solids: (a) the hydrogen atom, (b) silicon with its four outer shell electrons, and (c) representation of bonding in Si indicating the opposite spins of the shared electrons.

In compound semiconductors such as GaAs, the outer shells of Ga and As are

$$Ga : 4s^2 4p^1$$
$$As : 4s^2 4p^3 \tag{1.9}$$

Like Si atoms, each Ga or As atom in the diamond-like lattice structure shares its outer electron with the four nearest neighbors (Ga or As). The bonding is therefore of covalent nature. However, because Ga and As are drawn from different groups of the periodic table, there is a small degree of charge transfer and therefore the bonding is partially ionic or *heteropolar*. The ionic nature of the chemical bond increases in II–VI compounds, in which the constituent atoms are placed further apart in the periodic table. To summarize, it is fair to state that bonding in compound semiconductors is predominantly covalent and partly ionic. Typical characteristics of covalent crystals are hardness, derived from the strength of the paired bond, and brittleness, since the adjacent atoms must remain in accurate alignment.

In passing, we should mention *metallic bonding*, in which the valence electrons are shared by many atoms. This type of bonding is not directional, and the nearly free or free electrons contribute to the superior conduction properties of metals.

1.3 CRYSTALLINE NATURE OF SOLIDS

An aggregate of atoms that forms the solid state distinguishes itself from other forms such as the liquid or gaseous states principally in the number of atoms per unit volume. There are between 10^{22} and 10^{23} atoms/cm^3 in a solid compared to 10^{19} molecules/cm^3 in water vapor at $100°C$ and under a pressure of 1 atmosphere. Other important properties of solids are rigidity and mechanical strength. It is not our intention here to engage in a lengthy discourse on properties of crystalline solids, but to highlight the properties of the semiconductors we will be discussing throughout this text. It is essential, therefore, to learn a little about the crystalline nature of solids and how atoms are arranged in the lattice. A *crystal* or *crystalline* solid is distinguished from any other form of solid or condensed matter by two important facts: (1) the periodic arrangement of the atoms and (2) the binding forces that hold the atoms together. We will discuss the first aspect here. In a crystalline solid, the periodic arrangements of the atoms, usually in three dimensions, is repeated over the *entire* crystal. A unique feature of such a crystal is the translational symmetry of the atoms. In other words, the crystal appears identical at several equivalent regions defined by a basic periodicity. Such solids are termed *single crystals*. In contrast, solids that exhibit crystalline behavior over a small region and are divided by boundaries from other regions exhibiting a different periodicity and arrangement of atoms are called *polycrystalline*. Solids that have no periodic structure at all are called *amorphous*. The three types of solids are illustrated in Fig. 1.5. While single crystal materials have long-range order, polycrystalline and amorphous materials have short-range order. The periodic array of points, generally in three dimensions, representing the positions occupied by the atoms of the crystal is called a *lattice*.

A *unit cell* is a region of a crystal defined by vectors **a, b, c**, and angles α, β, and γ, which when translated by integral multiples of those vectors, reproduce a similar region of the crystal. This is the translational property mentioned above and is expressed by

$$\mathbf{r} = h\mathbf{a} + k\mathbf{b} + l\mathbf{c} \tag{1.10}$$

where h, k, and l are integers. The set of linearly independent vectors **a, b, c,** which can be used to define a unit cell are called *basis vectors*. The vector **r** is sometimes called the *translational vector*. **a**, **b**, and **c** can be interatomic distances, in which case they are called the *lattice constant*. They usually have slightly different values because of microscopic differences in the binding forces. Figure 1.6(a) illustrates a two-dimensional arrangement of atoms in which the atoms, basis vectors, and the translational property are illustrated. The unit cell in this case is defined by the vectors

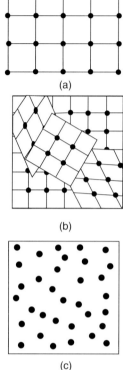

(a)

(b)

(c)

Figure 1.5 Schematic representation of atomic arrangement in different solids: (a) monocrystalline, (b) polycrystalline, and (c) amorphous. The first two are characterized by complete or partial ordering whereas amorphous materials have no periodicity at all.

a and **b**. A *primitive unit cell* is defined as the smallest unit cell in volume that can be defined for a given lattice. Similarly, the *primitive basis vectors* are a set of linearly independent vectors that define a primitive unit cell.

The vectors **a**, **b**, **c**, and angles α, β, γ define, in general, the parallelopiped shown in Fig. 1.6(b). Different combinations of $a, b, c,$ and α, β, γ give 14 different three-dimensional space lattices, called *Bravais lattices*. These are 1 triclinic, 2 monoclinic, 4 rhombic, 2 tetragonal, 1 hexagonal, 1 trigonal, and 3 cubic. It is the last classification, the cubic form, which is important, since Si and almost all III–V compound semiconductors crystallize in this form. The cubic form is characterized by $a = b = c$ and $\alpha = \beta = \gamma = 90°$, and its three classifications are simple or primitive cubic, face-centered cubic (FCC), and body-centered cubic (BCC).

The elemental semiconductors, Si and Ge, crystallize in the diamond structure formed by two interpenetrating FCC sublattices of atoms, shifted with respect to each other by one-fourth of the body diagonal. This is represented in Fig. 1.7(a). It is evident that the diamond lattice can be thought of as an FCC structure with an extra atom placed at a/4 + b/4 + c/4 from each FCC atom. This arrangement corresponds to the *tetrahedral* configuration in which each silicon atom in the center

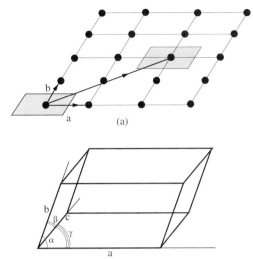

(a)

Figure 1.6 Representation of lattice atoms and unit cells in crystalline materials: (a) two-dimensional arrangement of atoms in which the basis vectors and translational property are illustrated and (b) parallelopiped indicating the basis vectors **a, b, c** and angles α, β, and γ.

(b)

of a tetrahedron (formed by four silicon atoms) forms four bonds with its nearest neighbors. In compound semiconductors, the chemical bonding is predominantly covalent and slightly ionic, leading to the tetrahedral bonding configuration as in Si and Ge. As a consequence, most III–V compound semiconductors crystallize in the *zincblende* structure. This is almost identical to the diamond form, with two interpenetrating FCC sublattices or cubes, except in this case each sublattice contains one kind of atom. Thus, for a compound $A_{III}B_V$, the two cubes are the A and B sublattices. The crystal structures of Si and GaAs are shown in Figs. 1.7(b) and 1.7(c), respectively. The diamond and zincblende structures have two atoms per lattice site and are therefore called diatomic lattices. In contrast, cadmium sulfide (CdS), which is a II–VI compound, crystallizes in the hexagonal or *wurtzite* form.

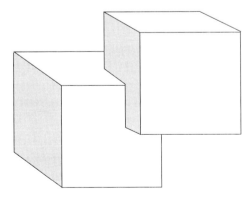

(a)

Figure 1.7 (a) Schematic of the diamond or zincblende structure, (b) the silicon crystal structure with $a = 5.43$ Å, and (c) the GaAs crystal structure with $a = 5.65$ Å (from S. M. Sze, *Physics of Semiconductor Devices*, 2nd ed., copyright ©1981. Reprinted by permission of Wiley, New York).

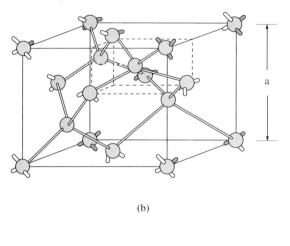

(b)

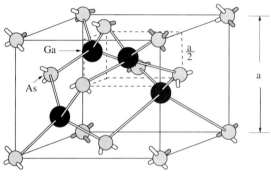

(c)

Figure 1.7 (*continued*)

The periodic arrangement of atoms in a crystal is of paramount importance, since it determines the crystal potential and bandstructure of the material, which in turn, determine the electronic and optical properties. Also, the physical arrangement of the atoms in the lattice together with their bonding determine the mechanical properties of the crystal.

In a cubic lattice the length of a side is called the lattice constant. There are three important parameters that describe a crystal lattice. These are the *coordination number*, the *packing fraction*, and the *nearest neighbor distance*. The coordination number is the number of nearest neighbor lattice sites. For example, in a metal with very non-directional bonding, the coordination number is expected to be very high. For the BCC lattice the coordination number is 8 and for the FCC structure the coordination number is 12. Again, for the FCC lattice, the nearest neighbor distance is one-half of a face diagonal, or $\frac{1}{2}(a\sqrt{2})$. Therefore the maximum required radius of an atom on the face to touch a corner atom of the same size is $\frac{1}{4}(a\sqrt{2})$. From this the packing fraction, as illustrated in the following example, can be calculated.

EXAMPLE 1.1
Objective.   To find the packing fraction of the BCC unit cell assuming spherical atomic shells.

The BCC unit cell has eight atoms in the corners of the cube and one atom in the center, a shown in Fig. 1.8. Each corner atom is, however, shared with seven neighboring cells. Thus, each cell contains $\frac{1}{8}$ of an atom (or sphere) at each corner and one atom at the body center of the cube, resulting in a total of two atoms per unit cell. The distance between the centers of two corner atoms is a. On the other hand, the distance between an atom at the corner and the atom at the body center is $\frac{\sqrt{3}}{2}a$. Therefore, the nearest neighbor distance in the BCC lattice is $\frac{\sqrt{3}}{2}a$ and the maximum radius for two atoms to touch each other is $\frac{\sqrt{3}}{4}a$.

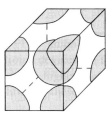

Figure 1.8   Packing of atoms as hard spheres in the BCC lattice.

Therefore, volume of each sphere $= \frac{4}{3}\pi \left(\frac{\sqrt{3}}{4}a\right)^3$

$$\text{Packing fraction} = \frac{\text{volume of each sphere} \times \text{number of spheres}}{\text{total volume of unit cell}}$$

$$= \frac{\frac{4}{3}\pi \left(\frac{\sqrt{3}a}{4}\right)^3 \times 2}{a^3}$$

$$= \frac{\sqrt{3}\pi}{8} = 0.68$$

1.3.1 Directions and Planes

In dealing with crystalline materials there is a need to form a basis that will enable us to refer to planes and directions within the lattice. This is made possible by the *Miller indices*, derived on the basis of a rectangular cartesian coordinate system. The steps to calculate the Miller indices are as follows:

  a. Take as the origin any atom in the crystal and erect coordinate axes from this atom in the direction of the basis vectors.

  b. Find the intercept of a plane belonging to a system, expressing them as integral multiples of the basis vectors along the crystal axes.

  c. Take the reciprocals of these numbers and reduce them to the smallest triad of integers h, k, and l having the same ratio. Then (hkl) is the Miller index of that system of parallel planes. The direction normal to the plane is the [hkl] direction.

EXAMPLE 1.2
Objective. To find the Miller indices of a plane that makes intercepts equal to 2, 3, and 4 times the basis vectors along the three crystal axes.

In terms of the basis vectors the intercepts are therefore 2a, 3b, and 4c, as shown in Fig. 1.9. We take the reciprocals of 2, 3, and 4 and obtain $\frac{1}{2}$, $\frac{1}{3}$, and $\frac{1}{4}$.

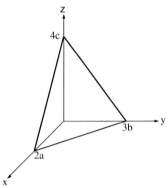

Figure 1.9 Representation for determination of Miller indices.

These fractions are then reduced to the smallest triad of integers having the same ratio. These are obtained as 6, 4, 3 by multiplying the fractions by 12. The Miller indices of the plane are (6, 4, 3).

The directions and planes of a cubic lattice are shown in Fig. 1.10. It is obvious that many planes in a lattice are equivalent. For example, the faces of a cube are equivalent, considering the orientation of the cube. Thus, the (100), (010), (001) . . . planes form the family of (100) planes. It also is apparent that parallel planes within a crystal have the same crystallographic direction and the same Miller indices. In the

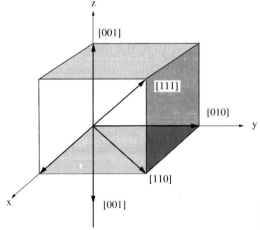

Figure 1.10 Planes and directions in a cubic crystal.

diamond and zincblende structures, depicted in Figs. 1.7(b) and (c), respectively, it is seen that the bonds between the nearest Si atoms, or that between the nearest Ga and As atoms are in the [111] directions. The bonds between the nearest equivalent atoms in GaAs, Ga, and Ga, or As and As, are in the [110] directions. The (110) planes in GaAs are called the cleavage planes. In Si they are the (111) planes. Usually GaAs-based or InP-based materials and heterostructures are grown in the [100] direction and are cleaved along the (011) or (01$\bar{1}$) planes. This technique is most commonly used to define laser cavities and the end facets.

The different planes of GaAs, visualized from the top, are shown in Fig. 1.11. On the (100) surface each Ga atom has two bonds with the As atoms below and two free or dangling bonds. The same is true for As atoms covering the surface. The (110) plane is the only one that has the same number of Ga and As atoms. Each surface atom has one bond with the layer below, two bonds with the in-plane nonequivalent atoms and one free or dangling bond. On the (111) surface as shown each Ga atom forms three bonds with the As atoms in the layer below and therefore has one dangling bond. Note that As atoms on the surface, on the other hand, have *one* bond with the Ga atoms below and three dangling bonds. The GaAs crystal lacks inversion symmetry, and therefore the [111] and [$\bar{1}\bar{1}\bar{1}$] directions are not equivalent. In the nomenclature commonly used the Ga to nearest As atom denotes the [$\bar{1}\bar{1}\bar{1}$] direction. The surface properties of a crystal are determined by the number of dangling bonds. The latter also determine the dopant incorporation behavior in crystals. In particular, *amphoteric* dopants, such as Si in GaAs and a few other III–V compounds can be made to preferentially incorporate in a Ga or As lattice site, depending on the orientation of the surface.

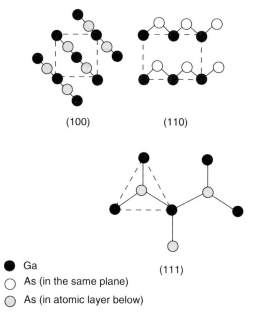

(100)   (110)

(111)

● Ga
○ As (in the same plane)
○ As (in atomic layer below)

Figure 1.11   Gallium arsenide crystal planes. Solid lines correspond to crystal bonds; dashed lines correspond to the intercepts of atomic planes with a unit cell (from M. Shur, *Physics of Semiconductor Devices*, ©1990. Reprinted by permission of Prentice Hall, Englewood Cliffs, NJ).

1.3.2 Reciprocal Lattice Vectors

An alternate elegant way to express the periodic properties of a crystal is through the *reciprocal lattice vectors*. If **a**, **b**, and **c** are the direct lattice vectors, which may be the basis vectors of the primitive cell, we can define the primitive translation vectors in the reciprocal lattice, **a***, **b***, **c***, by the relations

$$\mathbf{a}^* = 2\pi \frac{\mathbf{b} \times \mathbf{c}}{\mathbf{a} \cdot (\mathbf{b} \times \mathbf{c})} \; , \; \mathbf{b}^* = 2\pi \frac{\mathbf{c} \times \mathbf{a}}{\mathbf{a} \cdot (\mathbf{b} \times \mathbf{c})} \; , \mathbf{c}^* = 2\pi \frac{\mathbf{a} \times \mathbf{b}}{\mathbf{a} \cdot (\mathbf{b} \times \mathbf{c})} \tag{1.11}$$

It is obvious that the direct and reciprocal lattice vectors are related by the *Kronecker delta function.* In other words:

$$\mathbf{a}^* \cdot \mathbf{a} = \mathbf{b}^* \cdot \mathbf{b} = \mathbf{c}^* \cdot \mathbf{c} = 2\pi \tag{1.12}$$

while

$$\mathbf{a}^* \cdot \mathbf{b} = \mathbf{a}^* \cdot \mathbf{c} = 0, \; \mathbf{b}^* \cdot \mathbf{a} = \mathbf{b}^* \cdot \mathbf{c} = 0, \; \mathbf{c}^* \cdot \mathbf{a} = \mathbf{c}^* \cdot \mathbf{b} = 0 \tag{1.13}$$

The set of reciprocal lattice vectors define a reciprocal lattice and a vector in the reciprocal lattice is defined, with analogy to Eq. 1.10, by

$$r^* = h\mathbf{a}^* + k\mathbf{b}^* + l c^* \tag{1.14}$$

As shown in the example below, the vector **r*** is *normal* to the [hkl] plane.

EXAMPLE 1.3
Objective.   To show that the reciprocal lattice vector given by Eq. 1.14 is normal to the [hkl] plane.

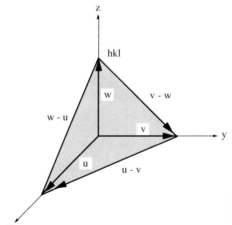

Figure 1.12 Diagram showing the intercept of the (hkl) plane with the crystal axes.

The intersection of the [hkl] plane with the x, y, and z-axes are shown in Fig. 1.12, with intercepts **u**, **v**, and **w**. The sides of the triangle formed by the intersection of the [hkl] plane with the coordinate axes are (**u**−**v**), (**v**−**w**), and (**w**−**u**). Using Eqs. (1.12) to (1.14), it can then be shown that

$$\mathbf{r}^* \cdot (\mathbf{u} - \mathbf{v}) = h\mathbf{a}^* \cdot \mathbf{u} - k\mathbf{b}^* \cdot \mathbf{v} = 0$$

and

$$\mathbf{r}^* \cdot (\mathbf{v} - \mathbf{w}) = 0$$

It follows from these equations that $\mathbf{r}^*$ is normal to the [hkl] plane, since any vector in this plane can be expressed as a linear combination of $(\mathbf{u} - \mathbf{v})$ and $(\mathbf{v} - \mathbf{w})$.

In general, the free surface of a crystal has fewer lattice points than the bulk. Therefore, it follows that in the reciprocal lattice there are more points. Since a true surface is two-dimensional, the reciprocal lattice points on the surface become infinite lines, or rods. These concepts are useful in monitoring the epitaxial growth of a crystal by in-situ measurements, as is done during molecular beam epitaxy.

To conclude this section, the tetrahedral radii of several elements, which determine the nearest neighbor distance, and the lattice constants of important semiconductors are given in Tables 1.1 and 1.2, respectively.

TABLE 1.1 COVALENT (TETRAHEDRAL) RADII (from M. Shur, *Physics of Semiconductor Devices*, Prentice Hall, Englewood Cliffs, NJ, 1990).

Element	Tetrahedral radius (Å)
Al	1.26
As	1.18
B	0.88
Bi	1.46
C	0.77
Ga	1.26
Cd	1.48
Ge	1.22
Hg	1.48
In	1.44
Mn	1.27
N	0.70
P	1.10
Sb	1.36
Si	1.17
Sn	1.40
Te	1.47
Zn	1.31

1.4 ALLOY SEMICONDUCTORS

The energy gaps and lattice constants of common binary III–V compounds were depicted in Fig. 1.2. Depending on the choice of the particular compound, the energy bandgap varies from 0.17 eV (InSb) to 2.46 eV (AlP). Some nitride compounds have bandgaps as large as 6 eV. Since the bandgap translates to the emission energy (wavelength) of a luminescent device, a wide emission spectrum is covered. In principle,

TABLE 1.2 LATTICE CONSTANTS, NEAREST-NEIGHBOR DISTANCES AND COVALENT RADII OF ELEMENTAL AND COMPOUND SEMICONDUCTORS (adapted from M. Shur, *Physics of Semiconductor Devices*, Prentice Hall, Englewood Cliffs, NJ, 1990).

Material	Lattice constant, a (Å) at $25°C$	Distance between nearest neighbors, $a\sqrt{3/4}$ (Å)	Sum of covalent radii (Å)
Si	5.4309	2.353	2.34
Ge	5.6461	2.450	2.44
$A_{III}B_V$			
AlAs	5.6611	2.430	2.44
AlP	5.451	2.360	2.36
AlSb	5.136	2.224	2.62
BAs	4.776	2.068	2.06
BN	3.615	1.565	1.58
BP	4.538	1.965	1.98
BSb	5.170	2.239	2.24
GaAs	5.6532	2.448	2.44
GaP	5.4495	2.360	2.36
GaSb	6.095	2.639	2.62
InAs	6.0584	2.623	2.62
InP	5.8687	2.540	2.54
InSb	6.479	2.805	2.80
$C_{II}D_{VI}$			
CdTe	6.482	2.807	2.95
HgS	5.841		
HgSe	6.084		
HgTe	6.462	2.798	2.95
ZnS	5.415		
ZnSe	5.653		
ZnTe	6.101	2.642	2.78

however, constraints are imposed by the requirements of lattice matching and the availability of suitable substrates. The only common substrates available with relatively low defect density are GaAs and InP. Almost concurrent with the development of binary III–V compounds was the development of *alloy* semiconductors. It was realized at an early stage that solid solutions of the binary compounds could form ternary or quaternary alloys. For example, a ternary III–V solid solution $A_x B_{1-x}C$ consists of x atoms of the element A and $(1-x)$ atoms of element B, randomly mixed in the group III sublattice and all the group V lattice sites are occupied by element C. The value of x can vary from 0 to 1. A typical example is $Al_x Ga_{1-x}As$, a ternary alloy that has emerged as being technologically very important. In the crystal lattice, Al atoms randomly replace Ga atoms in the group III sublattice. Another form of ternary material is $AB_{1-x}C_x$, where all the group III sublattice sites are occupied by element A and mixing or alloying takes place in the group V sublattice between x atoms of element C and $(1-x)$ atoms of element B. An example of this type of ternary compound is $GaAs_{1-x}P_x$, which is also technologically important for the manufacture of visi-

ble light-emitting diodes. From the combination of the binary compounds shown in
Fig. 1.2 there are 18 possible ternary derivatives. In a similar manner, quaternary compounds are formed by the mixing of four atomic species. Such a compound can consist
of two group III elements A and B randomly distributed in group III lattice sites, and
two group V elements C and D, randomly distributed in group V lattice sites, to give
the compound $A_x B_{1-x} C_y D_{1-y}$. An example of such a compound is $In_x Ga_{1-x} As_y P_{1-y}$.
The different alloy compositions of this material are extremely important for the realization of sources and detectors for optical fiber communication. If, on the other hand,
three elements A, B, and C mix in the group III sublattice, and only one element D
is present in the group V sublattice, the quaternary compound $A_x B_y C_z D$ is obtained.
The composition is more conveniently expressed as $(A_x B_{1-x})_y C_{1-y} D$, where x and
y can vary from zero to unity. An example is $(In_x Ga_{1-x})_y Al_{1-y} As$. From the point
of view of epitaxy, mixing in the group V sublattice with accurate control of alloy
composition is harder to achieve than mixing in the group III sublattice.

The existence of the random alloys, or solid solutions, is predicted by thermodynamic phase equilibria. For the $Al_x Ga_{1-x} As$ alloy system, for example, it is possible
to grow crystals over the entire composition range, from $x = 0$ to 1. However, for
certain compositions in some alloy systems, thermodynamics does not predict the existence of a completely random alloy with a uniform composition. Instead, they tend
to form *immiscible* solutions. In such a case, we say a *miscibility gap* exists in the
phase diagram, illustrated in Fig. 1.13. The tendency toward immiscibility increases
with increasing difference in the covalent radii of different atoms trying to occupy
the same sublattice, or lattice sites. There is another way to look at immiscibility.
Usually, in the immiscible solid solution, two or more compositions, or phases, co-
exist instead of one uniform composition. The different phases can be looked upon
as *clusters*, and the phenomenon can be looked upon as clustering. In a perfectly
random alloy, the cluster size is the size of the Wigner-Seitz cell. In a clustered alloy,
the cluster size can be much larger. Such clusters increase carrier scattering in the
material and degrade its transport properties.

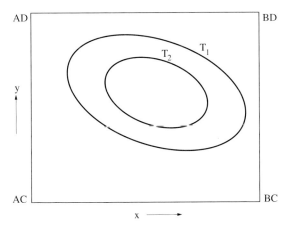

Figure 1.13 Schematic of
miscibility gap in a quaternary
compound $A_x B_{1-x} C_y D_{1-y}$. The gap
usually becomes smaller for higher
growth temperatures ($T_2 > T_1$).

In a ternary alloy, the lattice constant is linearly dependent on the composition, as expressed in Eq. 1.1. The linear relationship of the lattice constant generally holds for quaternary alloys too. However, other parameters, of a mixed alloy do not, in general, obey this linear relationship. The bandgap $\mathcal{E}_g$, for example, is usually given by an empirical relationship:

$$\mathcal{E}_g(x) = \mathcal{E}_{go} + bx + cx^2 \tag{1.15}$$

where $\mathcal{E}_{go}$ is the bandgap of the lower bandgap binary, b is a fitting parameter, and c is called the bowing parameter, which may be calculated theoretically or determined experimentally. As we shall soon see, it is not only the variation of bandgap with composition, but also the energy variation of the higher-lying bandstructure with composition, that is extremely important for the understanding of material properties. This variation for the $Al_xGa_{1-x}As$ system is depicted in Fig. 1.14. The figure shows the variation of the Γ, L, and X conduction bands with alloy composition x, and the energy disposition of the conduction valleys of the two end-point binaries. The relevant energies are approximately given by

$$\mathcal{E}_g^\Gamma(x) = 1.425 + 1.247x + 1.147(x - 0.45)^2$$

$$\mathcal{E}_g^X(x) = 1.9 + 0.125x + 0.143x^2$$

$$\mathcal{E}_g^L(x) = 1.708 + 0.642x \tag{1.16}$$

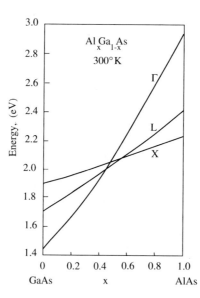

Figure 1.14 Compositional dependence of the direct (Γ) and indirect (X and L) conduction band minima in the $Al_xGa_{1-x}As$ mixed crystals.

The monotonic variation of bandgap with alloy composition allows us to realize *heterojunctions*, which is a junction of two semiconductors of unequal bandgap. Heterojunctions are of paramount importance in the design of high-performance electronic and optoelectronic devices. There is, however, a problem, which becomes evident upon examining Fig. 1.2. It is seen that, in general, the lattice constant of the alloy varies with composition, unless the end-point constituents have the same lattice constant. This does create a problem in realizing lattice-matched heterojunctions, since no two alloy compositions on the same substrate, with different energy bandgaps, would be lattice-matched to each other or to the substrate, thereby creating the possibility of dislocation generation. There are a few exceptions, and one of them is the $Al_xGa_{1-x}As$ system. GaAs and AlAs have a lattice mismatch of only 0.04%, and therefore all the intermediate alloy compositions are nearly perfectly lattice-matched. This heterostructure material system is widely used for the design and fabrication of near-infrared sources and detectors and is perhaps the most extensively investigated. Looking at Fig. 1.2, it is evident that the ternary alloy compositions $In_{0.53}Ga_{0.47}As$ ($\mathcal{E}_g = 0.74$ eV) and $In_{0.52}Al_{0.48}As$ ($\mathcal{E}_g = 1.45$ eV) are both lattice-matched to InP. The $In_{0.53}Ga_{0.47}As/In_{0.52}Al_{0.48}As$ heterojunction on InP has the advantage over the GaAs/AlGaAs one in that the energy bandgap of InGaAs is close to the region of low-loss and low-dispersion of optical fibers. In addition, it is seen that the quaternary alloys InGaAsP (with InP and $In_{0.53}Ga_{0.47}As$ as the end-point materials) and InGaAlAs (with $In_{0.53}Ga_{0.47}As$ and $In_{0.52}Al_{0.48}As$ as the end-point materials) are also lattice-matched to InP and are technologically very important for optical fiber communication. There are other alloy systems lattice-matched to GaSb substrates that will be important for far infrared sources and detectors, but some of them have miscibility gaps in the phase diagram. As we shall see in the next section, one does not always have to work under the constraints of lattice-matching, and use of *pseudomorphic* materials does alleviate this problem to some extent.

The process of random alloying causes a perturbation of the periodic potential of the lattice, which manifests itself as an additional scattering mechanism for carriers. The process is characterized by a disorder, or alloy-scattering potential, which is due to the deviation in electronegativity of atoms caused by the deviation of the covalent radius due to alloying. As seen earlier, III–V semiconductors have ionic bonding and have a net charge Q. As an example, the net charge in $Al_xGa_{1-x}As$ can be approximated as

$$Q_{Al_xGa_{1-x}As} = (1-x)Q_{GaAs} + xQ_{AlAs} \tag{1.17}$$

and the corresponding alloy-scattering potential can be approximated as

$$U_{Al_xGa_{1-x}As} = 0.46 + 0.01x. \tag{1.18}$$

1.5 LATTICE-MISMATCHED AND PSEUDOMORPHIC MATERIALS

It was mentioned in Sec. 1.4 that one of the constraints in heterojunction technology for optoelectronics is the lack of suitable lattice-matched substrates. Pioneering work by Frank and van der Merwe in the late 1940s established the possibility of the growth of lattice-mismatched crystalline monolayers. This was followed by a more complete analysis of the growth of lattice-mismatched layers by Matthews and Blakeslee, who pointed out that careful growth of an epitaxial layer whose lattice constant is close, but not equal to the lattice constant of the substrate can result in a coherent strain, as opposed to polycrystalline or amorphous incoherent growth.

Usually in lattice-matched epitaxial growth, the epitaxial film, and the substrate join along a plane boundary along which a common two-dimensional cell structure is maintained. This is shown in Fig. 1.15(a). In strained layer growth the situation is different, as illustrated in Figs. 1.15(b) and (c). These figures show the two-dimensional atomic arrangement of a substrate with lattice constant a_o and the epitaxial overlayer with lattice constant a. The lattice mismatch, or misfit, is defined as

$$\frac{\Delta a}{a} = \frac{a - a_o}{a}$$

$$= -f \tag{1.19}$$

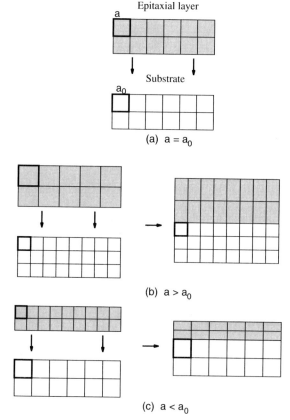

(a) $a = a_0$

(b) $a > a_0$

(c) $a < a_0$

Figure 1.15 Accommodation of lattice of epitaxial layer with that of substrate for different cases: (a) lattice-matched growth ($a = a_0$), (b) biaxial compressive strain ($a > a_0$), and (c) biaxial tensile strain ($a < a_0$).

It is clear that a match between the two crystalline structures can be achieved only if one or both crystals are elastically strained. If the substrate is much thicker than the growing epitaxial film, as is usually the case, then it is the atoms of the epitaxial layer that are displaced. If the strain is incorporated into the epitaxial crystal coherently, the lattice constant of the epitaxial layer in the direction parallel to the interface is forced to be equal to the lattice constant of the substrate. The lattice constant of the epitaxial layer perpendicular to the substrate will be changed by the Poisson effect. If the parallel lattice constant is forced to shrink, or a compressive strain is applied, the perpendicular lattice constant will grow. Conversely, if the parallel lattice constant of the epitaxial layer is forced to expand, under a tensile strain, the perpendicular lattice constant will shrink. These two cases are depicted in Fig. 1.15. This type of coherently strained crystal is called *pseudomorphic*, since it takes on the morphology or form, in this case the lattice constant, of the substrate.

In analyzing the effects of strain in a (001)-oriented epitaxial film, it is assumed that the strains are small, of the order of 1%–2%, and the cell size and shape do not change significantly, other than the tetragonal distortion. In the case of the growth of a strained epitaxial film of thickness d_1 with lattice constant a on a substrate of thickness d_2 with lattice constant a_o, where $a > a_o$, one can write the lattice constants of the strained film in the direction parallel to the interface as

$$a_\parallel = \frac{ad_1 + a_0 d_2}{d_1 + d_2} \tag{1.20}$$

In general, $d_2 \gg d_1$, and therefore

$$a_\parallel = a_0$$
$$= (1 + f)a \tag{1.21}$$

The lattice constant in the direction perpendicular to the interface is

$$a_\perp = (1 - \sigma_{ST} f)a \tag{1.22}$$

where σ_{ST} is a material parameter given by

$$\sigma_{ST} = \frac{c_{11}}{2c_{12}} \tag{1.23}$$

Here c_{11} and c_{12} are elastic constants of the epitaxial layer. For example, for GaAs c_{11} and c_{12} are 1.2×10^{12} and 5.5×10^{11} dynes/cm^2, respectively. We assume that the in-plane coordinate directions are x and y, and growth occurs along the z-direction. Equations 1.20 to 1.23 are only valid when the z-direction coincides with the (001) crystallographic axis. Strained growth places stress upon the epitaxial layer in the x and y directions. The stress components in the x and y directions are equal and force the lattice constants in both of the parallel directions to be equal. We can therefore write

$$f_{xx} = f_{yy} = f_\parallel = f \tag{1.24}$$

No compressive strain is placed upon the epitaxial layer in the z-direction, but the tensile strain in this direction will be nonzero. The strain in the z-direction can be determined by writing Hooke's law in terms of the elastic stiffness constants. Since the stress of interest yields a tetragonal distortion, the full elastic matrix need not be used as only the axial strain will be nonzero. We can therefore write for the stress, in matrix form:

$$\begin{bmatrix} S_{xx} \\ S_{yy} \\ S_{zz} \end{bmatrix} = \begin{bmatrix} c_{11} & c_{12} & c_{12} \\ c_{12} & c_{11} & c_{12} \\ c_{12} & c_{12} & c_{11} \end{bmatrix} \begin{bmatrix} f_{xx} \\ f_{yy} \\ f_{zz} \end{bmatrix} \tag{1.25}$$

where the c's are the elastic constants. Since the z-component of the stress, S_{zz}, is zero, we can combine Eqs. 1.24 and 1.25 to express the strain in the z-direction as

$$f_{zz} = -\frac{2c_{12}}{c_{11}} f \tag{1.26}$$

For a (001)-oriented crystal or in orientations in which the surface normal is an axis of symmetry, the strain tensor has only diagonal terms. In other words:

$$f \equiv \begin{bmatrix} f_{xx} & 0 & 0 \\ 0 & f_{yy} & 0 \\ 0 & 0 & f_{zz} \end{bmatrix} \tag{1.27}$$

The strain energy given by $\mathcal{E}_{st} = f^2 \left(c_{11} + c_{12} - \frac{2c_{12}^2}{c_{11}} \right) d$ increases linearly with the thickness of the film. However, if the coherent strain in the crystal becomes relatively large, either due to a large mismatch or a large thickness of the epitaxial layer, the strain energy stored in the crystal can be reduced by the formation of a network of dislocations at the interface. Figure 1.16 illustrates the formation of dislocations by missing atoms at the interface. The edge dislocations that are formed lie in the interfacial plane. The type of strain that we are discussing here is called *biaxial* in the plane of growth, having equal components in two perpendicular (in-plane) directions. Therefore, two perpendicular sets of parallel edge dislocations forming a square grid are required to accommodate the lattice mismatch. Once the misfit dislocations are generated, they usually propagate upward, with the growing crystal, as threading dislocations. As the threading dislocations propagate upward, they multiply. The interfacial misfit dislocations seen at the edge by cross-sectional transmission electron microscopy (TEM) are shown in Fig. 1.17(a). The dislocations in the interfacial plane, seen by plan-view TEM are depicted in Fig. 1.17(b). The network of threading dislocations and their multiplication mechanism, as observed by TEM, are shown in Figs. 1.17(c) and (d), respectively.

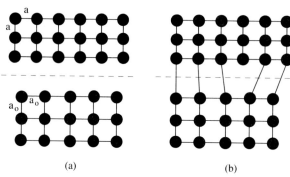

(a) (b)

Figure 1.16 Atomic arrangement showing the formation of an edge misfit dislocation in joining single cubic crystal epitaxial layer with lattice constant a and substrate with lattice constant a_0: (a) imaginary isolated case and (b) grown bi-crystal.

An important concept in strained layer epitaxy is that of critical layer thickness. *Critical thickness* arises because of a competition between strain energy and chemical energy. Below the critical thickness h_c, the minimum energy state of the bilayer system is achieved by strain. Above h_c, the minimum energy state is achieved by the formation of dislocations. The critical thickness is dependent upon the amount of the lattice mismatch and material parameters as well as the properties of the dislocations that form in the particular material. For the case of misfit dislocations forming at the

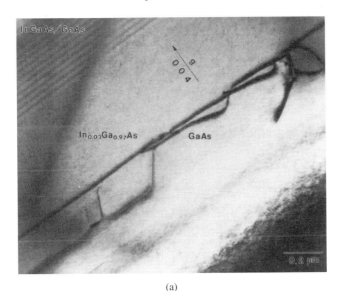

(a)

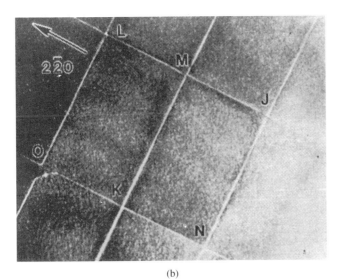

(b)

Figure 1.17 Misfit dislocations caused by molecular beam epitaxial growth of $In_{0.3}Ga_{0.7}As$ on GaAs ($f = 2\%$): (a) cross-section transmission electron micrograph (TEM) showing interfacial dislocation network, (b) plan-view TEM showing two-dimensional dislocation array, (c) dislocation multiplication and threading dislocation, and (d) detailed plan-view TEM showing dislocation multiplication mechanism. The measurements were made by the author and co-workers; K. H. Chang et al., *Journal of Applied Physics*, **66**, 2993 (1989).

(c)

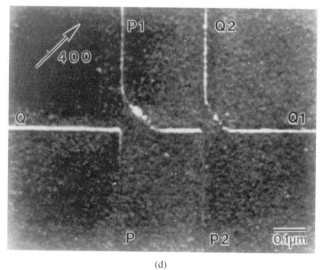

(d)

Figure 1.17 (*continued*)

interface when the critical thickness is exceeded, it has been shown that

$$h_c = \frac{a_o \left(1 - \frac{v_{PR}}{4}\right) \left[ln \left(\frac{h_c \sqrt{2}}{a_o}\right) + 1\right]}{2\sqrt{2}\pi f (1 + v_{PR})} \qquad (1.28)$$

It is evident that the larger the mismatch, the smaller the critical thickness (Problem 1.12). A much simpler approximate expression, $h_c \cong a/2f$, can be derived from energy minimization considerations (Problem 1.13). $\nu_{PR} = c_{12}/(c_{11} + c_{12})$ is the Poisson ratio. Its values are close to 1/3 for most semiconductors.

At this point it is important to understand and appreciate a crucial aspect of the use of strained layers in device structures. The use of strained semiconductors does remove the restrictions of lattice-matching and, in principle, offers an additional degree of freedom in bandgap engineering. Indeed, the use of strained layers was originally invoked as a technique to alter the bandgap. From the discussion above it is clear that such strained layers are in reality *strain-free* mismatched layers if their thickness is larger than h_c. Also, such mismatched layers have a large density of dislocations, sometimes as high as 10^8 to 10^9 cm^{-2}. Therefore, unless some special steps are taken to reduce the dislocation density, mismatched layers cannot be used in the active region of devices. Pseudomorphic layers, on the other hand, are coherently strained and dislocation-free and have, therefore, found important applications in electronic and optoelectronic device design. In such design, it has to be remembered that the pseudomorphic layers must have a thickness less than h_c. For misfits $\sim 1\%$, h_c is only a few hundred angstroms.

Compared to the bandstructure of lattice-matched III–V semiconductors, there are several changes due to biaxial strain. These include changes in the bandgaps, changes in the conduction band effective masses, and changes in the valence bandstructure. For unstrained materials, the equal energy surfaces for the Γ-valley are spherical in the materials that we are considering. Strain, however, lowers the symmetry of the crystal and changes the shape of the equi-energy surfaces into an ellipsoid. Thus, the biaxially strained Γ-valley must be characterized by a parallel and a perpendicular effective mass that will, in general, be different. Some recently calculated effective masses in strained $In_{1-x}Ga_x As$ on GaAs and InP substrates are shown in Table 1.3. It is seen from the table that the perpendicular effective mass tends to decrease very slowly as

TABLE 1.3 ELECTRON EFFECTIVE MASSES IN $In_xGa_{1-x}As$ GROWN PSEUDOMORPHICALLY ON GaAs AND $In_{0.53+x}Ga_{0.47-x}As$ GROWN PSEUDOMORPHICALLY ON InP (from M. Jaffe and J. Singh, *Journal of Applied Physics*, **65**(1), 329, 1989).

x	$In_xGa_{1-x}As$			$In_{0.53+x}Ga_{0.47-x}As$		
	$m^*_{unstrained}$	$m^*_{\|strained}$	$m^*_{\perp strained}$	$m^*_{unstrained}$	$m^*_{\|strained}*$	$m^*_{\perp strained}$
0.00	0.066	0.066	0.066	0.045	0.045	0.045
0.05	0.064	0.065	0.064	0.044	0.044	0.045
0.10	0.062	0.064	0.063	0.042	0.043	0.045
0.15	0.060	0.063	0.063	0.040	0.041	0.044
0.20	0.058	0.062	0.062	0.037	0.039	0.044
0.25	0.056	0.061	0.061	0.035	0.037	0.044
0.30	0.054	0.060	0.061	0.033	0.035	0.043
0.35	0.052	0.058	0.060	0.031	0.033	0.043
0.40	0.050	0.057	0.060	0.028	0.030	0.043

the In composition (and the strain) increases. The parallel effective mass decreases at a greater rate with excess In, although it does remain above the unstrained value.

The effect of biaxial strain on the valence bandstructure is far more drastic. This is because of the fact that the top of the valence band is doubly degenerate. The effect of a biaxial strain is, to a first order, to cause a splitting in the heavy-hole and light-hole states. The effects on the heavy- and light-hole bandgaps are expressed, for a (001)-oriented film, by

$$\mathcal{E}_{g_{HH}} = \mathcal{E}_{g_o} + \frac{1}{2}\delta\mathcal{E}_{sh} - \delta\mathcal{E}_{hy}$$

$$\mathcal{E}_{g_{LH}} = \mathcal{E}_{g_o} - \frac{1}{2}\delta\mathcal{E}_{sh} - \delta\mathcal{E}_{hy} \tag{1.29}$$

where $\delta\mathcal{E}_{sh}$ and $\delta\mathcal{E}_{hy}$ are the shear and hydrostatic components of the change in the band energies from the unstrained band energy position $\mathcal{E}_{go}$ and are given by

$$\delta\mathcal{E}_{sh} = -2b[(c_{11} + 2c_{12})/c_{11}]f$$

$$\delta\mathcal{E}_{hy} = -2a[(c_{11} - c_{12})/c_{11}]f. \tag{1.30}$$

Here a and b are the material deformation potentials. For GaAs, the shifts for the heavy- and light-hole gaps are $-5.96f$ eV and $-12.4f$ eV, respectively. Thus, in presence of biaxial tensile strain the light-hole is expected to be above the heavy-hole state, while for compressive strain the reverse is expected. This is shown in Fig. 1.18. The light-hole and heavy-hole bands couple very strongly and this coupling determines, to a large extent, the masses of the two bands and can be responsible for making the hole bands highly nonparabolic. Under sufficiently large biaxial compressive strain, the average effective hole masses in $In_xGa_{1-x}As$ on GaAs become

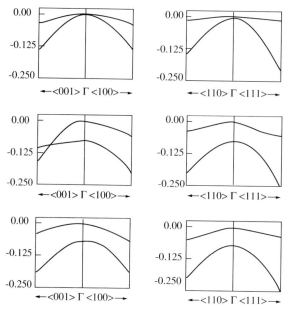

Figure 1.18 Hole band dispersions for (top) unstrained GaAs, (middle) GaAs under 1% biaxial tension, and (bottom) GaAs under 1% biaxial compression (courtesy of J. Singh, University of Michigan).

almost as small as the electron effective mass. The ability of biaxial strain to split heavy-hole and light-hole states can also be utilized to produce interesting and useful optical properties in strained materials.

We have considered, until now, the case of the "bulk"-strained layer. Optoelectronic devices such as lasers and modulators have *quantum wells* in the active region, and these wells could be pseudomorphically strained. We therefore have to know the joint effects of strain and quantum confinement. As we shall see in Chapter 2, discrete energy subbands are formed in a potential quantum well. The important consequence is that in pseudomorphic quantum wells, the effective mass of the bulk-strained crystal in the parallel direction, as given in Table 1.3, becomes the effective mass of the quantum subbands. Unlike the electron case, the hole dispersion can change significantly in a quantum well. Both strain and quantum confinement will alter the relative positions of the light- and heavy-hole subbands. For example, the effect of the quantum confinement alone is to split the degeneracy between the light- and heavy-hole subbands at the zone center. In a 100 Å lattice-matched $GaAs/Al_{0.3}Ga_{0.7}As$ quantum well, the splitting is approximately 10 meV. Biaxial compressive strain increases this splitting further, and tensile strain decreases it, since it moves the bands in the opposite directions. Change in the splitting changes the interaction between the light- and heavy-hole bands, thereby changing the curvature of the bands and the corresponding hole masses. In particular, as we shall see in Chapter 3, the oscillator strength for absorption and emission depends on the joint density of states, which is determined by the dispersion relations. As an example, for the case of *small* biaxial tensile strain, the light- and heavy-hole bands can become degenerate in a quantum well, resulting in a large in-plane hole mass and density of states. The resulting high-joint density of state for an upward carrier transition can give rise to enhanced optical absorption. This can be useful for the design of detectors and modulators.

In summary, it is important to bring out two aspects of strained layers. Thick mismatched layers, in which dislocations may be present, allow us to tune the bandgap and can also serve as buffer layers or even substrates for subsequent pseudomorphic active regions. In pseudomorphically strained layers, which are usually thin, both the bandgap and the bandstructure are altered, and these give rise to changes in electronic and optical properties.

1.6 TRANSMISSION MEDIA AND CHOICE OF MATERIALS

The optical part of the electromagnetic frequency spectrum is divided into three basic bands. These are the infrared, visible, and ultraviolet; and the wavelength and frequency are related by $c = \nu\lambda$, where c is the velocity of light (3×10^8 m/s), ν is the frequency in hertz cycles (sec), and λ is the wavelength of the electromagnetic wave (in meters). The optical spectrum with approximate wavelengths and corresponding frequencies is listed in Table 1.4. It is seen that the optical spectrum spans the wavelength range of 5×10^{-3} to 20 microns (μm). In contrast, audio-frequency waves have wavelengths of several thousand meters and radio frequency waves have wavelengths of the order of a meter.

TABLE 1.4 OPTICAL COLORS AND THEIR WAVELENGTHS (from R. G. Seippel, *Optoelectronics for Technology and Engineering*, Prentice Hall, Englewood Cliffs, NJ, 1989).

	Color	Wavelength[†] (μm)	Frequencies (Hz)
	Ultraviolet	0.005-0.39	6×10^{16}–7.69×10^{14}
	Violet	0.40–0.45	7.5–6.6×10^{14}
Visual	Blue	0.45–0.50	6.6–6.0×10^{14}
response	Green	0.50–0.57	6.0–5.27×10^{14}
	Yellow	0.57–0.59	5.27–5.01×10^{14}
	Orange	0.59–0.61	5.01–4.92×10^{14}
	Red	0.61–0.70	4.92–4.28×10^{14}
	Infrared	0.70–20	4.28×10^{14}–1.5×10^{13}

[†]Approximate only, with overlapping wavelengths and frequencies.

The optoelectronic devices that we will learn about in later chapters are used in communication, local area networks, as sensing devices and as switching and logic devices. In all these applications, light emission, detection, modulation and demodulation, switching, and signal processing are important functions. Depending on the application, light is transmitted through various media. In free-space communication, the medium is air. In guided wave systems, light is transmitted either by optical fibers or by waveguides made of dielectric materials including the compound semiconductors of interest. The choice of materials depends largely on the application and the transmission medium. As shown in Fig. 1.19, optical fibers have their lowest

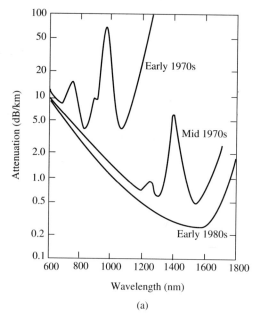

Figure 1.19 (a) Attenuation characteristics (from G. Keiser, *Optical Fiber Communication*, McGraw-Hill, Inc., New York, 1983) and (b) dispersion characteristics (after J. W. Fleming, *Electronics Letter*, **14** (11), 326, 1978) of silica optical fibers (from G. Keiser, *Optical Fiber Communications*, McGraw-Hill, Inc., New York, 1983).

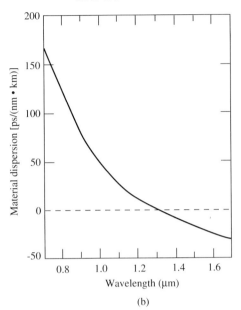

Figure 1.19   (*continued*)

loss at 1.55 μm. The lowest dispersion and optical amplification in fibers (with rare-earth doping) have been demonstrated at 1.3 μm. Therefore, for long distance optical communication using fibers, the materials of choice are InP-based lattice-matched materials, or GaAs- and InP-based mismatched materials. For short-distance optical fiber systems, GaAs/AlGaAs is used, since the fiber losses at 0.8 μm remain within acceptable limits. Fibers are now being developed for use at longer wavelengths, where there is a lot of potential for the Sb-containing compounds. Similarly, as semiconductor sources are developed for radiation in the visible part of the spectrum, using III–V and II–VI compounds, appropriate guiding materials will also be developed. For free-space communication, both long and short wavelength materials can be used, and other systems requirements and advantages determine the choice of materials. In what follows, we will briefly discuss the bandgaps and other important properties of binary III–V compounds and their ternary and quaternary derivatives. The bandgaps and lattice constants of the common III–V binary compounds are seen in Fig. 1.2. The compositional dependence of the energy gap of some important ternary derivatives are listed in Table 1.5.

Among the large bandgap III–V compounds, GaP and GaAs$_x$P$_{1-x}$ are important for the manufacture of visible LEDs. The compositional dependence of the direct and indirect energy gaps of the GaAs$_{1-x}$P$_x$ alloys is depicted in Fig. 1.20. These alloys and heterostructures with different compositions are grown on GaAs or GaP substrates. Ga$_{0.51}$In$_{0.49}$P ($\mathcal{E}_g = 1.96$ eV) and Al$_{0.51}$In$_{0.49}$P ($\mathcal{E}_g = 2.45$ eV) and their heterojunction lattice-matched to GaAs are important for the development of sources and detectors in the visible region of the spectrum. The GaAs-Ga$_{0.51}$In$_{0.49}$P heterojunction is important for bipolar devices and may be important for phototransistors.

TABLE 1.5 COMPOSITIONAL DEPENDENCE OF THE ENERGY GAP OF TERNARY III–V SEMICONDUCTORS AT 300°K[a] (from H. C. Casey and M. B. Panish, *Heterostructure Lasers*, Academic Press, New York, 1978).

Compound	Direct energy gap $\mathcal{E}_g$ (cV)	Indirect energy gap, $\mathcal{E}_g$ (eV)	
		X minima	L minima
$Al_xIn_{1-x}P$	$1.351 + 2.23x$	—	—
$Al_xGa_{1-x}As$	$1.425 + 1.247x + 1.147$ $\times(x - 0.45)^2$	$1.900 + 0.125x + 0.143x^2$	$1.708 + 0.642x$
$Al_xIn_{1-x}As$	$0.360 + 2.012x + 0.698x^2$	—	—
$Al_xGa_{1-x}Sb$	$0.726 + 1.129x + 0.368x^2$	$1.020 + 0.492x + 0.077^2$	$0.799 + 0.746x + 0.334x^2$
$Al_xIn_{1-x}Sb$	$0.172 + 1.621x + 0.43x^2$	—	—
$Ga_xIn_{1-x}P$	$1.351 + 0.643x + 0.786x^2$	—	—
$Ga_xIn_{1-x}As$	$0.36 + 1.064x$	—	—
$Ga_xIn_{1-x}Sb$	$0.172 + 0.139x + 0.415x^2$	—	—
GaP_xAs_{1-x}	$1.424 + 1.150x + 0.176x^2$	—	—
$GaAs_xSb_{1-x}$	$0.726 - 0.502x + 1.2x^2$	—	—
InP_xAs_{1-x}	$0.360 + 0.891x + 0.101x^2$	—	—
$InAs_xSb_{1-x}$	$0.18 - 0.41x + 0.58x^2$	—	—

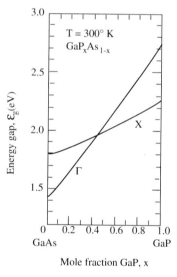

Figure 1.20 Compositional dependence of the direct-energy gap Γ and indirect-energy gap X for GaP_xAs_{1-x} (from M. R. Lorenz and A. Onton, *Proc. Int. Conf. Phys. Semiconduct.*, 10th, Cambridge, MA (S. P. Keller, J. C. Hensel, and F. Stern, eds.), 444, U.S. Atomic Energy Comm., Washington, D.C., 1970).

In the near infrared region, the most important and certainly the most extensively characterized semiconductors are GaAs, AlAs, and their ternary derivatives $Al_xGa_{1-x}As$. The mixed crystals span a wide range of bandgaps, dielectric properties, and electronic properties, and are lattice-matched to GaAs. The GaAs-$Al_xGa_{1-x}As$ heterojunction, being lattice-matched, has a low density of interface defects and dislocations and is therefore widely used for waveguiding and active optoelectronic devices for the near-infrared region of the spectrum. Although this wavelength range does not overlap with the desirable wavelengths of choice in optical fibers, the heterojunction still is used for sources and detectors due to the availability of high-quality GaAs substrates and the ability to epitaxially grow high-quality heterojunctions. GaAs-AlGaAs heterostructure lasers have very low threshold currents, and detectors made of these materials have high reponsitivity, indicating a very high radiative efficiency in these materials.

At longer wavelengths, the materials of importance for optoelectronic device applications are InP and the ternary and quaternary semiconductors lattice-matched to InP. These are $Ga_{0.47}In_{0.53}As$, $Al_{0.48}In_{0.52}As$ and the quaternary compounds InGaAlAs and InGaAsP spanning the bandgap regions in between. The compositional dependence of the energy gaps of $Ga_xIn_{1-x}As$ and $Al_xIn_{1-x}As$ are shown in Figs. 1.21 and 1.22, respectively. The $Ga_xIn_{1-x}As_yP_{1-y}/InP$ ($y = 2.2x$ for lattice-matched crystals) and the $Al_xGa_yIn_{(1-x-y)}As/Al_{0.48}In_{0.52}As$ heterojunctions are extensively used for lasers, light-emitting diodes, detectors, modulators, and solar cells. Since their bandgaps are compatible with the region of low-loss and low-dispersion of optical fibers, these materials and heterojunctions are important for optical fiber communi-

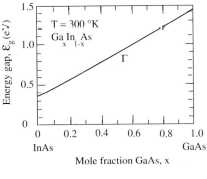

Figure 1.21 Compositional dependence of the direct-energy gap Γ for $Ga_xIn_{1-x}As$ (after B. J. Baliga et al., *Journal of Applied Physics*, **46**, 4608, 1975).

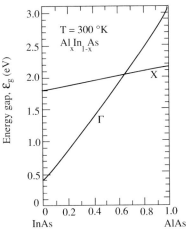

Figure 1.22 Compositional dependence of the direct-energy gap Γ and the indirect energy gap X for $Al_xIn_{1-x}As$ (from M. R. Lorenz and A. Onton, *Proc. Int. Conf. Phys. Semiconduct.*, 10th, Cambridge, MA (S. P. Keller, J. C. Hensel, and F. Stern, eds.), 444, U.S. Atomic Energy Comm., Washington, D.C., 1970).

cation. There are some additional advantages, which lead to applications of these heterojunctions to optoelectronic integrated circuits (OEICs). The electron mobilities in $Ga_{0.47}In_{0.53}As$ are very large (10,000 $cm^2/V.s$ at room temperature), and the Γ-L separation in (0.55 eV) is much larger than that in GaAs (0.28 eV). These attributes make this heterojunction ideal for making high-frequency modulation-doped field-effect transistors (MODFETs). In such devices a channel of carriers is created by transfer-doping from the wide-bandgap material into a triangular quantum well in the narrow-bandgap material at the heterojunction. The carriers in this quantum well have the properties of a two-dimensional electron gas. Thus, the InP-based heterojunctions can be used to make high-performance electronic *and* optoelectronic devices and both devices can then be placed on the same chip in an optoelectronic integrated circuit (OEIC). A photoreceiver circuit, for example, has at its front end a detector and an amplifier following it. Both devices can be epitaxially grown and integrated on the same chip. Similarly, a laser can be integrated with its driver (usually an FET or a bipolar transistor) and/or an external modulator on the same chip. We will learn more about these integrated devices in Chapter 12.

The smaller bandgap materials are useful for applications in the long wavelength range. The III–V compounds of interest here are the antimony(Sb)-bearing compounds and InAs and their ternary and quaternary derivatives. The $InAs_xSb_{1-x}$ alloys have the lowest bandgaps of all the known III–V compounds and offer the promise to be applicable in the 8–12 μm wavelength range. There are two problems, however, associated with the use of the Sb-bearing compounds. First, many of them have a miscibility gap in the phase diagram, and therefore the binary compounds and their alloys cannot be easily grown of the quality desirable for most optoelectronic devices. They can be easily grown by techniques such as molecular beam epitaxy (MBE), but the material may be clustered. The second problem associated with use of the Sb-bearing alloys is the lack of suitable substrates. InAs, InSb, and GaSb substrates are available, but not as readily as GaAs and InP. Moreover, all the ternary and quaternary alloys of interest are mismatched to these substrates. The only exceptions, as seen in Fig. 1.2, are the $Al_x Ga_{1-x}Sb$ alloys grown on GaSb, where the mismatch is very small. The GaSb-AlGaSb heterojunction may therefore become important.

We have, until now, confined our discussion only to lattice-matched materials and heterojunctions. As we have seen in Sec. 1.5, another degree of freedom in bandgap tailoring is obtained if lattice-matching requirements are forgone, or if thin, coherently strained active regions are used. The use of quantum wells provides additional tunability of the bandgap and therefore the wavelength of use. Also, in the context of this text, we have restricted ourselves to the III–V compounds. The II–VI compounds are also of immense importance for application to optoelectronic devices, both in the short- and long-wavelength regions. Visible lasers emitting in the blue region of the spectrum have been demonstrated, using CdZnSe/ZnSe junctions. Similarly, very far infrared detectors made of InAsSb/InSb and HgCdTe/CdTe heterojunctions are commercially available. These compounds and heterojunctions are synthesized by the various epitaxial techniques described in the next section.

1.7 CRYSTAL GROWTH

1.7.1 Introduction

An important difference between the growth of elemental and compound semiconductors is the difficulty in maintaining stoichiometry, the correct chemical ratio for
the compound's constituent elements, while at the same time controlling the thermal
parameters determining the crystalline perfection. These materials usually have one
component that is volatile and this component is lost more readily, at high temperatures. For example, in GaAs and InP, the group V components, As and P, are more
volatile than Ga and In, and therefore when the compounds are heated, the group
V components evaporate more readily, leaving behind group III-rich material. This
phenomenon is called incongruent evaporation and can pose special problems in the
growth and processing of these semiconductors. The single crystal materials that are
commonly used are divided into two categories: *substrate* or *bulk* crystals, and epitaxial layers. The substrate crystal is first formed and then the epitaxial layer is grown
on it. The substrate not only constitutes the seed crystal, but it plays an important role
in the fabrication of devices in that it provides the contact region, device isolation,
etc. In what follows, the techniques to synthesize the substrate materials and epitaxial
layers are briefly discussed.

1.7.2 Bulk Crystal Growth

Bulk crystals of compound semiconductors are produced by two common techniques.
Both are also known as melt growth techniques. In the first the starting materials and
seed crystal, are placed in a quartz tube, which is placed in a furnace. Hence, this
technique is also known as furnace growth. In the second technique the seed crystal
is dipped in a melt contained in a pressurized vessel and gradually pulled, while being
rotated at the same time. This results in single crystal growth of the substrate ingot.
In both cases growth is accomplished in a sealed environment to prevent the loss of
the more volatile constituents.

The furnace technique is illustrated in Fig. 1.23. The starting materials, polycrystalline GaAs and a seed crystal, are placed in a quartz boat, which itself is placed
in a sealed quartz tube or ampoule. Excess arsenic is provided in a second boat to
ensure stoichiometry during growth. Growth is achieved by the controlled cooling
of the molten starting materials, starting at the seed crystal. Either a fixed profile
hot zone is moved along the tube through the use of heating elements and a moving
boat or the temperature of the hot zone is varied with the heater in a fixed location.
The first is called the horizontal Bridgman (HB) technique and the latter is called the
gradient freeze method. In either technique the controlled cooling at the solid-melt
interface ensures single-crystal growth. The starting materials used are of high purity,
to ensure low impurity levels in the resulting ingot. The furnace method of growth is
relatively inexpensive and most bulk GaAs is currently produced by this technique. P-
containing compounds such as InP and GaP require a very large overpressure (10–20
atmospheres) and cannot therefore, be grown by this technique. There are, however,
some problems associated with the furnace growth technique. First, the lateral size

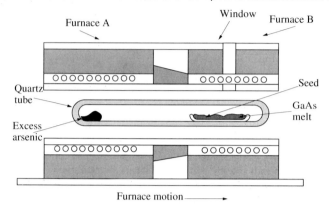

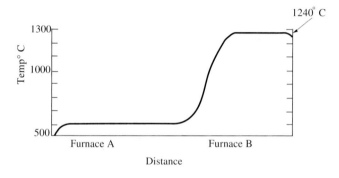

Figure 1.23 Schematics of a horizontal furnace and quartz tube with growth constituents for the growth of GaAs bulk crystals.

of the ingot is determined by the size of the boat. Being in intimate contact with the quartz crucible during growth, large amounts of Si can get incorporated in the growing crystal. Therefore, for the realization of high-resistivity materials, additional compensating acceptor dopants such as Cr may be needed. Finally, contact of the growing crystal with the walls can introduce stresses during solidification. In general, however, since thermal gradients are small, furnace grown materials tend to have lower dislocation densities.

GaAs and InP substrate crystals are also grown by the pulling technique, due originally to Czochralski. In this method the seed crystal is dipped into a solution and slowly withdrawn. For compound semiconductors the solution, contained in a crucible, is the molten compound. The melt may be first formed as polycrystalline material in a separate furnace. When a high-pressure crystal pulling apparatus is used, the melt can be synthesized in place in a chemical reaction between the constituent elements. The crucible is usually made of PBN instead of quartz to minimize contamination of the melt and ensure the production of undoped high-resistivity crystals. There are several advantages of this technique of bulk crystal growth. Orientation control is possible because the growing crystal has the same orientation as the seed crystal. There is no contact between the growing crystal and the crucible. Large diameter crystals can be made whose crystallographic quality is very good and its doping is controllable. There are, however, a few disadvantages. The diameter control of the growing crystal is very difficult, since it needs very accurate temperature control, to within $1°C$, at the solid-liquid interface. If the temperature of this interface drops, the growth rate increases and the diameter of the crystal increases. This results in more heat dissipation from the extended crystal surface. As a result the interface cools down further and the diameter continues increasing. To correct this situation, the pull rate is increased, which restores the correct temperature at the solid-liquid interface. This phenomenon can result in a crystal of uneven diameter. Finally, control of the ambient pressure is difficult. Since the growing crystal is at its melting point, the more volatile group V component will vaporize. For GaAs the minimum vapor pressure that needs to be maintained is 1 atmosphere at the melting point. For GaP the vapor pressure is 20 atmosphere at the melting point. If the growth ambient is sealed and the pressure inside the chamber is maintained to be the same as the vapor pressure of the more volatile component, then its loss by evaporation is eliminated. For the growth of GaAs inert Ba_2O_3, which melts at $700–800°C$, is placed on top of the melt and prevents the As from vaporizing. This encapsulant materials is a transparent solid at room temperature. In the growth of GaP, in addition to the Ba_2O_3 encapsulant, the chamber pressure is increased to 40 atmosphere by inserting a high-pressure inert gas such as Ne, Ar, N, etc. The contamination from the encapsulant or the gases is very small. The whole process just described is called the *liquid encapsulation Czochralski* (LEC) technique. It is a common technique for synthesizing bulk crystals of GaAs, InP, and GaP. The LEC apparatus is shown in Fig. 1.24.

After growth is completed the whole system is cooled and the crystal is removed from the puller. It is then ground into a round cylinder, flats ground on the cylinder to indicate the major and minor flats for orientation, and finally sliced into thin wafers, as shown in Fig. 1.25. The wafer surfaces are polished to a mirror-like finish to make them ready for epitaxial growth and device applications. A typical ingot diameter ranges from 3–5 in. and yields 50–150 wafers.

1.7.3 Epitaxial Material Growth

Epitaxy is the regular oriented growth of a single crystal with controlled thickness and doping, over a similar single crystal called the substrate. Originally epitaxy was used to provide materials characteristics superior to that in bulk crystals. Epitaxial techniques are now used to realize junctions, ultrathin layers, and multiquantum wells with unprecedential control over purity, doping, and thickness profiles. Very often epitaxy replaces techniques such as diffusion and implantation in the fabrication of device structures. There are many forms of epitaxy, but the most common ones are liquid phase epitaxy, vapor phase epitaxy, and molecular beam epitaxy. These epitaxial techniques have been described in detail in many texts and monographs. The purpose of the following brief description of each is to familiarize the reader with the basic techniques and their advantages and disadvantages.

1.7.3.1 Liquid Phase Epitaxy.   Historically, liquid phase epitaxy (LPE) was the first epitaxial process used for the growth of compound semiconductors. In this technique, crystal growth on a parent substrate results from the precipitation of a

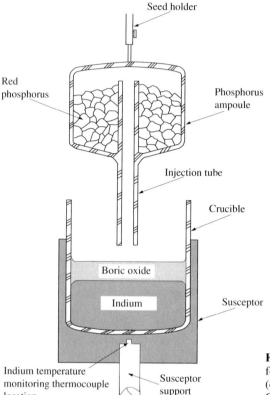

Seed holder

Red
phosphorus

Phosphorus
ampoule

Injection tube

Crucible

Boric oxide

Indium

Susceptor

Indium temperature
monitoring thermocouple
location

Susceptor
support

Figure 1.24   The LEC apparatus
for growing bulk InP crystals
(courtesy of G. A. Antypas and
Crystacomm, Inc.).

crystalline phase from a saturated solution of the constituents. The crystallographic
orientation of the grown layer is determined by that of the substrate. This technique
of growth is intrinsically capable of growing very pure films but has the limitation of
being unable to realize ultrathin layers.

In LPE, growth occurs from a solution at a temperature well below the melting point of the semiconductor. The principle of growth is based on a simple fact
that a mixture of the semiconductor with one of the constituent elements melts at a
much lower temperature than the semiconductor itself. As an example, if a solution
or mixture of GaAs and Ga (which is a liquid at temperatures slightly above room
temperature) is formed, then it is possible to recrystallize GaAs from the solution by
cooling it at temperatures from 600° to 800°C, which is much lower than 1237°C,
the melting point of GaAs. Usually molten Ga or In are used to form the melt, and
therefore LPE is commonly described as solution or melt growth. Growth occurs by
association, in which Ga and As combine in the liquid phase to form GaAs so that in
solution both GaAs molecules as well as Ga and As atoms are present. As the temperature of the solution approaches the melting point of GaAs, the structure of the liquid
becomes increasingly similar to that of the solid. This suggests that when the actual
crystallization of the solid starts from the melt, there is already a large percentage

(a)

(b)

Figure 1.25 (a) InP ingot grown by the LEC technique and (b) sliced wafers showing major and minor flats along the [110] directions (courtesy of G. A. Antypas and Crystacomm, Inc.).

of GaAs molecules present in the melt. It has been suggested that the following equilibrium exists in the solution between atoms (A and C) and molecules (AC):

$$AC \rightleftharpoons A + C \tag{1.31}$$

Liquid phase epitaxy is an equilibrium growth technique and the process of growth is governed by the laws of thermodynamic phase equilibria. Therefore, one refers to the liquidus-solidus phase diagram. For most binary III–V compounds, such as GaAs, the diagram is represented in its familiar schematic form in Fig. 1.26(a). Most of this diagram represents the equilibrium between the liquid phase and the *congruently melting* solid. Congruent melting occurs at a fixed temperature at which the solid and the liquid have the same composition. In Fig. 1.26(a), the liquid-solid phase equilibrium is represented by the upper curve. Since the melt or solution is rich in component A (group III), the left side of the curve is useful for liquid phase epitaxy. Growth can be initiated at the temperature T_2 and terminated at temperature T_1, both temperatures being much lower than the melting temperature (also called the temperature of *fusion*) T_F. The resulting crystalline solid may therefore be slightly rich in component A (e.g., $A_{0.5001}C_{.4999}$) instead of $A_{0.5}C_{0.5}$. This small departure from stoichiometry does not degrade the crystalline properties. In general, lower temperature growth is preferred, since diffusion processes are minimized and impurity and defect concentrations are lower. However, a lower limit of the growth temperature is set by surface diffusion. For growth of compounds such as GaAs and InP, typical growth temperatures range between 600° and 700°C. In actual practice, the melt consisting of component A saturated with component C by dissolution at temperature T_2 is brought in contact with the substrate and the temperature is gradually ramped down. The solid AC leaves the solution and recrystallizes on the substrate. As the cooling continues, the epitaxial crystal continues to grow. The rate of cooling decides the growth rate and the epitaxial layer thickness is determined by the total duration during which the melt is cooled and held on the substrate. Calculated and measured phase diagrams for some binary compounds are shown in Fig. 1.26(b). These curves

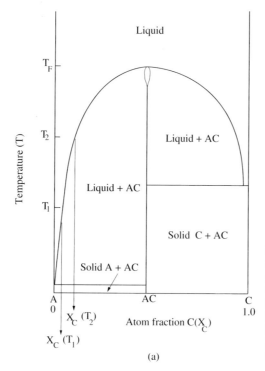

Figure 1.26 (a) Schematic representation of liquidus-solidus phase equilibria for a binary system AC and (b) calculated and measured liquidus composition as a function of reciprocal temperatures for different binary III–V systems (from M. B. Panish, *Journal of Crystal Growth*, **27**, 6, 1974).

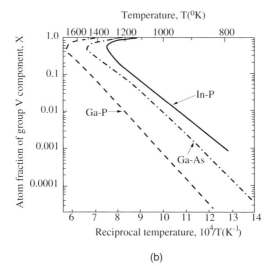

(b)

Figure 1.26   (*continued*)

give the equilibrium compositions in the liquid phase as a function of temperature, and are therefore called *liquidus* curves. They provide the parameters for actual LPE growth.

The apparatus for LPE growth is relatively simple and inexpensive. It consists of a graphite slider and boat assembly inside a quartz furnace tube with accurate temperature control. The melt is held in a groove, or *bin*, in the boat and the polished and cleaned substrate is held, with the growth surface pointing downward, in an opening in a slider. The slider can move along a groove cut along the length of the boat. As an illustration, consider the growth of GaAs. The saturated melt of GaAs in Ga is held at the appropriate temperature, and the slider is moved so that the substrate is positioned on top of the melt and is in contact with it. The melt temperature is then lowered at a predetermined rate during which the atoms from the melt crystallize on the substrate surface to form the epitaxial layer. After the required thickness of the layer is grown, the slider, with the substrate, is moved away from the melt and the whole system is cooled down.

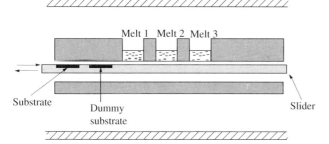

Figure 1.27   Schematics of the LPE growth technique showing boat and slider assembly having multiple bins for different melts. Such a system is suitable for heterostructure and junction growth.

For the growth of heterojunctions a boat with multiple bins, as shown in Fig. 1.27, is used. Each bin contains a melt with a different composition and the substrate is moved from one to the other to grow multiple layers, as is required for a laser. Several of these layers also needed to be doped to produce free carriers needed for the optical processes or for conduction. Impurity incorporation by doping is also determined by consideration of the equilibrium constant for phase equilibria. Thus, to dope a particular layer to a desired level, a predetermined amount of the dopant species in elemental form is dissolved in the appropriate melt. Common n-type dopants are Sn or Te, and the common p-type dopant is Zn.

In conclusion, it is fair to state that the LPE technique is characterized by simplicity of equipment and the general safety associated with it. Expensive vacuum equipment is not required. Also, the deposition rates are high and the crystal quality is very good. Computer-controlled equipment results in a high degree of reproducibility and thickness and composition control. However, due to the nature of growth from the melt, ultrathin layers, ≤ 1000 Å, or hyperabrupt junctions cannot be grown with precise control. The LPE technique is widely used to grow III–V, II–VI, and IV–VI compounds and laser structures with these materials.

1.7.3.2 Vapor Phase Epitaxy.   In this technique of epitaxial growth crystalline deposition of a semiconductor layer from a gaseous ambient results from a chemical reaction or decomposition. Vapor phase epitaxy (VPE), also termed chemical vapor deposition (CVD), is one of the most widely used techniques of crystal growth. CVD refers to the formation of a condensed phase from a gaseous medium of different chemical composition. It is distinguished from the physical vapor deposition processes such as sublimation, sputtering, or molecular beam epitaxy, where condensation occurs in the absence of a chemical change. In vapor phase epitaxy, a mixture of gases stream through a reactor and interact on a heated substrate to grow an epitaxial layer.

There are three different CVD techniques used for the growth of III–V and other compound semiconductors. These are the halide process, the hydride process and the organometallic chemical vapor deposition (OMCVD) technique. Chemical transport processes form a special class of CVD where initially a condensed phase (source) reacts with a transport agent to produce volatile species that, in another region and under different conditions, undergo chemical reactions to re-form the condensed phase. Chemical transport processes require both a source and a transport agent, whereas other CVD processes employ only gaseous reactants that are stable at room temperature and consequently require no separate condensed source. In any CVD process the chemical processes and apparatus must be selected with great care. Hydrogen is attractive as a carrier gas, since it can be purified to a high degree by palladium diffusion. In the halide or chloride process for the growth of GaAs, for example, the source materials are elemental Ga and $AsCl_3$, a liquid, both of which can be obtained in high purity. The $AsCl_3$ is put in bubblers through which hydrogen is passed. The resulting gas flows over the Ga source and then onto the substrate. The reactor tube is heated in the source (Ga boat) region to about $1,000°C$ and in the deposition (substrate) region to about 800°–850°C. When the AsCl$_3$-saturated hydrogen flows over the source region, the following reaction takes place:

$$AsCl_3(g) + \frac{3}{2}H_2(g) \rightharpoonup As(g) + 3HCl(g) \tag{1.32}$$

The HCl formed in this reaction reacts with the source Ga to produce GaCl, according to

$$Ga(l) + HCl(g) \rightharpoonup GaCl(g) + \frac{1}{2}H_2(g) \tag{1.33}$$

The following reaction then takes place over the heated substrate, resulting in crystalline growth:

$$GaCl(g) + As(g) + \frac{1}{2}H_2(g) \rightleftharpoons GaAs(s) + HCl(g) \tag{1.34}$$

This reaction is exothermic, and reversible to some extent. No growth occurs initially with a fresh Ga source, because the As dissolves in the molten Ga until it is saturated and GaAs precipitates are produced. So, in reality, the reverse of reaction (Eq. 1.34) proceeds to produce GaCl instead of reaction (Eq. 1.33). Usually Zn is used for p-type doping, and S or Se, in the form of H$_2$S or H$_2$Se are used as n-type dopants and controlled amounts of these are fed into the gas stream. Chloride-based VPE produces very pure layers, since the liquid source can be made very pure by distillation.

In the *hydride* process, gaseous AsH$_3$ or PH$_3$ are used as the sources for the group V elements. The hyrides decompose in the high temperature zone of the furnace according to

$$AsH_3(g) \rightharpoonup As + \frac{3}{2}H_2 \tag{1.35}$$

HCl reacts with the metallic group III source to produce the chloride according to Eq. 1.33 above, and finally, in the deposition or substrate region epitaxial growth occurs according to the reaction (Eq. 1.34). The hydride process allows control over the ratio of group III to group V vapor phase species. It is the common technique used in the manufacture of light-emitting diodes and detectors.

The OMCVD or organometallic vapor phase epitaxy (OMVPE) growth technique has emerged as a technologically important one for the production of single layers, heterojunctions and quantum well structures with excellent control over layer thickness, and doping and the achievement of hyperabrupt isotype and anisotype junctions. In this technique organometallic compounds such as metal alkyls and hydrides as vapors are transported to the growth chamber using a carrier gas such as hydrogen. The substrate wafer is mounted on a graphite holder or susceptor that is held at the growth temperature by rf heating. A simple OMCVD growth system is schematically shown in Fig. 1.28(a) and a present day multiwafer growth system is illustrated in Fig. 1.28(b). A chemical reaction occurs over the substrate to produce epitaxial growth. The basic reaction is an irreversible pyrolysis, in which intermediate compounds are formed, with the final compound depositing on the substrate surface. For example, for the growth of GaAs, the basic reaction is

$$Ga(CH_3)_3(l) + AsH_3(g) + H_2(g) \rightharpoonup GaAs(s) + 3CH_4(g) + H_2(g) \tag{1.36}$$

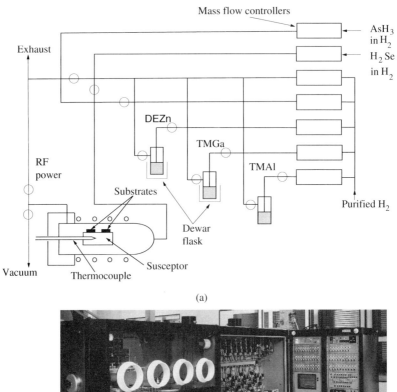

(a)

(b)

Figure 1.28 Organo-metallic vapor phase epitaxy (OMPVE) system: (a) schematics of a horizontal reactor with gas flow network and (b) a multi-wafer growth facility (courtesy of E. Woelk, AIXTRON GmbH, Aachen, Germany).

which is the analog of the silane reaction used to grow epitaxial silicon. OMVPE growth has a number of important and advantageous features. Since there is no source material with which equilibrium has to be reached, compounds such as GaAs can be grown with different As to Ga ratio simply by varying the relative vapor pressures of

AsH$_3$ and trimethyl gallium (TMG). This becomes important in the study of native defects and trap states, which might be related to stoichiometry. All the reactants and dopants can be in the vapor phase and only the substrate is heated. The equipment is simple and shorter turnaround times are possible. The absence of halides limits auto-doping and therefore abrupt p-n junctions can be grown. Essentially all the III–V compounds and their ternary and quaternary derivatives can be grown with this technique and II–VI compounds such as HgCdTe have also been grown, sometimes with photo-assisted reaction kinetics. Proper design of the gas flow system including precision mass flow controllers and evacuation systems for residual in-line gases makes it possible to grow ultrathin layers and abrupt junctions with precise doping control. As source purities continue to increase, compounds of very high purity can be realized by the OMCVD process. The process has also demonstrated great potential for growth on large-area substrates, selective-area growth, and patterned growth and shows great promise as a manufacturing process.

To summarize, the various VPE techniques have proved to be extremely important for the manufacture of optoelectronic devices. In the halide and hydride processes, the high growth temperature can result in higher impurity diffusion. This disadvantage is offset by the production capabilities of these techniques.

1.7.3.3 Molecular beam epitaxy.

In contrast with LPE and VPE growth of semiconducting crystals under quasi-equilibrium conditions, growth by molecular beam epitaxy is accomplished under nonequilibrium conditions and is principally governed by surface kinetic processes. Since the initial demonstration of the principle and process, MBE has come a long way and has emerged as a technique with unprecedented control over growth parameters. This has resulted from extensive research and is leading to a better understanding of the growth processes and to great advances in system design and reliability.

Molecular beam epitaxy is a controlled thermal evaporation process under ultrahigh vacuum conditions. The process is schematically shown is Fig. 1.29 for the

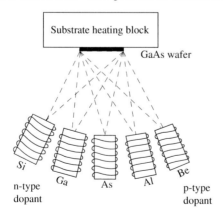

(a)

Figure 1.29 Schematics of the MBE (a) growth process and (b) growth chamber.

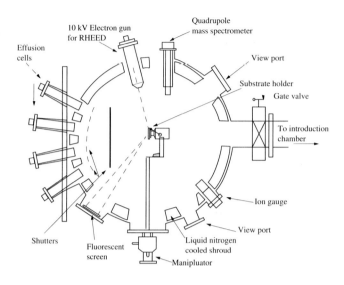

(b)

Figure 1.29 (*continued*)

growth of III–V compounds with the possibility of dopant incorporation. The cells are designed in such a way that realistic fluxes for crystal growth at the substrate can be realized, while the Knudsen effusion condition is maintained (i.e., the cell aperture is smaller than the mean free path of the vaporized effusing species within the cell). The flux density of a beam incident on the substrate surface is controlled by the temperature of the effusion cell provided the cell aperture is less than the mean free path of vapor molecules within the cell (i.e., Knudsen effusion). The flux F_i of species i per unit area at the substrate surface is given by

$$F_i = \frac{A_c P_i}{\pi d_s^2 \sqrt{2\pi m_i k_B T}} \cos \phi \quad (\text{atoms cm}^{-2}\text{s}^{-1}) \tag{1.37}$$

where A_c is the area of the cell aperture, P_i is the equilibrium vapor pressure of species i (in Torr) in the cell at the absolute temperature T, d_s is the distance of the orifice from the substrate, m_i is the mass of the effusion species, and ϕ is the angle between the beam and the normal of the substrate. The individual cells are provided with externally controlled mechanical shutters whose movement times are less than the time taken to grow a monolayer. Therefore, very abrupt composition and doping profiles are possible. Interfaces that are one monolayer abrupt can be obtained fairly easily.

The process of crystalline growth by MBE involves the adsorption of the constituent atoms or molecules, their dissociation and surface migration and, finally, incorporation, resulting in growth. From pulsed molecular beam experiments it has been observed that, below a substrate temperature of 480°C, Ga has a unity sticking coefficient on (100) GaAs. Above this temperature, the coefficient is less than unity. The absorption and incorporation of As$_2$ or As$_4$ molecules are more complex. It was found that, in general, As sticks only when a Ga adatom plane is already established.

The cation is in atomic form when it reaches the heated surface. It attaches randomly to a surface site and undergoes several kinetically controlled steps before it is finally incorporated. The As_2 or As_4 atom is first physisorbed into a mobile, weakly bound precursor state. As the state moves on the surface, some loss occurs due to reevaporation, and the rest are finally incorporated in paired Ga lattice sites by dissociative chemisorption. An important point to note in the MBE growth of III–V compounds is that the cation incorporation rate is close to unity. Under these conditions, the growing crystal will retain a smooth and atomically abrupt surface only if the cation surface migration rate is very high ($\geq 10^4$ hops/second). If the cations are given insufficient time or energy, then before a monolayer is completed (i.e., the cation is unable to move to an island step-edge) a new island will form on top of the former. Thus, the surface roughens and begins to resemble an assembly of three-dimensional islands. The growth rate R_g of III–V compounds is entirely controlled by the flux densities of the Gr III beams F_i:

$$R_g \infty \sum_{i=1}^{n} \alpha_i^s F_i \qquad (1.38)$$

where n is the number of the different Gr III elements and α_i^s are their respective sticking coefficients. The processes involved in cation and anion incorporation are schematically illustrated in Fig. 1.30. The important fact that emerged from the early kinetic studies is that stoichiometric GaAs can be grown over a wide range of substrate temperatures by maintaining an excessive overpressure of As_2 or As_4 over the Ga beam pressure. Substrate temperatures during growth are usually close to, and slightly higher than, the congruent evaporation temperature of the growing compound.

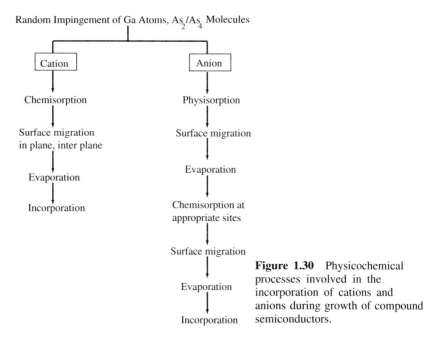

Figure 1.30 Physicochemical processes involved in the incorporation of cations and anions during growth of compound semiconductors.

A typical, present-day MBE growth facility consists of the UHV growth chamber into which the substrate is introduced through one or two sample-exchange loadlocks. The base pressure in the growth chamber is usually $\sim 10^{-11}$ torr, while the other chambers are at $\sim 10^{-10}$ torr. The effusion cells are made of pyrolithic BN. To obtain the dimers from the tetramer species of the group V elements, external cracking cells are sometimes incorporated. Cracking is enhanced in the presence of a loosely packed catalytic agent. Charge interlocks with auxiliary pumping are sometimes used for the more rapidly depleting group V species. The growth chamber and effusion cells are provided with liquid N_2 cryoshrouds, which are kept cold during growth. One of the systems being used in the author's laboratory is shown in Fig. 1.31.

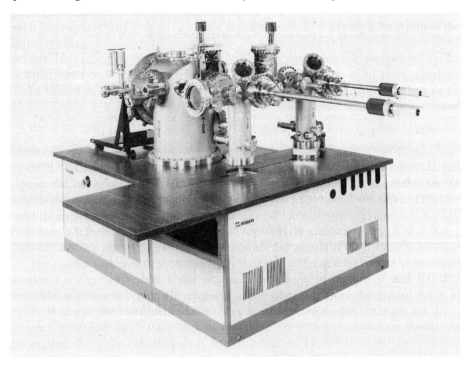

Figure 1.31   Photograph of an MBE growth facility in the Solid State Electronics Laboratory, University of Michigan (courtesy of RIBER SA).

Most growth systems are equipped with in situ surface diagnostic and analytical capabilities in the growth and auxiliary chambers. The most common facilities in the growth chamber are a quadruple mass spectrometer (or residual gas analyzer), which gives important information regarding the ambient in the growth chamber at all times, and a reflection high-energy electron diffraction (RHEED) system, which gives insight to the growth mechanism and surface reconstruction. The forward scattering angle of the RHEED makes it compatible with MBE. The RHEED pattern is formed by the elastic scattering of electrons from the periodic crystal lattice. The shallow grazing angle of the electron beam implies that it penetrates only the uppermost layers of the crystal. Thus, in the simplest sense, diffraction basically takes place from the first layer. Therefore, the RHEED pattern shows streaks, not spots, normal to the shadow edge. The incident electron beam does in actuality penetrate a little into the bulk crystal, which means at various azimuths, the RHEED pattern is a superposition of the bulk reconstruction and surface reconstruction.

1.7.3.4 Chemical beam epitaxy.   This is a technique similar to MBE except that all the sources are either alkyl or hydride. Growth is done under high vacuum in a conventional MBE growth system. Several variations of this technique are currently being used. For example, use of solid group III sources and hydrides for the group V is termed gas-source MBE (GSMBE). Use of alkyl group III sources and elemental group V sources is termed *metal-organic MBE* (MOMBE). Use of alkyl group III sources and hydride group V sources is called *chemical beam epitaxy* (CBE). All these techniques combine the advantages of MBE with semi-infinite sources. Use of gas sources, particularly for the group III elements, results is lower defect densities on the surface of the growing crystal.

1.7.4 Factors Controlling Heterointerface Quality

In the context of MBE, each semiconductor will have its own optimized growth temperature. For instance, Table 1.6 shows the physical constants of three important binary compounds. A comparison of the congruent temperature, melting point and heat of formation of the binaries reveals that the ratio of the magnitudes of these parameters are approximately 0.80/1.00/1.25 for InAs, GaAs, and AlAs, which are expected to be the ratio of their relative bond strengths. The bond energies are approximately equal to $c_{11}a^3/2$ and therefore of the order of 1 eV/bond. The ratio of optimized growth temperatures for these binaries should be in the ratio of

TABLE 1.6   PHYSICAL PARAMETERS FOR THE BINARY COMPOUNDS InAs, GaAs, AND AlAs USED TO ESTIMATE THE RATIO OF THEIR RELATIVE BOND STRENGTHS. THE RATIO OF BOND STRENGTHS CALCULATED FROM EACH PARAMETER ARE IN PARENTHESIS.

Parameter	InAs	GaAs	AlAs
Bandgap (300°K)	0.36 eV	1.43 eV	2.16 eV
Lattice Constant (300°K)	6.05838 Å	5.65315 Å	5.6622 Å
Congruent Temperature	380°C	630°C	850°C
Relative Bond Strength	(0.71)	(1.00)	(1.22)
Melting Point	942°C	1237°C	1600°C
Relative Bond Strength	(0.80)	(1.00)	(1.24)
Heat of Formation	115 kcal/mole	130 kcal/mole	150 kcal/mole
Relative Bond Strength	(0.89)	(1.00)	(1.1)

their relative bond strengths. The optimized growth temperature of any compound formed from a combination of binaries should be linearly interpolated using the binary data. In this context, we can think of a "growth window" for which optimized layer-by-layer growth for compound A takes place, and perhaps a different growth window for compound B. Forming a perfect heterointerface is dependent on compounds A and B having overlapping windows of growth. Often the windows do not overlap and novel growth techniques are required. The differences between growth temperatures of A and B will, of course, increase as their bond energy difference increases.

For typical growth, if compound A is the lower bandgap material and compound B the higher bandgap, then growth conditions are usually dictated by the growth window of compound A. This is because compound A, having the lower bandgap, is usually the active region of the device. A normal interface, with compound A in the bottom and B on top, is expected to be fairly good since the growth front of A would mostly determine the heterointerface. But growth of a single quantum well (SQW), would involve both normal and inverted interfaces. Since growth of compound B is not optimized, the inverted heterointerface could be quite rough. An example of a rough interface is shown in Fig. 1.32. It is important to add, however, that techniques such as MBE and gas-source MBE have demonstrated unprecedented thickness control and interface abruptness, as illustrated in Fig. 1.33.

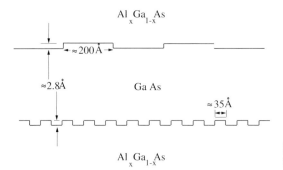

Figure 1.32   Schematic representation of a rough heterointerface.

1.7.5 Doping of Semiconductors

In an *intrinsic* or pure semiconductor, there are no extra atoms, other than those constituting the lattice structure. At $0°K$ in such a semiconductor, there are no *free carriers*, and hence electrical conduction processes are absent. As the temperature is raised, electron-hole pairs are created by thermal generation, but their numbers are still quite small. For more free carriers to be present in a semiconductor material, it has to be *doped* with an appropriate impurity, or element in the periodic table. Such a semiconductor is called an *extrinsic* semiconductor. Undoped materials grown by any of the epitaxial techniques described in the previous section are relatively pure, but have far more carriers than in intrinsic material. For example, the purest GaAs

Figure 1.33 High resolution (011) TEM cross-section showing a lattice image of an $In_{0.53}Ga_{0.47}As/InP$ quantum well structure grown by gas source MBE. The top well is ~ 3 molecular layers thick and the bottom well is ~ 9 molecular layers thick. Each bright spot has dimension along (100) direction corresponding to a pair of III–V atoms (courtesy of S. N. G. Chu, AT&T Bell Laboratories).

that can be grown by epitaxial techniques will have ~ 10^{13} cm^{-3} free carriers. These carriers arise from ubiquitous impurities that give rise to donor and/or acceptor levels. Such impurity species can originate from the source materials, carrier gases, process equipment, or even substrate handling.

Extrinsic semiconductors are made by adding dopants during growth. The dopant species are chosen from the appropriate column of the periodic table. Usually minute quantities of such dopants are required, ranging from one atom for every 10^3–10^{10} host lattice atoms. Depending on the type of dominant charge carrier in the semiconductor, electron or hole, extrinsic semiconductors are either n- or p-type. We will briefly discuss the process of doping in compound semiconductors. If atoms in the group III sublattice of a III–V compound are substituted by an element from column IV, for example, then an n-type semiconductor is obtained. The specific element is called a *donor*. The three outer or valence electrons of the donor atom substituting a group III host atom are shared with the valence electrons of the neighboring atoms and as a result, it has net positive charge. The fourth electron is not quite free, but rotates around the impurity atom in an orbit having a large radius. In other words, it traverses several lattice spacings in its orbit and is therefore quite delocalized in real space. The situation is similar to a single electron rotating around a hydrogen atom, and, by this analogy, the ionization energy, or binding energy of the donor atom can be expressed as

$$\mathcal{E}_D = -\frac{13.6}{\epsilon_r^2}\left(\frac{m_e^*}{m_o}\right) \quad (eV) \quad\quad (1.39)$$

where ϵ_r is the relative dielectric constant of the semiconductor, m_o is the free electron mass, and m_e^* is the electron conductivity effective mass. Remember that 13.6 eV is the binding energy of the electron in the hydrogen atom. The radius of the orbit of the fourth electron is again given by the hydrogenic model as

$$r_d = 0.05\epsilon_r \left(\frac{m_o}{m_e^*}\right) \quad (nm) \tag{1.40}$$

where 0.05 nm is the Bohr radius. It is clear that r_d is many times the Bohr radius, and this justifies use of the macroscopic parameters of the material, ϵ_r and m_e^*, in the above equations. In a typical III–V compound, $\mathcal{E}_D$ varies in the range 3–8 meV and is extremely small. Donor atoms give rise to an extra discrete energy in the energy gap very close to the conduction band edge, as shown in Fig. 1.34(a). As shown in this figure, $\mathcal{E}_C - \mathcal{E}_D$ is the donor ionization energy. In other words, only this small amount of energy is necessary to raise the loosely bound electron to the conduction band, where it is free to move in a large density of available states and contribute to electrical conduction. As an example, if the doping level is 1 atom in every 10^6 of host atoms, then the donor atom's density is 10^{16} cm^{-3} and there are that many extra electrons that can be ionized and exist in the conduction band.

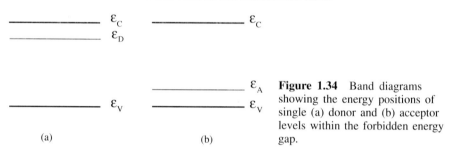

Figure 1.34   Band diagrams showing the energy positions of single (a) donor and (b) acceptor levels within the forbidden energy gap.

Similarly, if atoms in the group V sublattice of a III–V compound are substituted by elements from group IV or lower, then a p-type semiconductor is obtained. In this case the particular element is called an *acceptor*; for example, if the acceptor atom is a group IV element, it has four valence electrons. When it is in a substitutional site in the group V sublattice, there is one electron short for completing the five covalent bonds between it and the neighboring atoms. This incomplete bond itself is not a hole, since it is bound to the atom. However, at some temperature above 0°K, an electron from a covalent bond of a neighboring host atom can complete the impurity bonding, and the absence of this electron from the host atom creates a hole. In the band picture, acceptor impurities give rise to discrete states close to the valence band edge as shown in Fig. 1.34(b). Above 0°K, an electron is raised to the acceptor level, completing its bonds and simultaneously leaving a hole in the valence band for conduction. The ionization energy of the acceptor atom is given again by Eq. 1.39 with the electron effective mass replaced by the heavy hole effective mass. Therefore, the ionization energy of the doubly degenerate shallow acceptor levels is generally higher than that of shallow donors. In the bonding picture it can be imagined that the loosely bound "extra" hole with a positive charge rotates around the impurity atom core, which has a negative charge as in the hydrogen atom. The radius of this orbit can also be calculated using Eq. 1.40 after replacing the electron mass with the hole mass.

In considering dopant species for n- and p-type doping, the relative atomic radii of the host lattice atom and the impurity atom are important. Certain elements can act as *amphoteric* dopants in compound semiconductors. For example, Si, from column IV in the periodic table is an amphoteric dopant in GaAs. In other words, depending on the growth conditions, Si may incorporate in the group III sublattice replacing Ga atoms to act as a donor, or may incorporate in the group V sublattice replacing As atoms to act as an acceptor. Also in any real semiconductor there are unintentionally introduced impurities giving rise to donor and acceptor levels. Therefore, there may be multiple donor and acceptor levels. The electrons and holes generated from these levels *compensate* each other, giving rise to a *net* donor or acceptor density. However, when we dope a semiconductor for a specific application, the doping level is usually higher than the background unintentional doping levels. In fact, it is the objective of all epitaxial materials growth to grow undoped material which is nearly free from unwanted background impurities. We will learn more about compensation in Chapter 2. Energy levels produced in the bandgap of GaAs by different impurity species are shown in Fig. 1.35.

Doping is done differently in the different epitaxial processes. In liquid phase epitaxy, the pure element is usually dissolved in the appropriate melt, or solution. In molecular beam epitaxy, the dopant material in elemental form is evaporated from an effusion cell to codeposit with the constituent adatom species on the growing surface. It is necessary that the vapor pressure of the element is reasonably high. Sometimes, appropriate compounds, or *captive* sources, are evaporated. For example, PbSe and PbTe are evaporated as sources of the donor elements Se and Te in GaAs. In vapor phase epitaxy, the dopant species are transferred by a carrier gas to the growth zone. For example, for p-type doping of GaAs in the OMCVD process, H_2 carrier gas is bubbled through diethyl zinc in liquid form.

It should be mentioned that in addition to in-situ doping during epitaxy, ex-situ doping of wafers for device fabrication is accomplished by diffusion and ion implantation. The process of doping is one of the most essential steps in the fabrication of most devices. It is also important to realize that almost all devices have stringent tolerance limits on doping levels and doping profiles. The latter may require grading in the doping density as a function of depth. Techniques such as MBE and OMCVD can easily meet these requirements.

1.8 DEVICE PROCESSING

Most optoelectronic devices, such as lasers, light-emitting diodes, photodetectors, and modulators, are junction diodes. They therefore have diode configurations, with top and bottom ohmic contacts. Device structures are formed by ion implantation, diffusion, or one of the epitaxial techniques on high resistivity or conducting substrates. The latter is useful for devices that have top and bottom contacts. For certain high-speed devices, both contacts are formed on the top side and in this case a high-

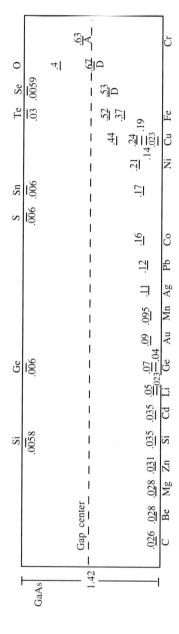

Figure 1.35 Energy levels of impurities in GaAs (from S. M. Sze, *Physics of Semiconductor Devices*, 2nd ed., Copyright ©1981. Reprinted by permission of Wiley, New York).

resistivity substrate is used. In what follows, the usual processing steps for fabricating most optoelectronic devices are described. For simplicity, many of the detailed steps and monitoring electrical measurements, that a complete process flow would contain, are being omitted here.

As a general rule, the device wafer is first cleaned with high-purity warm organic solvents before any processing. Usually the first processing step is to form top ohmic contacts. One of the unique distinctions of GaAs processing, from that of Si, is that chemical etching is rarely used to pattern metals. There are two reasons for this. First, many metal etchants will also etch GaAs. Second, GaAs processing tends to use multilayered metals, such as Au/Ge/Ni for n-type ohmic contacts and Ti/Pt/Au for rectifying or overlay metals, which are difficult to etch. As an alternative, metal patterning in GaAs processing usually employs the *lift-off* process. An organic compound called photoresist is applied to the wafer, baked, exposed through a patterned mask, and developed to define the desired patterns. The success of the lift-off process is highly sensitive to the edge profile of the patterned resist. Therefore, the resist is treated with a special solvent (e.g., chlorobenzene) before developing to provide an undercut resist edge profile. Metals are subsequently evaporated onto the wafer. A suitable solvent is used to dissolve the resist. This results in removal of the metals on top of the resist, but leaving those in contact with the wafer in selected areas. For most III–V compounds with reasonable doping ranges (1×10^{16} to 5×10^{17}), the most common metal systems used are Au/Ge/Ni and Au/Zn eutectics for n-type and p-type contacts, respectively. The contacts are annealed under appropriate ambient and temperature conditions to achieve good electrical characteristics. To make the bottom contacts, two additional processing steps are performed. The wafer is next cleaned and coated with photoresist, aligned and exposed through a mask to define the device areas. The patterned resist is used as the mask for etching to access the bottom contact layer. Either wet chemical etching (e.g., $H_gPO_4:H_2O_2:H_2O$) or dry reactive ion plasma etching can be used, depending on the desired etching profiles. Then the resist is stripped and similar lift-off process is repeated to form the bottom ohmic contacts.

Device isolation is usually provided by mesa-etching, although selective proton implantation can also be used with some advantages. Lithographically patterned resist is used as the etch mask and the wafer is etched down to the semi-insulating substrate or appropriate layer in the unmasked areas. This is generally accomplished by wet chemical etching. To protect the wafer surface (passivation) and to provide an inter-level dielectric, thin (~ 0.1–1μm) dielectric films such as silicon dioxide or silicon nitride are deposited, using plasma-enhanced chemical vapor deposition (PECVD) at a relatively low temperature ($\leq 200°$C). The films are then selectively etched with patterned resist as mask to open holes for accessing the ohmic contacts. Finally, the lift-off process is repeated to form thick ($\sim 1\mu$m) overlayer metallization (usually Ti/Pt/Au) for bonding pads and device interconnections.

Some extra steps are needed for laser diodes and modulators. To facilitate cleaving and to reduce heat dissipation in laser diodes, the wafer has to be thinned down to about 100μm. Then marks are scribed along the edges at desired lengths and laser

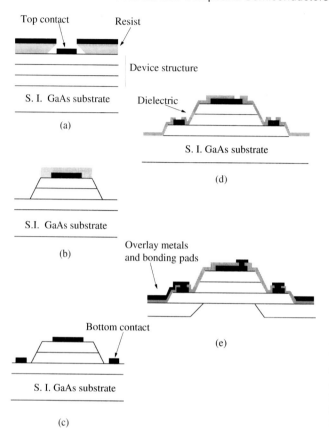

Figure 1.36 Typical processing for fabricating junction opto-electronic devices on semi-insulating substrates: (a) deposition of photoresist and metals and lift-off for formation of top contact, (b) mesa-etching, (c) formation of bottom contact following steps described in (a), (d) mesa etching for device isolation and deposition of dielectric passivation layer, and (e) overlay metallization for interconnects and bonding pads, substrate thinning and backhole formation (if necessary).

bars are cleaved. To avoid substrate absorption in modulators, the wafer is first thinned and then back holes are etched through the substrate for each device using photoresist or metals as mask. Infrared back illumination, which is transparent to the GaAs substrate, is used to help align the back-holes with the devices. The typical processing steps are illustrated in Fig. 1.36 with the help of a device schematic.

PROBLEMS

1.1 Discuss, with suitable diagrams, the difference between ionic and covalent bonding.

1.2 Draw a cubic unit cell and sketch the (111) and (211) planes. Also indicate the [110] direction.

1.3 Given that the unit cell edge of a cubic crystal, $a = 2.62$ Å, at what angles can the [100] and [110] reflections of monochromatic X-rays be observed for $\lambda = 1.54$ Å?

1.4 On a planar projection of a cubic lattice with dots to represent lattice sites, draw lines to represent the intersection of (100), (110), and (210) planes. If the lattice constant is a, what is the separation between successive intersection lines in the three cases?

1.5 Determine the packing fraction of

    (a) simple cubic lattice

    (b) face-centered cubic lattice

    (c) diamond lattice

assuming that the atoms are rigid spheres.

1.6 What is the number of (a) nearest neighbors, and (b) second nearest neighbors for the FCC, BCC, and diamond crystal lattices?

1.7 The lattice constants of Si and GaAs are 5.43 and 5.65 Å, respectively. The atomic weights of Si, Ga, and As are, respectively, 28.1, 69.7, and 74.9. Avogadro's number is 6.02×10^{23}. Calculate the densities of Si and GaAs. [Hint: Si has 8 atoms per unit cell and GaAs has 4 Ga atoms and 4 As atoms per unit cell.]

1.8 **(a)** Find the number of surface atoms on the (100) surface of InP.

    (b) Calculate the distance between nearest Al neighbors in a AlAs crystal.

1.9 Using Eqs. 1.1 and 1.2, calculate lattice constant and dielectric constant of

    (a) $In_{0.53}Ga_{0.47}As$

    (b) $InAs_{0.4}P_{0.6}$

    (c) $In_{0.8}Ga_{0.2}As_{0.4}P_{0.6}$

1.10 From Fig. 1.14 it is evident that in $Al_{0.35}Ga_{0.65}As$, the X- and L-minima are very close to the Γ minimum. As a result, if the electrons in the Γ minimum of this alloy are given a small amount of energy, they may transfer to the X- and L-valleys. Do you know of, or can you suggest some consequences and applications of this phenomenon? Since a momentum change is also involved in such a transfer, where is this momentum derived from?

1.11 It is decided to use InP-based materials and heterojunctions for a laser to emit at 1.3 μm for a optical fiber communication link. With reference to Fig. 1.2, mention the approximate compositions of low- and high-bandgap materials that can form a suitable heterojunction and the low-bandgap material is the active material for the laser.

1.12 Using Eq. 1.28, plot the variation of h_c with composition x for the epitaxial growth of $In_x Ga_{1-x}As$ ($0 \le x \le 1$) on GaAs substrate.

[Given: The Poisson ratio $\sigma_{PR} = \frac{1}{3}$.]

1.13 At the critical thickness, it can be assumed that a two-dimensional dislocation network with equal spacings is produced at the substrate-epilayer interface. The dislocation energy can therefore be represented by a line of atoms with broken bonds. Show, from energy minimization conditions, that the critical thickness $h_c \cong a/2f$, where the symbols have their usual meanings. Calculate h_c for $In_xGa_{1-x}As$ ($0 \leq x \leq 1$) on GaAs and compare the results with those of Problem 1.12.

1.14 What crystal growth technique(s) would you employ if you were asked to grow the following:

(a) a very pure layer of GaAs 20 μm thick

(b) a GaAs/AlGaAs heterostructure, where each layer is 0.2 μm thick

(c) a 50-period multiquantum well in which the well and barrier regions are each 20 monolayers thick

READING LIST

CASEY, H. C., and PANISH, M. *Heterostructure Lasers: Parts A and B*. Academic Press, New York, 1978.

KITTEL, C. *Introduction to Solid State Physics*, 6th ed. Wiley, New York, 1986.

KRESSEL, H., and BUTLER, J. K. *Semiconductor Lasers and Heterojunction LEDs*. Academic Press, New York, 1977.

LAUDISE, R. A. *Growth of Single Crystals*. Prentice Hall, Englewood Cliffs, NJ, 1970.

NARAYANAMURTI, V. "Crystalline Semiconductor Heterostructures." *Physics Today*, **37**, No. 10 (October 1984): 24–32.

PARKER, E. H. C., ed. *The Technology and Physics of Molecular Beam Epitaxy*. Plenum Press, New York, 1985.

RAZEGHI, M. *The MOCVD Challenge; Vol. I. A Survey of GaInAsP-InP for Photonic and Electronic Applications*. Adam Hilger, Philadelphia, 1989.

STREETMAN, B. G. *Solid State Electronic Devices*, 3rd ed. Prentice Hall, Englewood Cliffs, NJ, 1990.

WILLIAMS, R. *Modern GaAs Processing Methods*, 2nd ed. Artech House, Boston, 1990.

2

Electronic Properties of Semiconductors

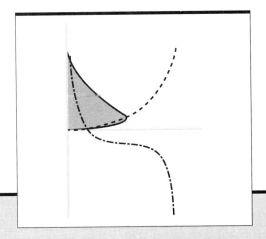

Chapter Contents

2.1 INTRODUCTION

The band theory of solids has been discussed at great lengths in many texts. The theory encompasses the basic principles of quantum mechanics and embraces the concepts of the exclusion principle and the Bohr theory. It also overcomes the limitations of the *free-electron* model in explaining phenomena in covalently bonded solids such as semiconductors. The free-electron model is adequate to explain conduction and related phenomena in metals. In this model, the nearly free valence electrons are shared by all the atoms in the solid. Thus, there is a "sea" of free electrons swimming around and these electrons see a nearly constant smeared-out potential. A model that is more successful in explaining the electronic properties of semiconductors is the *one-electron* model of a solid. Here it is assumed that a conduction electron obeying the Pauli exclusion principle is not entirely free, and does not see a constant potential. In semiconductor crystals there exist a periodic potential with the periodicity of the lattice. The periodic potential arises from the periodic charge distribution of the ion cores of host atoms at the lattice sites combined with the average nearly constant and much weaker potential distribution of the other electrons in the solid. It is the quantum-mechanical interaction of an electron with the periodic potential of the lattice that gives rise to the band theory of solids. An important outcome of theory is the formation of allowed energy bands in which electrons can exists, satisfying Pauli's exclusion principle, separated by *forbidden* energy regions. The concepts of a *hole*, effective mass of carriers, and ultimately the conduction properties also arise from the band theory.

2.2 CARRIER EFFECTIVE MASSES AND BANDSTRUCTURE

By taking into consideration the fact that the electron is acted upon by not only an external field, but also by forces arising from the periodic potential of the lattice, it can be shown that the equation of motion for an electron in a perfect lattice with no scattering is

$$F = m_e^* \frac{d\vartheta_g}{dt} \tag{2.1}$$

where F is the force acting on the electron, ϑ_g is the group velocity of the wavepacket describing the electron motion, and m_e^* is the electron *effective mass*, given by

$$m_e^* = \hbar^2 \left(\frac{d^2\mathcal{E}}{dk^2} \right)^{-1} \tag{2.2}$$

Equation 2.2 implies that the motion of electrons can be treated in a semiclassical manner provided quantum-mechanical interactions with the lattice are accounted for. The function $\left(\frac{d^2\mathcal{E}}{dk^2} \right)^{-1}$ is plotted against k, the wave vector, in Fig. 2.1, which is not the true picture in real materials. It is seen that the value is positive for small electron energies. As the electron energy increases, the value of the function becomes

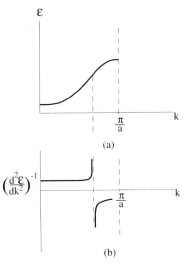

Figure 2.1 Plots of (a) $\mathcal{E}$ and (b) $\left(\frac{d^2\mathcal{E}}{dk^2}\right)^{-1}$ versus wavevector k as obtained from the solution of the Schrödinger equation for a one-dimensional lattice. $k = \pi/a$ denotes the boundary of the first Brillouin zone. The value of k at which $\left(\frac{d^2\mathcal{E}}{dk^2}\right)^{-1}$ and the effective mass attain negative values corresponds to the point of maximum slope of the $\mathcal{E}(k)$ curve.

indeterminate and then becomes *negative*. This is the region of the $\mathcal{E} - k$ curve, which is concave downward. Quantum mechanically, a carrier with a negative mass and negative charge is interpreted as an equivalent charged particle with a positive mass and *positive* charge, and this is the *hole*. In a nearly filled valence band, the hole can be regarded as the absence of an electron, and under the influence of an external electric field holes move in a direction opposite to that of electrons. This is also consistent with their positive charge. In a semiconductor at $0°$K, the conduction band is empty and the valence band is completely filled with electrons. Neither bands can conduct electricity. As the temperature is raised, some electrons may overcome the forbidden energy gap and reach the conduction band, leaving a few holes near the top of the valence band. This is the intrinsic pair generation process, which can also be achieved by photoexcitation. Under an external field, both types of carriers can move in opposite directions and conduct. As expected from this discussion, the conduction band is concave upward, and the valence band is concave downward.

It is now appropriate to discuss the energy band diagrams and effective masses in a little more detail. The real $\mathcal{E} - k$ diagrams of elemental and compound semi-conductors were illustrated in Fig. 1.1. Such bandstructures can be calculated by a variety of techniques, a common one being based on the *pseudopotential* method. In this approach the atomic potential in the crystal is approximated by a more tractable model potential that leads to the same energy levels. Experimentally, the bandstructure can be determined from cyclotron resonance and Shubnikov–de Haas measurements, which determine the effective masses, measurements of impurity binding energies, and optical absorption measurements.

In a direct bandgap semiconductor such as GaAs, whose simplified bandstructure is shown in Fig. 2.2, the lowest Γ minimum in the conduction band is at the (000) point in the first Brillouin zone. This is the *zone center* in k-space. Under normal

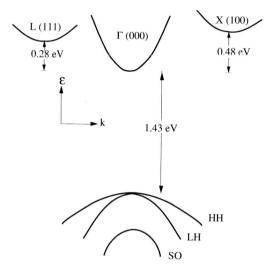

Figure 2.2 Simplified $\mathcal{E} - k$ diagram showing the various conduction band minima and valence band maxima in GaAs.

conditions, electrons in the conduction band of GaAs will occupy this minimum. The electron energy in this band is related to the wave vector by

$$\mathcal{E}(k) = \frac{\hbar^2 k^2}{2m_e^*} \tag{2.3}$$

In this case m_e^* is a scalar quantity and Eq. 2.3 corresponds to a band with a spherical constant energy surface, illustrated in Fig. 2.3(a). For the L and X minima, which are along the (111) and (100) directions, respectively, Eq. 2.3 is not valid, since the constant energy surfaces are ellipsoids of revolution as shown in Figs. 2.3(b) and (c). Also, there are more than one *equivalent* minima for each. There are three equivalent X-minima and four equivalent L-minima. The energy-wave vector dependency now becomes

$$\mathcal{E}(k) = \frac{\hbar^2}{2}\left(\frac{k_x^2}{m_x} + \frac{k_y^2}{m_y} + \frac{k_z^2}{m_z}\right) \tag{2.4}$$

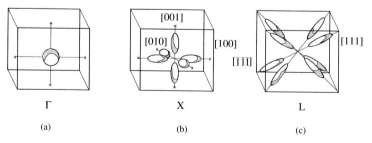

Figure 2.3 Constant energy surfaces for Γ, L, and X minima in GaAs and other zincblende semiconductors.

Here the effective mass depends on the direction of **k** and as such is a tensor quantity, $\mathbf{m_{ij}}$, defined by

$$\mathbf{m_{ij}} = \hbar^2 \left(\frac{\partial^2 \mathcal{E}}{\partial k_i \partial k_j} \right)^{-1} \tag{2.5}$$

Because of the rotational symmetry of the ellipsoids of equal energy, Eq. 2.4 can be simplified to

$$\mathcal{E}(k) = \frac{\hbar^2}{2} \left(\frac{k_l^2}{m_l} + \frac{2k_t^2}{m_t} \right) \tag{2.6}$$

where m_l and m_t are defined as the longitudinal and transverse effective mass and k_l and k_t are the longitudinal and transverse components of the wavevector, respectively. For the Γ minimum in GaAs or other III–V compounds, $m_x = m_y = m_z = m_l = m_t$. At this point it is important to distinguish two effective masses of carriers, the density-of-states effective mass, m_d, and the conductivity or mobility effective mass, m_c. The former is used in the density of states function, as we shall soon see. The latter enters into the expression for mobility and for impurity level ionization energy. The density of states effective mass for electrons is given by

$$m_{de} = (m_x m_y m_z)^{1/3} \tag{2.7}$$

For ellipsoidal bands, as is the case for the X and L valleys,

$$m_{de} = (m_l m_t^2)^{1/3} \tag{2.8}$$

The conductivity or mobility effective mass is given by

$$\frac{1}{m_{ce}} = \frac{1}{3} \left(\frac{1}{m_x} + \frac{1}{m_y} + \frac{1}{m_z} \right)$$

$$= \frac{1}{3} \left(\frac{1}{m_l} + \frac{2}{m_t} \right) \tag{2.9}$$

The different effective masses for GaAs, InAs and InP are listed in Table 2.1.

The valence bandstructure is more complicated. As can be seen in Fig. 2.2, there is a degeneracy at the zone center between the light- and heavy-hole bands and there is the split-off band. Actually, the degeneracy is fourfold due to the opposite spins in each band. The energy-momentum relationship for the degenerate bands is

$$\mathcal{E}(k) = -\frac{\hbar^2}{2m_0} \left[Ak^2 \pm \left\{ B^2 k^4 + C^2 (k_x^2 k_y^2 + k_y^2 k_z^2 + k_z^2 k_x^2) \right\}^{1/2} \right] \tag{2.10}$$

where m_0 is the free-electron mass. The plus and minus signs correspond to heavy- and light-hole bands, respectively, and A, B, and C are constants. The hole effective mass is given by a combination of the light- and heavy-hole effective masses according to

$$m_{dh} = \left(m_{lh}^{*3/2} + m_{hh}^{*3/2} \right)^{2/3} \tag{2.11}$$

TABLE 2.1 ENERGY GAPS AND TRANSVERSE AND LONGITUDINAL ELECTRON EFFECTIVE MASSES FOR SOME IMPORTANT III-V BINARY COMPOUNDS.

	ε_Γ (eV)	ε (eV)	ε_X (eV)	$m_e^{\Gamma*}(m_o)$[a]	$m_e^{L*}(m_o)$[a]	$m_e^{X*}(m_o)$[a]	$m_{hh}^{*}(m_o)$[b]	$m_{lh}^{*}(m_o)$[b]	$m_{sh}^{*}(m_o)$[c]
GaP	2.24	2.75	2.38	0.126	1.493(l) 0.142(t)	1.993(l) 0.250(t)	0.79	0.14	0.24
GaAs	1.42	1.71	1.91	0.063	1.538(l) 0.127(t)	1.987(l) 0.229(t)	0.48	0.09	0.15
AlAs	2.95	2.67	2.20	0.149	1.386(l)	0.813(l)	0.76	0.15	
InAs	0.35	1.45	2.14	0.031	1.565(l) 0.124(t)	3.619(l) 0.271(t)	0.60	0.03	0.089
InP	1.35	2.0	2.3	0.082	1.878(l) 0.153(t)	1.321(l) 0.273(t)	0.85	0.09	0.17

[a] M. V. Fischetti, *IEEE Trans. Electron Devices*, **38** (3), 634–649, 1991.

[b] M. Shur, *Physics of Semiconductor Devices*, Prentice Hall, Englewood Cliffs, NJ, 1990.

[c] Split-off hole mass, from G. P. Agrawal and N. K. Dutta, *Long Wavelength Semiconductor Lasers*, Van Nostrand Reinhold, New York, 1986.

(l) and (t) denote longitudinal and transverse effective masses, respectively.

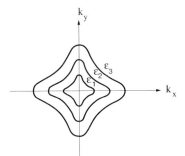

Figure 2.4 Two-dimensional constant energy contours for warped constant energy surfaces.

For example, from Table 2.1, in GaAs $m^*_{hh} = 0.48m_o$ and $m^*_{lh} = 0.09m_o$, resulting in $m_{dh} = 0.51m_o$. Equation 2.10 also indicates that both hole bands have "warped" constant energy surfaces, as shown in Fig. 2.4 in two dimensions. Finally, the split-off valence band also has a spherical constant energy surface, and its energy-wave vector dependence is given by

$$\mathcal{E}(k) = \mathcal{E}_{so} - \frac{\hbar^2 k^2}{2m^*_h} \tag{2.12}$$

where $\mathcal{E}_{so}$ is the energy corresponding to the top of the split-off band. Another important point regarding the bandstructure needs to be mentioned at this point. Equation 2.3, for example, suggests a parabolic energy band. This is true only for low energies of carriers. At higher values of energy, there is a substantial deviation from parabolicity. This is usually accounted for by substituting for energy the function

$$\gamma(\mathcal{E}) = \mathcal{E}(1 + \alpha_{np}\mathcal{E}) \tag{2.13}$$

where α_{np} is a nonparabolicity parameter related to the bandgap $\mathcal{E}_g$ and the effective mass $m^*_{e,h}$. Thus, for the Γ minimum in a direct bandgap semiconductor,

$$\mathcal{E}(1 + \alpha_{np}\mathcal{E}) = \frac{\hbar^2 k^2}{2m^*_e} \tag{2.14}$$

In GaAs $\alpha_{np} = 0.6 eV^{-1}$ and in $In_{0.5}Ga_{0.5}As$ $\alpha_{np} = 1.0$ eV^{-1}.

2.3 EFFECT OF TEMPERATURE AND PRESSURE ON BANDGAP

As the temperature of a semiconductor is varied, several important effects take place in the lattice. The lattice will expand or contract, and the oscillations of the lattice atoms around their mean positions will increase or decrease. The electron-lattice interaction also changes with temperature. Impurity potentials of shallow donors and acceptors are also affected, although the ionization energy of the impurities remain fairly constant. The experimentally observed macroscopic effect is a change in the bandgap with temperature. The dependence varies in different temperature ranges and approximately obeys the empirical relation, first proposed by Varshni:

$$\mathcal{E}_g(T) = \mathcal{E}_g(0) - \frac{AT^2}{T + B} \tag{2.15}$$

where $\mathcal{E}_g(0)$ is the energy bandgap at $0°K$ and A and B are constants, which have been determined from fitting of measured data. The values of A, B, and $\mathcal{E}_g(0)$ for GaAs and InP are listed in Table 2.2, and $\mathcal{E}_g(T)$ for GaAs is plotted in Fig. 2.5.

TABLE 2.2   PARAMETERS FOR THE VARSHNI EQUATION.

Semiconductor	$\mathcal{E}_g\ (0°K)$ (eV)	A (eV/K²)	B (°K)
GaAs	1.519	5.405×10^{-4}	204
InP	1.421	4.906×10^{-4}	327

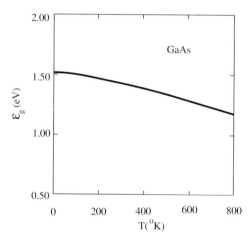

Figure 2.5 Temperature dependence of the fundamental (Γ) bandgap for GaAs using the parameters listed in Table 2.2.

   In Chapter 1 we have seen the effects of biaxial strain on the properties of a semiconductor. The built-in biaxial strain is caused by the lattice mismatch between the growing layer and the substrate. Such biaxial strain in the plane of the layer can, of course, be applied to the crystal after growth by applying *uniaxial* stress normal to the surface with a diamond anvil pressure cell. Sometimes it is more interesting to know the effects of *hydrostatic* pressure on a semiconductor crystal, when equal stress or strain is applied to the crystal in all directions. Hydrostatic pressure can be applied by immersing the crystal in a suitable pressure-transmitting medium, which itself is compressed with a piston or anvil arrangement.

   Compressive hydrostatic pressure decreases the interatomic spacing, which results in an increase in the fundamental energy bandgap. In general, all energy levels in a semiconductor are shifted, and the change can be expressed by the equation

$$\mathcal{E}(P) = \mathcal{E}(0) + \mathcal{E}_1 \Delta a \tag{2.16}$$

which denotes a linear change in the energy level with change in lattice constant Δa. This relation would be valid only for small changes in Δa. $\mathcal{E}(0)$ is the energy level without the application of hydrostatic pressure, and $\mathcal{E}_1$ is a pressure coefficient or deformation potential. The change in the energy gap of a semiconductor can then be expressed by the equation

$$\mathcal{E}_g(P) = \mathcal{E}_g(0) \pm (\mathcal{E}_{1C} + \mathcal{E}_{1V})\Delta a \tag{2.17}$$

where $\mathcal{E}_{1C}$ and $\mathcal{E}_{1V}$ are the respective pressure coefficients for the conduction and valence bands. Different valleys have different pressure coefficients and some may even have negative values. For example, in the conduction band of GaAs, the Γ valley increases in energy at a faster rate than the L valleys, and the X-valleys decrease in energy. The rate of change with pressure for the three sets of minima are expressed by the equations

$$\frac{d(\mathcal{E}_\Gamma)}{dP} = 12.6 \times 10^{-6} eV/bar$$

$$\frac{d(\mathcal{E}_L)}{dP} = 5.5 \times 10^{-6} eV/bar$$

$$\frac{d(\mathcal{E}_X)}{dP} = -1.5 \times 10^{-6} eV/bar \tag{2.18}$$

where 1 bar (pressure) = 10 newton/cm^2. Thus, the different valleys cross over, and beyond a certain magnitude of hydrostatic pressure *GaAs becomes an indirect bandgap semiconductor*. In fact, the change in the conduction bandstructure of GaAs as a function of hydrostatic pressure, is almost identical to the change produced by alloying with Al, as shown in Fig. 1.14. It is clear that the processes of alloying and application of hydrostatic pressure are equivalent. This is somewhat predictable, since in the process of alloying, depending on the relative atomic sizes, the host lattice can be compressed or dilated. It is instructive to note that as the $\Gamma - L$ and $\Gamma - X$ energy separations decrease, due to alloying or hydrostatic pressure, there will be transfer and redistribution of carriers among the direct and indirect minima. The mobility and equilibrium carrier concentration will now be average quantities, determined by the parameters of all the bands. Since the effective mass and density of states are higher in the X and L minima, the average measured mobility will be lower than that in GaAs. Also, the concentration of electrons in the Γ-minimum will decrease. The distribution of electrons between the different conduction minima can be simply expressed by Boltzmann statistics, to be discussed in Sec. 2.5. The total electron concentration $n_T = n_\Gamma + n_L + n_X$.

A change of the bandgap due to hydrostatic pressure causes changes in other properties of the materials. For example, the intrinsic carrier concentration, as we shall see, depends exponentially on the energy bandgap. A change in the bandgap will cause a change in the dielectric constant, which indirectly alters the ionization energies of impurity atoms and the binding energy of *excitons*. We will study the properties of excitons in more detail in Chapter 3. Thus, the fundamental properties of semiconductors can be studied by the application of hydrostatic pressure.

To summarize, we have learned about the effects of temperature and pressure on the bandstructure of compound semiconductors, which in turn change the optical properties of these materials. Electric field also has a profound effects on the optical processes, which we shall learn about in Chapter 3. A magnetic field changes the energy levels, due to its effect on carrier motion. It usually causes the energy levels to split. Depending on the magnitude of the field at which such splitting takes place, the effects are known as *Zeeman* splitting or *Landau* splitting.

2.4 CARRIER SCATTERING PHENOMENA

A charged carrier, electron or hole, in a semiconductor is usually not stationary. In equilibrium, it has a thermal velocity $\vartheta_{th} = \left(\frac{3k_BT}{m^*}\right)^{1/2}$, which is derived from its energy distribution. This motion is usually random and the *net* displacement is zero. The carriers can also move in a specified direction under the influence of an externally applied bias, or due to a concentration gradient. Such motion within the lattice can be greatly restricted by impurities, defects, and other imperfections. In addition, at any finite temperature, the atoms in the lattice are vibrating about their mean position in the lattice. The quanta of lattice vibrations are called *phonons*, which are quantum mechanical particles. In a diatomic lattice, such as GaAs and InP and their derivatives there are *acoustic* and *optical* phonons. Their dispersion relation is derived in Appendix 2. The names are associated with their oscillation frequencies. In a monatomic lattice such as Si, there are only acoustic phonons. The probability of finding a phonon of frequency ω, is given by the function

$$f_{BE}(\omega) = \frac{1}{\left[\exp\left(\frac{\hbar\omega}{k_BT}\right) - 1\right]} \qquad (2.19)$$

which is the *Bose-Einstein* distribution function.

An interesting relation exists between the optical phonon frequencies and the dielectric constants in a crystal. At lower frequencies in the optical spectrum, such as for infrared light, the atomic sublattices respond to the variations caused by the electromagnetic wave, and the corresponding change of dipole moment can be accounted for by the static dielectric constant ϵ_s. At higher frequencies, for example for visible light, the atoms can no longer respond to the fast-varying electric field, and thus the high-frequency dielectric constant ϵ_∞ is determined by electronic motion alone. Usually ϵ_∞ is smaller than ϵ_s and the relationship

$$\frac{\omega_{L0}}{\omega_{T0}} = \left(\frac{\epsilon_s}{\epsilon_\infty}\right)^{1/2} \qquad (2.20)$$

was derived by Liddane, Sachs, and Teller. The equation is therefore known as the *Liddane-Sachs-Teller relationship*.

Electrons and holes moving within a crystal are scattered by phonons, or lattice vibrations, which we just discussed. There are other sources of carrier scattering, such as scattering by charged impurities and by other defects within the crystal. In general, anything that perturbs the periodic crystal potential in the lattice, which in turn alters the bandedge potentials, will scatter carriers. We will derive a simple relation between the relaxation time for carrier scattering and the resulting mobility of the carriers.

We define a scattering cross section $\sigma_s(\theta, \phi)d\Omega$, which is the probability that an electron is scattered from $(\theta, \phi) = 0$ to some angle (θ, ϕ) within an incremental solid angle $d\Omega$ [Fig. 2.6(a)]. The total cross section is then

$$\sigma_{st} = \int \sigma_s(\theta, \phi)d\Omega \tag{2.21}$$

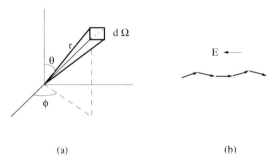

(a)

(b)

Figure 2.6 (a) Scattering geometry in polar coordinates, and (b) motion of an electron under the influence of an electric field. The motion is a superposition of drift and random motion with thermal velocity. As the field increases, the drift component becomes more dominant.

We also define a mean free time τ_C between successive collisions such that

$$\tau_C = \frac{l}{\vartheta} \tag{2.22}$$

where l is the mean free path and ϑ is the mean velocity. Consider n electrons moving with velocity ϑ in a given direction. The number of collisions, dn, in time dt is proportional to n and dt, so that

$$dn = Cndt \tag{2.23}$$

where C is a constant of proportionality. We define

$$C = \frac{1}{\tau_C} \tag{2.24}$$

where τ_C is defined as the *relaxation time*. Combining Eqs. 2.23 and 2.24,

$$\frac{dn}{n} = \frac{dt}{\tau_C} \tag{2.25}$$

which, on integration gives

$$n = n^o e^{-t/\tau_C} \tag{2.26}$$

where $n = n^o$ at $t = 0$. The probability that an electron has not made a collision is

$$\frac{n}{n^o} = e^{-t/\tau_C} \tag{2.27}$$

The mean time between collisions is

$$\bar{t} = \frac{1}{\tau_C} \int_0^\infty t e^{-t/\tau_C} = \tau_C \tag{2.28}$$

The mean free path can also be defined as

$$\frac{1}{l} = N_{sc}\sigma_{st} \tag{2.29}$$

where N_{sc} is the density of scattering centers. Therefore, from Eq. 2.22,

$$\tau_C = \frac{1}{N_{sc}\sigma_{st}\vartheta} \tag{2.30}$$

Consider now an electron under the influence of an electric field and suffering collisions, as depicted in Fig. 2.6(b). At time $t = 0$, its velocity is ϑ_0 and the velocity ϑ at time t, when it suffers collision, is given by

$$\vartheta = \vartheta_0 - \frac{qE\tau_C}{m_e^*} \tag{2.31}$$

where E is the applied field. This equation must be averaged over all time knowing that $\frac{1}{\tau_C}e^{-t/\tau_C}$ is the probability that a collision will occur after t seconds. Thus, the time-averaged velocity is given by

$$\bar{\vartheta} = \bar{\vartheta}_0 - \frac{qE}{\tau_C m_e^*} \int_0^\infty t e^{-t/\tau_C} dt$$

$$= \bar{\vartheta}_0 - \frac{qE\tau_C}{m_e^*} \tag{2.32}$$

If the collisions are truly random, $\bar{\vartheta}_0 = 0$ and the mean drift velocity is given by

$$\bar{\vartheta} = \bar{\vartheta}_D = -\frac{qE\tau_C}{m_e^*} \tag{2.33}$$

The magnitude of the mean drift velocity per unit field is defined as *mobility*, such that for electrons

$$\mu_e = \frac{\bar{\vartheta}_D}{E} = -\frac{q\tau_{Ce}}{m_e^*} \tag{2.34}$$

and for holes

$$\mu_h = \frac{q\tau_{C_h}}{m_h^*} \tag{2.35}$$

Therefore, through the effective masses, the carrier mobilities depend on the dispersion curve.

In a very pure crystal, the mobility is limited at high temperatures by carrier-lattice, or phonon scattering. The lattice vibrations depend on the temperature. As a phonon moves through the crystal, the bandgap develops a periodic perturbation, which can be approximated by a periodic potential distribution. For such carrier-phonon scattering the mobility is approximately proportional to $T^{-3/2}$ at low temperatures. As the temperature of the crystal is lowered, the rate of phonon scattering decreases and the mobility increases. However, even in sufficiently pure crystals, there are impurities and other electrically active defects. As the temperature is lowered, the carriers move more slowly through the crystal and therefore the probability of collision with such charged or neutral (at very low temperatures) impurity centers increase. To a first approximation, the mobility limited by ionized impurity scattering is proportional to $T^{3/2}$ and to N_i^{-1} where N_i is the density of impurity centers. The total mobility, as a function of temperature, is then given by *Mattheisen's rule*, as

$$\frac{1}{\mu} = \frac{1}{\mu_I} + \frac{1}{\mu_P} \tag{2.36}$$

where μ_I is the mobility limited by impurity scattering and μ_P is the mobility limited by phonon scattering. The total mobility is schematically shown in Fig. 2.7(a) and for high-purity GaAs grown by vapor phase epitaxy (VPE) in Fig. 2.7(b). The mobilities of electrons and holes in GaAs measured at 300°K, as a function of impurity density, are shown in Fig. 2.8.

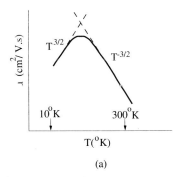

(a)

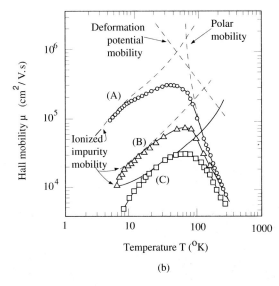

(b)

Figure 2.7 (a) Approximate temperature dependence of mobility in a very pure semiconductor sample. The mobility is principally limited by impurity scattering and optical phonon scattering; (b) mobility measured as a function of temperature in a very pure sample of GaAs grown by vapor phase epitaxy (from M. Shur, *Physics of Semiconductor Devices*, ©1990. Reprinted by permission of Prentice Hall, Englewood Cliffs, NJ; the data are compiled from C. M. Wolfe, et al., *Journal of Applied Physics*, **41**, 3088, 1970).

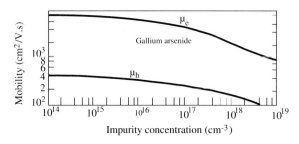

Figure 2.8 Variation of electron and hole mobilities in GaAs as a function of doping level (from B. G. Streetman, *Solid State Electronic Devices*, 3rd ed., ©1990. Reprinted by permission of Prentice Hall, Englewood Cliffs, NJ).

In addition to these two dominant carrier-scattering mechanisms, there are other sources of scattering in a real crystal. In particular, in alloy semiconductors there is a dominant effect called *alloy scattering*, which limits the mobility. Alloy scattering arises from the random positioning of the substituting atom species in the relevant sublattice and the consequent perturbation of the crystal potential. The magnitude of alloy scattering and the corresponding relaxation time are expressed by an alloy scattering potential, such as that expressed by Eq. 1.18.

Before concluding this section, some comments will be made on the measurement of mobility and effective mass in semiconductors. Both measurements can be made with suitable samples with the application of a magnetic field. The mobility of carriers is usually determined from Hall measurements [Fig. 2.9(a)], in which the motion of

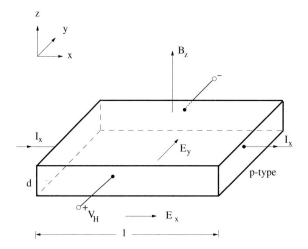

Lorentz Force : $F_y = q\,(E_y - v_x B_z)$

Hall Coefficient: $R_H = \dfrac{1}{qp} = \dfrac{dV_H}{I_x B_z}$

Hall Voltage: $\mu_H = \dfrac{R_H}{\rho}$

(a)

Figure 2.9 Schematics of (a) Hall measurements in a p-type sample and (b) the Haynes-Shockley measurement in a n-type sample. Note that the former is used to measure majority-carrier parameters and the latter determines minority-carrier parameters.

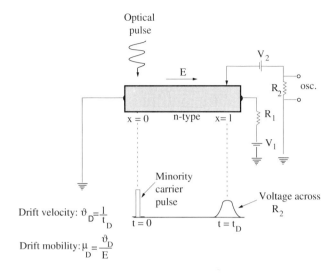

Drift velocity: $\vartheta_D = \dfrac{1}{t_D}$

Drift mobility: $\mu_D = \dfrac{\vartheta_D}{E}$

(b)

Figure 2.9   (*continued*)

the carriers across the sample is altered by the Lorentz force due to the magnetic field. This creates a *Hall voltage* in the sample, which can be related to the mobility. It is important to note that this mobility is the *Hall mobility*, μ_H, which can be different from the drift or conductivity mobility, μ_d. The ratio μ_H/μ_d is usually close to unity in direct bandgap semiconductors, but can be larger than unity in indirect bandgap semiconductors. The drift mobility can be determined by the well-known *Haynes-Shockley experiment* [Fig. 2.9(b)]. The effective mass of carriers is measured by the cyclotron resonance measurement. In this experiment, the frequency at which there is a strong absorption of electromagnetic signals passing through a sample due to absorption between Landau levels created by a magnetic field is measured. If this resonance frequency is ω, the effective mass is simply given by $m^* = qB/\omega$, where B is the applied magnetic field.

EXAMPLE 2.1
Objective.   The electron mobility in a very pure binary semiconductor is 200,000 cm²/V.s at 4.2°K. What is the mobility at 300°K?

Since the semiconductor is very pure, impurity scattering can be neglected and so can most of the other scattering mechanisms. Also, in a binary material, alloy scattering is absent. Therefore, carrier mobility is limited only by lattice or optical phonon scattering, and for this the mobility as a function of temperature is expressed as

$$\mu_P = CT^{-3/2} \tag{2.37}$$

where C is a constant. It should be cautioned that this relation is not strictly valid at high temperatures, where optical phonon scattering exhibits a more complex temperature dependence.

It follows that

$$2 \times 10^5 = C(4.2)^{-1.5} \tag{2.38}$$

from which we get

$$C = 1.72 \times 10^6 \left(\frac{cm^2 . K^{3/2}}{V.s} \right) \tag{2.39}$$

and

$$\mu(300° K) = 1.72 \times 10^6 \times (300)^{-3/2}$$
$$= 331 \text{ cm}^2/\text{V.s} \tag{2.40}$$

2.5 SEMICONDUCTOR STATISTICS

2.5.1 Energy Distribution Functions

Free carriers, electrons and holes, are essential for the operation of active semiconductor devices. We have seen that these free carriers are introduced in a semiconductor by the process of *doping*. Depending on the temperature of the crystal, the carriers are distributed in energy in the dopant energy levels and the respective bands. The number of carriers at any energy level will then depend on the number of available states at that energy and the energy distribution of the carriers. The two important functions that determine carrier distribution in a semiconductor are the *energy distribution function* and the *density of states function*.

It must be remembered that in a semiconductor, the occupancy of states by carriers, either in the conduction or valence band, or in the impurity levels is determined by Pauli's exclusion principle—that is, no two electrons with the same spin are allowed to occupy the same state. The carriers are also indistinguishable. These constraints imply that energy distribution functions such as the *Boltzmann distribution* cannot be used to determine the statistics of electrons and holes in a semiconductor. The distribution that appropriately describes the occupation of states in a semiconductor is the *Fermi-Dirac distribution function*, $f(\mathcal{E})$, expressed as

$$f(\mathcal{E}) = \frac{1}{1 + \exp\left(\frac{\mathcal{E} - \mathcal{E}_F}{k_B T}\right)} \tag{2.41}$$

where $\mathcal{E}$ is the energy of the state or level whose occupation we are interested in and could be an energy in the conduction a valence bands, or the energy of an impurity level within the forbidden energy gap. The energy $\mathcal{E}_F$ is defined as the *Fermi level*. The derivation of Eq. 2.41 can be found in texts in solid-state physics and statistical mechanics. The form of the Fermi function as a function of energy at $0°K$ and higher temperatures is illustrated in Fig. 2.10. These curves bring out several important characteristics of the function. First, it is seen that the maximum value of the function is unity. Therefore, the probability of occupation of an energy level can never exceed unity, or not more than one electron can occupy the same quantum state. At $\mathcal{E} = \mathcal{E}_F$ the value of the function goes from unity to zero at $0°K$ and at any temperature T, $f(\mathcal{E}) = \frac{1}{2}$ at $\mathcal{E} = \mathcal{E}_F$. Thus, the Fermi energy can be defined as that energy level up

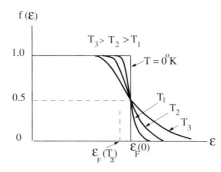

Figure 2.10 Schematic representation of the Fermi-Dirac distribution function at 0°K and higher temperatures. Also indicated is the lowering of the Fermi energy $\mathcal{E}_F$ with increase of temperature.

to which all levels are occupied and above which all levels are empty at 0°K, and at any temperature $T > 0°K$ the probability of occupation at the Fermi level is 1/2.

EXAMPLE 2.2
Objective.   To show that the Fermi-Dirac distribution function is symmetrical about $\mathcal{E} = \mathcal{E}_F$ at any temperature T.

Assume equal increments of energy $\Delta\mathcal{E}$ above and below $\mathcal{E}_F$, as shown in Fig. 2.11. With respect to Eq. 2.41, the probability of a level being full at $(\mathcal{E}_F + \Delta\mathcal{E})$ is

$$f(\mathcal{E}_F + \Delta\mathcal{E}) = \frac{1}{1 + \exp[(\mathcal{E}_F + \Delta\mathcal{E} - \mathcal{E}_F)/k_B T]}$$

$$= \frac{1}{1 + \exp[\Delta\mathcal{E}/k_B T]} \tag{2.42}$$

Similarly, the probability of a level being empty at $(\mathcal{E}_F - \Delta\mathcal{E})$ is

$$[1 - f(\mathcal{E}_F - \Delta\mathcal{E})] = 1 - \frac{1}{1 + \exp[(\mathcal{E}_F - \Delta\mathcal{E} - \mathcal{E}_F)/k_B T]}$$

$$= \frac{1}{1 + \exp[\Delta\mathcal{E}/k_B T]} \tag{2.43}$$

Thus, $f(\mathcal{E})$ is symmetrical about $\mathcal{E} = \mathcal{E}_F$.

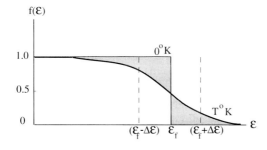

Figure 2.11 Diagram indicating the symmetry of the Fermi-Dirac distribution function around $\mathcal{E} = \mathcal{E}_F$.

The energy position of the Fermi energy level in a semiconductor is closely related to the doping type and level. As we shall soon see, in an n-type semiconductor the Fermi level is close to the conduction band edge, while in a p-type semiconductor, $\mathcal{E}_F$ is close to the valence band edge. It may also be noted that no temperature dependence of $\mathcal{E}_F$ has been shown in Fig. 2.10. In fact, $\mathcal{E}_F$ does have a temperature dependence and its value decreases with increase of temperature.

In Eq. 2.41 expressing the Fermi-Dirac distribution function, if $(\mathcal{E} - \mathcal{E}_F) \gg k_B T$, the function may be approximated by

$$f(\mathcal{E}) \cong \exp\left[-\left(\frac{\mathcal{E} - \mathcal{E}_F}{k_B T}\right)\right] \tag{2.44}$$

Alternately,

$$f(\mathcal{E}) = A \exp\left(-\frac{\mathcal{E}}{k_B T}\right) \tag{2.45}$$

where $A = \exp(\mathcal{E}_F / k_B T)$. Equation 2.45 is of the form of the Boltzmann distribution function. Such conditions can occur in a semiconductor with very low doping, such that $(\mathcal{E} - \mathcal{E}_F)$ has a large value, or if carriers have large energy, due to an applied electric field or due to heating of the sample. In either case, carriers occupy higher energy regions of the band, where the number of available states may be much larger than the number of carriers. Thus, the situation becomes similar to that of a Boltzmann gas, and the restrictions set by the exclusion principle is relaxed. The condition expressed by Eq. 2.45 is called the Boltzmann approximation, and is very useful in calculating carrier statistics in nondegenerate semiconductors.

2.5.2 Density of States Function

We will now discuss the density of states function, which is, as mentioned in Sec. 2.5.1, as important as the distribution function in determining carrier statistics. The density of states function $N(\mathcal{E})d\mathcal{E}$ gives the number of available quantum states in the energy interval between $\mathcal{E}$ and $\mathcal{E} + d\mathcal{E}$. In what follows, we will derive an expression for $N(\mathcal{E})$, first for the case of a three-dimensional *bulk* semiconductor, and then for the case of a two-dimensional density of carriers, as found in a *quantum well*. Such structures, as we shall see, are extremely important for the design of lasers and modulators.

2.5.2.1 Three-Dimensional Density of States.   We consider a cubic region of the crystal with dimensions L along the three perpendicular directions and impose the condition that the electron wavefunctions become zero at the boundaries of the cube defined by values of x, y, and z equal to 0 and L. The boundary conditions are satisfied by a wavefunction of the form

$$\Psi_k(r) = U_k(r) \sin k_x x \sin k_y y \sin k_z z \tag{2.46}$$

and the boundary conditions lead to

$$k_x L = 2\pi n_1$$

$$k_y L = 2\pi n_2$$

$$k_z L = 2\pi n_3 \tag{2.47}$$

where n_1, n_2, and n_3 are integers. Therefore, each allowed value of k with coordinates k_x, k_y, and k_z occupies a volume $(2\pi/L)^3$ in **k**-space. In other words, the density of allowed points in **k**-space is $V/(2\pi)^3$, where $V = L^3$ is the crystal volume.

We first derive the three-dimensional density of states. The volume in **k**-space defined by vectors **k** and **k** + d**k** is $4\pi k^2 dk$. Hence, the total number of states with k-values between k and $k + dk$ is

$$dN = 4\pi V k^2 dk/(2\pi)^3 \tag{2.48}$$

Taking into account the two possible values of spin,

$$dN = 8\pi V k^2 dk/(2\pi)^3 \tag{2.49}$$

For electrons in the conduction band of a semiconductor

$$k^2 = \frac{2m_{de}(\mathcal{E} - \mathcal{E}_c)}{\hbar^2} \tag{2.50}$$

which, by differentiation, leads to

$$2kdk = \frac{2m_{de}d\mathcal{E}}{\hbar^2} \tag{2.51}$$

Substituting Eqs. 2.50 and 2.51 into Eq. 2.49, we obtain

$$dN = \frac{8\pi V[2m_{de}(\mathcal{E} - \mathcal{E}_c)/\hbar^2]\left(\frac{2m_{de}d\mathcal{E}}{\hbar^2}\right)}{\left\{2[2m_{de}(\mathcal{E} - \mathcal{E}_c)/\hbar^2]^{1/2}(2\pi)^3\right\}} \tag{2.52}$$

and for unit volume of the crystal,

$$N(\mathcal{E})d\mathcal{E} = M_c \frac{\sqrt{2}}{\pi^2}\left(\frac{m_{de}}{\hbar^2}\right)^{3/2}(\mathcal{E} - \mathcal{E}_c)^{1/2}d\mathcal{E} \tag{2.53}$$

The form of this function is schematically shown in Fig. 2.12. Here M_c is the number of equivalent minima in the conduction band. A similar equation holds for the density of states in the valence band.

2.5.2.2 Quantum-Sized Effects and Two-Dimensional Density of States.
In most cases, we deal with the electronic and optical properties in bulk materials. In such cases, the electron mean free path or de Broglie wavelength ($\sim$ 200–400 Å) is much smaller than the sample dimensions. When the dimensions of the semiconductor becomes comparable to the mean free path of carriers, *quantum-sized effects* (QSEs) become important and they dominate the electronic properties of the

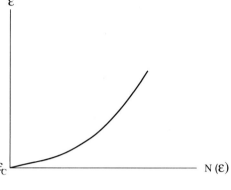

Figure 2.12 A plot of the three-dimensional density-of-states function in accordance with Eq. 2.53. The energy $\mathcal{E}_C$ represents the bottom of the conduction band.

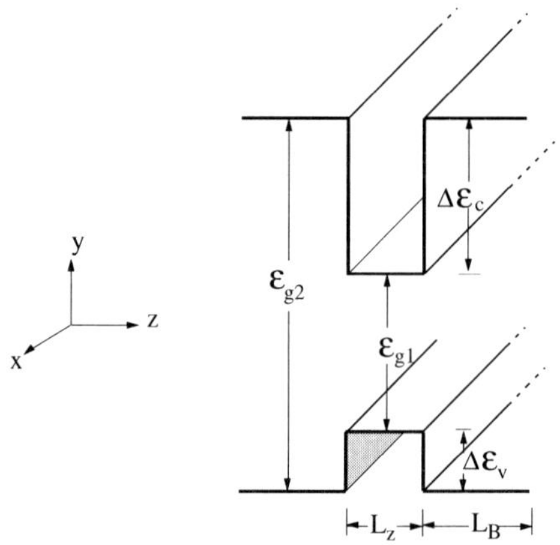

Figure 2.13 Schematic representation of a single-quantum well (SQW) in three dimensions formed by two semiconductors with bandgaps $\mathcal{E}_{g1}$ and $\mathcal{E}_{g2}$. L_z and L_B are well and barrier widths, respectively. In the x–y plane the material dimensions have no limits. $\Delta\mathcal{E}_C$ and $\Delta\mathcal{E}_V$, which are called conduction and valence band offsets, form the energy barriers to electrons and holes in their respective wells. Also, in this figure, band-bending due to formation of the heterojunctions is ignored.

material. Of particular interest are quantum well structures, shown in Fig. 2.13. In such structures, two semiconductors of different bandgap energies and of thicknesses comparable to the electron mean free path alternate to form a synthetically modulated structure. If the narrow gap semiconductor layer is small enough, the motion of carriers in the direction perpendicular to the heterointerfaces is quantized. If the thickness of the wide-bandgap layers is thin enough, such that carriers may tunnel through, the discrete energy levels formed in the quantum wells due to size quantization splits into minibands. Such a composite material is called a *superlattice*, which has electronic and optical properties that are unique to the whole structure. For example, carriers are transported through such material by miniband conduction. Also, the effective bandgap of the entire composite structure is approximately equal to the energy separation between the electron and hole minibands. Usually in compound semiconductor materials with $\mathcal{E}_g \cong 1eV$, such coupling between wells occurs for barrier thicknesses ≤ 50 Å. For barrier thicknesses larger than 50 Å, the individual wells are largely uncoupled and the electronic properties in a single-quantum well (SQW) or multiquantum well (MQW) are quite similar. The structures that we have just described are called *compositional superlattices*, since they are formed by a periodic modulation of composition (bandgap). Another modulated structure is a *doping* or *n-i-p-i* superlattice, so called because it is formed by alternate p- and n-doped layers separated by i-regions of the same semiconductor. This superlattice has several unusual and interesting properties. Electrons and holes are confined in their respective "wells," but the subbands are separated in real space. As a result, carrier recombination lifetimes can be quite large and also tunable by a transverse applied bias.

Returning to the case of the quantum well structure formed by bandgap modulation, in the well region the quantized subband electron energies $\mathcal{E}_l$ are given by, with analogy to the problem of a particle in a finite well,

$$\mathcal{E}_l = \frac{\pi^2 \hbar^2 l^2}{2m_e^* L_z^2}, \qquad l = 1, 2, 3, \cdots \qquad (2.54)$$

where $\mathcal{E}_l$ is measured from the edge of the conduction band in the well region and L_z is the well thickness. A similar equation is true for hole subbands in the valence band. It is useful to note that the criterion for quantization is that the energy difference between the subbands should be larger than the subband energy $k_B T$. As an example, for GaAs, with $m_e^* = 0.063m_o$, the levels are quantized at T = 300°K for $L_z < 150$ Å. In other words, if L_z is > 150 Å, the electrons in the well begin to assume bulk-like properties. In the direction parallel to the heterointerfaces, the wave function is described as a two-dimensional Bloch function with the dispersion relation

$$\mathcal{E} = \mathcal{E}_l + \frac{\hbar^2}{2m_e^*}(k_x^2 + k_y^2) \qquad (2.55)$$

with $\mathcal{E}_l$ given by Eq. 2.54. Thus, each quantum level given in Eq. 2.54 corresponds to an energy subband shown in Fig. 2.14. In the quantum well formed in the valence

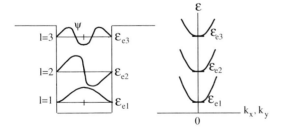

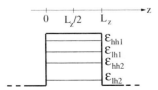

Figure 2.14 Electron and hole energy subbands in a quantum well. Also shown are the electron wavefunctions Ψ and the electron dispersion relations in the *x–y* plane.

band of the well material, the degeneracy between the light- and heavy-hole states at the zone center is lifted. For example, in a 100 Å GaAs-Al$_{0.3}$Ga$_{0.7}$As quantum well, their energy separation at k = 0 is ∼ 10 meV. Therefore, separate heavy-hole (hh) and light-hole (lh) confined states are formed. In Eq. 2.54, $\mathcal{E}_l$ is $\mathcal{E}_{e1}, \mathcal{E}_{e2}, \mathcal{E}_{e3} \cdots$ for electron confined states or $\mathcal{E}_{hh1}, \mathcal{E}_{hh2} \cdots$ and $\mathcal{E}_{lh}, \mathcal{E}_{lh2} \cdots$ for heavy-hole and light-hole confined states.

For the two-dimensional case, the density of states for each subband in a quantum well can be calculated, as follows. As before,

$$k_x = \frac{2\pi}{L_x}, k_y = \frac{2\pi}{L_y}, \text{ and } k_z = \frac{2\pi}{L_z} \tag{2.56}$$

and it is assumed that $L_z \ll L_x, L_y$. In a quantum well, L_z represents the width of the well. Therefore, each allowed value of k, or mode, occupies a volume in k-space equal to $(2\pi)^3/L_x L_y L_z$. In the present case we have to determine the number of modes included between k and $k + dk$ keeping $k_z = \pi/L_z$ constant. The mode numbers can be treated as continuous variables in the $k_x - k_y$ plane. As before, the density of allowed points in k-space $= L_x L_y L_z/(2\pi)^3$. The volume in k-space between k and $k + dk = 2\pi k_\| dk_\| \left(\frac{2\pi}{L_z}\right)$, where $k_\|$ is the wavevector in the $k_x - k_y$ plane. However, since

$$k^2 = \left(\frac{2\pi}{L_z}\right)^2 + k_\|^2 \tag{2.57}$$

it follows that

$$k dk = k_\| dk_\| \tag{2.58}$$

The total number of states between k and $k + dk$ is

$$dN = 2\left[\frac{L_x L_y L_z}{(2\pi)^3} 2\pi k dk \left(\frac{2\pi}{L_z}\right)\right] \tag{2.59}$$

which takes into account the two possible spin values. On simplification,

$$dN = (L_x L_y L_z)\frac{1}{\pi^2} k dk \left(\frac{\pi}{L_z}\right). \tag{2.60}$$

In this case,

$$k^2 = \frac{2m_{de}(\mathcal{E} - \mathcal{E}_l)}{\hbar^2}, l = 1, 2, \cdots \tag{2.61}$$

On substitution in Eq. 2.60 the density of states in the energy interval $d\mathcal{E}$ is

$$N(\mathcal{E})d\mathcal{E} = \frac{m_{de}}{\pi \hbar^2 L_z} d\mathcal{E} \tag{2.62}$$

provided $\mathcal{E} > \mathcal{E}_1 = [\hbar(\pi/L_z)]^2/2m_{de} (l = 1)$. The two-dimensional density of states is then

$$N(\mathcal{E})d\mathcal{E} = \frac{m_{de}}{\pi \hbar^2} d\mathcal{E} \tag{2.63}$$

Therefore, each of the two-dimensional bands gives rise to a band density of states that is independent of energy. As can be seen from Fig. 2.15, the states of the first subband overlap with the states of the second subband for energies larger than the second subband level, and so on. As a result, the cumulative density of states for a series of subbands will be steplike in character.

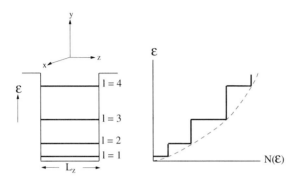

Figure 2.15 Energy subbands and the two-dimensional density of states formed in a quantum well in the z-direction. In the x–y plane, the density of states is three-dimensional and is given by the dashed curve.

Equation 2.63 also gives the density of states functions for the light and heavy holes, by choosing the appropriate effective masses. If one assumes zero-phonon transitions, then electrons in the $l = 1$ (or 2, 3, $\cdots$) electron subband recombines with $l = 1$ (or 2, 3, $\cdots$) heavy holes, or with $l = 1$ (or 2, 3, $\cdots$) light holes. However, it is important to remember that in a real structure imperfections can change the selection rules. There is an important consequence of the of the steplike density of states of the carriers in laser operation. It is easy to see that a large density of electrons, all at a fixed energy (say $\mathcal{E}_{e1}$) can recombine with a similar density of holes at a fixed subband energy.

2.5.3 Density of Carriers in Intrinsic and Extrinsic Semiconductors

At $0°K$ or very low temperatures, the valence band is filled with electrons and the conduction band is empty. Under such conditions there can be no electrical conduction. For the latter to occur, a covalent bond has to be broken, for which the minimum energy required is the bandgap energy $\mathcal{E}_g$. This energy can be provided by heat, electric field, optical excitation, and a variety of other techniques. In an intrinsic semiconductor, without any dopant atoms, the breaking of a covalent bond creates an electron-hole pair, and under an electric field, the two carriers move in opposite directions to give rise to *two-carrier transport*. In an intrinsic semiconductor, therefore, under thermal equilibrium conditions,

$$n_o = p_o = n_i \tag{2.64}$$

where n_i is the *intrinsic carrier concentration*. Usually, in semiconductors with $\mathcal{E}_g \sim 1\,eV$, n_i is too small for any practical use. Therefore, doping is used to increase n, or p, and such semiconductors are called *extrinsic semiconductors*. In the following,

we establish relations that give the density of carriers in intrinsic and extrinsic semiconductors under thermal equilibrium conditions. For the conduction and valence bands,
the three-dimensional density of states functions can be rewritten from Eq. 2.53 as

$$N(\mathcal{E}) = \frac{4\pi (2m_{de})^{3/2}(\mathcal{E} - \mathcal{E}_C)^{1/2}}{h^3} d\mathcal{E} \tag{2.65}$$

and

$$N(\mathcal{E}) = \frac{4\pi (2m_{dh})^{3/2}(\mathcal{E}_V - \mathcal{E})^{1/2}}{h^3} d\mathcal{E} \tag{2.66}$$

respectively, where the energy of holes is measured from the edge of the valence
band. The parameter M_c is excluded in these equations. An important detail may
be mentioned here. The expressions for the density of states in Eqs. 2.65 and 2.66
include a factor of 2 to include the two possible spins of the carriers. However, when
calculating upward or downward transition probabilities as in a band-to-band radiative
transition, the density of states function forms a product in an integral and the factor 2
is dropped because the spin of the carrier remains unchanged during such a transition.
Also, when calculating the density of states in the Γ, X, or L conduction minima, the
appropriate value of $m_{de}^{\Gamma}, m_{de}^{X}$, or m_{de}^{L} is used. The density of electrons and holes in
the conduction and valence bands are given, respectively, by

$$n_o = \int_{\mathcal{E}_C}^{\infty} N(\mathcal{E}) f(\mathcal{E}) d\mathcal{E} \tag{2.67}$$

and

$$p_o = \int_{-\infty}^{\mathcal{E}_V} N(\mathcal{E})[1 - f(\mathcal{E})] d\mathcal{E} \tag{2.68}$$

which, on substitution of Eqs. 2.41, 2.65, and 2.66 lead to

$$n_o = N_C F_{1/2}(\eta) \tag{2.69}$$

where

$$N_C = 2 \left(\frac{2\pi m_{de} k_B T}{h^2} \right)^{3/2} \tag{2.70}$$

and

$$p_o = N_V F_{1/2}(\eta) \tag{2.71}$$

where

$$N_V = 2 \left(\frac{2\pi m_{dh} k_B T}{h^2} \right)^{3/2} \tag{2.72}$$

N_C and N_V are the *effective density of states* in the conduction and valence bands,
respectively. In order to calculate these parameters numerically, it is useful to express
them in terms of the variables, effective mass, and temperature, in the following form:

$$N_{C(V)} = 2.5 \times 10^{19} \left(\frac{m_d}{m_0} \frac{T}{300} \right)^{3/2} (cm^{-3}) \tag{2.73}$$

The quantity $F_{1/2}(\eta)$ is called the Fermi integral. For example, for electrons

$$F_{1/2}(\eta) = \frac{2}{\sqrt{\pi}} \int_0^{\infty} \frac{x^{1/2} dx}{1 + \exp(x - \eta)} \tag{2.74}$$

where

$$x = \frac{\mathcal{E} - \mathcal{E}_C}{k_B T} \tag{2.75}$$

and

$$\eta = \frac{(\mathcal{E}_F - \mathcal{E}_C)}{k_B T} \tag{2.76}$$

These quantities can be calculated from the equations given in Appendix 3.

In the case of a lightly doped or nondegenerate semiconductor, where the Fermi level is within the forbidden energy gap and $|\mathcal{E}_{C(V)} - \mathcal{E}_F| \geq 3k_B T$, the Boltzmann approximation of Eq. 2.45 is valid. In this case Eq. 2.69 becomes

$$n_o = N_C e^{-\left(\frac{\mathcal{E}_C - \mathcal{E}_F}{k_B T}\right)} \frac{2}{\sqrt{\pi}} \int_0^\infty x^{1/2} e^{-x} dx \tag{2.77}$$

The integral in Eq. 2.77 is of the form of a *gamma function* and is equal to $\Gamma\left(\frac{3}{2}\right)$ whose numerical value is $\sqrt{\pi}/2$. Therefore, it follows that

$$n_o = N_C e^{-(\mathcal{E}_C - \mathcal{E}_F)/k_B T} \tag{2.78}$$

Similarly

$$p_o = N_V e^{-(\mathcal{E}_F - \mathcal{E}_V)/k_B T} \tag{2.79}$$

Note that in this case $\eta \ll -1$ and $F_{1/2}(\eta) = \exp(\eta)$. These equations give the energy separation between the Fermi energy level and the bandedges in terms of the free-carrier concentrations and the effective density of states.

EXAMPLE 2.3
Objective.   To estimate the density electrons in the conduction band of intrinsic Si ($\mathcal{E}_g = 1.1$ eV) and diamond ($\mathcal{E}_g = 5.6$ eV) at room temperature.

We shall use Eq. 2.78 to solve this problem. For intrinsic material $(\mathcal{E}_C - \mathcal{E}_F) \cong \mathcal{E}_g/2$.

For Si, using Table 2.3,
$$n_o = 2.9 \times 10^{19} e^{-1.1/0.052}$$
$$= 1.9 \times 10^{10} cm^{-3} \tag{2.80}$$

For diamond, N_C is not given, but as we shall see, it is not necessary. In this case
$$n_o = N_C e^{-5.6/0.052} = 0 \tag{2.81}$$

The significance of these results are that as the bandgap of a semiconductor increases, or if the temperature is lowered, the semiconductor will behave more and more as an insulator.

In an n-type semiconductor in which all the donor atoms are ionized, such that $n_o = N_D$,

$$\mathcal{E}_C - \mathcal{E}_F = k_B T \ln \frac{N_C}{N_D} \tag{2.82}$$

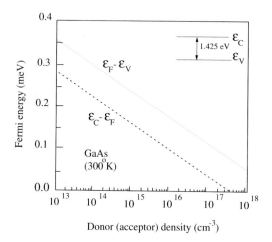

Figure 2.16   The energy position of the Fermi levels in GaAs at 300°K as a function of doping level.

The energy position of the Fermi level with respect to the bandedges in GaAs for different concentrations of shallow donor and acceptor impurities is shown in Fig. 2.16. From Eqs. 2.78 and 2.79 for a non-degenerate semiconductor

$$n_o p_o = N_C N_V e^{-\mathcal{E}_g / k_B T} = n_i^2 \tag{2.83}$$

where $\mathcal{E}_g = \mathcal{E}_C - \mathcal{E}_V$ is the bandgap energy and n_i is the intrinsic carrier concentration, which can be expressed as

$$n_i = 2 \left(\frac{2\pi k_B T}{h^2} \right)^{3/2} (m_{de} m_{dh})^{3/4} \exp\left(-\frac{\mathcal{E}_g}{2k_B T} \right) \tag{2.84}$$

The relation $n_o p_o = n_i^2$, also known as the *law of mass action*, holds only in a nondegenerate semiconductor and not for a degenerate semiconductor produced, for example, by heavy doping. In an intrinsic semiconductor, Eq. 2.64 is valid. In a doped semiconductor, n_0 and p_0 are the densities of free carriers in the bands resulting mainly from the ionization of impurity levels. From Eq. 2.84 it is seen that n_i has a negative exponential relation to the bandgap and is mainly related to temperature by an exponential relation, although there is a weaker $T^{3/2}$ dependence. It may also be noted that n_i is independent of the position of the Fermi level. Values of N_C, N_V, and n_i at 300°K for a few important semiconductors are listed in Table 2.3. The

TABLE 2.3   VALUES OF N_C, N_V, AND n_i AT 300°K FOR A FEW SEMICONDUCTORS.

cm^{-3}	Si	GaAs	InP
N_C	2.9×10^{19}	4.4×10^{17}	5.4×10^{17}
N_V	1.1×10^{19}	8.2×10^{18}	1.2×10^{19}
n_i	1.0×10^{10}	5.0×10^6	1.2×10^7

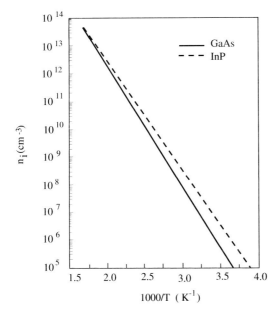

Figure 2.17 Intrinsic carrier concentration for GaAs and InP as a function of inverse temperature.

variation of n_i in GaAs and InP as a function of inverse temperature is also shown in Fig. 2.17.

EXAMPLE 2.4

Objective.   To find the position of the Fermi level in an intrinsic semiconductor.

Since the bandgap energy $\mathcal{E}_g$ does not depend on the impurity concentration in a nondegenerate semiconductor and since n_i is independent of $\mathcal{E}_F$, which is affected by the doping level, it follows that Eq. 2.84 is equally valid for intrinsic and extrinsic semiconductors. In an intrinsic semiconductor, since $n_0 = p_0$, it follows from Eqs. 2.78 and 2.79,

$$m_{de}^{3/2} \exp[-(\mathcal{E}_C - \mathcal{E}_F)/k_B T] \;=\; m_{dh}^{3/2} \exp[-(\mathcal{E}_F - \mathcal{E}_V)/k_B T] \tag{2.85}$$

from which it follows that

$$\mathcal{E}_F \;=\; \frac{\mathcal{E}_g}{2} - \frac{3}{4} k_B T \ln\left(\frac{m_{de}}{m_{dh}}\right)$$

$$= \mathcal{E}_{Fi} \tag{2.86}$$

Equation 2.86 gives the position of the Fermi level in an intrinsic semiconductor, $\mathcal{E}_{Fi}$, which is also defined as the intrinsic Fermi level. If the electron and hole effective masses are equal, then the Fermi level is at the middle of the forbidden energy gap. Two important points may be noted here. First, the Fermi energy has a temperature dependence, as mentioned earlier. Second, it has been assumed that the value of the effective mass is constant in the respective bands, which is not strictly true, but is a reasonable assumption.

The functions $f(\mathcal{E})$ and $N_{C,V}(\mathcal{E})$ are illustrated in Fig. 2.18 for an n-type semiconductor. Also plotted are the carrier distributions $n_o(\mathcal{E})$ and $p_o(\mathcal{E})$ in the respective bands.

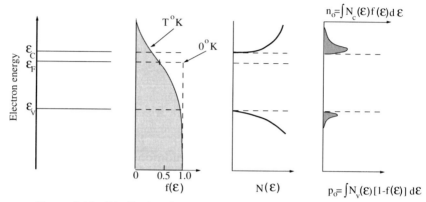

Figure 2.18 Distribution functions, density-of-states functions, and carrier distributions (in energy) in a n-type nondegenerate semiconductor.

2.5.4 Compensation in Semiconductors

We have been considering the cases of semiconductors doped with either donors or acceptors, but not both. More often than not, in real semiconductors, unintentional impurity species are present in addition to the intentional ones. Thus, there is some concentration of donor and acceptor impurities present in most semiconductors. Of course, the aim in all epitaxial growth techniques is to keep the density of these ubiquitous impurity species down to a negligible value. As a result, in an n-type semiconductor, the net electron concentration in the conduction band is given by

$$n_o = N_D - N_A \qquad (2.87)$$

where N_D is the density of intentionally introduced donor impurities and N_A is the density of unintentional acceptor impurities present in the sample. Similarly, in a p-type sample

$$p_o = N_A - N_D \qquad (2.88)$$

Another point made by Eqs. 2.87 and 2.88 is that the conductivity type of a semiconductor is determined by the impurity type which has the larger concentration. A semiconductor in which the number of free carriers produced by one type of dopant is reduced by the presence of the other type of dopant is said to be *compensated*. The

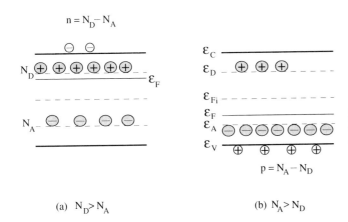

Figure 2.19 Schematic band diagram showing the process of compensation in (a) n-type semiconductor and (b) p-type semiconductor. The clear symbols denote free carriers in the bands, and the shaded symbols denote ionized charge.

process of compensation in n- and p-type semiconductors is illustrated in Fig. 2.19. The *compensation ratio* is defined as

$$\beta_r = \frac{N_A}{N_D} \qquad \text{(n-type semiconductor)}$$

$$= \frac{N_D}{N_A} \qquad \text{(p-type semiconductor)} \qquad (2.89)$$

Ideally, it is desirable to have $\beta_r \leq 0.1$. For some applications, compensation is intentionally introduced in the sample to reduce its conductivity, such as in the bulk growth of semi-insulating substrate materials.

We will next briefly review the *charge neutrality* condition and the temperature-dependent density of carriers in a doped semiconductor. We consider a nondegenerately doped n-type semiconductor with a small density of compensating acceptors also present. The situation is illustrated in Fig. 2.19(a). For the crystal to remain electrically neutral, the sum of all the positive charges must be equal to the sum of the negative charges. This condition is aptly expressed by the equation

$$n_o + N_A^- = p_o + N_D^+ \qquad (2.90)$$

On the left-hand side, the negative charge density is made up of the free-conduction band electrons and the charge of the immobile ionized acceptors. On the right-hand side, the positive charge is made up of the free-valence band holes and the charge of the immobile ionized donors. Before proceeding further, it is useful to know the form of the Fermi functions for the donor and acceptor doping levels. For the donor energy level

$$f(\mathcal{E}_D) = \frac{N_D - N_D^+}{N_D} = \frac{1}{1 + g_D^{-1} \exp[(\mathcal{E}_D - \mathcal{E}_F)/k_B T]} \qquad (2.91)$$

and for the acceptor energy level

$$f(\mathcal{E}_A) = \frac{N_A^-}{N_A} = \frac{1}{1 + g_D \exp[(\mathcal{E}_A - \mathcal{E}_F)/k_B T]} \qquad (2.92)$$

where g_D is the degeneracy factor and has a value of 2. Substituting Eqs. 2.87, 2.88, 2.91, and 2.92 in the charge neutrality condition leads to

$$N_C \exp\left(-\frac{\mathcal{E}_C - \mathcal{E}_F}{k_B T}\right) + \frac{N_A}{1 + 2\exp[(\mathcal{E}_A - \mathcal{E}_F)/k_B T]}$$

$$= N_V \exp\left(-\frac{\mathcal{E}_F - \mathcal{E}_V}{k_B T}\right) + \frac{N_D}{1 + 2\exp[(\mathcal{E}_F - \mathcal{E}_D)/k_B T]} \tag{2.93}$$

This equation can be solved numerically and an accurate value of $\mathcal{E}_F$ can be obtained by an iterative procedure. Once $\mathcal{E}_F$ is known, the values of n_o and p_o can be determined from Eqs. 2.78 and 2.79. The Fermi energy level can be uniquely determined for a semiconductor at any temperature for a given N_D (or N_A) by a simple graphical method shown in Fig. 2.20, which is also known as a Shockley diagram. The diagram is illustrated for Si at 300°K with $N_D = 10^{16}$ cm^{-3}. N_C, N_V, and n_i are calculated for a particular T. The lines $n_o(\mathcal{E}_F)$ and $p_o(\mathcal{E}_F)$ join at n_i. For an n-type semiconductor the intersection of $N_D = 10^{16}$ (cm^{-3}), and the $n_o(\mathcal{E}_F)$ lines determines the position of the Fermi level. If the temperature, doping type and level changes, a new value for $\mathcal{E}_F$ will be obtained.

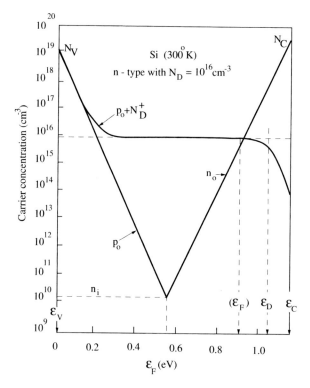

Figure 2.20 Graphical method, first described by W. Shockley, to determine the Fermi energy level in a semiconductor (from S. M. Sze, *Physics of Semiconductor Devices*, 2nd ed., ©1981. Reprinted by permission of Wiley, New York).

A great deal of insight of the carrier statistics can be obtained by making some reasonable approximations. The energies of shallow donors and acceptors in GaAs and InP and compounds with similar bandgaps are in the range of 4–30 meV. They are therefore completely ionized at 300°K, and the value of $k_B T$ at this temperature is 26 meV. Under such conditions of full ionization

$$N_A^- = N_A \tag{2.94}$$

and

$$N_D^+ = N_D \tag{2.95}$$

Equation 2.90, can therefore be rewritten as

$$n_o + N_A = p_o + N_D. \tag{2.96}$$

Elimination of p_o by using the relation $n_o p_o = n_i^2$ yields the quadratic equation

$$n_o^2 - (N_D - N_A)n - n_i^2 = 0 \tag{2.97}$$

whose allowable solution is

$$n_o = \frac{N_D - N_A}{2} + \left[\left(\frac{N_D - N_A}{2} \right)^2 + n_i^2 \right]^{1/2} \tag{2.98}$$

Similarly, from the quadratic equation with p as a variable,

$$p_o = -\frac{N_D - N_A}{2} + \left[\left(\frac{N_D - N_A}{2} \right)^2 + n_i^2 \right]^{1/2} \tag{2.99}$$

where it is assumed that for the n-type semiconductor $N_D > N_A$. The free-carrier concentration in a semiconductor as a function of temperature is very important for many device applications. This parameter can be obtained from Eqs. 2.98 and 2.99.

At very high temperatures, where the intrinsic generation of electron-hole pairs is a dominant process, $n_i \gg (N_D - N_A)/2$, and from Eqs. 2.98 and 2.99,

$$n_o = p_o \cong n_i(T) \tag{2.100}$$

This range of temperature is known as the *intrinsic* region. In the intermediate temperature range, where $n_i \ll (N_D - N_A)/2$,

$$n_o = N_D - N_A \tag{2.101}$$

and

$$p_o = \frac{n_i^2}{N_D - N_A} \tag{2.102}$$

At low temperatures, if we make the simplifying assumption that $N_A = 0$ (no compensation), then Eq. 2.93 can be solved to yield

$$n_o = \sqrt{\frac{N_C N_D}{2}} \exp \left(-\frac{\mathcal{E}_c - \mathcal{E}_D}{2k_B T} \right) \tag{2.103}$$

and

$$p_o = \sqrt{\frac{2N_C}{N_D}} N_V \exp \left(-\frac{\mathcal{E}_C + \mathcal{E}_D - 2\mathcal{E}_V}{2k_B T} \right) \tag{2.104}$$

In the intermediate- and low-temperature regions, the free carriers are generated by the ionization of impurity centers, and they therefore represent the *extrinsic* region. If $N_A \neq 0$, then Eqs. 2.103 and 2.104 are of a different form. The important difference is that the denominators in the exponential terms are $k_B T$ instead of $2k_B T$. Thus, compensation changes the slope, or the activation energy of the plot of $\ln[n(p)]$ versus $1/T$ at low temperatures. The variation of $\ln n$ with inverse temperature in an n-type semiconductor is illustrated in Fig. 2.21(a).

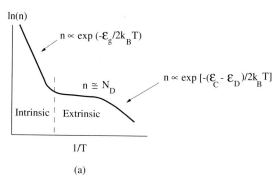

(a)

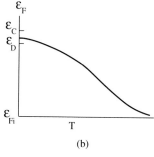

(b)

Figure 2.21 Schematic variation of (a) electron concentration with inverse temperature and (b) Fermi level with temperature in an n-type semiconductor.

Finally, before concluding this discussion, it is useful to see the change in the energy position of the Fermi level as the temperature of the sample changes. In the intrinsic region $\mathcal{E}_F = \mathcal{E}_{Fi}$, which is near mid-bandgap. In the intermediate-temperature range

$$\mathcal{E}_F = \mathcal{E}_c - k_B T \ln \frac{N_C}{N_D - N_A} \qquad (2.105)$$

As the temperature is lowered further, $\mathcal{E}_F$ moves closer to $\mathcal{E}_C$ and at some temperature, when the donor centers are filled with electrons again, crosses the donor level. The variation of $\mathcal{E}_F$ with T in an n-type semiconductor is shown in Fig. 2.21(b).

2.5.5 Consequences of Heavy Doping: Bandtail States

Up to this point we have discussed the case of lightly doped, *nondegenerate*, semiconductors in which the Fermi level is several times $k_B T$ away from the bandedges. In devices, the active region usually has a high density of carriers, produced by doping, injection, or photoexcitation. Under these conditions the semiconductor can become degenerate and $|\mathcal{E}_F - \mathcal{E}_{C(V)}| \leq 3k_B T$. As the doping is increased several things happen. First, the Fermi level approaches the band. At the same time, the density of states functions are also perturbed, resulting in *bandtail states*. This results in bandgap narrowing, an important phenomenon in laser operation, and the Fermi level is within the conduction or valence bands, signifying that the ionization energy of the impurity centers, donors, or acceptors goes to zero. When this occurs, the free-carrier concentration becomes temperature-independent and all the impurity centers are ionized.

Referring to the discussions in Sec. 2.5.3, it can be shown (Problem 2.20) that when a semiconductor becomes degenerate the value of the Fermi integral $F_{1/2}(\eta)$ is given by

$$F_{1/2}(\eta) \cong \frac{4\eta^{3/2}}{3\sqrt{\pi}} \tag{2.106}$$

and the degenerate doping densities are given by

$$n'_o = N_D \cong \frac{1}{3\pi^2}\left[\frac{2m_{de}(\mathcal{E}_F - \mathcal{E}_C)}{h^2}\right]^{3/2} \tag{2.107}$$

or

$$p'_o = N_A \cong \frac{1}{3\pi^2}\left[\frac{2m_{dh}(\mathcal{E}_V - \mathcal{E}_F)}{\hbar^2}\right]^{3/2} \tag{2.108}$$

Note also that $n'_o p'_o \neq n_i^2$.

EXAMPLE 2.5
Objective.   To derive an expression for the product (np) for a degenerate n-type semiconductor.

For a degenerate n-type semiconductor, since $p \ll n$, the charge neutrality condition reduces to

$$n'_o \cong N_D \tag{2.109}$$

Using Eqs. 2.69 and 2.106,

$$n'_o \cong N_D = \frac{4N_C}{3\sqrt{\pi}}\left(\frac{\mathcal{E}_F - \mathcal{E}_C}{k_B T}\right)^{3/2} \tag{2.110}$$

and the expression for p_o is given by Eq. 2.79. Therefore,

$$n'_o p_o = N_D N_V \exp[(\mathcal{E}_V - \mathcal{E}_F)/k_B T] \tag{2.111}$$

Using the substitution

$$\mathcal{E}_V - \mathcal{E}_F = \mathcal{E}_C - \mathcal{E}_F - \mathcal{E}_g \tag{2.112}$$

it follows that

$$n'_o p_o = N_D N_V \exp\left\{-\frac{1}{k_B T}\left[\mathcal{E}_g + \left(\frac{3N_D\sqrt{\pi}}{4N_C}\right)^{2/3} k_B T\right]\right\} \tag{2.113}$$

An important consequence of high doping is that the resultant free carriers screen the Coulomb interaction between the ionized impurity centers and the carriers. In effect, the range of the Coulomb interaction decreases with increasing concentration. A *screening length* is therefore defined to describe this situation, which is inversely proportional to the free-carrier concentration. The screening length, or Debye length, given by

$$L_D = \left[\frac{k_B T \epsilon_s \epsilon_o}{q^2 n}\right]^{1/2} \tag{2.114}$$

is plotted in Fig. 2.22 for GaAs. In Eq. 2.114, ϵ_s is the static dielectric constant and for the case of holes n is replaced by p.

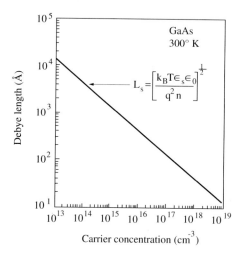

Figure 2.22 Screening or Debye length as a function of carrier concentration.

The onset of degeneracy and the narrowing of the bandgap are related to a concentration-dependent density-of-states function. The calculation of this density of states is outside the scope of this text, but the various models that have been used are qualitatively discussed in the following. In the bandtail model proposed by Kane,[†]

[†]E. O. Kane, *Phys. Rev.*, **131**, 79, 1963.

the carriers are assumed to have very low kinetic energy, so that they can follow the fluctuations in the crystal potential due to the high density of impurity atoms. These potential fluctuations are assumed to be Gaussian in form. In Kane's model, the extent of bandtailing is overestimated, unless a large effective mass is assumed, since the carriers are assumed to have no kinetic energy. In the model proposed by Halperin and Lax,[†] the kinetic energy of the carriers is taken into account, but the potential fluctuation is non-Gaussian. Stern[‡] calculated the density of states by using a model that is similar to the Halperin-Lax model, but with a Gaussian potential. However, the density of states calculated by these models does not merge with a parabolic density of states at higher energies, and usually a smooth interpolation is necessary. The density of states calculated by the different models is illustrated in Figs. 2.23(a) and (b).

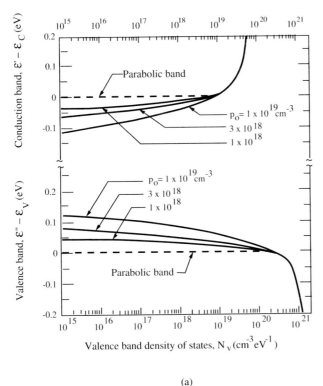

Figure 2.23 (a) Bandtail density of states in GaAs at 297°K as given by the Kane model and (b) densities of states in the conduction and valence bands for GaAs with 2×10^{18} net acceptors per cm^3. The upper curves show the densities of states in the unperturbed bands, the lower curves show the densities of states in the bandtail as calculated from the theory of Halperin and Lax, and the curves which join them are the interpolated Kane functions used as the densities of states in the Stern model (from H. C. Casey, Jr. and F. Stern, *Journal of Applied Physics*, **47**, 631, 1976).

(a)

[†]B. I. Halperin and M. Lax, *Phys. Rev.*, **148**, 722, 1966.
[‡]H. C. Casey and F. Stern, *J. Appl. Phys.*, **47**, 631, 1976.

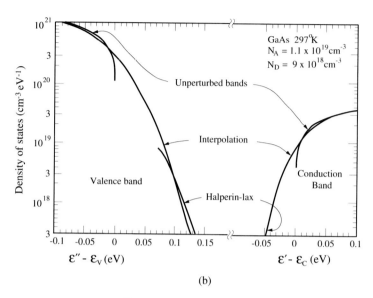

(b)

Figure 2.23 (*continued*)

A consequence of the tail states due to heavy doping is bandgap narrowing, which is seen in Fig. 2.24. Experimentally, such narrowing is observed as a shift of the bandedge absorption and photoluminescence peak to lower energies. This phenomenon is extremely important in the operation of lasers. High doping densities also result in a broadening of the luminescence peak. From experimental observations, the dependence of bandgap on doping level in GaAs is found to be given by[†]

$$\mathcal{E}_g = 1.424 - 1.6 \times 10^{-8} \, (p^{1/3} + n^{1/3})(eV) \qquad (2.115)$$

Thus, in a p-type sample for $p = 3 \times 10^{19} cm^{-3}$, $\mathcal{E}_g \simeq 1.37 eV$, and $\mathcal{E}_A \cong 0$. Optical transitions involving bandtail states will be discussed in Chapter 3.

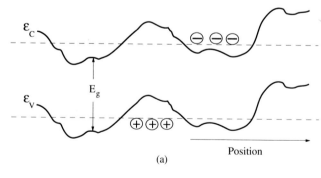

(a)

Position

Figure 2.24 Schematic illustration of (a) perturbation of conduction and valence bandedges due to strong charge-induced local fields and (b) bandgap narrowing in the case of heavy doping. Note that the perturbation of the bandedges causes an effective *spatial* separation of electrons and *holes*, thereby changing carrier lifetimes.

[†]F. Stern, *J. Appl. Phys.*, **47**, 5382, 1976.

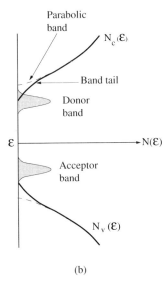

(b)   **Figure 2.24**   (*continued*)

2.6 CONDUCTION PROCESSES IN SEMICONDUCTORS

In order for a semiconductor material to conduct, electrons and holes must be in motion in their respective bands. As we have seen, the first requirement is that there should be a partially filled band, since a completely filled or a completely empty band will not conduct. It is also necessary that the carrier motion has a net direction, and for this an external force is needed. Without such external stimulus, the carriers have a scattering limited thermal velocity, which is not directional. Electrons and holes, moving in their respective bands collide with other carriers, impurity centers, and phonons, and their motion is redirected. Since electrons and holes are charged particles, an externally applied electric field can move carriers in a band in the direction of the electric field. Such motion is called *drift*, and is illustrated in Fig. 2.25(a). Note the bending of the bands and the motion of electrons and holes in opposite directions. Electrons and holes, like neutral particles, can also acquire directional motion due to a concentration gradient, and such motion is termed *diffusion*. The process is illustrated in Fig. 2.25(b).

We will first discuss the process of drift, which arises due to the force $|qE|$ applied by an externally applied electric field E on the carriers. Electrical conduction by electrons and holes in their respective bands is similar to conduction in metals by free electrons. Therefore, the current due to electrons in the conduction band is given by

$$J_{dr} = -nq\vartheta$$

$$= \sigma E \quad (A/cm^2) \tag{2.116}$$

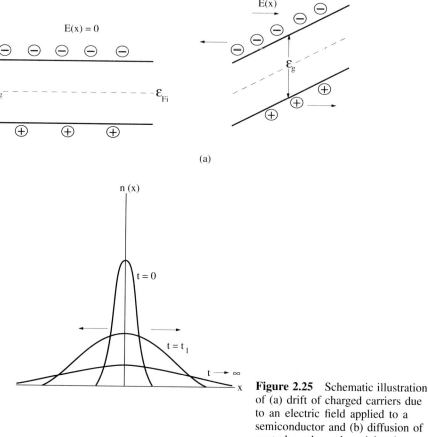

Figure 2.25 Schematic illustration of (a) drift of charged carriers due to an electric field applied to a semiconductor and (b) diffusion of neutral or charged particles due to a concentration gradient.

which is essentially Ohm's law. Here σ is the conductivity of the sample, E is the applied electric field, and $\vartheta = \vartheta_D$ is the average scattering-limited *drift* velocity of the electrons. If the average time between collisions is τ_C, then the average rate of change of momentum due to collisions is $m\vartheta_D/\tau_C$. The equation of motion of an electron subject to an electric field in the x-direction is then given by

$$-qE = m_e^* \frac{d\vartheta_{Dx}}{dt} + \frac{m_e^* \vartheta_{Dx}}{\tau_C} \tag{2.117}$$

where the first term on the right is the energy gained from the field and the second term is the energy lost due to collisions. Solution of the differential equation leads to

$$\vartheta_{Dx} = -\frac{q\tau_C E}{m_e^*} \left(1 - e^{-t/\tau_C}\right) \tag{2.118}$$

and, from Eq. 2.116, the current density is given by

$$J_{dr} = \frac{nq^2\tau_C E}{m_e^*}\left(1 - e^{-t/\tau_C}\right) \tag{2.119}$$

Equations 2.118 and 2.119 indicate that ϑ_{Dx} and J_{dr} rise exponentially with time to a constant value in a time comparable to τ_C, which is defined as a *relaxation time*. Physically, it is the time taken by the system to relax back to thermal equilibrium after the field is switched to zero. Thus,

$$\vartheta_{Dx} = \vartheta_{D0}e^{-t/\tau_C} \tag{2.120}$$

and in the time τ_C the current also reduces to zero. The *steady-state* values of velocity and current, with reference to Eq. 2.34, are given by

$$\bar{\vartheta}_{Dx} = \mu_e E \tag{2.121}$$

and

$$J_x = nq\mu_e E \tag{2.122}$$

if τ_C is not a function E, which is usually a valid assumption. It follows that

$$\sigma = nq\mu_e = \frac{nq^2\tau_C}{m_e^*} \quad (\text{ohm.cm})^{-1} \tag{2.123}$$

The equations derived above are equally valid for hole transport in the valence band. Therefore, the total current density due to drift of electrons and holes is given by

$$J_{dr} = q(n\mu_e + p\mu_h)E \tag{2.124}$$

and the conductivity is given by

$$\sigma = q(n\mu_e + p\mu_h) \tag{2.125}$$

For doped semiconductors in which the impurity levels are fully ionized, n and p are replaced by N_D and N_A, respectively.

EXAMPLE 2.6
Objective.   A sample of GaAs is doped with 10^{15} donors/cm^3 and 5×10^{14} acceptors/cm^3. To find the total conduction current density for an applied field of 100V/cm if $\mu_e = 5000$ cm^2/V.s and $\mu_h = 400$ cm^2/V.s.

We are required to find the drift current density expressed by Eq. 2.124. The conductivity of the sample is

$$\sigma = 1.6 \times 10^{-19} \left[(10^{15} \times 5000) + (5 \times 10^{14} \times 400) \right]$$
$$= 0.832 \ (ohm.cm)^{-1} \tag{2.126}$$

and

$$J_{dr} = 0.832 \times 100$$
$$= 83.2 \ A/cm^2 \tag{2.127}$$

Diffusion arises from a nonuniform density of carriers—electrons and holes. Thus, in the absence of any other processes such as drift, the carriers will diffuse from a region of high density to a region of low density. The process is identical for neutral and charged particles. To analyze the process of diffusion we consider an arbitrary distribution of electrons shown in Fig. 2.25(b). The force of diffusion acting on each electron is given by

$$F_{diff} = -\frac{1}{n} \frac{dP}{dx} \tag{2.128}$$

where the negative sign signifies that the carriers move in a direction opposite to the concentration gradient. Here

$$P = nk_B T \tag{2.129}$$

is the force per unit area acting on the distribution of electrons. It must be remembered that the motion of carriers by diffusion is limited by collisions and scattering. Thus, F_{diff} is equivalent to the force exerted by an electric field, $|eE|$. The velocity due to diffusion is therefore given by

$$\vartheta_{diff} = -\frac{\tau_{Ce}}{m_e^*} \frac{1}{n} \frac{dP}{dx} \tag{2.130}$$

and taking into account Eq. 2.129,

$$\vartheta_{diff} = -\frac{\tau_{Ce} k_B T}{m_e^*} \frac{1}{n} \frac{dn}{dx} \tag{2.131}$$

This leads to the well-known equation for diffusion

$$\vartheta_{diff} = -\frac{D_e}{n} \frac{dn}{dx} \tag{2.132}$$

where D_e is the *diffusion coefficient* for electrons, given by

$$D_e = \frac{\tau_{Ce} k_B T}{m_e^*} \tag{2.133}$$

The current due to diffusion of electrons is expressed as

$$J_{diff}^{e} = -nq\vartheta_{diff} = qD_e\frac{dn}{dx} \tag{2.134}$$

Similarly for holes

$$J_{diff}^{h} = -qD_h\frac{dp}{dx} \tag{2.135}$$

where D_h is the diffusion coefficient for holes. The positive and negative signs in Eqs. 2.134 and 2.135 signify the direction of current with respect to the concentration gradient. These are illustrated in Figs. 2.26(a) and (b). Thus, for electrons having a positive concentration gradient the diffusion velocity is in the negative x direction and the diffusion current is in the positive x direction. For holes having a positive concentration gradient, the hole diffusion velocity and current are both in the negative x direction. Combining Eqs. 2.34 and 2.133, the diffusion constant for electrons can be expressed as

$$D_e = -\frac{\mu_e k_B T}{q} \quad (cm^2/s) \tag{2.136}$$

Similarly,

$$D_h = \frac{\mu_h k_B T}{q} \quad (cm^2/s) \tag{2.137}$$

from which

$$\left|\frac{D_{e(h)}}{\mu_{e(h)}}\right| = \frac{k_B T}{q} \tag{2.138}$$

This is known as the *Einstein relation*. At room temperature $D/\mu = 26$ mV.

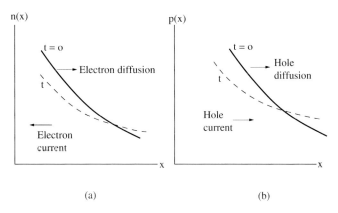

(a)   (b)

Figure 2.26   Diffusion of (a) electrons and (b) holes due to concentration gradient and the corresponding current directions.

If an electric field is present in addition to a concentration gradient in a semiconductor, the total current density for electrons and holes are given by

$$J_e = qn\mu_e E + qD_e \frac{dn}{dx}$$

$$J_h = qp\mu_h E - qD_h \frac{dp}{dx} \tag{2.139}$$

in which the first term on the right arises from drift and the second term arises from diffusion. The total current density is the sum of the contributions due to electrons and holes

$$J(x) = J_e(x) + J_h(x) \tag{2.140}$$

Equations 2.139 and 2.140 are called the current density equations. Some important insights regarding the transport processes can be obtained from Eq. 2.139. The drift term is proportional to the carrier density. Since the density of minority carriers in a semiconductor can be many orders of magnitude smaller than the density of majority carriers, the process of drift is essentially controlled by majority carriers. The process of diffusion, on the other hand, depends on the concentration gradient. Since the density of minority carriers is small, a concentration gradient of minority carriers is easily produced, by current injection in a p-n junction or by intrinsic photoexcitation. Thus, diffusion is essentially controlled by the density of minority carriers.

The process of diffusion has some interesting and useful consequences. At equilibrium there is no net current flow. Thus, if a concentration gradient is somehow induced in the material, a diffusion current is produced. Since no net current can flow, the diffusion current is exactly balanced by a drift current due to a *built-in* electric field that is accommodated by band-bending. Assuming a p-type material, the built-in field is given by Eq. 2.139 as

$$E = \frac{D_h}{\mu_h} \frac{1}{p} \frac{dp}{dx} \tag{2.141}$$

and a similar expression is true for an n-type material. The bending of the bands constitutes a potential gradient, as shown in Fig. 2.25(a). It may be noted in passing that an effective way of introducing a built-in electric field in a region of a semiconductor is to grade the composition. For example, if the region is graded from GaAs to $Al_{0.3}Ga_{0.7}As$, the bandgap increases from 1.43 to 1.85 eV and a potential gradient exists in the conduction band, and drift of electrons is induced.

In optoelectronic devices such as p-i-n photodiodes and modulators, electrons and holes travel with their *saturation velocity* ϑ_S under the influence of the applied bias. It is important to know what this velocity is. In what follows, the mechanism of velocity saturation in semiconductors is briefly discussed. In equilibrium, electrons will normally occupy the lowest energy levels in the conduction band. With the application of an electric field, electrons gain in energy and momentum and move to higher energies within the band. Eventually, the carriers will suffer collisions with the lattice and lose all their energy in the process. The average, or maximum, drift velocity is given by $\bar{\vartheta}_D = \vartheta_S = p_{max}/2m_e^* = (\hbar\omega_p/2m_e^*)^{1/2}$, where p_{max} is the maximum value of the electron momentum and ω_p is the phonon frequency. As an example, in Si $\hbar w_p = 63$ meV and $m_e^* = 0.43m_0$, from which $\bar{\vartheta}_D = 1.08 \times 10^7$ cm/sec at room temperature. In all semiconductors, carriers attain saturated velocities at high temperatures. The velocity-field characteristics of Si, GaAs, In$_{0.53}$Ga$_{0.47}$As and InP are shown in Fig. 2.27. Note that for the compound semiconductors, the electron velocity, instead of monotonically attaining a saturated velocity, first reaches a peak value and then decreases and attains a saturation value. This behavior can be understood on the basis of the conduction bandstructure of these direct bandgap semiconductors, shown in Fig. 2.2. As the applied electric field increases, electrons can gain enough energy from it to attain the energy corresponding to the L minima, or even higher. Collision with phonons can result in the scattering of the carriers to the L minima. Usually polar optical phonons are involved in this process. Since the effective mass in the L minimum is almost eight times that in the Γ minimum, the density of states in these minima is much higher than that in the Γ minimum and therefore the probability of the electrons remaining in the L minima is much higher than that in the Γ minimum. However, in the L minimum the mobility of carriers become much smaller, since the effective mass is higher. As the field continues to increase, the velocity-field characteristics will now be controlled by the transport parameters of the L minima. Thus, at low fields the velocity is given by $\vartheta = \mu_\Gamma E$ and after carrier transfer $\vartheta = \mu_L E$. In the region in between, there is a *negative differential mobility* given by $\mu = -d\vartheta_D/dE$. In real semiconductors the drift velocity does not continue to increase after carrier transfer, but saturates due to reasons outlined

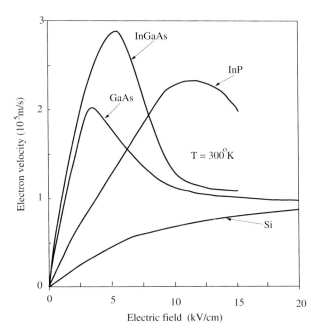

Figure 2.27 Electron drift velocity as a function of electric field at room temperature in elemental and compound semiconductors (from M. Shur, *Physics of Semiconductor Devices*, ©1990. Reprinted by permission of Prentice Hall, Englewood Cliffs, NJ).

earlier. The phenomenon just described is commonly known as the *Gunn effect*. The important parameters in the velocity-field characteristics are the critical field E_C, which marks the onset of the carrier transfer process and the peak and saturation velocities.

2.7 BULK AND SURFACE RECOMBINATION PHENOMENA

2.7.1 Introduction

In a semiconductor, minority carriers are generated by intrinsic photoexcitation or by injection across a forward-biased p-n junction. Since the density of majority carriers is not usually affected, these are termed minority-carrier generation processes. The *excess* minority carrier, after living a mean lifetime, generally recombines with a majority carrier and the pair is dissipated. In an n-type semiconductor, one would be concerned with the generation and recombination of holes, and the net rate of recombination is given by

$$R = \frac{1}{\tau_h}(p - p_o) \tag{2.142}$$

where τ_h is the hole lifetime and p and p_o are the nonequilibrium and equilibrium hole concentrations. The recombination can be *radiative* or *nonradiative*. We will study radiative processes, in which a photon is emitted, in Chapter 3. In this section we will discuss nonradiative recombination, in which a phonon is usually emitted.

2.7.2 Recombination-Generation via Defects or Levels in the Bandgap

The radiative processes of interest to us usually involve the absorption or emission of a photon with energy close to the bandgap. An example is a band-to-band downward transition of an electron, in which a photon is emitted. There is a small probability that during such a downward transition phonons may be emitted, in which case the recombination becomes a nonradiative process. It is more likely, however, that such nonradiative recombination will take place via levels within the bandgap of the semiconductor, as shown in Fig. 2.28. In evaluating the quality of a semiconductor for device applications, it has been found that centers with deep energy levels in the forbidden energy gap of large bandgap semiconductors play an important role. Deep levels essentially act as carrier recombination or trapping centers and adversely affect device performance. Deep levels can be produced by a variety of defects that include substitutional and interstitial impurity atoms, lattice vacancies, or complex defects formed by a combination of two types of defects. As seen from Fig. 1.35 many substitutional impurity species give rise to deep levels in a semiconductor. Note that the probability of the involvement of a phonon is very high in such transitions, which make them nonradiative.

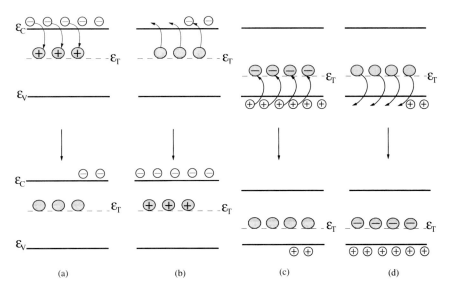

Figure 2.28   Illustration of (a) electron capture, (b) electron emission, (c) hole capture, and (d) hole emission. The deep levels in (a) and (b) are electron traps $(s_e \gg s_h)$ and those in (c) and (d) are hole traps $(s_h \gg s_e)$.

The processes illustrated in Fig. 2.28 are (a) electron capture, (b) electron emission, (c) hole capture, and (d) hole emission. Representing the rates, r, of these processes with the corresponding subscripts, they can be expressed as

$$r_a = c_n n N_T (1 - f(\mathcal{E}))$$
$$= \vartheta_{th} s_e n N_T (1 - f(\mathcal{E}))$$
$$r_b = e_n N_T f(\mathcal{E})$$
$$r_c = \vartheta_{th} s_h p N_T f(\mathcal{E})$$
$$r_d = e_p N_T (1 - f(\mathcal{E})) \tag{2.143}$$

Here $c_{n(p)}$ and $e_{n(p)}$ are the carrier capture and emission rates, respectively, at the deep level or *trap*, N_T is the trap concentration, n and p are the free-carrier concentrations in the respective bands, ϑ_{th} is the thermal velocity of the carriers, and $f(\mathcal{E})$ is the Fermi function at the trap level. The parameters s_e and s_h are the electron and hole capture cross sections at the trap level. Note that the units of $e_n (s^{-1})$ and c_n $(cm^3 s^{-1})$ are different. The various rates r are expressed in $(cm^{-3} s^{-1})$. Under equilibrium conditions, and with no generation of carriers by photoexcitation or other means (generation rate G = 0),

$$r_a = r_b \tag{2.144}$$

and substitution of Eq. 2.143 leads to

$$e_n = \vartheta_{th}^e s_e N_c e^{-(\mathcal{E}_C - \mathcal{E}_T)/k_B T} \tag{2.145}$$

Similarly,

$$r_c = r_d, \tag{2.146}$$

which leads to

$$e_p = \vartheta_{th}^h s_h N_V e^{-(\mathcal{E}_T - \mathcal{E}_V)/k_B T} \tag{2.147}$$

In Eqs. 2.145 and 2.147 $\mathcal{E}_T$ is the energy position of the trap level.

If it is assumed that carriers are also generated at a rate G, then the semiconductor is under nonequilibrium conditions. In the pair-generation process, an electron is raised from the valence band to the conduction band, leaving behind a hole. We also consider *steady-state* conditions, under which the rate at which carriers enter a band is equal to the rate at which they leave the band. Therefore, for an n-type semiconductor,

$$\frac{dn}{dt} = G - (r_a - r_b) = 0 \tag{2.148}$$

and

$$\frac{dp}{dt} = G - (r_c - r_d) = 0 \tag{2.149}$$

Under steady-state nonequilibrium conditions,

$$r_a - r_b = r_c - r_d \tag{2.150}$$

It is necessary to determine the distribution function $f(\mathcal{E})$ under these nonequilibrium conditions, when n and p depend on the generation rate G, and consequently $f(\mathcal{E})$ depends on G. Substitution of Eqs. 2.143, 2.145, and 2.147 into Eq. 2.150 and simplification leads to

$$f(\mathcal{E}_T) = \frac{s_e n + s_h N_\vartheta e^{-(\mathcal{E}_T - \mathcal{E}_V)/k_B T}}{s_e[n + N_c e^{-(\mathcal{E}_C - \mathcal{E}_T)/k_B T}] + s_h[p + N_V e^{-(\mathcal{E}_T - \mathcal{E}_V)/k_B T}]} \tag{2.151}$$

This expression for $f(\mathcal{E}_T)$ can be substituted into the expressions for the individual rates r_a, r_b, r_c, and r_d. Therefore, the net rate of recombination through deep-level traps under steady-state nonequilibrium conditions is given by

$$R = r_a - r_b = r_c - r_d$$

$$= \frac{s_e s_h \vartheta_{th} N_T [np - n_i^2]}{s_e[n + N_c e^{-(\mathcal{E}_C - \mathcal{E}_T)/k_B T}] + s_h[p + N_V e^{-(\mathcal{E}_T - \mathcal{E}_V)/k_B T}]} \tag{2.152}$$

The capture cross section is a measure of how close to a trap center a carrier has to come to get captured. Usually, for an *electron trap* $s_e \gg s_h$ and for a *hole trap* $s_h \gg s_e$. For normal traps, $s_{e(h)} \sim 10^{-15} - 10^{-13}$ cm^2. However, when traps behave more like true nonradiative *recombination centers*,

$$s_e = s_h = s_r \tag{2.153}$$

Therefore, for nonradiative recombination via recombination centers,

$$R = s_r \vartheta_{th} N_T \frac{np - n_i^2}{n + p + 2n_i \cosh\left(\frac{\mathcal{E}_T - \mathcal{E}_{Fi}}{k_B T}\right)} \tag{2.154}$$

where $\mathcal{E}_{Fi}$ is the *intrinsic level* and

$$N_c e^{-(\mathcal{E}_C - \mathcal{E}_T)/k_B T} = n_i e^{-(\mathcal{E}_{Fi} - \mathcal{E}_T)/k_B T}$$

$$N_V e^{-(\mathcal{E}_T - \mathcal{E}_V)/k_B T} = n_i e^{-(\mathcal{E}_T - \mathcal{E}_{Fi})/k_B T} \tag{2.155}$$

Equations 2.152 and 2.154 represent the *Shockley-Read-Hall theory of recombination.*[†][‡] Some important points may be noted with respect to Eq. 2.154. The quantity $(np - n_i^2)$ is a measure of the deviation from equilibrium conditions. It is this deviation that acts as the source of carrier recombination. In deriving Eq. 2.154 it was assumed that the defects behave as true recombination centers that are usually mid-bandgap and for which $s_e \cong s_h$. If this is not true, the recombination rate will decrease. For example, if $(\mathcal{E}_T - \mathcal{E}_{Fi})$ increases (i.e., the center moves away from mid-bandgap), then $s_e \gg s_h$ or $s_h \gg s_e$. This condition holds for true trapping centers or shallow donor and acceptor levels. Finally, it is important to note that true recombination centers can also act as generation centers. For low-level injection in an n-type semiconductor, $n \gg p$ and $n \gg n_i e^{(\mathcal{E}_T - \mathcal{E}_{Fi})/k_B T}$. With these approximations,

$$R = \frac{s_h \vartheta_{th} N_T [np - n_i^2]}{n} \tag{2.156}$$

Remembering that $n_0 p_0 = n_i^2$, where n_0 and p_0 are equilibrium values, and assuming $n \cong n_0$,

$$R = s_h \vartheta_{th} N_T [p - p_0]$$

$$= \frac{1}{\tau_h}(p - p_0) \tag{2.157}$$

where

$$\tau_h = (s_h \vartheta_{th} N_T)^{-1} \tag{2.158}$$

It is seen that the lifetime is not a function of the majority carrier density, n. The rate-limiting step in the recombination process is the concentration of minority carriers.

2.7.3 Surface Recombination

All the bulk properties of a semiconductor come to an abrupt halt at a surface. The surface usually consists of *dangling bonds* or bonds that are satisfied by atoms other than the host atoms in the bulk. A common element is oxygen, and therefore a native oxide is quickly formed on a semiconductor surface. The dangling bonds and bonding with foreign atoms give rise to a high density of defects at the surface of a semiconductor. As a result there is a distribution of defect states in the bandgap at the surface, as shown in Fig. 2.29, and the Fermi level is *pinned* by the overall charge state at the surface rather than by charge neutrality in the bulk. Due to the

---

[†] R. N. Hall, *Phys. Rev.*, **87**, 387, 1952.
[‡] W. Shockley and W. T. Read, *Phys. Rev.*, **87**, 835, 1952.

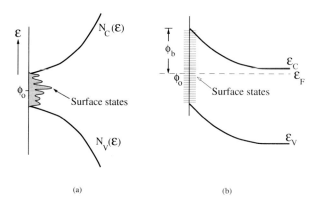

(a)   (b)

Figure 2.29   (a) Distribution of surface states in the bandgap of a semiconductor and (b) band-bending caused by Fermi level pinning at the surface. Φ_0 is called the *neutral* level. In (b) the acceptor-like surface states are occupied with electrons above Φ_0 and the surface has a net negative charge, which balances the positive charge in the depletion layer of the n-type semiconductor.

large density of such surface states, there is an enhanced recombination at the surface of the semiconductor. The surface state density N_{ST} is usually characterized by a delta function at the surface, as shown in Fig. 2.30. When light falls on such a surface, most of it can recombine at the surface even before reaching the bulk. This is extremely detrimental to the operation of most optoelectronic devices, and special treatment of the semiconductor surface is usually necessary. Due to the large density of recombination centers at the surface, the resulting distribution of excess minority carriers in the semiconductor is as shown in Fig. 2.30. It is assumed that

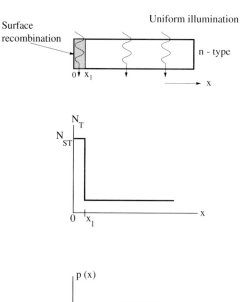

Figure 2.30   Enhanced surface recombination and resulting distribution of excess minority carriers in the presence of surface states.

the surface-state density N_{ST} extends to a thickness x_1 into the material. Then the surface recombination rate R_S is given, by analogy with Eq. 2.157, as

$$R_S = s_h \vartheta_{th} N_{ST} x_1 [p(0) - p_0] \tag{2.159}$$

where it is again assumed that the material is n-type and holes are minority carriers. Under steady-state conditions, R_S must be equal to the flux of minority carriers into the surface region. Thus,

$$D_h \frac{\partial p}{\partial x}\bigg|_{x=0} = s_h \vartheta_{th} N_{ST} x_1 [p(0) - p_0]$$
$$= s_R [p(0) - p_0] \tag{2.160}$$

where

$$s_R = s_h \vartheta_{th} N_{ST} \tag{2.161}$$

is the *surface recombination velocity*. It is a measure of the density of surface recombination rate or the density of defects responsible for it.

Surface recombination can be minimized either by *passivating* the surface with a dielectric such as silicon dioxide or silicon nitride or by having a lattice-matched heterojunction at the free surface. In both cases the wider bandgap material on top of the free surface not only minimizes surface recombination, but also serves as a *window layer*, so that in a device such as a detector or solar cell light can be absorbed in the active region of interest. To conclude this section, it may be noted that bulk and surface nonradiative recombination are extremely detrimental to the operation of optoelectronic devices. These centers, sometimes called "killer centers," provide a nonradiative shunt path through which the excess carriers are dissipated. We will study the important *radiative* processes in a semiconductor in the next two chapters.

PROBLEMS

2.1 A hypothetical energy band can be fitted approximately to the expression

$$\mathcal{E}(k) = \mathcal{E}_0 [1 - \exp(-2a^2 k^2)]$$

where a is the lattice constant of the crystal. Calculate

(a) The effective mass at $k = 0$

(b) The value of k for maximum electron velocity

(c) The effective mass at the edge of the Brillouin zone

2.2 A series of Si-doped n-type $Al_x Ga_{1-x} As (0 \leq x \leq 1)$ samples, each 2 μm thick, were grown on semi-insulating GaAs substrates. The doping level was held approximately constant. When the equilibrium electron concentration

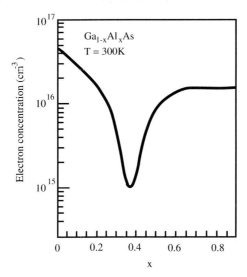

Figure P.2.2

was measured for each sample, by a technique such as Hall measurement, it showed a strong dependence on composition, as shown in Fig. P.2.2. Explain this experimentally observed variation of the electron concentration with x, taking into account the conduction bandstructure of the alloy system.

2.3   The variation of the electron effective mass in the Γ minimum of the $Al_x Ga_{1-x} As$ alloy system may be calculated from theory as

$$\frac{m_0}{m_\Gamma^*} = 1 + \mathcal{E}_{p\Gamma}\left(\frac{2}{\mathcal{E}_\Gamma} + \frac{1}{\mathcal{E}_\Gamma + \Delta}\right)$$

where m_0 is the free electron mass, $\mathcal{E}_\Gamma$ is the energy of the Γ minimum, $\mathcal{E}_{p\Gamma}$ and Δ are energies related to the momentum matrix element and the spin orbit splitting of the valence band. Using the information in Fig. 1.14, determine and plot the values of m_Γ^*/m_0 for $x = 0$ (GaAs) and at intervals of $x = 0.1$. Which effective mass have you determined (m_c or m_d) and why? With reference to Fig. 1.14, comment on the variation of m_X^* in the alloy system.

[Take $\mathcal{E}_{p\Gamma} = 7.51$ eV and $\Delta = 0.341$ eV]

2.4   Can the variation of bandgap of GaAs and other similar compounds with pressure be effectively used to design a pressure sensor? What are the merits and drawbacks of such a device?

2.5   **(a)**   For a semiconductor at $300°$K, the relaxation time for scattering by lattice waves is 10^{-12} sec and that for ionized impurity scattering is 10^{-13} s. At what temperature does the mobility attain its maximum value? The effective mass of the charge carrier is $0.5\ m_0$.

(b) In a cubic semiconductor crystal with ellipsoidal equal energy surfaces, a mobility of 1000 cm^2/V.s is measured for electrons, for which the average scattering relaxation time is 10^{-13} s. If the transverse effective mass is 0.5 m_0, what is the longitudinal effective mass?

2.6 Derive Eq. 2.8 for the density of states effective mass.

2.7 Determine the position of the Fermi level in an n-type GaAs sample with a shallow donor density of 5×10^{15} cm^{-3}.

2.8 A semiconductor contains two independent donor levels with densities $N_{D1} = 1 \times 10^{16}$ cm^{-3} and $N_{D2} = 2 \times 10^{16}$ cm^{-3} and located at 20 meV and 200 meV below the conduction bandedge. Write the charge neutrality equation and plot the dependence of the electron concentration on inverse temperature. How will the profile change if the sample is compensated to some extent with the presence of a shallow acceptor level with density $N_A = 0.9 \times 10^{16}$ cm^{-3}?

2.9 Explain in detail the phenomenon of impurity compensation in a semiconductor. If a semiconductor exhibits a nearly intrinsic resistivity, what can be said of its purity?

2.10 Show that the maximum in the electron distribution of an n-type nondegenerate semiconductor occurs at an energy $k_B T/2$ above the bottom of the conduction band.

2.11 Au and Cu give rise to two very effective recombination centers in Si at 0.03 eV and 0.01 eV, respectively, from the intrinsic Fermi level. Calculate the variation in the recombination rate R, normalized to its maximum value, as a function of $(\mathcal{E}_T - \mathcal{E}_{Fi})/k_B T$, for the following conditions: $p \ll n, n = 10^{16}$ cm^{-3}, $pn - n_i^2 = 1.5 \times 10^{31}$ cm^{-6}, and $\tau_h = 10^{-7}$s.

Repeat this for the case of generation where $pn - n_i^2 = -2.1 \times 10^{20}$ cm^{-6}, $\tau_h = 10^{-7}$s, and p and $n \ll n_i$.

Plot your results on the same graph for a variation of $(\mathcal{E}_T - \mathcal{E}_{Fi})/k_B T$ from 0 to 20 and comment on your results.

2.12 **(a)** In an n-type semiconductor at 300°K, 30% of the donor atoms are ionized. Does the Fermi level lie above or below the donor level? Which way does $\mathcal{E}_F$ move if the temperature increases? Derive the necessary relationship.

(b) Show that a semiconductor has its minimum conductivity when

$$n = n_i \left(\frac{\mu_h}{\mu_e} \right)^{1/2} \quad \text{and} \quad p = n_i \left(\frac{\mu_e}{\mu_h} \right)^{1/2}$$

where n_i is the intrinsic carrier concentration, n and p are the electron and hole concentrations and μ_e and μ_h are the electron and hole mobilities.

2.13 A hypothetical "intrinsic" sample of InP is doped with acceptor impurities from one side such that a doping profile $N_A = N_A(0)e^{-bz}$ is obtained. Here z represents the direction into the sample and $N_A(0)$ is the concentration on the surface ($z = 0$). Note also that for $z \to \infty$, $N_A \to n_i$, the intrinsic carrier concentration.

   (a) Sketch the doping profile and derive an expression for the electric field that will develop under thermal equilibrium conditions for $N_A \gg n_i$ due to this doping profile.

   (b) Evaluate the value of the electric field when $b = (2\mu m)^{-1}$.

2.14 What is the representation of the Fermi-Dirac distribution function in *momentum* space at low temperatures? Comment on what happens to this distribution at very high temperatures.

2.15 **(a)** What is the Fermi energy level? In a semiconductor the Fermi level is 40 meV below the edge of the conduction band and the bandgap is 1.3 eV. What would you do to bring the Fermi level to mid-bandgap? Will the resistivity of the semiconductor change? If so, how?

   (b) An n-type direct bandgap semiconductor has $\mathcal{E}_g = 1.43$ eV at $300°$K. It has a shallow donor level with a binding energy of 15 meV and a deep trap level at an energy of 600 meV from the conduction band-edge. Draw the $\mathcal{E} - \mathbf{k}$ diagram showing the conduction and valence bands and the donor and trap levels.

   [Hint: remember the uncertainty principle.]

2.16 A one-dimensional quantum confined structure is called a quantum wire. In such a wire, carrier motion is free along the length of the wire and is quantized in the two transverse directions. With assumptions similar to that made in Sec. 2.5.2 (i.e., the carrier energy in a subband is parabolic), find the density of states function $N(\mathcal{E})d\mathcal{E}$ in the subbands of such a wire.

2.17 A sample of InP is doped with Si (a shallow donor) and C (a shallow acceptor). Using the values given in Table 2.3, find the electron and hole concentrations, n and p, respectively, for the following doping concentrations at $300°$K:

   (a) $N_A = 3 \times 10^{16}$ cm^{-3}, $N_D = 4 \times 10^{16}$ cm^{-3}.

   (b) $N_A = 7 \times 10^{16}$ cm^{-3}, $N_D = 2 \times 10^{16}$ cm^{-3}.

   Show the variation of the Fermi level with temperature in both cases.

2.18 Find the relation among the Fermi energy level, the hole concentration, and the temperature for a two-dimensional hole gas.

2.19 **(a)** Derive expressions for the ratio of the electron concentrations n_X/n_Γ and n_L/n_Γ in a multivalleyed semiconductor.

   (b) Calculate the values of the expressions derived in (a) for GaAs in which the $\Gamma - L$ and $\Gamma - X$ energy band separations are 0.28 and 0.48 eV, respectively.

2.20   Derive Eq. 2.106 for a degenerate semiconductor.

[Hint: In this case $\eta \gg 1$.]

READING LIST

CASEY, H. C., and PANISH, M. B. *Heterostructure Lasers, Parts A and B*. Academic Press, New York, 1978.

MCKELVEY, J. P. *Solid State and Semiconductor Physics*. Krieger Publishing Co., Malabar, FL, 1982.

NAG, B. R. *Theory of Electrical Transport in Semiconductors*. Pergamon Press, Oxford, 1972.

SHOCKLEY, W. *Electrons and Holes in Semiconductors*. D. van Nostrand, Princeton, NJ, 1950.

SHUR, M. *Physics of Semiconductor Devices*. Prentice Hall, Englewood Cliffs, NJ, 1990.

SINGH, J. *Physics of Semiconductors and Their Heterostructures*. McGraw-Hill, New York, 1993.

SZE, S. *Physics of Semiconductor Devices*. 2nd ed. Wiley, New York, 1981.

WANG, S. *Fundamentals of Semiconductor Theory and Device Physics*. Prentice Hall, Englewood Cliffs, NJ, 1989.

3

Optical Processes in Semiconductors

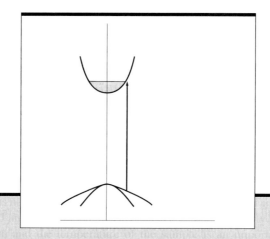

Chapter Contents

3.1 ELECTRON-HOLE PAIR FORMATION AND RECOMBINATION

The operation of almost all optoelectronic devices is based on the creation or annihilation of electron-hole pairs. Pair formation essentially involves raising an electron in energy from the valence band to the conduction band, thereby leaving a hole behind in the valence band. In principle, any energetic particle incident on a semiconductor, which can impart an energy at least equal to the bandgap energy to a valence band electron, will create pairs. With respect to the bonding in the lattice, this process is equivalent to breaking a covalent bond. The simplest way to create electron-hole pairs is to irradiate the semiconductor. Photons with sufficient energy are absorbed, and these impart their energy to the valence band electrons and raise them to the conduction band. This process is, therefore, also called *absorption*. The reverse process, that of electron and hole recombination, is associated with the pair giving up its excess energy. Recombination may be *radiative* or *nonradiative*. In a nonradiative transition, the excess energy due to recombination is usually imparted to phonons and dissipated as heat. In a radiative transition, the excess energy is dissipated as photons, usually having energy equal to the bandgap (i.e., $\hbar\omega = \mathcal{E}_g$). This is the *luminescent* process, which is classified according to the method by which the electron-hole pairs are created. *Photoluminescence* involves the radiative recombination of electron-hole pairs created by injection of photons. *Cathodoluminescence* is the process of radiative recombination of electron-hole pairs created by electron bombardment. *Electroluminescence* is the process of radiative recombination following injection with a p-n junction or similar device.

In a semiconductor in equilibrium (i.e., without any incident photons or injection of electrons), the carrier densities can be calculated from an equilibrium Fermi level by using Fermi-Dirac or Boltzmann statistics outlined in Sec. 2.5.3. When excess carriers are created by one of the techniques described above, nonequilibrium conditions are generated and the concept of a Fermi level is no longer valid. One can, however, define nonequilibrium distribution functions for electrons and holes as

$$f_n(\mathcal{E}) = \frac{1}{1 + \exp\left(\frac{\mathcal{E} - \mathcal{E}_{fn}}{k_B T}\right)} \tag{3.1}$$

$$f_p(\mathcal{E}) = \frac{1}{1 + \exp\left(\frac{\mathcal{E} - \mathcal{E}_{fp}}{k_B T}\right)} \tag{3.2}$$

These distribution functions define $\mathcal{E}_{fn}$ and $\mathcal{E}_{fp}$, the *quasi-Fermi levels* for electrons and holes, respectively. In some texts they are referred to as IMREFs (Fermi spelled backward). When the excitation source creating excess carriers is removed, $\mathcal{E}_{fn} = \mathcal{E}_{fp} = \mathcal{E}_F$. The difference ($\mathcal{E}_{fn} - \mathcal{E}_{fp}$) is a measure of the deviation from equilibrium. As with equilibrium statistics, we obtain for the nondegenerate case

$$f_n(\mathcal{E}) \cong \exp\left(\frac{\mathcal{E}_{fn} - \mathcal{E}}{k_B T}\right) \tag{3.3}$$

$$f_p(\mathcal{E}) \cong \exp\left(\frac{\mathcal{E} - \mathcal{E}_{fp}}{k_B T}\right) \tag{3.4}$$

and the nonequilibrium carrier concentrations are given by

$$n = N_C \exp\left(\frac{\mathcal{E}_{fn} - \mathcal{E}_C}{k_B T}\right) \tag{3.5}$$

$$p = N_V \exp\left(\frac{\mathcal{E}_V - \mathcal{E}_{fp}}{k_B T}\right) \tag{3.6}$$

The concept of quasi-Fermi levels is extremely useful, since it provides a means to take into account changes of carrier concentration as a function of position in a semiconductor. As we shall see in Chapter 4, in a p-n junction under forward bias a large density of excess carriers exist in the depletion region and close to it on either side. The concentration of these carriers can be determined from the appropriate quasi-Fermi levels. A junction laser is operated under such forward bias injection conditions to create population inversion. To consider a simple example, assume that an n-type semiconductor with an equilibrium electron density n_o ($= N_D$, the donor density) is uniformly irradiated with intrinsic photoexcitation (above-bandgap light) so as to produce Δn electron-hole pairs with a generation rate G. The nonequilibrium electron and hole concentrations are given by

$$n = \Delta n + n_o \tag{3.7}$$

$$p = \Delta n + n_i^2/n_o \tag{3.8}$$

Using Eqs. 3.3–3.8, Fig. 3.1 illustrates the change in the energy position of the quasi Fermi levels in GaAs as the generation rate changes.

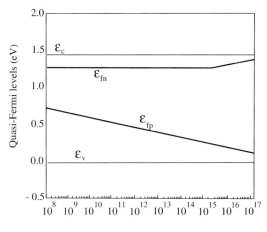

Figure 3.1 caption (to the right of figure):

Figure 3.1 Energy position of the electron and hole quasi-Fermi levels as a function of pair generation rate in GaAs at room temperature. It is assumed that the sample is n-type with $N_D = 10^{15}$ cm^{-3} (from M. Shur, *Physics of Semiconductor Devices*, ©1990. Reprinted by permission of Prentice-Hall, Englewood Cliffs, New Jersey).

The excess carriers created in a semiconductor must eventually recombine. In fact, under steady-state conditions the recombination rate must be equal to the generation rate:

$$G = R \tag{3.9}$$

Generation and recombination processes involve transition of carriers across the energy bandgap and are therefore different for direct and indirect bandgap semiconductors, as illustrated in Fig. 3.2. In a direct bandgap semiconductor, as shown in Fig. 3.2(a), the valence band maximum and the conduction band minimum occur at the zone center ($\mathbf{k} = 0$) and an upward or downward transition of electrons does not require a change in momentum or the involvement of a phonon. Therefore, in direct bandgap semiconductors such as GaAs, an electron raised to the conduction band, say, by photon absorption, will dwell there for a very short time and recombine again with a valence band hole to emit light of energy equal to the bandgap. Thus, the probability of *radiative recombination* is very high in direct bandgap semiconductors. The processes are quite different in an indirect bandgap semiconductor. Considering the band diagrams shown in Fig. 3.2(b) and (c), since the conduction band minima are not at $\mathbf{k} = 0$, upward or downward transition of carriers require a change in momentum, or the involvement of a phonon. Thus, an electron dwelling in the conduction band minimum, at $\mathbf{k} \neq 0$, cannot recombine with a hole at $\mathbf{k} = 0$ until a phonon with the right energy and momentum is available. Both phonon emission or absorption processes can assist the downward transition. In order for the right phonon collision to occur, the dwell time of the electron in the conduction band increases. Since no crystal is perfect, there are impurities and defects in the lattice that manifest themselves as traps and recombination centers. It is most likely that the electron and hole will recombine nonradiatively through such a defect center, and the excess energy is dissipated into the lattice as heat. The competing nonradiative processes reduce the probability of radiative recombination in indirect bandgap materials such as Si, Ge, or

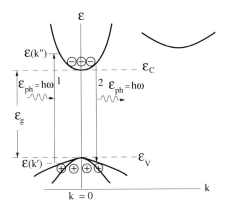

(a)

Figure 3.2(a) Illustration of band-to-band of absorption and recombination processes in (a) direct bandgap semiconductor and (b) and (c) indirect bandgap semiconductor.

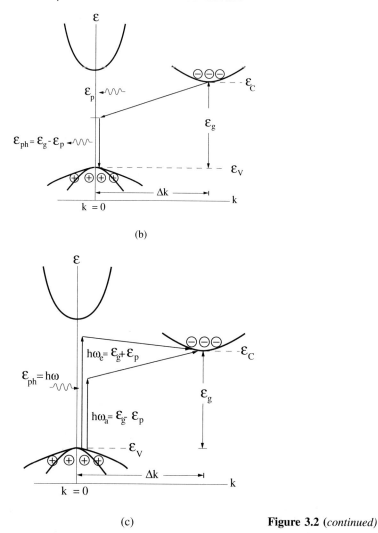

(b)

(c)   **Figure 3.2** (*continued*)

GaP. These semiconductors are therefore, in general, not suitable for the realization of light sources such as light-emitting diodes and lasers.

3.1.1 Radiative and Nonradiative Recombination

For continuous carrier generation by optical excitation or injection, a quasi equilibrium or steady state is produced. Electrons and holes are created and annihilated in pairs and, depending on the injection level, a steady-state excess density $\Delta n = \Delta p$ is established in the crystal. This equality is also necessary for the maintenance of overall charge neutrality. When the excitation source is removed, the density of excess carriers returns to the equilibrium values, n_o and p_o. The decay of excess carriers usually follows an exponential law with respect to time $\sim \exp(-t/\tau)$, where τ is defined as the lifetime of excess carriers. The lifetime is determined by a combination of intrinsic and extrinsic parameters, and the performance characteristics of most optoelectronic devices depend on it. In the discussion that follows, we will be concerned mainly with *bulk* recombination processes. It is important to remember that, depending on the semiconductor sample and its surface, there can be a very strong *surface* recombination component which depends on the density of surface states.

In general, the excess carriers decay by radiative and/or nonradiative recombination, in which the excess energy is dissipated by photons and phonons. The former is of importance for the operation of luminescent devices. Nonradiative recombination usually takes place via surface or bulk defects and traps (Fig. 3.3), as discussed in Chapter 2, and reduces the radiative efficiency of the material. Therefore the total lifetime τ can be expressed as

$$\frac{1}{\tau} = \frac{1}{\tau_r} + \frac{1}{\tau_{nr}} \tag{3.10}$$

where τ_r and τ_{nr} are the radiative and nonradiative lifetimes, respectively. Also, the total recombination rate R_{total} is given by

$$R_{\text{total}} = R_r + R_{nr} = R_{sp} \tag{3.11}$$

where R_r and R_{nr} are radiative and nonradiative recombination rates per unit volume, respectively, and R_{sp} is called the *spontaneous* recombination rate, to distinguish

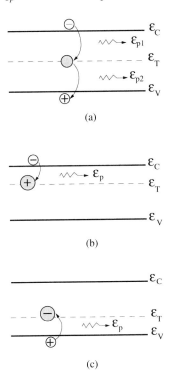

(a)

(b)

(c)

Figure 3.3 Nonradiative recombination at (a) recombination center, (b) electron trap, and (c) hole trap. The excess carrier energy in all cases is dissipated by single or multiple phonons.

R_{total} from the *stimulated* recombination rate to be defined later in Chapter 6. The internal quantum efficiency or radiative recombination efficiency is defined as

$$\eta_r = \frac{R_r}{R_r + R_{nr}} \tag{3.12}$$

For an exponential decay process, $\tau_r = \Delta n / R_r$ and $\tau_{nr} = \Delta n / R_{nr}$ where Δn is the excess electron concentration. Therefore,

$$\eta_r = \frac{1}{1 + \tau_r / \tau_{nr}} \tag{3.13}$$

To achieve high internal quantum efficiency, the ratio τ_r / τ_{nr} should be as small as possible, or τ_{nr} should be as large as possible. The value of τ_{nr} is controlled by the properties of defects, which produce levels in the bandgap of a semiconductor. The excess energy of carriers recombining at these levels is dissipated by phonons. Another nonradiative process is *Auger recombination*, to be discussed in Sec. 3.8. It also follows from Eq. 3.9 that under steady-state conditions $\Delta n = G\tau_r$.

3.1.2 Band-to-Band Recombination

The simplest carrier decay process is spontaneous band-to-band recombination, whose rate, without momentum conservation, is given by

$$R_{sp} = B_r np \tag{3.14}$$

where B_r is defined as the coefficient for band-to-band recombination in units of $cm^3.s^{-1}$. B_r is related to the transition probability P to be discussed in the next section. In terms of the equilibrium and excess carrier densities,

$$R_{sp} = B_r(n_o + \Delta n)(p_o + \Delta p) \tag{3.15}$$

where $\Delta n = \Delta p$. The spontaneous radiative recombination rate for excess carriers can be expressed as

$$R_{sp}^{ex} = \frac{\Delta n}{\tau_r} \tag{3.16}$$

and therefore

$$R_{sp} = R_{sp}^{o} + R_{sp}^{ex} \tag{3.17}$$

where

$$R_{sp}^{o} = B_r n_o p_o \tag{3.18}$$

is the spontaneous recombination rate in thermal equilibrium. From Eqs. 3.15 and 3.18,

$$R_{sp} = B_r \left[n_o p_o + \Delta n(n_o + p_o) + (\Delta n^2) \right] \tag{3.19}$$

$$R_{sp}^{ex} = B_r \Delta n \left[n_o + p_o + \Delta n \right] \tag{3.20}$$

and

$$\tau_r = \frac{1}{B_r(n_o + p_o + \Delta n)} \tag{3.21}$$

When $\Delta n \gg n_o, p_o$, which is relevant to laser operation,

$$\tau_r \cong \frac{1}{B_r \Delta n} \tag{3.22}$$

This is the *bimolecular* recombination regime, when the lifetime changes with Δn. At low injection levels, such that $\Delta n < n_o, p_o$

$$\tau_r \cong \frac{1}{B_r(n_o + p_o)} \tag{3.23}$$

which remains constant, being determined by the background carrier concentrations. For an intrinsic semiconductor under low-level injection, since $n_o = p_o = n_i$,

$$\tau_r = \frac{1}{2B_r n_i} \tag{3.24}$$

Eq. 3.22 is valid for $\Delta n \leq 10^{18}$ cm^{-3}. For higher values of Δn,

$$\tau_r \cong \tau_o \tag{3.25}$$

which is usually constant for any material. For example for GaAs, $\tau_o \cong 0.5$ ns.

The value of the recombination coefficient depends on the bandgap and whether the semiconductor has a direct or an indirect bandgap. Direct bandgap semiconductors usually have values of B_r ranging from 10^{-11} to 10^{-9} cm^3.s^{-1} and indirect bandgap semiconductors have values of B_r ranging from 10^{-15} to 10^{-13} cm^3.s^{-1}.

EXAMPLE 3.1
Objective.   To calculate τ_r in GaAs having $n_o = 10^{14}$ cm^{-3} under high- and low-level injections for $B_r = 7 \times 10^{-10}$ cm^3/s.

At a high injection level of 10^{18} cm^{-3}, $\tau_r = (7 \times 10^{-10} \times 10^{18})^{-1} s = 1.43$ns.

At a low injection level of 10^{16} cm^{-3}, $\tau_r = 143$ ns for the same value of B_r. This value of τ_r is almost an order of magnitude larger than that measured in pure GaAs samples. Therefore, the value of B_r used here is only valid for the high-level injection case and should be larger for the case of low-level injection.

3.2 ABSORPTION IN SEMICONDUCTORS

3.2.1 Matrix Elements and Oscillator Strength for Band-to-Band Transitions

The operation of optical devices that we will describe and discuss in this text depends on the upward and downward transitions of carriers between energy bands. These transitions result in absorption or emission of light, which is electromagnetic energy. The measurement of absorption and emission spectra in semiconductors constitutes an important aspect of materials characterization. They provide information not only on the bandgap, but the measurements also provide information on direct and indirect transitions, the distribution of states, and the energy position of defect and impurity levels. The absorption spectrum spans a wide energy (or wavelength) range, extending from the near-bandgap energies to the low-energy transitions involving free carriers and lattice vibrations. In the context of this text the more important ones are the near-bandgap transitions.

The process of photon absorption results in the transition of an electron from a lower energy state to a higher energy state, the simplest form of which may be a direct transition from the valence to the conduction band. The different possible transitions are outlined in this chapter. In what follows, the process of band-to-band transitions in semiconductors, in which photons are absorbed, is analyzed.

The energy-momentum diagrams of a direct and an indirect semiconductor were shown in Figs. 3.2(a) and (b), respectively. Considering the case of an electron raised from the top of the valence band to the bottom of the conduction band due to absorption of a photon in a direct transition, there is no change in momentum. Strictly, there is a small change in $\mathbf{k}$ due to the finite momentum of the photon, which is equal to h/λ. For most III–V semiconductors $\lambda \sim 1\mu$m, and the resultant momentum change is very small. An indirect transition due to the absorption of a photon is illustrated in Fig. 3.2(c). Since a large change in momentum is involved in this case, the transition can occur only by the emission or absorption of a phonon. The process can be described by the equation

$$\mathcal{E}_{ph} \pm \mathcal{E}_p = \mathcal{E}_g \tag{3.26}$$

and the change in momentum is given by

$$\Delta \mathbf{k} = \mathbf{k}_p \tag{3.27}$$

where $\mathbf{k}_p$ is the wavevector of the phonon and $\mathcal{E}_{ph}$ and $\mathcal{E}_p$ are the photon and phonon energies, respectively. Therefore, an optical or acoustic phonon with the right energy and momentum must be involved in an indirect transition.

The wavelength dependence of direct transitions, for the case of absorption, is illustrated in Fig. 3.2(a), where we consider a transition away from the zone center. The top of the valence band is taken as the zero of energy. The transition occurs in energy from $\mathcal{E}(\mathbf{k}')$ to $\mathcal{E}(\mathbf{k}'')$ where

$$\mathcal{E}(\mathbf{k}') = -\frac{\hbar^2 \mathbf{k}'^2}{2m_h^*} \tag{3.28}$$

and

$$\mathcal{E}(\mathbf{k}'') = \mathcal{E}_g + \frac{\hbar^2 \mathbf{k}''^2}{2m_e^*}. \tag{3.29}$$

Here $\mathcal{E}_g$ is the direct bandgap at $\mathbf{k} = 0$. The energy of the absorbed photon is $\mathcal{E}_{ph} = \mathcal{E}(\mathbf{k}'') - \mathcal{E}(\mathbf{k}')$, which is the requirement for energy conservation. Remember that such a transition will take place only if the level at $\mathcal{E}(\mathbf{k}')$ is filled and that at $\mathcal{E}(\mathbf{k}'')$ is empty. Also, for momentum conservation $\mathbf{k}'$ must be nearly equal to $\mathbf{k}''$. This is called the *k-selection rule*.

The matrix element and probability of an optical transition from $\mathcal{E}(\mathbf{k}')$ to $\mathcal{E}(\mathbf{k}'')$ can be calculated by considering first-order time-dependent perturbation theory. The time-independent form of the Schrödinger equation is

$$\mathbf{H}_0\Psi \;=\; \mathcal{E}(\mathbf{k}')\Psi \tag{3.30}$$

where $\mathbf{H}_0$ is the Hamiltonian of the unperturbed system. In the case of a perturbation $\mathbf{H}_1$, which in our context is light or electromagnetic radiation, causing a carrier transition from a state at $\mathcal{E}(\mathbf{k}')$ to a state at $\mathcal{E}(\mathbf{k}'')$, the time-dependent Schrödinger equation can be expressed as

$$(\mathbf{H}_0 + \mathbf{H}_1)\Psi \;=\; j\hbar\frac{d\Psi}{dt} \tag{3.31}$$

with

$$\Psi \;=\; \sum_m A_m(t)\Psi_m e^{-j\mathcal{E}_m(\mathbf{k}')t/\hbar} \tag{3.32}$$

It may be noted that $|A_m(t)|^2$ is the transition probability. The calculation of the matrix element of an optical transition has been described in detail in a few texts and is not repeated here. The matrix element for direct transitions, where the condition

$$\mathbf{k}'' - \mathbf{k}' \;=\; \mathbf{k} \cong 0 \tag{3.33}$$

is satisfied, is given by

$$
\begin{aligned}
H_{k''k'} &= \int \Psi_{k''}^*\mathbf{H}_1\Psi_{k'}d\mathbf{r} \\
&= \frac{jq\hbar A}{2m_0}\int u_C^*(\mathbf{r},\mathbf{k}'')[\mathbf{a}_0\cdot\nabla u_V(\mathbf{r},\mathbf{k}') \;+\; j(\mathbf{a}_0\cdot\mathbf{k}')u_V(\mathbf{r},\mathbf{k}')]d\mathbf{r}
\end{aligned} \tag{3.34}
$$

where u_V and u_C are the Bloch functions corresponding to the valence and conduction bands, respectively, $\mathbf{A}$ is the magnetic vector potential of the electromagnetic wave and $\mathbf{a}_0$ is a polarization unit vector. The first term represents the matrix element for *allowed* direct transitions and is usually much larger than the second term, which represents *forbidden* transitions. If $\mathbf{k}' = \mathbf{k}''$, the matrix element of the forbidden transition is zero. However, because of the small change in momentum due to the small but finite momentum of the photon, the matrix element of the forbidden transition has a finite value. It can be shown that the transition probability per unit volume per unit time for an allowed direct transition is given by[†]

$$P(\hbar\omega) \;=\; \frac{q^2|\mathbf{A}|^2(2m_r^*)^{3/2}p_{CV}^2}{4\pi m_0^2\hbar^4}(\hbar\omega - \mathcal{E}_g)^{1/2} \tag{3.35}$$

where p_{CV} is the matrix element of the momentum operator (or the momentum matrix element) and m_r^* is the reduced mass given by

$$m_r^* \;=\; \frac{m_e^*m_h^*}{m_e^* + m_h^*} \tag{3.36}$$

[†]R. H. Bube, *Electronic Properties of Crystalline Solids*, Academic Press, New York, 1974.

Equation 3.35 contains the *joint density of states* to be discussed in Chapter 5. When the **k**-selection rule is obeyed, $|p_{CV}|^2 = 0$ unless $\mathbf{k}' = \mathbf{k}''$. Equation 3.35 is an important relationship and shows that the transition probability of a direct allowed transition varies as $(\hbar\omega - \mathcal{E}_g)^{1/2}$. The transition probability includes the summation over all filled valence band states and empty conduction band states and over all $\mathbf{k}'$ and $\mathbf{k}''$ values that satisfy energy and momentum conservation. In the form expressed in Eq. 3.35, it is assumed that the semiconductor is at $0°K$ when the valence band is completely filled and the conduction band is empty.

Since the absorption of a photon of energy $\hbar\omega$ is involved in a direct transition, it is important to calculate the absorption coefficient α. Assume that a monochromatic photon flux $\Im_i$, given by

$$\Im_i = \frac{|\mathbf{S}|}{\hbar\omega} \quad \text{(photons/cm}^2\text{.s)} \tag{3.37}$$

is incident on the crystal. Here $|\mathbf{S}|$ is the radiation energy crossing unit area in unit time, or the *Poynting vector*. The transmitted intensity $\Im_d$ is then

$$\Im_d = \frac{|\mathbf{S}|}{\hbar\omega} - P(\hbar\omega)d \tag{3.38}$$

where d is the thickness of the sample. The second term represents the number of photons absorbed per unit time per unit area, normal to the incident light in a thickness d. Equation 3.38 can be written as

$$\Im_d = \frac{|\mathbf{S}|}{\hbar\omega}e^{-\alpha d}$$

$$\cong \frac{|\mathbf{S}|}{\hbar\omega}(1 - \alpha d) \tag{3.39}$$

for small αd. Thus,

$$\alpha(\hbar\omega) = \frac{P\hbar\omega}{|\mathbf{S}|} \tag{3.40}$$

The average value of the Poynting vector over a period of the electromagnetic wave can be expressed as

$$|\mathbf{S}| = \frac{1}{2}n_r\epsilon_0c\omega^2|\mathbf{A}^2| \tag{3.41}$$

where n_r is the refractive index of the crystal. Substitution of Eqs. 3.35 and 3.41 into Eq. 3.40 leads to

$$\alpha(\hbar\omega) = C_1 n_r^{-1}\left(\frac{2m_r^*}{m_0}\right)^{3/2}\frac{f_{CV}}{\hbar\omega}(\hbar\omega - \mathcal{E}_g)^{1/2} \tag{3.42}$$

where

$$C_1 = \frac{q^2 m_0^{1/2}}{4\pi\hbar^2\epsilon_0 c} \tag{3.43}$$

and

$$f_{CV} = \frac{2p_{CV}^2}{m_0} \qquad (3.44)$$

Expressing $\hbar\omega$ and $\mathcal{E}_g$ in eV,

$$\alpha(\hbar\omega) = 2.64 \times 10^5 n_r^{-1} \left(\frac{2m_r^*}{m_0}\right)^{3/2} \frac{f_{CV}}{\hbar\omega}(\hbar\omega - \mathcal{E}_g)^{1/2} \quad (cm^{-1}) \qquad (3.45)$$

f_{CV} is called the *oscillator strength* for the transition. It has a value approximately equal to 20 eV in most semiconductors. Therefore, for GaAs ($f_{CV} = 23$ eV), we get from Eq. 3.45,

$$\alpha(\hbar\omega) = 5.6 \times 10^4 \frac{(\hbar\omega - \mathcal{E}_g)^{1/2}}{\hbar\omega} \quad (cm^{-1}) \qquad (3.46)$$

The value of α expressed in Eqs. 3.42 or 3.45 corresponds to a fixed photon energy $\hbar\omega$ when the semiconductor is at $0°$K and the values of the Fermi-Dirac distribution functions in the conduction and valence bands are zero and unity, respectively. To express the temperature dependence of α, one must include the Fermi functions in the summation over all energies used to calculate the probability function $P(\hbar\omega)$. As a result, the absorption coefficient expressed by Eq. 3.42 or 3.45 must include a factor $[f_p(\mathcal{E}(\mathbf{k}')) - f_n(\mathcal{E}(\mathbf{k}''))]$, where f_n and f_p are the Fermi functions in the respective bands. It is also assumed that the semiconductor is very pure. Impurity atoms will induce scattering, which will relax the momentum conservation requirements. In addition, impurity levels will give rise to bandtail states that will result in a finite value of α for photon energies $\hbar\omega < \mathcal{E}_g$. The absorption coefficient α expressed in Eq. 3.46 corresponds to a measured value for photon energies $\hbar\omega > \mathcal{E}_g$ under the ideal conditions described above.

Similarly, starting with the matrix element for the direct forbidden transition in Eq. 3.34, it can be shown that the transition probability is given by

$$P(\hbar\omega) = \frac{q^2|\mathbf{A}|^2}{12\pi m_0^2 \hbar^4}(2m_r^*)^{5/2} f_{CV}'(\hbar\omega - \mathcal{E}_g)^{3/2} \qquad (3.47)$$

It is important to note that the probability is proportional to $(\hbar\omega - \mathcal{E}_g)^{3/2}$. The absorption coefficient is given by

$$\alpha(\hbar\omega) = C_2 n_r^{-1} \left(\frac{2m_r^*}{m_0}\right)^{5/2} \frac{f_{CV}'}{\hbar\omega}(\hbar w - \mathcal{E}_g)^{3/2} \qquad (3.48)$$

where

$$C_2 = \frac{q^2 m_0^{1/2}}{6\pi\hbar^2\epsilon_0 c} \qquad (3.49)$$

f_{CV}' is the oscillator strength for the forbidden transition, and its value is much less than unity. Again, if $\hbar\omega$ and $\mathcal{E}_g$ are expressed in eV,

$$\alpha(\hbar\omega) = 1.76 \times 10^5 \frac{n_r^{-1}}{\hbar\omega} \left(\frac{2m_r^*}{m_0}\right)^{5/2} f_{CV}'(\hbar\omega - \mathcal{E}_g)^{3/2} \quad (cm^{-1}) \qquad (3.50)$$

From experimental results it is evident that the direct optical transition corresponding to the absorption of a photon with $\hbar\omega > \mathcal{E}_g$ is dominantly observed in most direct-bandgap semiconductors.

EXAMPLE 3.2
Objective.   To calculate α for the allowed transitions in GaAs at a photon energy $\hbar\omega = 1.52$ eV.

Assuming $\mathcal{E}_g = 1.5$ eV at $0°$K, from Eq. 3.46 $\alpha = 5.2 \times 10^3$ cm^{-1}.

The following general comments may be made regarding the relations given above. For very small values of $(\hbar\omega - \mathcal{E}_g)$ and in a pure semiconductor, *excitons* are generally formed due to the Coulomb interaction between electrons and holes. The expressions for the matrix element, transition probability, and absorption coefficient given above are only for band-to-band transitions and do not consider exciton-related processes. Second, the equations are valid insofar as the parabolic band approximation is true, and are therefore generally not valid for higher-lying regions of the direct band. For large values of photon energy, contributions from the satellite valleys can become important. In most III–V semiconductors, the valence band is degenerate at the zone center. Strictly, the effect of such degenerate bands will contribute to the absorption. For a heavily doped semiconductor, the absorption edge moves to higher energies, as illustrated in Fig. 3.4. In addition, bandtail states are formed. It is important to note that the matrix element for transitions between bandtail states is different from that involving free-electron and hole states. Consequently, the **k**-selection rule does not apply.

With variation of temperature, the variation of the absorption coefficient follows the variation of the bandgap with temperature given by Eq. 2.15. Finally, it should

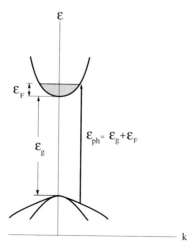

Figure 3.4 Simplified illustration of absorption of photons of energy larger than the bandgap in a degenerately doped n-type semiconductor.

be remembered that the process of radiation is complementary to absorption and is governed by similar equations for transition probability and oscillator strength.

3.2.2 Indirect Intrinsic Transitions

The momentum or wavevector change required in an indirect transition may be provided by single or multiple phonons, although the probability of the latter to occur is very small. As seen in Chapter 2, there are optical and acoustic phonons. Each of these has transverse and longitudinal modes of vibrations, with characteristic energy and momentum. The indirect transition process is illustrated in Fig. 3.2(c). Conservation of momentum requires

$$\mathbf{k}'' \pm \mathbf{k}_p = \mathbf{k}' + \mathbf{k}_{ph} \tag{3.51}$$

where $\mathbf{k}''$ and $\mathbf{k}'$ are the electron wavevectors for the final and initial states, $\mathbf{k}_p$ is the wavevector of the phonon, and $\mathbf{k}_{ph}$ is the wavevector of the absorbed photon. Since the latter is small, the conservation of momentum for an indirect transition can be expressed as

$$\mathbf{k}'' - \mathbf{k}' = \mp \mathbf{k}_p \tag{3.52}$$

Similarly, the conservation of energy for the two cases of phonon emission and absorption can be expressed as ·

$$\hbar\omega_e = \mathcal{E}_C - \mathcal{E}_V + \mathcal{E}_p \tag{3.53}$$

$$\hbar\omega_a = \mathcal{E}_C - \mathcal{E}_V - \mathcal{E}_p \tag{3.54}$$

where the left-hand side represents the energy of the photon absorbed. Note that in the first case of phonon emission, the energy of the absorbed photon could be equal to the direct gap at or very near $\mathbf{k} = 0$. From this energy state the electron finally reaches the indirect valley by phonon scattering. The intermediate energy state of the electron is termed a *virtual* state, in which the carrier resides until a phonon of the right energy and momentum is available for the scattering process. Indirect transition probabilities involving virtual states can be calculated using a second-order time-dependent perturbation theory. However, there is a process to slightly counterbalance the low transition probability, which is often overlooked. From Eqs. 3.53 and 3.54 it is evident that the initial and final states of the electron in the valence and conduction bands, respectively, can have an energy range given by $\hbar(\omega_{ph} \pm \omega_p)$, where ω_p and ω_{ph} correspond to the angular frequencies of the phonon and photon, respectively. The total probability is obtained by a summation over these energy states, as long as each particular transition conserves energy between initial and final states.

For a transition with phonon absorption,

$$\alpha_a(\hbar\omega) \propto \frac{(\hbar\omega - \mathcal{E}_g \mid \mathcal{E}_p)^2}{e^{\mathcal{E}_p/k_B T} - 1} \tag{3.55}$$

for a photon energy $\hbar\omega > (\mathcal{E}_g - \mathcal{E}_p)$. Similarly, for a transition with phonon emission the absorption coefficient is given by

$$\alpha_e(\hbar\omega) \propto \frac{(\hbar\omega - \mathcal{E}_g - \mathcal{E}_p)^2}{1 - e^{-\mathcal{E}_p/k_B T}} \tag{3.56}$$

for $\hbar\omega > (\mathcal{E}_g + \mathcal{E}_p)$. Since for $\hbar\omega > (\mathcal{E}_g + \mathcal{E}_p)$ both phonon emission and absorption are possible, under these conditions

$$\alpha(\hbar\omega) = \alpha_a(\hbar\omega) + \alpha_e(\hbar\omega) \tag{3.57}$$

The temperature dependence of the absorption coefficient is illustrated in Fig. 3.5. At very low temperatures, the density of phonons available for absorption becomes small and therefore α_a is small. With increase of temperature, α_a increases. The shift of the curves to lower energies with increase of temperature reflects the temperature dependence of $\mathcal{E}_g$. In fact, the plots of $\sqrt{\alpha_e}$ and $\sqrt{\alpha_a}$ extrapolate to the energy axis at $(\mathcal{E}_g + \mathcal{E}_p)$ and $(\mathcal{E}_g - \mathcal{E}_p)$, respectively. This is a convenient technique to experimentally determine the bandgap.

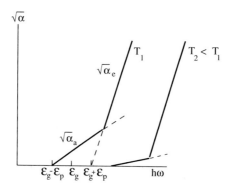

Figure 3.5 Energy-dependent absorption coefficient due to phonon emission and absorption as a function of temperature.

3.2.3 Exciton Absorption

In very pure semiconductors, where the screening effect of free carriers is almost absent, electrons and holes produced by the absorption of a photon of near-bandgap energy pair to form an *exciton*. This is the free exciton. The binding energy of the exciton, $\mathcal{E}_{ex}$, is calculated by drawing analogy with the Bohr atom for an impurity center, and is quantized. It is therefore expressed as

$$\mathcal{E}_{ex}^l = \frac{-m_r^* q^4}{2(4\pi\epsilon_r\epsilon_o\hbar)^2} \cdot \frac{1}{l^2}, \quad l = 1, 2, 3, \cdots$$

$$= \frac{-13.6\, m_r^*}{l^2} \frac{m_r^*}{m_0} \left(\frac{1}{\epsilon_r}\right)^2 (eV). \tag{3.58}$$

Here m_r^* is the reduced effective mass of the exciton given by Eq. 3.36 and l is an integer.

The optical excitation and formation of excitons usually manifest themselves as a series of sharp resonances (peaks) at the low energy side of the band edge in the absorption spectra of direct bandgap semiconductors. The total energy of the exciton is given by

$$\mathcal{E}_{ex} = \frac{\hbar^2 \mathbf{k}_{ex}^2}{2(m_e^* + m_h^*)} - \mathcal{E}_{ex}^l \tag{3.59}$$

where the first term on the right is the kinetic energy of the exciton. The kinetic energy contributes to a slight broadening of the exciton levels. For a direct transition conservation of momentum requires that $\mathbf{k}_{ex} \cong 0$. This is because the electron and hole must move in the same direction. Usually a sharp line transition is observed for direct excitonic transitions, which broadens with increase of temperature. Data on the excitonic absorption in pure GaAs are shown in Fig. 3.6.

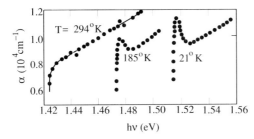

Figure 3.6   Absorption coefficient measured near the band edge of GaAs at T = 294, 185, and 21°K. The two absorption peaks at $h\nu$ slightly below the respective band gap at 185 K and 21 K are due to bound excitons (from M. D. Sturge, *Physical Review*, **127**, 768, 1962).

In indirect bandgap semiconductors excitons may also be formed with the absorption or emission of a phonon. In this case the center of gravity of the exciton may have a finite momentum $\hbar\mathbf{k}_{ex}$, conserved by an interacting phonon. Again, transverse and longitudinal acoustic and optical phonons may participate. An increase in absorption coefficient is obtained near the band edge due to exciton absorption, given by

$$\hbar\omega = \mathcal{E}_g \mp \mathcal{E}_p - \mathcal{E}_{ex} \tag{3.60}$$

where the two signs of the second term on the right hand side correspond to the cases of phonon absorption or emission. Exciton-related transitions are seen in the absorption spectra of an indirect bandgap semiconductor as a large number of steps near the absorption edge. Note that steps are observed instead of peaks, as in direct bandgap semiconductors, because the interacting phonons allow the carrier transition between states with equal $d\mathcal{E}/dk$ in the valence and conduction bands at energies greater than in direct band gap semiconductors where usually excitons are formed at the zone center ($d\mathcal{E}/dk = 0$).

Excitons are formed in very pure semiconductors at low temperatures. In fact, excitons were not observed in semiconductors until epitaxial techniques enabled the growth of very pure crystals. In such crystals, the very few unintentional impurities that are present—donors and acceptors—are neutral. If an electric field is applied, it can ionize these impurities, and the additional charge modifies the bandedge potential. This is seen in the experimental absorption spectra as a change in the slope of the absorption edge. In addition, the ionized carriers screen the Coulomb interaction between the electrons and holes, thereby inhibiting or preventing the formation of excitons. This is observed in the experimental absorption spectra as a disappearance of the excitonic resonances, peaks or steps, as the case may be.

3.2.4 Donor-Acceptor and Impurity-Band Absorption

Intentionally or unintentionally, both donors and acceptors are simultaneously present in a semiconductor, and any semiconductor is usually always compensated to some degree. Depending on the temperature and the state of occupancy of the impurity levels, it is possible to raise an electron from the acceptor to the donor level by absorbing a photon. This process is shown in Fig. 3.7. The energy of the photon absorbed is given by

$$\hbar\omega = \mathcal{E}_g - \mathcal{E}_D - \mathcal{E}_A + \frac{q^2}{\epsilon_0\epsilon_r r} \tag{3.61}$$

where the last term on the right-hand side accounts for the Coulomb interaction between the donor and acceptor atoms in substitutional sites, which results in a lowering of their binding energies. This can be understood as follows. Assume that at very low temperatures the donor and acceptor atoms are neutral. If they are brought closer together, the additional orbiting electron of the donor becomes "shared" by the acceptor, as in a covalent bond, and both become more ionized, resulting in a lowering of their binding energy. Also, it is important to remember that since the donor and acceptor atoms are located at discrete substitutional sites in the lattice, r varies in finite increments, being the smallest for nearest neighbors. Therefore, for the ground state of the impurities, the energies $\mathcal{E}_D$ and $\mathcal{E}_A$ correspond to the most distant pairs and $\hbar\omega \cong \mathcal{E}_g - \mathcal{E}_D - \mathcal{E}_A$. For fully ionized impurities, such as for nearest neighbors, the excited states may lie within the respective band and it is possible that $(q^2/\epsilon_0\epsilon_r r) > \mathcal{E}_D + \mathcal{E}_A$. At low temperatures the absorption resonances modify the bandedge absorption, with the lowest energy transitions for the most distant pairs and higher-energy transitions for nearer pairs. However, because the resonances occur so close to the absorption edge, they are not always very clearly defined. The pair transitions are more clearly identified in emission experiments.

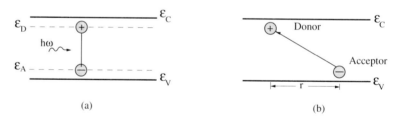

(a)   (b)

Figure 3.7 Illustration of photon absorption due to a donor-acceptor transition. The separation between the impurity centers, r, is shown in (b).

High-energy (near-bandgap) transitions can occur between ionized impurity levels and the opposite bandedge, as illustrated in Fig. 3.8. The photon energy absorbed is $\hbar\omega \cong \mathcal{E}_g - \mathcal{E}_b$, where $\mathcal{E}_b$ is the binding energy of the donor or acceptor level. It should be noted that the impurity levels need to be ionized. Since the transition occurs between a discrete impurity level and a band of energies, the transitions are observed as shoulders on the low-energy side of the absorption edge. In the emission spectra, these transition are observed as peaks. As in a band-to-band transition, phonons need to be involved in impurity-band transitions in an indirect bandgap semiconductor for momentum conservation.

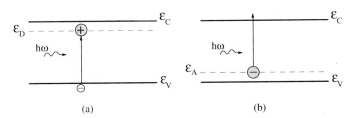

Figure 3.8   Illustration of (a) donor-band and (b) acceptor-band absorption transitions.

The absorption spectrum is largely altered if the doping level is increased and gradually taken to the point of degeneracy. For example, in a degenerately doped n-type semiconductor, the Fermi level $\mathcal{E}_{fn}$ is above the conduction bandedge. If the semiconductor is direct bandgap, as shown in Fig. 3.4, then, for the conservation of momentum, the transition resulting from the absorption of a photon will involve states in the conduction band that are at or higher than $\mathcal{E}_g + \mathcal{E}_{fn}$. This shift of the absorption to higher energies due to doping-induced band-filling is called the *Burstein-Moss shift* An indirect bandgap semiconductor will be similarly affected, except that phonons need not be involved in the transition. Momentum is conserved by impurity scattering.

Degeneracy in semiconductors not only pushes the Fermi level into the band, but also results in a shrinkage of the bandgap. This effect is more commonly known as *bandtailing*, which results in an exponentially increasing absorption edge with photon energy as shown is Fig. 3.9 for GaAs.

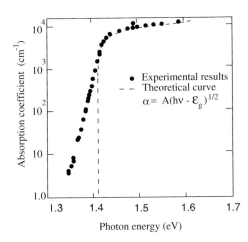

Figure 3.9   Absorption edge of GaAs at room temperature (from T. S. Moss, *Journal of Applied Physics*, **32**, 2136, 1961).

3.2.5 Low-Energy (Long-Wavelength) Absorption

Several types of transitions involving shallow impurity levels, bandedges, split bands, and free carriers give rise to resonances at very small energies in the absorption spectra. These are observed as steps or peaks in the long-wavelength region of the absorption spectra. The different processes are briefly described below.

3.2.5.1 Impurity-Band Transitions.

We have seen impurity band transitions that have energies close to the bandgap. These higher-energy impurity-band transitions usually require that the impurity levels are ionized (or empty). At low temperatures, when these shallow impurity levels are usually filled with their respective carriers, these carriers can be excited to the respective bandedge by a photon (Fig. 3.10). For this absorption process the energy of the photon must be at least equal to the ionization energy of the impurity. This energy usually corresponds to the far infrared region of the optical spectrum. Peaks corresponding to such impurity-band transitions have been observed in many semiconductors.

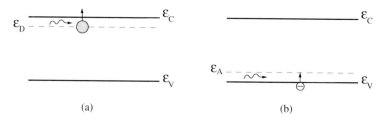

Figure 3.10   Low-energy (a) donor-band and (b) acceptor-band absorption transition.

3.2.5.2 Intraband Transitions.

At the zone center the valence band structure of most semiconductors consists of the light-hole (LH), the heavy-hole (HH) bands, and the split-off (SO) band. The three subbands are separated by spin-orbit interaction. In a p-type semiconductor the valence band is filled with holes and the occupancy of the different bands depend on the degree of doping and the position of the Fermi level. Absorption of photons with the right energy can result in transitions from LH to HH, SO to HH, and SO to LH bands, depending on the doping and temperature of the sample. These transitions have been observed experimentally. They are normally not observed in n-type semiconductors.

3.2.5.3 Free-Carrier Absorption.

As the name suggests, this mechanism involves the absorption of a photon by the interaction of a free carrier within a band, which is consequently raised to a higher energy. The transition of the carrier to a higher energy within the same valley must conserve momentum. This momentum change is provided by optical or acoustic phonons, optical phononse, or by impurity scattering. Free-carrier absorption usually manifests in the long-wavelength region of the spectrum as a monotonic increase in absorption with a wavelength dependence of the form λ^p, where p ranges from 1.5 to 3.5. The value of p depends on the nature of the momentum-conserving scattering (i.e., the involvement of acoustic phonons, optical phonons, or ionized impurities). The absorption coefficient due to free-carrier absorption can be expressed as

$$\alpha = \frac{Nq^2\lambda^2}{4\pi^2 m^* n_r c^3 \epsilon_0} \left\langle \frac{1}{\tau} \right\rangle \tag{3.62}$$

where N is the free-carrier concentration, n_r is the refractive index of the semiconductor, and $\left\langle \frac{1}{\tau} \right\rangle$ is the average value of the inverse of the relaxation time of the scattering process.

As a concluding note to this section, the absorption coefficients for different elemental and III–V compound semiconductors at room temperature are shown in Fig. 3.11.

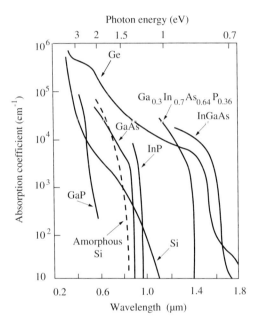

Figure 3.11   Near-bandgap absorption spectra of different semiconductors (from M. Shur, *Physics of Semiconductor Devices,* ©1990. Reprinted by permission of Prentice Hall, Englewood Cliffs, New Jersey).

3.3 EFFECT OF ELECTRIC FIELD ON ABSORPTION: FRANZ-KELDYSH AND STARK EFFECTS

The change in absorption in a semiconductor in the presence of a strong electric field is the *Franz-Keldysh effect,* which results in the absorption of photons with energies less than the bandgap of the semiconductor. The energy bands of a semiconductor in the presence of an electric field E and with an incident photon of energy $\hbar\omega < \mathcal{E}_g$ are shown in Figs. 3.12(a) and (b). It is important to note that at the classical turning points marked A and B, the electron wavefunctions change from oscillatory to decaying behavior. Thus, the electron in the energy gap is described by an exponentially decaying function $u_k e^{jkx}$, where **k** is imaginary. With increase of electric field, the distance AB decreases and the overlap of the wavefunctions within the gap increases.

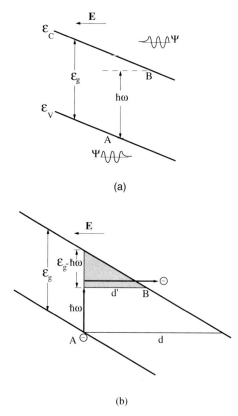

(a)

(b)

Figure 3.12 (a) Bending of bands due to an applied electric field and (b) absorption of photon with $\hbar\omega < \mathcal{E}_g$ due to carrier tunneling (Franz-Keldysh effect).

In the absence of a photon, the valence electron has to tunnel through a triangular barrier of height $\mathcal{E}_g$ and thickness d, given by $d = \mathcal{E}_g/qE$. With the assistance of an absorbed photon of energy $\hbar\omega < \mathcal{E}_g$, it is evident that the tunneling barrier thickness is reduced to $d' = (\mathcal{E}_g - \hbar\omega)/qE$, the overlap of the wavefunctions increases further, and the valence electron can easily tunnel to the conduction band. The net result is that a photon with $\hbar\omega < \mathcal{E}_g$ is absorbed. One has to keep in mind, of course, the conservation of momentum in these transitions and in this case the transverse component of momentum is conserved. The Franz-Keldysh effect is therefore, in essence, photon assisted tunneling.

It can be shown that the electric-field dependent absorption coefficient is given by[†]

$$\alpha = K(E')^{1/2}(8\beta)^{-1}\exp\left(-\frac{4}{3}\beta^{3/2}\right) \tag{3.63}$$

Here $E' = \left(\dfrac{q^2 E^2 \hbar^2}{2m_r^*}\right)^{1/3}$, $\beta = \dfrac{\mathcal{E}_g - \hbar\omega}{E'}$, and K is a material-dependent parameter that has

---

[†]S. Wang, *Fundamentals of Semiconductor Theory and Device Physics*, Prentice Hall, Englewood Cliffs, NJ, 1989.

a value of 5×10^4 cm^{-1} (eV)$^{-1/2}$ in GaAs. Although not derived here, the various terms in Eq. 3.63 can be examined qualitatively. The exponential term is the transmission coefficient (or tunneling probability) of an electron through a triangular barrier of height ($\mathcal{E}_g - \hbar\omega$) and can be obtained from the well-known Wentzel-Kramers-Brillouin (WKB) approximation. The other factors are related to the upward transition of an electron due to photon absorption. Substituting appropriate values for the different parameters, it is seen that in GaAs $\alpha = 4cm^{-1}$ at a photon energy of $\mathcal{E}_g - 20$ meV with electric field $E \sim 10^4$ V/cm. This value of absorption coefficient is much smaller than the values of α at the band edge at zero field. Therefore, the Franz-Keldysh effect will be small unless $E \geq 10^5$ V/cm.

The *Stark effect* refers to the change in atomic energy upon the application of an electric field. The electric field affects the higher-order, or outer, orbits of the precessing electrons so that the center of gravity of the elliptical orbit and the focus are displaced with respect to each other and linearly aligned in the direction of the electric field. As a result, there is a splitting of the energy of the outer $2s$ or $2p$ states, and the energy shift is simply given by $\Delta\mathcal{E} = qdE$, where d is the eccentricity of the orbit. This is the *linear Stark effect*. The effect of the electric field on ground state orbits also leads to an energy shift of the state, and this is the *quadratic* or *second-order Stark effect*.

3.4 ABSORPTION IN QUANTUM WELLS AND THE QUANTUM-CONFINED STARK EFFECT

In a bulk semiconductor the exciton binding energy is given by Eq. 3.58. For example in GaAs, upon substitution of the effective mass and dielectric constant, a ground-state binding energy of 4.4 meV is obtained. This is comparable to the thermal broadening of ~ 4 meV produced by optical phonon scattering and inhomogeneous broadening, to be discussed in Chapter 6. In other words, the exciton dissociates in a very short time (a few hundred femtoseconds) and is hardly detected in the absorption spectra at room temperature, except in very pure samples.

The situation is drastically altered in a quantum well. In a single-quantum well (SQW) or multiple-quantum well (MQW) with thick barriers (≥ 100 Å), electrons and holes are confined in the region defined by the well width, and the overlap of their wavefunctions is increased. This results in an increase in the oscillator strength of the interband transitions between the discrete electron and hole energy bound states, which are produced by the size quantization. Consequently, strong resonances corresponding to the heavy-hole and light-hole transitions are seen near the bandedge of the well material even at room temperature. Shown in Fig. 3.13 are the measured and calculated absorption spectra of an In$_{0.53}$Ga$_{0.47}$As (100 Å)/In$_{0.52}$Al$_{0.48}$As (100 Å) MQW lattice matched to InP, in which the resonances for $l = 2$ and 3, are also clearly seen. It may be noted that distinct resonances are seen for heavy-hole and light-hole transitions. This is because a splitting between the HH and LH bands occurs at the zone center due to the difference in the energy eigenvalues resulting from different hole masses.

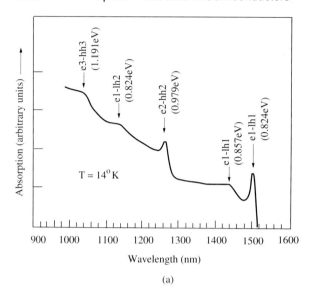

(a)

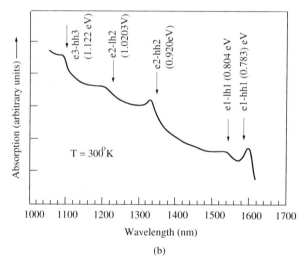

(b)

Figure 3.13   (a) Absorption spectrum of 40-period lattice-matched InGaAs/InAlAs MQW (measured by the author and co-workers) at 14°K; and (b) the calculated transitions based on a finite square well model at room temperature (from S. Gupta et al., *Journal of Applied Physics*, **69**, 3219, 1991).

In a two-dimensional quantum well the exciton is compressed like a pancake. However, since typical well dimensions are ~ 100 Å and the exciton diameter is ≥ 300 Å, there is some penetration of the exciton wavefunction into the barrier material. This is depicted in Fig. 3.14. In the limit, for small well widths, the situation becomes similar to a three-dimensional solid. It has been shown that for a purely two-dimensional exciton, its binding energy is four times the bulk value, given by Eq. 3.58. However, because of the extension of its wavefunction into the barrier, the binding energy in practical SQW and MQW structures ranges between $2\mathcal{E}_{ex}$ and $3\mathcal{E}_{ex}$. Since this binding energy is much larger than the thermal broadening, the exciton

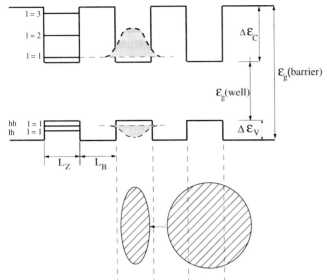

Figure 3.14 A typical multiquantum well and the compression of the bulk exciton in the well region.

resonances are clearly seen in the absorption spectra even at room temperature. The coefficient for intersubband absorption in a quantum well can be expressed in cgs units as[†]

$$\alpha_{2D}(\hbar\omega) = \frac{4\pi^2 q^2 \hbar}{n_r c m_0^2 \hbar\omega} p_{CV}^2 a_p \frac{N_{2D}(\hbar\omega - \mathcal{E}_g)}{L_z}$$

$$= 1.77 \times 10^{-28} \frac{a_p}{n_r} \left(\frac{f_{CV}}{\hbar\omega} \right) \frac{N_{2D}}{L_z} \tag{3.64}$$

where L_z is the width of the well, $N_{2D} = m_r^*/\pi\hbar^2$ is the 2-D density of states, a_p is a factor due to polarization dependence of the matrix elements (in GaAs $a_p = 1/2$). This equation is valid for a single pair of conduction and valence subbands. Note also that the value of α is constant and will remain so till the next subband energies are reached. This is evident in Fig. 3.13. It may seem from Eq. 3.64 that the absorption coefficient can be increased by decreasing L_z. However, as L_z is decreased, the electron and hole wavefunctions spread outside the well and the overlap decreases. Therefore, for every material system there is an optimum well size. For example, for GaAs this $L_z \sim 50\text{Å}$ and for $In_{0.53}Ga_{0.47}As$ $L_z \sim 80\text{Å}$.

The ground-state wavefunctions of the electron and hole subband with no applied transverse field (in the direction perpendicular to the layers) are shown in Fig. 3.15(a). With the application of an electric field several things happen. The electron and hole wavefunctions are separated and pushed toward opposite sides of the well, as

---

[†]J. Singh, *Physics of Semiconductors and Their Heterostructures*, McGraw-Hill, New York, 1993.

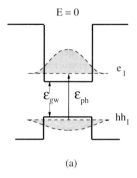

(a)

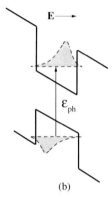

(b)

Figure 3.15 Absorption in a quantum well in the (a) absence and (b) presence of a transverse electric field.

shown in Fig. 3.15(b). The reduced overlap results in a corresponding reduction in absorption and in luminescence. The probability of carriers tunneling out of the wells also increases, resulting in a decrease in carrier lifetimes and a broadening of the absorption spectra. The transition energy is given by

$$\mathcal{E}_{ph} = \mathcal{E}_e + \mathcal{E}_h + \mathcal{E}_{gw} - \mathcal{E}_{ex} \tag{3.65}$$

where $\mathcal{E}_e$ and $\mathcal{E}_h$ are the electron and hole subband energies. With the application of moderate electric fields ($10^4 - 10^5$ V/cm), there is little change in $\mathcal{E}_{ex}$ and a very small change in $\mathcal{E}_{gw}$ due to the Stark effect in the well material. However, due to the modification of the envelope functions, there is a reduction in $\mathcal{E}_e$ and $\mathcal{E}_h$, the subband energies. This results in a shift of the absorption spectrum to lower energies, including the heavy- and light-hole resonances. The shifts are much larger than the Stark shift in bulk materials and is ~ 20 meV for $E = 10^5$ V/cm in a 100 Å GaAs/Al$_{0.3}$Ga$_{0.7}$As quantum well. Experimental data are shown in Fig. 3.16 and the phenomenon is known as the *quantum confined Stark effect* (QCSE). As we shall see in Chapter 11, the effect can be used for the design and realization of very efficient light modulators.

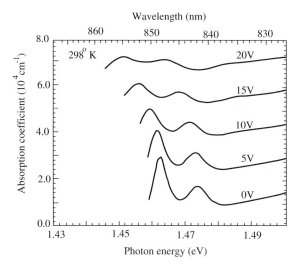

Figure 3.16   Room temperature optical absorption coefficients measured by the author and co-workers in a 100-Å GaAs/Al$_{0.3}$Ga$_{0.7}$As MQW structure at various bias levels (from J. Singh et al., *Journal of Lightwave Technology*, **6**, 818, ©1988 IEEE).

3.5 THE KRAMER-KRÖNIG RELATIONS

The complex refractive index of a semiconductor is given by

$$\mathbf{n}_c = \mathbf{n}_r + j\mathbf{k}_a \qquad (3.66)$$

where $\mathbf{n}_r$ and $\mathbf{k}_a$ are the real and imaginary parts of $\mathbf{n}_c$. It may be noted that $\mathbf{k}_a$ is an attenuation or damping factor and is a measure of the loss in power of a wave propagating through the semiconductor. Therefore, $\mathbf{k}_a$ is directly proportional to the absorption coefficient α of the material. If a plane wave described by

$$\mathbf{E} = \mathbf{E}_0 \exp\left[j\omega\left(\frac{x}{\vartheta} - t\right)\right]$$

$$= \mathbf{E}_0 \exp\left[j\omega\left(\frac{n_r x}{c} - t\right)\right] \exp\left(-\frac{\omega \mathbf{k}_a x}{c}\right) \qquad (3.67)$$

is propagating in the x-direction in a semiconductor with a velocity $\vartheta = \frac{c}{n_c}$, then it can be shown (Problem 3.7) that the absorption coefficient is given by

$$\alpha = \frac{2\omega \mathbf{k}_a}{c} = \frac{4\pi \nu \mathbf{k}_a}{c} \qquad (3.68)$$

In a material whose conductivity $\sigma \to 0$, the refractive index is related to the dielectric constant by the relation

$$n_r \cong \sqrt{\epsilon_r} \qquad (3.69)$$

where ϵ_r is the static dielectric constant ϵ_s and

$$k_a \cong 0 \qquad (3.70)$$

It is also useful to know that the refractive index is inversely related to the bandgap of a semiconductor.

The complex dielectric constant of a material is given by

$$\epsilon_r(\omega) = \epsilon_r'(\omega) + j\epsilon_r''(\omega) \qquad (3.71)$$

In time-invariant form, the electric field $\mathbf{E}$ and the electric flux density $\mathbf{D}$ are related by

$$\mathbf{D} = \epsilon_0 \left(1 + \chi^e\right) \mathbf{E}$$

$$= \epsilon_r \epsilon_o \mathbf{E} \tag{3.72}$$

where χ^e is the electric *susceptibility*. The temporal response of $\mathbf{D}$ to a change or switching of $\mathbf{E}$ must include the change of polarization with time, and can be expressed by a causality relation of the type

$$\mathbf{D}(t) = \epsilon_o \epsilon'_\infty \delta(t)\mathbf{E}(t) + \int_{-\infty}^{t} \epsilon_o f(t - t')\mathbf{E}(t')dt' \tag{3.73}$$

in which the integral represents the response of the system at time t to the applied field $\mathbf{E}$ at a previous time t'. Note that $\epsilon(t)$ has a δ-function singularity at $t = 0$. ϵ'_∞ is the high-frequency dielectric constant, and $\epsilon''_\infty = 0$. Now the following Fourier transforms can be written for $\mathbf{D}$ and $\mathbf{E}$:

$$\mathbf{D}(t) = \frac{1}{2\pi} \int_{-\infty}^{\infty} \mathbf{D}(\omega)e^{-j\omega t}d\omega \tag{3.74}$$

and

$$\mathbf{E}(t) = \frac{1}{2\pi} \int_{-\infty}^{\infty} \mathbf{E}(\omega)e^{-j\omega t}d\omega \tag{3.75}$$

Substituting these in Eq. 3.73 one gets

$$\int_{-\infty}^{\infty} \left[\mathbf{D}(\omega) - \epsilon_o(\epsilon'_\infty + f(\omega))\mathbf{E}(\omega)\right] e^{-j\omega t}d\omega = 0 \tag{3.76}$$

with

$$f(\omega) = \frac{1}{2\pi} \int_{-\infty}^{\infty} f(t)e^{j\omega t}dt \tag{3.77}$$

Equation 3.76 must be valid for all values of t. Therefore, the relation

$$\mathbf{D}(\omega) = \epsilon_o \left(\epsilon'_\infty + f(\omega)\right) \mathbf{E}(\omega) \tag{3.78}$$

is valid between the Fourier components, so that

$$\epsilon_r(\omega) = \epsilon'_\infty + \frac{1}{2\pi} \int_{-\infty}^{\infty} f(t)e^{j\omega t}dt$$

$$= \epsilon'_\infty + f(\omega) \tag{3.79}$$

From Eq. 3.79, by application of the Cauchy theorem to the function $[\epsilon_r(\omega) - \epsilon'_\infty]/(\omega' - \omega)$, the following relations can be derived (Problem 3.14):

$$\epsilon'_r(\omega) = \epsilon'_\infty + \frac{2}{\pi} P \int_0^{\infty} \frac{\omega' \epsilon''_r(\omega')d\omega'}{\omega'^2 - \omega^2} \tag{3.80}$$

$$\epsilon''_r(\omega) = -\frac{2\omega}{\pi} P \int_0^{\infty} \frac{\left[\epsilon'_r(\omega') - \epsilon'_\infty\right]d\omega'}{\omega'^2 - \omega^2} \tag{3.81}$$

where P is the principal value of the Cauchy integrals. These integrals are known as the Kramer-Krönig relations. A more relevant form is the relation between refractive index and absorption coefficient. By analogy with Eq. 3.80,

$$n_r(\mathcal{E}) - 1 = \frac{2}{\pi} P \int_0^\infty \frac{\mathcal{E}' k_a(\mathcal{E}')}{\mathcal{E}'^2 - \mathcal{E}^2} d\mathcal{E}' \tag{3.82}$$

and, by virtue of the relation in Eq. 3.68,

$$n_r(\mathcal{E}) - 1 = \frac{ch}{2\pi^2} P \int_0^\infty \frac{\alpha(\mathcal{E}')}{\mathcal{E}'^2 - \mathcal{E}^2} d\mathcal{E}' \tag{3.83}$$

which enables the determination of the refractive index from the absorption spectrum. The dielectric constant and refractive index of some important binary III–V compounds are given in Table 3.1.

TABLE 3.1 DIELECTRIC CONSTANT AND REFRACTIVE INDEX IN SOME BINARY III-V COMPOUNDS.

Material	Static Dielectric Constant (ϵ_s)	High-Frequency Dielectric Constant (ϵ_∞)	Refractive Index (n_r) at Bandgap Energy
AlAs	10.06	8.5	3.17
GaP	11.11	9.11	3.45
GaAs	13.18	10.89	3.66
InP	12.56	9.61	3.45
InAs	15.15	12.3	3.52

3.6 RADIATION IN SEMICONDUCTORS

3.6.1 Relation between Absorption and Emission Spectra

In the last few sections of this chapter we have studied the various absorption processes. In these a photon is absorbed in the semiconductor, as a result of which an electron is usually raised from a lower-energy filled state to a higher-energy empty state, and in most cases the energy difference between the two states is equal to the energy of the absorbed photon. If the higher-energy level to which the electron is raised to is not the equilibrium state, then it will make a downward transition to the lower-energy empty state and emit electromagnetic radiation in the process. The energy of the radiation is very close to the energy difference between the higher and lower energy states. These are radiative transitions.

In principle, the reverse of all the absorption processes we have considered can occur to produce radiation. However, there is an important difference between absorption and emission spectra. While the absorption process can couple a broad energy range of filled and empty states (with momentum conservation) to produce a broad absorption spectrum, the emission process usually couples a narrow band of nonequilibrium filled states with a narrow band of empty states, to give a narrow emission spectrum. Therefore, a shoulder in the absorption spectrum can very well correspond to a narrow emission peak. Also, it is essential for the semiconductor to have a nonequilibrium population in the higher-energy states to produce a spontaneous emission spectrum. Depending on how the nonequilibrium state is produced, we defined different types of luminescence in Sec. 3.1.

The absorption and spontaneous emission spectra are related by the principle of detailed balance as calculated by van Roosbroeck and Shockley. At thermodynamic equilibrium, the rate of spontaneous photon emission $R_{sp}(\nu)$ at frequency ν in an interval $d\nu$ is given by

$$R_{sp}(\nu)d\nu = P_{abs}(\nu)\varphi(\nu)d\nu \qquad (3.84)$$

where $P_{abs}(\nu)$ is the probability of absorbing a photon of energy $h\nu$ per unit time, and $\varphi(\nu)d\nu$ is the radiation density of frequency ν in an interval $d\nu$. This is obtained from Planck's radiation law (Appendix 4) as:

$$\varphi(\nu)d\nu = \frac{8\pi\nu^3 n_r^3}{c^3} \frac{1}{\exp\left(\frac{h\nu}{k_B T}\right) - 1} d\nu \qquad (3.85)$$

The absorption probability $P(\nu)$ can be calculated in the following way. If the absorption coefficient of the photon is $\alpha(\nu)$ and it travels with a velocity $v = c/n_r$ in the material with refractive index n_r, then the mean lifetime of the photon is given by $\tau(\nu) = 1/\alpha(\nu)v$ and the absorption probability is given by

$$P(\nu) = \frac{1}{\tau(\nu)} = \alpha(\nu)v = \alpha(\nu)\frac{c}{n_r} \qquad (3.86)$$

Substituting Eqs. 3.85 and 3.86 into Eq. 3.84, we get

$$R_{sp}(\nu)d\nu = \frac{\alpha(\nu)8\pi\nu^3 n_r^2}{c^2\left[\exp(h\nu/k_B T) - 1\right]} d\nu \qquad (3.87)$$

which expresses the desired relation between absorption and emission spectra. Substitution of Eq. 3.68 leads to

$$R_{sp}(\nu)d\nu = \frac{32\pi^2 k_a(\nu)n_r^2 \nu^4}{c^3\left[\exp(h\nu/k_B T) - 1\right]} d\nu \qquad (3.88)$$

The total emission rate per unit volume is obtained by integrating Eq. 3.87 over all frequencies, or energies, as

$$R_{sp} = \frac{8\pi n_r^2 (k_B T)^4}{c^2 h^4} \int_0^\infty \frac{\alpha(\nu)u^3}{e^u - 1} du \qquad (3.89)$$

where $u = h\nu/k_B T$. Although derived for thermodynamic equilibrum, Eqs. 3.87 and 3.89 express the fundamental relation between absorption and emission spectra for any means of excitation. This formulation is valid for any transition between a higher-energy and a lower-energy state. The relation between emission and absorption spectra is schematically illustrated in Fig. 3.17.

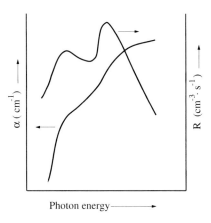

Figure 3.17 Schematic illustration of relation between emission and absorption spectra.

3.6.2 Stokes Shift in Optical Transitions

The Stokes shift is a difference in transition energy of the emission and absorption spectra resulting from defects in the material and, in general, partial nonradiative decay. The process can be understood with respect to the configuration coordinate diagram shown in Fig. 3.18. The lower and upper curves represent the energies of the lower and upper states of an optical transition as a function of distance that could be the ground and excited states of an impurity atom, a host lattice atom, or a deep-level trap. The point labeled 1 in the lower curve is the minimum energy, or equilibrium, position in the ground state. Due to photon absorption of energy $\hbar\omega_1$, an electron may be raised to the point 2 in the upper state ($\Delta \mathbf{k} = 0$), which is not the minimum energy configuration. The displacement Δx of the excited state may be caused by the defect or impurity potential. Therefore, the system relaxes to the state 3, which is at the lowest energy. This phenomenon is usually termed *lattice relaxation* and a phonon is involved. After living a mean lifetime the excited carrier returns to the ground state

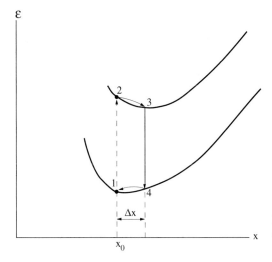

Figure 3.18 Configuration coordinate diagram illustrating the Stokes shift in a semiconductor.

by radiative recombination at the point 4, which is again not the minimum energy configuration. The energy of the emitted photon is $\hbar\omega_2$. Therefore, the system relaxes again with phonon participation to the point 1. Both optical and/or acoustic phonons may be involved but usually optical phonons produce the largest change in energy per unit displacement. The energy difference of the photon absorbed and emitted, $(\hbar\omega_1 - \hbar\omega_2)$ is called the Stokes shift or Frank-Condon shift. This degradation of optical energy arises directly from imperfections in the material or interfaces, such as in a heterostructure or quantum well.

3.6.3 Near-Bandgap Radiative Transitions

3.6.3.1 Exciton Recombination.   We have seen in Sec. 3.2.3 that electrons and holes produced by the absorption of a photon of near-bandgap energy can pair to form an exciton. Recombination of the electron-hole pair results in a narrow and sharp peak in the emission spectra. The energy of the emitted photon is

$$\hbar\omega = \mathcal{E}_g - \mathcal{E}_{ex} \tag{3.90}$$

where $\mathcal{E}_{ex}$ is quantized. In other words, in very pure crystals emission lines corresponding to the ground state and higher-order states may be seen. The process is shown in Fig. 3.19(a). In indirect bandgap semiconductors, a phonon needs to be involved, in the transition for momentum conservation, as shown in Fig. 3.19(b). Therefore, the probability of exciton recombination transitions is very low in indirect bandgap materials.

In semiconductors with impurities—donors and/or acceptors—present, the free exciton couples with the impurity atoms to produce *bound excitons*. Bound excitons produce sharp peaks at photon energies lower than that of the free exciton. Also, the linewidth of the bound exciton resonances are much smaller than that of the free excitons—almost by a factor of 10. In most semiconductors, free and bound exciton resonances are seen simultaneously in the emission spectra. Figure 3.20

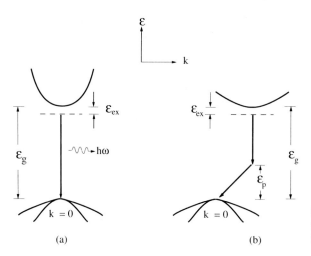

Figure 3.19   Exciton recombination in (a) direct bandgap and (b) indirect bandgap semiconductor.

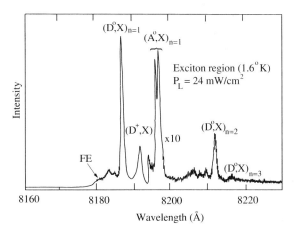

Figure 3.20 Exciton-related recombinations seen in the photoluminescence spectrum of a high-purity GaAs sample grown by the hydride VPE process. FE denotes free exciton and $D° − X$ and $A° − X$ are neutral donor-bound and acceptor-bound excitonic transitions, respectively. $D^+ − X$ denotes an ionized donor-bound excitonic transition (from B. J. Skromme et al., *Journal of Electronic Materials*, **12** (2), 433, ©1983 IEEE).

shows the low-temperature photoluminence spectrum of high-purity GaAs grown by VPE. In addition to the free-exciton peak, a host of peaks attributed to excitons bound to impurities and electrically active defects are seen in the energy range 1.4–1.51 eV.

3.6.3.2 Band-to-Band Recombination.

If the temperature of the sample is high enough so that $k_B T > \mathcal{E}_{ex}$, or if there are sufficient number of free carriers in the semiconductor producing local fields to dissociate the exciton, then most photogenerated carriers exist as separate electrons and holes in the bands. Most of these free carriers live a mean lifetime and then recombine radiatively. In direct gap semiconductors the process is complementary to the absorption process and electrons recombine with holes with momentum conservation. The energy position of the emission peak depends on the temperature and intensity of excitation. At low temperature and low excitation intensity the recombination is characterized by a single peak with the peak energy or the low energy cut-off at $\hbar\omega = \mathcal{E}_g$. As the temperature or excitation energy is increased, electrons and holes are filled at higher energies in the respective bands and these recombine to produce photons of higher energy. In the emission spectrum this is seen as a temperature or intensity-dependent tail on the high-energy side. Similarly, as the doping of the sample is increased, the whole curve may move to lower energies due to bandtailing effects. Some experimental curves obtained for GaAs are shown in Fig. 3.21. One may wonder at this point about radiative direct transitions involving the light-hole band that is at a higher (hole) energy for $\mathbf{k} \neq 0$. Since the hole mass is considerably lower in this band, the density of states and transition probability is much lower, and therefore transitions involving light holes are normally not seen in emission spectra of bulk semiconductors.

In indirect bandgap semiconductors radiative transitions from the conduction band to the valence band take place with the help of phonons to conserve momentum. Usually the process of phonon emission is most likely and the probability of phonon emission remains high even at very low temperatures.

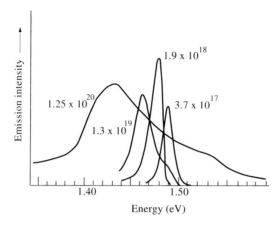

Figure 3.21 Cathodoluminescence spectra of Zn-doped GaAs at 4.2°K (from J. I. Pankove, Proceedings of the International Conference on the Physics of Semiconductors, Kyoto, 1966, *Journal of the Physical Society of Japan*, **21**, Supplement, 1966).

3.6.3.3 Donor-Acceptor and Impurity-Band Transitions.

Intentional and unintentional donor and acceptor levels in semiconductors give rise to radiative transitions. In this section we will restrict the discussion to *shallow*, hydrogenic donors and acceptors, for which the energy of the emitted photons is close to the fundamental bandgap. It may be remembered that in GaAs typical donor energies are between 4 and 8 meV and typical acceptor energies are between 25 and 40 meV. As in the case of absorption, the energy of the emitted photon is given by Eq. 3.61. In general, the donor-acceptor (D-A) transition gives rise to a broad peak in the emission spectrum.

The other important near-bandgap transitions are band-to-impurity transitions, which are complementary to the absorption process. In semiconductors where the donor and acceptor binding energies are nearly equal (due to equality of effective masses), it is not easy to distinguish between the two types of impurity-band transitions (donor-band and acceptor-band). In this case the conductivity type of the material has to be known and the temperature of the sample is an important factor. For indirect impurity-related transitions a phonon emission process is involved, and in this case the emitted photon energy is given by $\hbar\omega = \mathcal{E}_g - \mathcal{E}_i - \mathcal{E}_p$ where $\mathcal{E}_i = \mathcal{E}_D$ or $\mathcal{E}_A$.

It is important to know the difference in transition probability between the D-A and impurity-band transitions, provided there are carriers available for both to occur. These transition probabilities can be determined from quantum mechanical calculations, as outlined for the case of photon absorption in Sec. 3.2. The important parameter is the carrier lifetime for the relevant process, which is crucial for the operation of injection lasers. From these calculations one gets the impurity-band carrier lifetimes of the order of several nanoseconds, while that for the band-to-band transitions varies in the range of several hundred picoseconds to one nanosecond. What it amounts to roughly is that if there are electrons in the conduction band and donor level, and there are holes in the valence band and acceptor level, the probability of the band-to-band transition is approximately four times that of the impurity-band transitions.

Donor-acceptor or impurity-band transitions can be selectively observed in the luminescence spectra by causing selective occupation of the bandedge or impurity levels. This can be done by altering the temperature of the sample or by changing the excitation intensity. Consider a GaAs sample that has both donor ($\mathcal{E}_D \simeq 5$ meV) and acceptor ($\mathcal{E}_A = 30$ meV) impurities. At very low temperatures ($\sim 4°$K) both donor and acceptor levels are occupied with electrons and holes, respectively, and the prominent peak seen in the luminescence spectrum is due to a D-A transition. If the temperature is raised slightly, to $20°$K, some of the donor atoms are ionized and the electrons from these levels are raised to the conduction band. The acceptor level will still remain filled with holes. A shoulder develops to the high-energy side of the D-A peak in the luminescence spectrum, which corresponds to band-to-acceptor te (B-A) transitions. As the temperature is raised, the B-A transition becomes more prominent and the D-A transition is quenched. In fact, the energy separation between the two peaks corresponds to the donor ionization energy. The D-A and B-A peaks are schematically shown in Fig. 3.22. Selective occupation of donor levels and the conduction bandedge can also be achieved at a fixed (low) temperature by varying the excitation intensity. At low excitation (photon density) levels the D-A transition is the prominent one. As the excitation intensity is increased, the conduction band is filled with electrons and the B-A transition becomes more prominent.

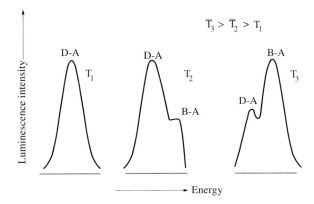

Figure 3.22 Schematic illustration of the evolution of band-acceptor and donor-acceptor transitions in the PL spectra of a semiconductor at varying temperatures.

As a concluding note, it should be mentioned that impurity-band transitions, involving the impurity level *closest* to the bandedge, can give rise to low-energy transitions. The transition energy corresponds to the impurity binding energy. Peaks believed to be due these transitions have been observed in the far-infrared region of the emission spectra of semiconductors. However, considering the transition probabilities it is debatable whether such transitions are radiative or nonradiative. In other words, instead of emitting a photon, single or multiple phonons may be emitted. In the latter case the excess energy is dissipated as heat in the lattice.

3.7 DEEP-LEVEL TRANSITIONS

Deep levels in the forbidden energy gap of semiconductors essentially act as carrier recombination or trapping centers and adversely affect device performance. Native defects in the lattice, such as vacancies, interstitial or substitutional impurities, or impurity-vacancy complexes can give rise to deep levels in semiconductors. It is generally believed that the excess energy of carriers recombining at these levels is carried away by single or multiple phonons. They were therefore also called "killer centers." Some deep levels due to impurities and defects are, however, radiative. Examples are Cu (0.18 and 0.41 eV) and Mn (0.10 eV) in GaAs, and a native defect in the same semiconductor called the EL2 level, which has an energy of ~ 0.8 eV below the conduction bandedge. Similar deep radiative transitions have been observed in $Al_xGa_{1-x}As$. It is important to realize that although a radiative transition is obtained at an energy equal to the ionization energy of the deep level, such transitions constitute a degradation of the radiative efficiency at near-bandgap energies. Also, it may be noted that the Franck-Condon shift of deep levels increases as the ionization energy of the level increases. This is because the increased localization of the carrier trapped in the potential of the impurity leads to a stronger interaction with the surrounding ions and therefore a larger lattice relaxation.

3.8 AUGER RECOMBINATION

Recombination via deep levels provides a nonradiative shunt path for the dissipation of radiative energy corresponding to the bandgap. Another important nonradiative process, which becomes important in heavily doped semiconductors, is *Auger recombination*. The Auger process is a three-carrier nonradiative recombination process, in which the excess energy released by the recombination of an electron-hole pair is transferred as kinetic energy of a third free carrier, which is raised in energy deep into the respective band. This carrier finally thermalizes back to the bottom of the band. The various possible Auger processes in semiconductors containing excess electrons and holes, are shown in Fig. 3.23. The carrier-concentration-dependent radiative recombination rate leading to spontaneous emission can be expressed as

$$R(n) = An + Bn^2 + Cn^3 \qquad (3.91)$$

In this equation the first term on the right accounts for Shockley-Read-Hall recombination at defects and traps and $A = s\vartheta_{th}N_T$. The second term accounts for spontaneous

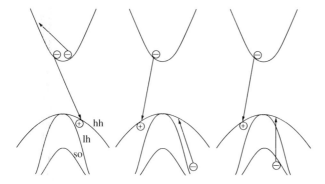

Figure 3.23 Different possible band-to-band Auger recombination processes in a direct bandgap semiconductor. Similar Auger transitions are also possible for impurity-band or D-A recombinations.

radiative recombination, and the third term accounts for Auger recombination, which plays a significant role in the operation of junction lasers. It may be noted that the second and third terms on the right-hand side of Eq. 3.91 are valid for a Boltzmann distribution.

The efficiency of Auger recombination depends on the concentration of excess carriers, which is proportional to $(k_B T / \mathcal{E}_g)^{3/2} \exp(-\mathcal{E}_g / k_B T)$. Therefore, the rate of the Auger process increases with increasing temperature and with decreasing bandgap. Auger processes are more dominant in narrow bandgap semiconductors, and play an important role in limiting the performance of junction lasers made with these materials. It may also be noted that Auger recombination is, in a sense, the reverse of the impact ionization process to be discussed in Chapter 8.

3.9 LUMINESCENCE FROM QUANTUM WELLS

If excess carriers are produced in a quantum well by any of the excitation techniques, they recombine to produce a narrow emission peak. The energy position of this peak is given by

$$\hbar\omega \cong \mathcal{E}_{gw} + \mathcal{E}_{e1} + \mathcal{E}_{hh1} \qquad (3.92)$$

where $\mathcal{E}_{gw}$ is the bandgap of the well material and $\mathcal{E}_{e1}$ and $\mathcal{E}_{hh1}$ are the electron and hole subband energies, respectively. The density of states of the heavy-hole band is much larger than that of the light-hole band, and therefore the heavy holes are primarily involved in the recombination process. Remember that we are considering nonresonant excitation, in which an excess electron-hole population is generally created in the barriers, from which they quickly diffuse and thermalize to the lowest bound states in the well region. If resonant excitation is used, then radiative transitions corresponding to the light-hole bands and higher-order bands can be seen in the luminescence spectrum.

For very large well sizes (≥ 200 Å), the luminescence spectrum begins to resemble that of the bulk well material. As the well size is reduced, quantization effects become important. At $L_z = 100$ Å for GaAs/AlGaAs quantum wells, for example, most of the wavefunctions corresponding to the bound states are still confined to the well. For smaller well sizes, a substantial part of the wavefunctions is in the barriers. Therefore, for well sizes of 100 Å or less, the luminescent properties of the quantum well are determined by the quality of the well material, the perfection of the interface, and the quality of the barrier material. In fact, the width of the luminescence spectrum can be analyzed to understand the optical quality of the quantum well.

3.10 MEASUREMENT OF ABSORPTION AND LUMINESCENCE SPECTRA

The measurement of absorption coefficient of a semiconductor is schematically shown in Fig. 3.24. Light from a variable wavelength light source, or monochromator, is incident normally on the sample. The light is partly absorbed and the intensity of the transmitted light is measured by a photodetector. The absorption coefficient $\alpha(\hbar\omega)$ is

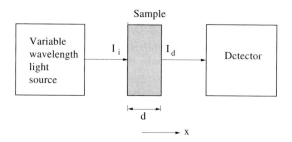

Figure 3.24 Schematic for optical absorption (transmission) experiment.

related to the relative decrease of light intensity, $\Im$, as it passes through the absorption medium, according to

$$\alpha = -\frac{1}{\Im}\frac{d\Im}{dx} \qquad (3.93)$$

or

$$-\frac{d\Im(x)}{dx} = \alpha\Im(x) \qquad (3.94)$$

The solution to this equation is

$$\Im(x) = \Im_i e^{-\alpha x} \qquad (3.95)$$

where $\Im_i$ is the intensity of the incident light. If the sample thickness is d, then the intensity of the transmitted light falling on the detector is

$$\Im_d = \Im_i e^{-\alpha d} \qquad (3.96)$$

The value of α will depend on the photon energy or wavelength. For a semiconductor of bandgap $\mathcal{E}_g$, photons of energy $\hbar\omega < \mathcal{E}_g$ pass unattenuated through the sample and $\alpha \cong 0$. As shallow impurity levels and the bandgap energy are approached, α begins to increase and rises abruptly at bandgap energy. The threshold of the absorption profile gives the value of $\mathcal{E}_g$.

Photoluminescence is the optical radiation emitted by a material (in addition to the thermal equilibrium blackbody radiation) resulting from its nonequilibrium state caused by an external light excitation. Three distinct processes take place to result in the light emission from the system: (1) absorption of exciting light and thus creation of electron-hole pairs, (2) partial radiative recombination of these electron-hole pairs, and (3) escape of this radiation from the system. Since the electron-hole pairs are generated by the absorbed exciting light, their highest concentration is near the illuminated surface of the sample; the resulting carrier distribution is both inhomogeneous and nonequilibrium. To regain homogeneity and equilibrium, the excess carriers will diffuse away from the surface while recombining radiatively and nonradiatively. The emitted radiation is subject to self-absorption; thus, it will not propagate far from the emission region. Most of the processes are therefore occurring in a region within an absorption length of the illuminated surface, and the emitted light escapes mostly through the nearby illuminated surface.

A typical photoluminescence measurement system consists of an excitation source, a variable-temperature cryostat and sample holder assembly, a high-resolution scanning spectrometer, and a detection system; a schematic diagram is shown in Fig. 3.25. The excitation source for most III–V semiconductors can be an argon-ion laser capable of delivering 0–15 mW of power at 5145 Å. A narrow band interference filter is used between the laser and the sample because of the rather intense radiation from the argon plasma. The laser beam is typically focused to a 100–500 μm diameter spot on the sample surface. By varying the laser power and using neutral density filters, it is possible to vary the power density on the sample in the range 10–10^3 W/cm^2. In order to get the maximum amount of information from the radiative relaxation process, it is necessary to cool the sample to cryogenic temperatures. This is accomplished by mounting the sample with cryogenic thermally conductive grease on a gold-coated copper cold finger of a liquid He cryostat equipped with optical windows. A Lambertian distribution of the luminescence from the sample is collected by a suitable lens and is focused on the slit of a high-resolution scanning spectrometer with a resolution

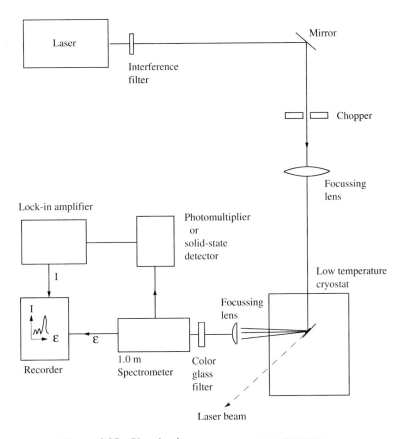

Figure 3.25   Photoluminescence measurement system.

$\sim$ 1–2 Å. The laser beam is incident at an angle on the sample so that the reflected beam does not go into the collecting lens. For GaAs and GaAs-based materials such as AlGaAs a liquid nitrogen-cooled photomultiplier tube is used to detect the luminescence signal. For longer wavelength (smaller bandgap) materials, Ge and CdS detectors are generally used.

3.11 TIME-RESOLVED PHOTOLUMINESCENCE

With intrinsic photoexcitation electron-hole pairs are generally created at energies much higher than the bandgap and the luminescence arises in the semiconductor after the following processes have occurred: (1) a rapid loss (in approximately 5–20 ps) of energy of the excess carriers by emitting phonons, (2) thermalization of the carriers, and (3) radiative and nonradiative recombination of the carriers. The recombination process is characterized by the rate equation

$$\frac{dn}{dt} = G - \frac{n}{\tau} \tag{3.97}$$

where n is the total number of excess carriers in the well and G is the generation rate. The time constant τ is related to the radiative and nonradiative time constants τ_r and τ_{nr} by Eq. 3.10. Recombination times in bulk semiconductors can vary between 500 ps and a few nanoseconds. In quantum wells, a faster recombination of the electron-hole pairs is observed, compared to bulk semiconductors, due to the enhanced overlap of electron and hole wavefunctions. Typical recombination times, depending on well width, are in the range 300–800 ps. The time resolved photoluminescence technique is used to measure these times. Tunneling of carriers out of a quantum well through the barrier will affect these recombination times. The tunneling rates will depend on transverse electric fields, due to the change the latter causes to the shape of the barrier, and these tunneling rates can be very high. Tunneling times as low as 100 ps is observed for barriers $\sim$ 50–100 Å. Time-resolved photoluminescence spectroscopy is useful to investigate all such dynamical processes in the temporal range of 10 ps up to a few nanoseconds.

A typical time-resolved luminescence measurement system is similar to that for steady-state luminescence, except in the excitation source and detection system. Excitation is provided by 1–5 picosecond pulses from a tunable dye laser, Ti: sapphire laser, or tunable color center laser depending on the material under test. These lasers are usually pumped by a mode-locked Nd:YAG laser, which has a repetition rate of 76 MHz. Typical output power characteristics of this laser are 10 W average power, 100 ps pulse width. The output is frequency doubled in a KDP crystal, giving an average power of 0.9–1.0 W. The dye laser cavity is folded into a Z configuration, with cavity length equal to that of the Nd:YAG pump laser, so that its repetition rate is also 76 MHz. The wavelength is tuned with a birefringent filter, which consists of three flat and parallel crystalline quartz plates placed inside the dye laser cavity at Brewster angle. The birefringent filter has low loss for linearly polarized light at

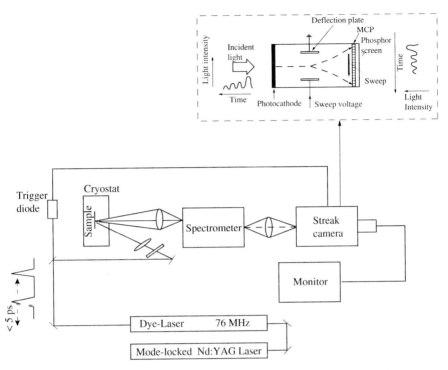

Figure 3.26   Arrangement for picosecond luminescence measurements with a synchronously pumped dye laser and a synchroscan streak camera.

a particular wavelength. The laser output wavelength is tuned. The output power of the dye laser or color center laser varies in the range 10–100 nW.

The luminescence from the sample is collected by a suitable lens and is focused on the slit of a high-resolution auto-scan spectrometer. The analyzed spectrum is detected by a streak camera with a typical temporal resolution of 10 ps. Figure 3.26 shows the operating principle of the streak tube, which forms the heart of the streak camera. The light pulse to be measured is projected on to the slit of the streak camera. The slit image of the incident light is focused on the photocathode of the streak tube via a relay lens where the photons are converted into electrons. The electrons are accelerated by the strong electrostatic field between the photocathode and the mesh-electrode, and conducted into the deflection field beyond which they are swept at high speed in a direction perpendicular to the slit length by applying a deflection voltage synchronized with the arrival of the electrons to the deflection field. The electrons are then multiplied in a microchannel plate (MCP) by a factor of approximately 3×10^3. Electrons exiting the MCP then bombard the phosphor screen of the steak tube and are converted to the optical image (called "streak image"). As a result of this structure and the sweeping system used, the time at which electrons were released from the photocathode surface can be determined by their deflected angle (vertical position on the phosphor screen). Therefore, the time axis of the incident light corresponds to the vertical axis on the phosphor screen, and the intensity of the incident light can be determined by the density of the streak image. Since it is necessary that the timing of the high-speed deflection is synchronized to the arrival time of electrons at the deflection field, part of the incident light is usually focused onto a PIN photodiode to generate a trigger signal for the sweeping. Typical time-resolved luminescence data obtained from measurements on GaAs-AlGaAs multiquantum wells are shown in Fig. 3.27. The excitation energy is above the conduction bandedge of the AlGaAs barrier. The excess carrier therefore quickly reach the well. This contributes to the observed rise time. The decay time reflects radiative and nonradiative recombination in the well and some tunneling into the barriers. The different dynamical processes for electrons are illustrated in Fig. 3.28.

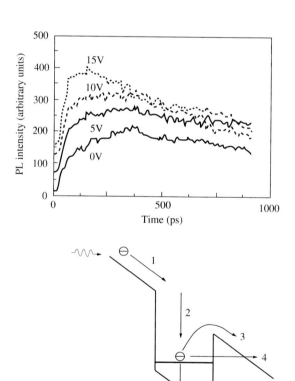

Figure 3.27 Temporal profile of the low temperature luminescence (T = 10°K) intensity from a 100 Å GaAs quantum well with 50 Å $Al_{0.3}Ga_{0.7}$ As barrier at different applied biases due to photoexcitation with 10 ps optical pulses of energy higher than the AlGaAs bandgap. The decay times are proportional to carrier tunnel rates from the quantum wells (N. Debbar, *Investigation of Optical Properties of Quantum Wells for Optoelectronic Device Applications*, Ph.D. Thesis, University of Michigan, 1989).

Figure 3.28 Carrier dynamics in a quantum well upon photoexcitation at an energy above the AlGaAs bandgap. The different processes are (1) drift and diffusion, (2) thermalization in the well by phonon scattering, (3) thermionic emission, (4) tunneling, and (5) radiative and nonradiative recombination. Photoexcited holes undergo the same processes.

PROBLEMS

3.1 An n-type sample of GaAs has 10^{14} cm^{-3} electrons in the conduction band in equilibrium. Intrinsic photoexcitation is used to create 10^{20} cm^{-3}.s^{-1} electron-hole pairs in the sample and the radiative lifetime is 1 ns. Assuming that there is no nonradiative recombination in the sample, find the energy position of the hole quasi-Fermi level in the energy bandgap under photoexcitation conditions. Comment on the position of the electron quasi-Fermi level. (Remember that under steady state conditions, Eq. 3.9 is valid.)

3.2 Absorption data taken from the literature for Ge at 300°K and 77°K are given in Fig. P3.2. Analyze the 300°K data to obtain the value of the direct bandgap, the indirect bandgap, and the phonon energy participating in the indirect transitions. Which category of phonons participate in these transitions?

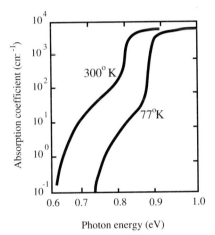

Figure P3.2

3.3 Alternate multiple layers of $n(5 \times 10^{18}$ cm$^{-3})$ and $p(5 \times 10^{18}$ cm$^{-3})$-type GaAs, each 500 Å thick, are grown by molecular beam epitaxy. What does the band diagram of this composite material look like? What are its special properties? What will the absorption and luminescence spectra of the modulated structure look like? Comment on its radiative lifetime.

3.4 Photons with an energy $\hbar\omega$ slightly less than the bandgap $\mathcal{E}_g$ are incident on a semiconductor material. Can an absorption edge spectrum still be obtained? If so, mention the mechanisms by which this might be possible in the case of a direct bandgap semiconductor and an indirect bandgap semiconductor.

3.5 Intrinsic light is incident at one end of a 10-cm-long semiconductor sample. If 15% of the light incident on the sample is absorbed per centimeter, calculate the fraction of the incident light that is transmitted through the sample and the absorption coefficient of the material.

3.6 Discuss, with figures, the process of electroabsorption in bulk semiconductors, multiquantum wells, and superlattices. Remember that in the last case the quantum wells are coupled (thin barriers).

3.7 Derive Eq. 3.68.

3.8 From the data of Fig. 3.11, calculate the bandedge emission spectrum for GaAs at $300°$K.

3.9 As we shall see in Chapter 11, the refractive index n_r is uniquely defined for transverse electric (TE) and transverse magnetic (TM) mode propagations. In fact, $n_r^{TE} - n_r^{TM} = \Delta n_r$ is directly proportional to the electro-optic coefficients of the crystal. By approximating the heavy-hole and light-hole excitonic resonances in the absorption spectrum of a semiconductor as δ-functions, show that Δn_r is directly related to the energy $(\mathcal{E}_{lh} - \mathcal{E}_{hh})$.

3.10 Discuss and illustrate the changes in the luminescence spectrum of n-type GaAs $(N_D = 10^{15}$ cm$^{-3})$ at $4.2°$K when

  (a) the donor density is gradually increased

  (b) when the compensation in the sample is gradually increased

3.11 Calculate the value of α for direct allowed transitions in InP at $2°$K. Does this value agree with the data of Fig. 3.11 after taking the temperature dependence of bandgap in account?

3.12 Derive the relation between the real and imaginary parts of the complex refractive index and the conductivity σ.
[Hint: You need to start with Maxwell's equations.]

3.13 Describe, with diagrams, the following processes:

  (a) Auger recombination

  (b) Shockley-Read-Hall recombination

  (c) radiative recombination

How would you minimize the rates of each of these processes?

3.14 Derive Eqs. 3.80 and 3.81.

READING LIST

AGRAWAL, G. P., and DUTTA, N. K. *Long Wavelength Semiconductor Lasers.* Van Nostrand Reinhold, New York, 1986.

KRESSEL, H., and BUTLER, J. K. *Semiconductor Lasers and Heterojunction LEDs.* Academic Press, New York, 1977.

PANKOVE, J. I. *Optical Processes in Semiconductors.* Prentice Hall, Englewood Cliffs, NJ, 1971.

SINGH, J. *Physics of Semiconductors and Their Heterostructures.* McGraw-Hill, New York, 1993.

WANG, S. *Fundamentals of Semiconductor Theory and Device Physics.* Prentice Hall, Englewood Cliffs, NJ, 1989.

4

Junction Theory

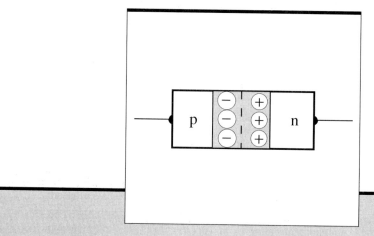

Chapter Contents

4.1 INTRODUCTION

At the present time, p-n junctions have wide applications in the design and realization of a host of electronic devices. This is because these junction diodes can, like vacuum tubes, perform functions such as rectification and switching. When incorporated in a bipolar transistor, amplification can be obtained. It will be seen in the following chapters that p-n junctions are also crucial for the operation of most optoelectronic devices. For example, in a photodiode, photogenerated carriers are separated and collected by a reverse bias. In a solar cell, the photogenerated carriers create a short-circuit current or an open-circuit voltage. In lasers and light-emitting diodes a forward bias is used to inject and create a nonequilibrium population of minority carriers, which then recombine radiatively with majority carriers. This chapter will therefore outline the basic principles of p-n junctions in their simplest form as anisotype homojunctions. The latter usually signifies a junction formed by the same semiconductor doped n- and p-type in different regions. Two other kinds of junctions have assumed roles of great importance in present-day devices. These are metal-semiconductor junctions and heterojunctions. A heterojunction is formed with two dissimilar semiconductors (unequal bandgaps), which can be either of the same doping type (isotype) or of the opposite doping type (anisotype). The basic principles of these junction will also be outlined in this chapter.

4.2 P-N JUNCTIONS

The simplest semiconductor junction diode is formed by "bringing together" a p-type and an n-type semiconductor. The subject of p-n junctions has been extensively treated and widely discussed in a number of texts. Therefore, only the aspects relevant to optoelectronic devices will be described here. These are junction formation, electrostatic properties, and current voltage characteristics. Also of importance are practical graded junctions and the case of high-level injection.

4.2.1 Junction Formation

The three common techniques of junction formation at the present time are epitaxial growth, diffusion and ion implantation. The principles and techniques of epitaxial growth have been outlined in Chapter 1. As an example, to form a GaAs p-n junction by liquid phase epitaxy (LPE), an appropriately doped substrate is moved successively over melts that have dissolved dopant species, usually in their elemental form. Typical n- and p-type dopants for GaAs and AlGaAs are Te and Zn. To have n-type GaAs the Ga-rich growth melt is nearly saturated with As and has an appropriate amount of Te, whose incorporation may be described by

$$Te(l) + V_{As} \rightleftharpoons Te(s)^+ + e^- \tag{4.1}$$

In this reaction Te reacts with an As vacancy in the liquid phase to give an ionized substitutional donor on an As site, $Te(s)^+$, and a free electron. The amount of Te to be dissolved in the melt to obtain the required doping density is determined by the solid solubility isotherm, which is independently determined by measuring dopant concentrations in the liquid phase (at a constant temperature) and in the solid crystal. The distribution coefficients of Te in GaAs as a function of temperature are shown in Fig. 4.1.

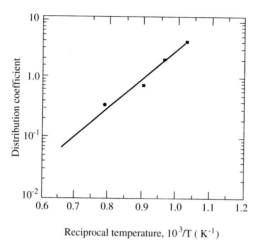

Figure 4.1 The distribution coefficient of Te in GaAs as a function of temperature (from H. C. Casey and M. B. Panish, *Heterostructure Lasers, Part B: Materials and Operating Characteristics*, copyright © 1978. Reprinted by permission of Academic Press, New York). The data points are adapted from H. C. Casey, Jr. et al., *J. Phys. Chem. Solids*, Pergamon Press, **32**, 571–580, 1971 and C. S. Kang and P. E. Greene, *Gallium Arsenide: 1968 Symp. Proc.*, 18, Inst. of Physics and Phys. Soc., London, 1969.

The incorporation of dopants during molecular beam epitaxy (MBE) is achieved by co-evaporating the dopant species, usually in elemental form, with the main flux species. In the case of GaAs, for example, a p-n junction is realized from effusing fluxes of Ga, As, Si(or Sn) (n-type), and Be (p-type). The appropriate dopant cells can be shuttered on or off to form the junction. The effusion cell temperature and dopant flux are decided by the equilibrium vapor pressure curve of the dopant species. Figure 4.2 shows the measured p-type concentration due to Be doping in GaAs as a

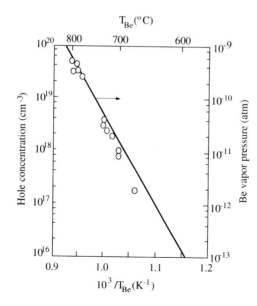

Figure 4.2 Hole concentration versus Be oven temperature for GaAs layers grown by MBE. The solid line represents the Be equilibrium vapor pressure curve (from M. Ilegems, *Journal of Applied Physics*, **48**, 1278, 1977).

function of Be cell temperature. The data agree fairly well with the Be vapor pressure curve, also shown in the same figure, indicating a unity incorporation coefficient. Most dopants exhibit similar behavior. Similarly, in organo-metallic vapor phase epitaxy (OMVPE), the dopant source, usually in the form of a liquid, is vaporized and introduced to the reaction mixture by hydrogen carrier gas. n- and p-type materials are obtained by switching the dopant flows on and off by mass flow controllers.

A mention should be made regarding the abruptness of junctions obtained by the three epitaxial techniques. In liquid phase epitaxy there is an inherent formation of a compensated region around the metallurgical interface due to melt dissolution, creating a doping profile shown in Fig. 4.3(a). Very abrupt junctions, as shown in Fig. 4.3(b), can be obtained by MBE, since the fluxes are terminated by mechanical shutters. Similarly, abrupt junctions can be obtained by OMVPE with mass flow controllers and concurrent pumping to avoid memory effects in the gas lines.

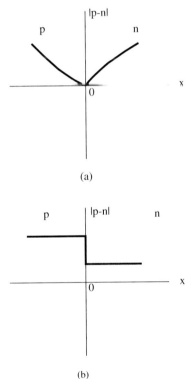

Figure 4.3 (a) Nonabrupt doping profile created near the junction by melt dissolution and (b) abrupt doping profile created by MBE or OMVPE.

Diffusion of impurities is a common technique of doping in semiconductors, more so in Si than in III–V semiconductors. Compound semiconductor solar cells and the heavily doped contact regions of lasers are commonly formed by the diffusion process. Diffusion is a thermally activated random motion process, characterized by a diffusion coefficient, D. The values of D for common impurities in GaAs are shown in Fig. 4.4. The process of diffusion doping involves the substitution of a host lattice atom by

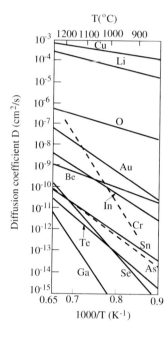

Figure 4.4 Diffusion constant of several impurity species in GaAs (from M. Zambuto, *Semiconductor Devices*, McGraw-Hill, 1989). The data points are adapted from D. L. Kendall, *Diffusion, Semiconductors and Semimetals*, **4**, Academic Press, New York, 163, 1968.

the impurity atom. In other words, the lattice atom has to move from the lattice site to make room for the diffusing impurity atom, which is until then in a nonlattice or *interstitial* site. This substitution typically involves an energy barrier of 1 eV and therefore temperatures around 1000°C are necessary to complete the diffusion process in a time of 1 hour or so.

There are two types of diffusion processes, determined by the amount of source (impurity) atoms. In essence there are correspondingly two types of boundary conditions and two types of solutions to the diffusion equation (Fick's second law):

$$\frac{\partial N}{\partial t} = D \frac{\partial^2 N}{\partial x^2} \tag{4.2}$$

where N is the concentration of the impurity. In the first, *limited source diffusion*, a fixed quantity of the impurity species is deposited in a thin layer on the wafer surface. This is modeled as an impulse function with the magnitude equal to the dose Q. The solution is given by the Gaussian distribution:

$$N(x, t) = (Q/\sqrt{\pi D t}) \exp\left[-(x/2\sqrt{D t})^2\right] \tag{4.3}$$

which is illustrated in Fig. 4.5(a). With increasing time, at any fixed temperature, the diffusion front moves into the wafer with a corresponding decrease of the surface concentration so that the area under the curves (dependent on the dose) remains constant. The second type is *constant source diffusion* or *error function diffusion*, where the impurity concentration is constant over time. In other words, an unlimited supply of dopant species is provided. Under this boundary condition the solution to

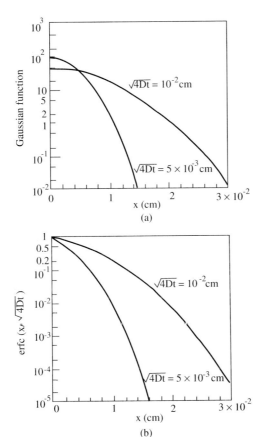

Figure 4.5 (a) Gaussian distribution and (b) complementary error function distribution for two different values of $\sqrt{4Dt}$ (from S. Wang, *Fundamentals of Semiconductor Theory and Device Physics*, Prentice Hall, Englewood Cliffs, NJ, 1989).

the diffusion equation is given by

$$N(x, t) = N_o \text{ erfc } (x/2\sqrt{Dt}) \tag{4.4}$$

where N_o is the impurity concentration at $x = 0$ (wafer surface). This solution is illustrated in Fig. 4.5 (b), where the impurity profile moves deeper into the wafer with increasing time, increasing temperature, or a combination of both. As an illustration, the impurity concentration profile, formed by either type of diffusion, and the net concentration profile of a p-n junction formed by diffusing acceptor impurities into an n-type wafer are shown in Fig. 4.6. At any given temperature, the upper limit to the amount of impurity that can be absorbed by the wafer is called the *solid solubility limit* for that species. Also, it should be remembered that at these high doping concentrations, only a fraction of the dopant species is electrically active (i.e., they are in substitutional lattice sites).

The third common technique of junction formation is *ion implantation* in which high-energy ions are projected onto the wafer surface. This technique of doping is more commonly used for the fabrication of electron devices such as metal-semicon-

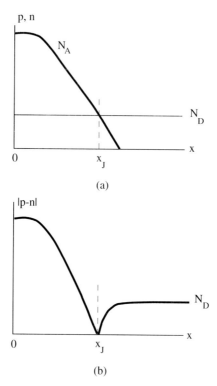

(a)

(b)

Figure 4.6 (a) Diffusion profile of acceptor species into an n-type semiconductor and (b) the corresponding free-carrier profile of the p-n junction. The point x_J corresponds to the metallurgical junction.

ductor field-effect transistors (MESFETs). It is also useful for dopant species that have very low vapor pressure and therefore cannot be effused during MBE. The ions are directed in a *nonchanneling* direction to ensure that their collision with the host lattice and their energy determine the average penetration depth, also called the *projected range*. A typical ion-implanter consists of a source chamber, accelerator, mass separator, and target chamber. The implant profile obeys a Gaussian law. Typical implant profiles for the case of single and multiple implants are shown schematically in Fig. 4.7. The latter is used for obtaining large junction depths with a uniform doping profile. The damage caused by the high-energy ions during implantation makes the

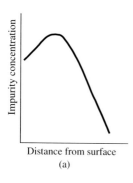

(a)

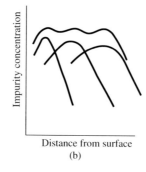

(b)

Figure 4.7 Schematic representation of implantation profile of dopant species in a semiconductor for (a) single implant and (b) multiple implants at successively higher energies. The typical range of energies in GaAs is 50–200 keV.

material almost amorphous. To restore single crystal behavior, the implanted wafer is usually annealed, typically at temperatures ranging from 700° to 1000°C for periods ranging from 1/2 to 1 hour. Presently a technique of *flash lamp annealing* is also used in which the wafer is annealed at 1000°–1100°C for a few seconds in an inert atmosphere with or without encapsulation. A last point to note is that for most device fabrication, diffusion or implantation is done through on oxide mask to delineate the device area.

4.2.2 Electrostatics of the p-n junction: Contact Potential and Space Charge

A p-n junction is schematically shown in Fig. 4.8, which also depicts the equilibrium impurity concentrations and carrier densities in the n- and p-type semiconductors. The p- and n-type semiconductors have large concentrations of free holes and free electrons, respectively. Once they are joined, there is a large concentration gradient of holes from the p- to the n-side across the metallurgical junction and vice versa for electrons. Diffusion of free carriers takes place by virtue of this gradient. There are, however, two important points to take note of. First, we are dealing with the diffusion of charged particles, and second, electrons leaving the n-region leave behind positively ionized donor atoms and similarly holes leaving the p-region leave behind negatively ionized acceptor atoms. There is therefore the formation of two regions, consisting of *immobile* charge of opposite types, on the two sides of the metallurgical junction. As carrier diffusion proceeds, the extent (length) of these regions increases. The two

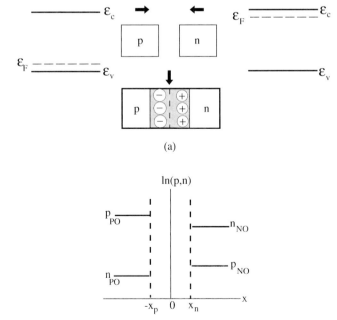

(a)

(b)

Figure 4.8   Electrostatics of the p-n junction: (a) formation of depletion region, (b) majority and minority carrier densities in neutral region, (c) space charge density, (d) electric field profile, and (e) potential profile.

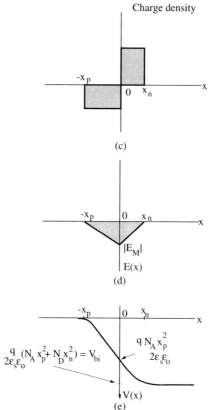

Figure 4.8 (*continued*)

regions of immobile charges result in an electric field, which is directed from the positive charge toward the negative charge. As this field builds up, it opposes further diffusion of carriers across the junction. In essence, the electric field generates drift currents of electrons and holes that exactly match the corresponding diffusion currents in equilibrium. This is consistent with the fact that under equilibrium conditions, with no applied bias, there is no *net* current flow. In equilibrium, the ionized region consisting of positive and negative charges on opposite sides is called the *depletion*, *space charge*, or *transition* region. The total depletion layer width $W = x_n + x_p$, where x_n and x_p are the space charge layer thicknesses in the two regions. The extent of the depletion region on each side is inversely dependent on the doping on that side. As a typical example, in GaAs doped with $n = 10^{15}$ cm^{-3}, the space charge region is approximately 0.4 μm thick.

The built-in electric field is a maximum at the junction and goes to zero at the edges of the depletion region on both sides. By virtue of this electric field E a potential gradient develops, according to

$$V_{bi} = -\int E\,dx \qquad (4.5)$$

The bands bend in the depletion region by an energy qV_{bi}, and this is called the contact or *built-in potential*. Looking at the picture simply, majority carriers diffusing due to contact formation are finally prevented from doing so because they cannot surmount this energy barrier. It is important to remember that minority carriers can easily go down the potential hills to the opposite sides, but these two components, due to electrons from the p-side and holes from the n-side, must exactly balance the majority carrier diffusion currents under equilibrium conditions. We are initially assuming an *abrupt* junction, in which a step change in impurity type occurs at the junction. Referring to Fig. 4.8, and the Gauss law, it follows that

$$qN_A x_p = qN_D x_n = \epsilon_s \epsilon_o E \tag{4.6}$$

where ϵ_s is the static dielectric constant of the semiconductor and N_A and N_D are the doping densities on the p- and n-sides respectively. Also, full ionization of the dopants are assumed for this analysis. Equation 4.6 signifies that the total ionized negative charge per unit area on the p-side must be equal to the total ionized positive charge per unit area on the n-side, and this is known as the *depletion approximation*. From Poisson's equation for the abrupt junction,

$$-\frac{d^2 V}{dx^2} = \frac{dE}{dx} = \frac{q}{\epsilon_s \epsilon_o} \left[p(x) - n(x) + N_D^+(x) - N_A^-(x) \right] \tag{4.7}$$

Therefore,

$$-\frac{d^2 V}{dx^2} \cong \frac{q}{\epsilon_s \epsilon_o} N_D \,, 0 < x \leq x_n \tag{4.8}$$

and

$$-\frac{d^2 V}{dx^2} \cong \frac{q}{\epsilon_s \epsilon_o} N_A \,, -x_p \leq x < 0 \tag{4.9}$$

Upon integration, the electric field profile is given by

$$E(x) = \frac{qN_D}{\epsilon_s \epsilon_o}(x - x_n) \qquad \text{(n-side)}$$

$$= -\frac{qN_A}{\epsilon_s \epsilon_o}(x + x_p) \qquad \text{(p-side)} \tag{4.10}$$

Integrating Eqs. 4.8 and 4.9 twice gives

$$V(x) = E_M \left(x - \frac{x^2}{2W} \right) \tag{4.11}$$

where E_M is the maximum field at the junction. From Eq. 4.11, the contact potential distribution is obtained for different values of x, as indicated in Fig. 4.8, as

$$V(x) = -\frac{qN_A}{2\epsilon_s \epsilon_o}(x_p + x)^2 \qquad \text{(p-side)}$$

$$= -\frac{qN_D}{2\epsilon_s \epsilon_o}(x - x_n)^2 \qquad \text{(n-side)} \tag{4.12}$$

and it follows that

$$V_{bi} = \frac{q}{2\epsilon_s \epsilon_o} \left[N_A x_p^2 + N_D x_n^2 \right]$$ (4.13)

At the metallurgical junction,

$$V(x) = -\frac{q N_A x_p^2}{2\epsilon_s \epsilon_o} = -\frac{q N_D x_n^2}{2\epsilon_s \epsilon_o}$$ (4.14)

4.2.2.1 The Built-in Voltage.   In equilibrium, when there is no further movement of carriers and the built-in voltage V_{bi} is established, the electron and hole currents have to be zero. That is, the drift and diffusion components of current for both electrons and holes must cancel. From the current density equations (Eq. 2.139), neglecting generation and recombination processes in the relatively thin depletion region,

$$J_e = q \left(n\mu_e E + D_e \frac{dn}{dx} \right) = 0$$

$$J_h = q \left(p\mu_h E - D_h \frac{dp}{dx} \right) = 0$$ (4.15)

where, carrier flow is assumed in the x-direction. For zero net *hole* current, $E = \frac{D_h}{p\mu_h} \frac{dp}{dx}$ and remembering Eq. 4.5, we get

$$V_{bi} = \frac{D_h}{\mu_h} \int \frac{1}{p} dp$$ (4.16)

At any point x,

$$-V(x) = \frac{D_h}{\mu_h} \left| \ln p \right|_{p_{PO}}^{p(x)}$$ (4.17)

Thus,

$$V_{bi} = \frac{D_h}{\mu_h} \ln \frac{p_{PO}}{p_{NO}}$$ (4.18)

Using the Einstein relation, Eq. 2.138,

$$V_{bi} = \frac{k_B T}{q} \ln \left(\frac{p_{PO}}{p_{NO}} \right)$$ (4.19)

Similarly, starting with the equation for electrons,

$$V_{bi} = \frac{k_B T}{q} \ln \left(\frac{n_{NO}}{n_{PO}} \right)$$ (4.20)

from which it follows,

$$n_{PO} p_{PO} = n_{NO} p_{NO}$$ (4.21)

The quantities in Eq. 4.21 represent equilibrium majority and minority carrier concentrations on both sides of the junction. Since in equilibrium the relation $n_o p_o = n_i^2$ holds, Eq. 4.19 or 4.20 may be rewritten as

$$V_{bi} = \frac{k_B T}{q} \ln \left(\frac{n_{NO} p_{PO}}{n_i^2} \right) \tag{4.22}$$

Again, assuming full ionization of the dopant impurity levels on either side (i.e., $p_{PO} = N_A$ and $n_{NO} = N_D$), one gets the familiar form,

$$V_{bi} = \frac{k_B T}{q} \ln \left(\frac{N_A N_D}{n_i^2} \right) \quad (V) \tag{4.23}$$

In this equation V_{bi} is expressed in terms of known and measurable parameters of the materials forming the junction. The energy band diagrams of p- and n-type semiconductors and a p-n junction are depicted in Fig. 4.9(a).

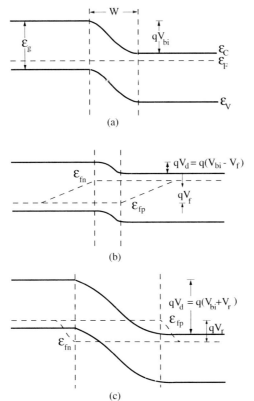

Figure 4.9 Energy band diagrams of a p-n junction (a) in equilibrium, (b) under forward bias, and (c) under reverse bias. The quasi-Fermi levels for the last two cases are also shown.

An alternate and simple way of deriving an expression for V_{bi} is as follows. For Fermi energies in the n- and p-type materials much larger than $k_B T$, the Boltzmann distribution function is valid. Assume the equilibrium majority and minority concentrations in the n- and p-sides to be n_{NO}, p_{NO}, and p_{PO}, p_{NO}, respectively. Then with respect to Fig. 4.9(a),

$$q V_{bi} = \mathcal{E}_g - (\mathcal{E}_C - \mathcal{E}_F) - (\mathcal{E}_F - \mathcal{E}_V) \tag{4.24}$$

It follows from Eqs. 2.78 and 2.79,

$$q V_{bi} = \mathcal{E}_g + k_B T \ln \frac{n_{NO}}{N_C} + k_B T \ln \frac{p_{PO}}{N_V} \tag{4.25}$$

Again, substituting for the first term on the r.h.s. from Eq. 2.83,

$$
\begin{aligned}
q V_{bi} &= k_B T \left(\ln \frac{N_C N_V}{n_i^2} + \ln \frac{n_{NO}}{N_C} + \ln \frac{p_{PO}}{N_V} \right) \\
&= k_B T \ln \left(\frac{n_{NO}\, p_{PO}}{n_i^2} \right) \\
&= k_B T \ln \left(\frac{N_A N_D}{n_i^2} \right)
\end{aligned}
\tag{4.26}
$$

which is of the same form as Eq. 4.23. It is obvious that as N_A and N_D tend to n_i, V_{bi} tends to zero. On the other hand as N_A and N_D increase, V_{bi} increases to a maximum value very close to the bandgap energy $\mathcal{E}_g$ under equilibrium conditions.

Finally, a simple way to look at the formation of the potential barrier is to consider the motion of the carriers themselves, when the junction is formed. Holes diffusing from the p-side leave it negatively charged, thereby raising all energy levels. Similarly, electrons migrating in the opposite direction cause all levels in the n-type material to be lowered. As a consequence, the potential hill of height V_{bi} is formed. From the energy levels shown in Fig. 4.9(a) and the Boltzmann distribution it follows that

$$n_{NO} = N_C \exp\left[-(\mathcal{E}_g - \mathcal{E}_F)/k_B T \right] \tag{4.27}$$

and

$$n_{PO} = N_C \exp\left[-\left\{ (\mathcal{E}_g + q V_{bi}) - \mathcal{E}_F \right\}/k_B T \right] \tag{4.28}$$

Therefore,

$$\frac{n_{NO}}{n_{PO}} = \exp\left(\frac{q V_{bi}}{k_B T} \right) \tag{4.29}$$

or

$$V_{bi} = \frac{k_B T}{q} \ln\left(\frac{n_{NO}}{n_{PO}} \right) \tag{4.30}$$

which is of the same form as Eq. 4.20.

4.2.2.2 Depletion Layer Width.   With respect to Fig. 4.8, in the space charge region where Poisson's equation is valid, the space charge density is equal to $q N_D$ and $q N_A$ on the n- and p-sides, respectively. Outside the depletion region, in neutral material on either side, the space charge density is zero. For an *abrupt* junction, the maximum electric field E_M is given by

$$\epsilon_s \epsilon_o E_M = q N_D\, x_n = q N_A x_p \tag{4.31}$$

The area enclosed by the electric field profile is the built-in voltage, V_{bi}, which can therefore be expressed as

$$|-V_{bi}| = \frac{E_M}{2}(x_n + x_p) \tag{4.32}$$

It follows that

$$W = x_n + x_p$$

$$= \sqrt{\frac{2\epsilon_s\epsilon_o}{q}\left(\frac{1}{N_D} + \frac{1}{N_A}\right)|-V_{bi}|} \tag{4.33}$$

If $N_A \gg N_D$, as in a one-sided step junction,

$$W = \sqrt{\frac{2\epsilon_s\epsilon_o}{qN_D}\left|-V_{bi}\right|} \tag{4.34}$$

Similarly, for $N_D \gg N_A$, N_D is replaced by N_A in the denominator of Eq. 4.34. Hence, the depletion layer extends more into the side with the lower doping. The variation of W with doping in Si, Ge and GaAs one-sided abrupt junctions is shown in Fig. 4.10.

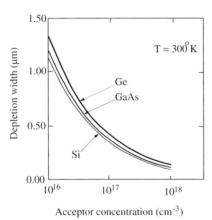

Figure 4.10 Depletion layer width of n^+-p junction for GaAs, Si, and Ge vs. the acceptor density at zero external bias. The donor density in the n^+ region is 5×10^{17} cm^{-3} and T = 300K (from M. Shur, *Physics of Semiconductor Devices*, Prentice Hall, Englewood Cliffs, NJ, 1990).

Until now we have assumed the depletion approximation to be true. In other words, the free-carrier densities abruptly end as step functions at the depletion layer boundaries on both sides. In practice, the free carriers have distributions as shown in Fig. 4.11, which in turn produce a space charge distribution at the depletion layer edge. For example, on the n-side, the space charge density is given by $q[N_D - n(x)]$ instead of qN_D. A similar equation holds for the p-side. As a result, it can be shown

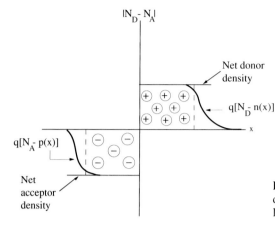

Figure 4.11   Space-charge distribution in a p-n junction with large free-carrier densities.

that (Problem 4.5)

$$W = \sqrt{\frac{2\epsilon_s \epsilon_o}{q N_B} \left(V_{bi} - \frac{2 k_B T}{q} \right)}$$

$$= L_D \sqrt{2 \left(\frac{q V_{bi}}{k_B T} - 2 \right)} \tag{4.35}$$

where $N_B = N_D$ or N_A and L_D is the Debye length given by Eq. 2.114. The effect is therefore also called *Debye tailing*.

EXAMPLE 4.1

Objective.   To calculate the extent of the depletion region on the n- and p-sides of an abrupt p-n junction in terms of the total depletion width W and the doping densities N_A and N_D.

From Eq. 4.31 it follows that

$$x_n = \frac{\epsilon_s \epsilon_o E_M}{q N_D}, \quad x_p = \frac{\epsilon_s \epsilon_o E_M}{q N_A} \tag{4.36}$$

The total depletion layer width

$$W = x_n + x_p$$

$$= \frac{\epsilon_s \epsilon_o E_M}{q} \left(\frac{1}{N_D} + \frac{1}{N_A} \right) \tag{4.37}$$

It follows from Eqs. 4.36 and 4.37

$$x_n = \frac{W N_A}{N_A + N_D}, \quad x_p = \frac{W N_D}{N_A + N_D} \tag{4.38}$$

4.2.2.3 Junction Capacitance.   The depletion layer of a p-n junction can be viewed as a parallel plate capacitor with its capacitance given by

$$C = \frac{\epsilon_s \epsilon_o A}{W} \tag{4.39}$$

where A is the junction area and W is the depletion-layer width. Using Eq. 4.33,

$$C = \frac{\epsilon_s \epsilon_o A}{\left[\frac{2\epsilon_s \epsilon_o}{q} (V_{bi} - V_a) \frac{N_A + N_D}{N_A N_D} \right]^{1/2}} \tag{4.40}$$

where V_a is the applied bias. For an asymmetric junction where $N_A \gg N_D$,

$$C \cong \frac{\epsilon_s \epsilon_o A}{\left[\frac{2\epsilon_s \epsilon_o}{q N_D} (V_{bi} - V_a) \right]^{1/2}} \tag{4.41}$$

These equations represent the bias dependence of the capacitance of a p-n junction. Eq. 4.41 can be rewritten in the form

$$\frac{1}{C^2} = \frac{2}{q \epsilon_s \epsilon_o N_D A^2} (V_{bi} + V_a) \tag{4.42}$$

Therefore, a linear plot of $1/C^2$ versus V_a can be obtained from direct measurements. The slope of this line is inversely proportional to the donor concentration N_D in a p^+-n junction, and its intercept on the voltage axis gives the value of V_{bi}. There are other techniques, however, to determine the built-in voltage of a junction, such as current-voltage and photoemission measurements.

EXAMPLE 4.2
Objective.   To show that the peak field E_M is determined by the doping density of the lightly doped side of a p-n junction.

From Eqs. 4.34 and 4.38

$$x_n = \left[\frac{2\epsilon_s \epsilon_o V_{bi}}{q N_D} \left(\frac{1}{1 + N_D/N_A} \right)^2 \right]^{1/2} \tag{4.43}$$

For a p^+-n junction in which $N_A \gg N_D$,

$$x_n \cong \left(\frac{2\epsilon_s \epsilon_o V_{bi}}{q N_D} \right)^{1/2}$$

and

$$E_M = \frac{q}{\epsilon_s \epsilon_0} N_D x_n$$

$$= \left(\frac{2q V_{bi} N_D}{\epsilon_s \epsilon_o} \right)^{1/2} \tag{4.44}$$

Therefore, for $N_A \gg N_D$, E_M is proportional to $N_D^{1/2}$. A similar result holds for a n^+-p junction, where $E_M \propto N_A^{1/2}$.

4.2.3 Current-Voltage Relationship

There are four components of current that can flow across the junction. These are $J_h^{p \to n}$, $J_h^{n \to p}$, $J_e^{n \to p}$, and $J_e^{p \to n}$. $J_h^{n \to p}$ and $J_e^{p \to n}$ are due to thermal generation of minority carriers (electron-hole pairs) that travel down the hill by drift motion. $J_h^{p \to n}$ and $J_e^{n \to p}$ depend on the injection of majority carriers to the opposite side and their subsequent diffusion and are therefore exponentially dependent on applied bias. Under equilibrium conditions, the electron and hole currents must exactly balance. Thus, $J_h^{p \to n} = J_h^{n \to p}$ and $J_e^{n \to p} = J_e^{p \to n}$.

4.2.3.1 Junction under Forward Bias.

Equilibrium is established at zero bias due to formation of a potential barrier, which restricts the motion of majority carriers. For any net current to flow, the barrier height V_{bi} has to be reduced, and this is achieved by applying a forward bias to the junction. A bias V_a applied to the diode will almost completely drop across the depletion region because of its high resistivity. So the effective bias across the depletion region for a forward bias V_f is $(V_{bi} - V_f)$ and the width of this region is also reduced. $(V_{bi} - V_f)$ is defined as the diffusion potential V_d. The band diagram of a forward-biased junction is shown in Fig. 4.9(b). Current flows across the junctions as drift and diffusion currents are no longer balanced. Majority carriers become minority carriers upon crossing to the other side. It is the change in the minority carrier density on both sides that is significant. The majority carrier densities are hardly affected. The diffusion of carriers cause excess minority carrier densities $\delta n(x)$ and $\delta p(x)$ on the p- and n-sides, respectively, as shown in Fig. 4.12. These excess carrier densities can be determined from a solution of the appropriate *continuity* equation (see Eqs. 8.59-8.62), governing the transport of *minority* carriers. For example, for holes in the n-side, this equation at steady state,

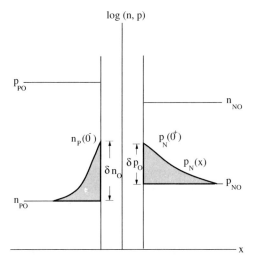

Figure 4.12 Injected carrier densities in a forward-biased junction diode.

neglecting drift, is of the form

$$-\frac{p - p_o}{\tau_h} + D_h \frac{d^2 p}{dx^2} = 0 \qquad (4.45)$$

where p_o is the equilibrium concentration and τ_h is the hole recombination lifetime in the n-region. The solution is of the form

$$\delta p(x) = \delta p_o \, e^{-x/L_h} \qquad (4.46)$$

where $L_h = \sqrt{\tau_h D_h}$ is the diffusion length for holes and $\delta p_0 = \delta p(0)$ is the excess hole density at the edge of the depletion region. With applied forward bias there is injection of both majority and minority carriers across the junction. With reference to Fig. 4.12,

$$p_{NO} = p_{PO} \, e^{-q V_{bi}/k_B T} \qquad (4.47)$$

which signifies that the minority carrier density in one side of a junction is given by the majority carrier density of the other side and the built-in potential. In the following it is assumed that the depletion regions on both sides of the junction have negligible width compared to that of the neutral regions. The minority carrier density $p_N(0^+)$ at the edge of the depletion region on the n-side is given by

$$p_N(0^+) = p_{NO} \, e^{q V_f/k_B T} \qquad (4.48)$$

By substitution of Eq. 4.47,

$$p_N(0^+) = p_{PO} \, e^{-q(V_{bi} - V_f)/k_B T} \qquad (4.49)$$

Therefore, the excess minority carrier density at the edge of the depletion region is

$$\delta p(0) = p_N(0^+) - p_{NO} = p_{NO} \left(e^{q V_f/k_B T} - 1 \right) \qquad (4.50)$$

and from Eq. 4.46,

$$\delta p(x) = p_{NO} \left(e^{q V_f/k_B T} - 1 \right) e^{-x/L_h} \qquad (4.51)$$

The diffusion current resulting from these excess carriers is given by

$$J_h(x) = -q D_h \frac{d}{dx} \delta p(x) \qquad (4.52)$$

At the origin, or the edge of the depletion region,

$$J_h(0) = J_h(0^+) = \frac{q D_h p_{NO}}{L_h} \left(e^{q V_f/k_B T} - 1 \right) \qquad (4.53)$$

Similarly, the current resulting from the diffusion of electrons on the p-side is

$$J_e(0) = J_e(0^-) = \frac{q D_e n_{PO}}{L_e} \left(e^{q V_f/k_B T} - 1 \right) \qquad (4.54)$$

And the total current due to forward bias injection is then

$$J = J_h(0^+) + J_e(0^-) = q \left(\frac{D_h p_{NO}}{L_h} + \frac{D_e n_{PO}}{L_e} \right) \left(e^{q V_f/k_B T} - 1 \right) \qquad (4.55)$$

It is important to note the following. First, $J_h(0^+) = J_h^{p \to n} - J_h^{n \to p}$, and $J_e(0^-) = J_e^{n \to p} - J_e^{p \to n}$. Second, that the prefactor in Eq. 4.55 before the exponential term is made up of parameters related to minority carriers.

4.2.3.2 Junction Under Reverse Bias.   The band-bending in the junction under reverse bias is shown in Fig. 4.9(c). Again, the applied bias V_r drops almost entirely across the depletion region. The potentials on the p-side are raised and those on the n-side are lowered, resulting in a potential hill of increased height ($V_{bi} + V_r$), which is now the diffusion potential V_d. Consequently, the depletion layer width is also increased. In terms of Eq. 4.55 above, if V_f is replaced by $-V_r$ and $|V_r|$ is large,

$$J \cong -q \left(\frac{D_h p_{NO}}{L_h} + \frac{D_e n_{PO}}{L_e} \right) = J_s \qquad (4.56)$$

where J_s is termed the reverse saturation current of the diode. In terms of drift and diffusion current components, $J_s = J_h^{n \to p} + J_e^{p \to n}$. In other words, the reverse saturation current is composed of the drift current due to the thermally generated minority carriers across the depletion region. The injection of majority carriers and the associated diffusion currents are negligible, due to the large potential barrier. The current-voltage relation of a diode can therefore be expressed as

$$J = J_s \left(e^{q V_a / k_B T} - 1 \right) \qquad (4.57)$$

where V_a is the applied bias. Typically, for $|V_r|$ larger than a few millivolts, $J = -J_s$. A typical current-voltage plot of a p-n junction is shown in Fig. 4.13(a). The following may be noted about the current-voltage characteristics. Under forward bias greater than a few millivolts (usually $> 3k_B T$)

$$J \cong J_s \, e^{q V_a / k_B T} \qquad (4.58)$$

An *ideality factor* n_f is usually incorporated in the exponent as

$$J \cong J_s \, e^{q V_a / n_f k_B T} \qquad (4.59)$$

For an ideal diode n_f should be close to unity. Due to recombination and generation processes in the depletion region of real diodes $n_f > 1$. In fact, at low applied biases, n_f is usually larger than unity. Also, for large forward biases, there is a saturation effect, shown by the dashed line in Fig. 4.13(b), in the characteristics of real diodes. This is due to ohmic losses due to the finite resistance of the neutral n- and p-regions.

EXAMPLE 4.3
Objective.   To find the *electron* current in the n-type region of a p-n junction under forward bias.

The total junction current I is given by Eq. 4.55. The hole current on the n-side is obtained from Eqs. 4.51 and 4.52 as

$$I_h(x) = q A \frac{D_h}{L_h} p_{NO} e^{-x/L_h} \left(e^{q V_f / k_B T} - 1 \right) \qquad (4.60)$$

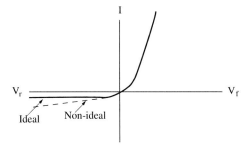

(a)

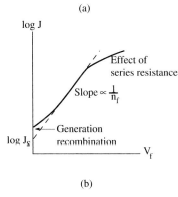

(b)

Figure 4.13 Schematic representation of (a) diode current-voltage characteristics and (b) forward-bias current-voltage characteristics in detail.

Therefore, the electron current in the n-region is

$$I_e(x) = I - I_h(x)$$

$$= qA\left[\frac{D_h}{L_h}\left(1 - e^{-x/L_h}\right)p_{NO} + \frac{D_e}{L_e}n_{PO}\right]\left(e^{qV_f/k_BT} - 1\right) \tag{4.61}$$

where A is the area of the diode. This current is the sum of two components. The first accounts for the supply of electrons for recombination with injected holes, and the second is related to the injection of electrons across the junction into the p-side. A similar expression can be written for the hole current on the p-side.

4.2.3.3 Current Densities across the Junction.

Consider an abrupt p-n junction, in which the dopings on the p- and n-sides are $\sim 10^{16}$ cm^{-3}. Then the minority carrier densities in equilibrium are $\sim 10^4$ cm^{-3}, assuming the semiconductor to have a bandgap ~ 1 eV. We will examine the current components flowing across the junction under forward bias injection conditions. We will analyze the injection of holes in the n-side of the junction, and the situation on the p-side is exactly complementary. Consider the edge of the depletion region on the n-side to be the origin of the distance coordinates. At the injecting end ($x = 0$), there is a large concentration and concentration gradient of excess injected holes, whose density is given by Eq. 4.46. At this end, therefore, the drift component of current is zero and the hole current $J_h(x)$ is essentially a diffusion current, $J_{diff}^h(x)$. Therefore,

$$J_{diff}^h(x) = -q D_h \frac{d(\delta p)}{dx} = \frac{q D_h \delta p_o}{L_h} e^{-x/L_h} \qquad (4.62)$$

Assuming the total current flowing through the diode is J_o, we know that

$$J_h(0) = J_o = \frac{q D_h \delta p_o}{L_h} \qquad (4.63)$$

and

$$J_{diff}^h(x) = J_o e^{-x/L_h} \qquad (4.64)$$

Since the total current $J_T = J_e(x) + J_h(x) = J_o$, the nature of $J_h(x)$ and $J_e(x)$ are as shown in Fig. 4.14. Here $J_e(x)$ is the current carried by electrons. On examining the nature of this electron current, it is clear that at the far end, where $J_e(x)$ tends to saturate, it is a drift current, since for diffusion a concentration gradient is necessary. However, closer to the origin, the electron current has both drift and diffusion components. To calculate the diffusion component of the electron current, $J_{diff}^e(x)$, it is necessary to know $\delta n(x)$, which is not simple. However, for charge neutrality to hold and to avoid the generation of high fields in neutral material, the condition $\delta n(x) \cong \delta p(x)$ must be valid. Thus,

$$J_{diff}^e(x) = q D_e \frac{d(\delta p)}{dx} = -\beta_D J_o e^{-x/L_h} \qquad (4.65)$$

and

$$\beta_D = \frac{D_e}{D_h} = \frac{\mu_e}{\mu_h} \cong 2 - 10 \qquad (4.66)$$

in most semiconductors. Referring to Fig. 4.14, in the region OA there is an electron diffusion current, which is of the same order of magnitude as the hole-diffusion current. Therefore, $J_h(x)$ is entirely a diffusion current and $J_e(x)$ has both drift and diffusion components. The drift component of the electron current

$$J_{dr}^e = q n \mu_e E = \sigma E \qquad (4.67)$$

which is Ohm's law. Here σ is the conductivity of the n-region. Adding all the current components,

$$\sigma E - \beta_D J_o e^{-x/L_h} + J_o e^{-x/L_h} = J_o \qquad (4.68)$$

from which

$$E = \frac{J_o}{\sigma} + (\beta_D - 1)\frac{J_o}{\sigma} e^{-x/L_h} \qquad (4.69)$$

or

$$E = E_\infty \left[1 + (\beta_D - 1) e^{-x/L_h} \right] \qquad (4.70)$$

where E_∞ is the electric field at the region far from $x = 0$, where the total current is predominantly an electron current. The electric field, excess carrier density, and electron and hole-current densities are shown in Fig. 4.14. Similar results can be obtained by considering the injection of electrons into the p-side. The carrier distributions and

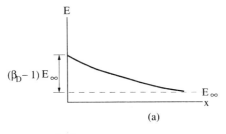

(a)

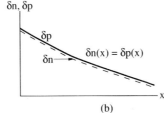

(b)

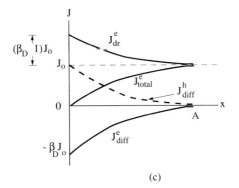

(c)

Figure 4.14   (a) The electric field, (b) excess carrier density, and (c) electron and hole current densities in the neutral n-region of a p-n junction under forward bias.

current densities in forward- and reverse-biased junctions are shown in Figs. 4.15(a) and (b).

The nature of the electric field profile $E(x)$, shown in Fig. 4.14, merits some discussion. Far from the junction the equilibrium value E_∞ results from the applied bias and in neutral material E_∞ will have a small value. As the junction is approached $E(x)$ increases. The extra field is required to transport electrons from the bulk toward the junction so that the electron density is approximately equal to the injected hole density and the excess holes can recombine.

4.2.3.4 Temperature Dependence of the Reverse Saturation Current.
In accordance with Eq. 4.56,

$$J_s = \frac{q D_h \, p_{NO}}{L_h} + \frac{q D_e \, n_{PO}}{L_e} \qquad (4.71)$$

To examine the temperature dependence of J_s we consider a p^+-n one-sided abrupt junction. The arguments presented below hold equally for a two-sided junction. For

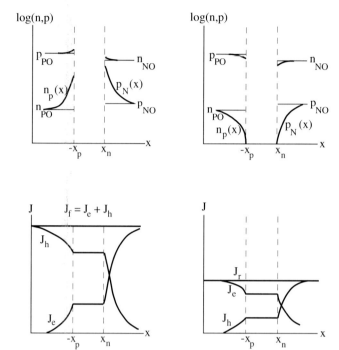

Figure 4.15 Carrier distributions and current densities in (a) forward-biased and (b) reverse-biased p-n junction.

the p^+-n junction $p_{NO} \gg n_{PO}$ and the second term in Eq. 4.71 can be neglected. In the first term D_h, p_{NO} and L_h are all functions of temperature. Thus,

$$J_s \cong \frac{q D_h \, p_{NO}}{L_h} = q \sqrt{\frac{D_h}{\tau_h}} \frac{n_i^2}{N_D} \tag{4.72}$$

where N_D is the donor density on the n-side and for full ionization $n_{NO} = N_D$. Assuming that the ratio D_h/τ_h has a temperature dependence of the form T^γ, where γ is an integer, and knowing that $n_i \propto [T^{3/2} e^{-\mathcal{E}_g/2k_B T}]$,

$$J_s(T) \propto T^{(3+\gamma/2)} e^{-\mathcal{E}_g/k_B T} \tag{4.73}$$

The temperature dependence of the $T^{(3+\gamma/2)}$ term is much weaker than the exponential term. Therefore, the slope of the plot of $\ln J_s$ (T) versus inverse temperature is determined by the bandgap $\mathcal{E}_g$. Under reverse bias $|J_r| \cong J_s \propto e^{-\mathcal{E}_g/k_B T}$. Under forward bias $J_f \simeq J_s e^{q V_f/k_B T}$ and the forward bias current is proportional to $e^{[-(\mathcal{E}_g - q V_f)/k_B T]}$.

4.2.4 Quasi-Fermi Levels and High-Level Injection
Until now, we have mostly considered the motion and distribution of carriers in the neutral n- and p-regions of the diode, *outside* the depletion region. As seen in the

previous sections, the current-voltage characteristics can be determined by considering the motion of these carriers. For the sake of completeness, it is important to know what goes on within the depletion region. Under forward bias, as majority carriers are injected to the opposite side, these carriers drift through the depletion region, suffering some recombination. The lost carriers are replenished at the contacts, thereby increasing the diode current. Similarly, under reverse bias, there is generation of carriers in the depletion region. It is important to realize that within the depletion region there is a state of *quasiequilibrium*. It is easy to see the situation under forward bias, when there is a dynamic distribution of a large number of free carriers, both electrons and holes, whose densities as a function of distance cannot be determined by the equilibrium Fermi levels in the neutral materials outside the depletion region. The situation is similar to that created by shining light on a semiconductor, which produces an extra population of electron-hole pairs. Their density also, cannot be determined by the equilibrium Fermi level in the dark. In Chapter 3, the concept of quasi-Fermi levels was introduced to calculate the nonequilibrium density of photogenerated carriers. The carrier densities in terms of the quasi-Fermi levels $\mathcal{E}_{fn}$ and $\mathcal{E}_{fp}$ are given by Eqs. 3.5 and 3.6. Similar conditions exist in the depletion region under forward bias, as seen in Fig. 4.9(b). At equilibrium, $\mathcal{E}_{fn} = \mathcal{E}_{fp} = \mathcal{E}_F$. In terms of the intrinsic Fermi level $\mathcal{E}_{Fi}$,

$$n_i = N_C \exp\left(\frac{\mathcal{E}_{Fi} - \mathcal{E}_C}{k_B T}\right)$$

$$= N_V \exp\left(\frac{\mathcal{E}_V - \mathcal{E}_{Fi}}{k_B T}\right) \tag{4.74}$$

Combining Eqs. 3.5, 3.6, and 4.74, we get

$$n = n_i \exp\left(\frac{\mathcal{E}_{fn} - \mathcal{E}_{Fi}}{k_B T}\right) \tag{4.75}$$

$$p = n_i \exp\left(\frac{\mathcal{E}_{Fi} - \mathcal{E}_{fp}}{k_B T}\right) \tag{4.76}$$

These are general relations that hold for any semiconductor region under nonequilibrium conditions. Note that n and p are different from the equilibrium concentrations n_o and p_o. It follows from Eqs. 4.75 and 4.76,

$$np = n_i^2 \exp\left(\frac{\mathcal{E}_{fn} - \mathcal{E}_{fp}}{k_B T}\right) \tag{4.77}$$

Therefore, $\mathcal{E}_{fn}$ and $\mathcal{E}_{fp}$ are separate energy levels that define the nonequilibrium electron and hole populations, respectively.

Under forward bias $(\mathcal{E}_{fn} - \mathcal{E}_{fp}) > 0$ and $pn > n_i^2$ and for a reverse bias [Fig. 4.9(c)] $(\mathcal{E}_{fp} - \mathcal{E}_{fn}) > 0$ and $pn < n_i^2$. We have seen earlier from Eq. 4.49 that the minority carrier concentration at the edge of the depletion region on either side of a forward-biased junction is equal to the corresponding majority carrier concentration on the other side of the depletion layer, reduced by an exponential factor

$\exp[q(V_{bi} - V_a)/k_B T]$, where $(V_{bi} - V_a)$ is the height of the barrier overcome by the carriers in crossing the space charge region. Thus, the quasi-Fermi levels do not vary in crossing the depletion layer. In other words,

$$\mathcal{E}_{fn}(-x_p) = \mathcal{E}_{fn}(x_n) \tag{4.78}$$

$$\mathcal{E}_{fp}(-x_p) = \mathcal{E}_{fp}(x_n) \tag{4.79}$$

where $(x_n + x_p)$ is the total depletion layer width. Similar arguments can be presented for the reverse-biased junction. It is interesting to follow the variation of the quasi Fermi levels beyond the depletion layer edges. For example, the density of excess holes on the n-side, shown in Fig. 4.12, is given by Eq. 4.46. As $p_N(x)$ approaches p_{NO}, $\mathcal{E}_{fp}$ varies linearly with x_n according to Eqs. 4.46 and 4.76. This linear variation continues for several diffusion lengths. In fact, when $\mathcal{E}_{fp}$ crosses the intrinsic level $\mathcal{E}_{Fi}$, the excess hole density is only n_i. $\mathcal{E}_{fn}(x_p)$ varies in a similar manner.

Under high-level injection conditions, the injected minority carrier densities become comparable to the majority carrier densities, and a substantial portion of the applied bias drops across the regions outside the space charge layer, which itself shrinks. Therefore, both drift and diffusion components of current become important. The current density equations (2.139) combined with Eqs. 4.75 and 4.76 and the fact that $E = -\frac{1}{q}\frac{d\mathcal{E}_{Fi}}{dx}$, give

$$J_e(x) = q\mu_e n \left(\frac{1}{q}\frac{d\mathcal{E}_{Fi}}{dx}\right) + \mu_e k_B T \left(\frac{n}{k_B T}\left[\frac{d\mathcal{E}_{fn}}{dx} - \frac{d\mathcal{E}_{Fi}}{dx}\right]\right)$$

$$= n\mu_e \frac{d\mathcal{E}_{fn}}{dx} \tag{4.80}$$

and

$$J_h(x) = p\mu_h \frac{d\mathcal{E}_{fp}}{dx} \tag{4.81}$$

Thus, the currents flowing through the diode are determined by the *gradients* of the quasi-Fermi levels. Appreciable gradients of these Fermi levels develop within the depletion region and outside, to account for the spatially dependent concentration of carriers. If the quasi Fermi levels are constant throughout, there is no current flow. From Figs. 4.9(b) and (c) it is clear that the separation between the quasi Fermi levels increases monotonically from zero to a maximum value of V_a. When the separation is zero, $pn \cong n_i^2$ and when it is nonzero

$$pn \leq n_i^2 \, e^{q V_a/k_B T} \tag{4.82}$$

If $p \cong n$, then

$$p = n = n_i \, e^{q V_a/2k_B T} \tag{4.83}$$

This condition holds for high-injection conditions also.

EXAMPLE 4.4

Objective.   A GaAs p-n junction diode has $N_D = 2 \times 10^{16}$ cm^{-3} on the n-side and $N_A = 3 \times 10^{19}$ cm^{-3} on the p-side. To calculate the forward bias at which the injected hole concentration at the edge of the depletion region on the n-side becomes equal to the majority carrier (electron) concentration.

The condition of high-level injection is $p_N(0^+) = n_{NO}$. Using this equality and Eq. 4.49,

$$n_{NO} = p_{PO} e^{-q(V_{bi} - V_f)/k_B T} \tag{4.84}$$

Substituting the values of N_A, N_D and n_i (for GaAs at 300K) = 6.5 $\times 10^6$ cm^{-3} in Eq. 4.23 gives $V_{bi} = 1.35$V. Substituting this value in Eq. 4.84 and knowing that $n_{NO} = N_D$, $p_{PO} = N_A$, we get $V_f = 1.1$V.

4.2.5 Graded Junctions

Practical p-n junctions made by diffusion are seldom abrupt. As we have seen earlier in Sec. 4.1, this is due to the profiles of the diffusing species. Consequently, a graded (doping) region is created on both sides of the metallurgical junction. With ion-implantation, by greater control of the implant profiles, it is possible to achieve near-abrupt behavior, and epitaxial techniques such as molecular beam epitaxy allow us to realize ideal diodes. Nonetheless, it is important to know the parameters for graded junctions. The analysis of such junctions is briefly outlined. We will consider the simplest case, the *linearly* graded junction. In this case, the net doping on both sides of the junction, and consequently the ionized charge densities in the depletion region, vary linearly. The electrostatic parameters of a linearly graded junction are depicted in Fig. 4.16.

For an impurity concentration gradient a_I [cm^{-4}] we have, in the depletion region,

$$-\frac{\partial^2 V}{\partial x^2} = \frac{\partial E}{\partial x} = \frac{q}{\epsilon_s \epsilon_o}(p - n + a_I x)$$

$$\cong \frac{q}{\epsilon_s \epsilon_o} a_I x, \qquad -\frac{W}{2} \leq x \leq \frac{W}{2} \tag{4.85}$$

where W is the width of the depletion region. Therefore,

$$E(x) = -\frac{q a_I}{\epsilon_s \epsilon_o} \frac{(W/2)^2 - x^2}{2} \tag{4.86}$$

and the maximum field E_M at $x = 0$ is given by

$$|E_M| = \frac{q a_I W^2}{8 \epsilon_s \epsilon_o} \tag{4.87}$$

The built-in potential is given by

$$V_{bi} = \frac{q a_I W^3}{12 \epsilon_s \epsilon_o} \tag{4.88}$$

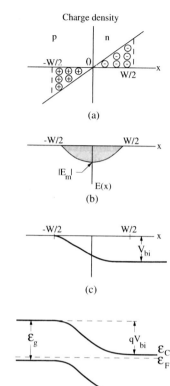

Figure 4.16 Electrostatics of linearly graded junction in thermal equilibrium: (a) space-charge distribution, (b) electric field profile, (c) potential profile, and (d) energy-band diagram.

from which

$$W = \left(\frac{12\epsilon_s\epsilon_o V_{bi}}{qa_I} \right)^{1/3} \tag{4.89}$$

Since the value of the impurity concentration at the edges of the depletion region are $a_I W/2$, by drawing analogy with Eq. 4.23, the built-in potential V_{bi} can be approximately written as

$$V_{bi} = \frac{k_B T}{q} \ln \left[\frac{(a_I W/2)^2}{n_i^2} \right]$$

$$= \frac{k_B T}{q} \ln \left[\frac{a_I W}{2n_i} \right]^2 \tag{4.90}$$

The depletion-layer capacitance as a function of applied bias is given by

$$C = \frac{\epsilon_s\epsilon_o}{W} = \left[\frac{qa_I(\epsilon_s\epsilon_o)^2}{12(V_{bi} \pm V_a)} \right]^{1/3} \quad (\text{Farad.cm}^{-2}). \tag{4.91}$$

The difference of these expressions, with those for abrupt junctions, should be noted.

4.2.6 AC Operation of Diodes: Diffusion Capacitance

The equivalent circuit of a reverse biased p-n junction is shown in Fig. 4.17(a). Here r_b is the total resistance of the bulk n- and p-regions, C_j is the bias-dependent depletion layer capacitance, and g_d is the conductance (usually small) of the junction due to the flow of the small reverse current. r_b and g_d are small compared to C_j. Under forward bias, however, the small-signal equivalent circuit changes to that shown in Fig. 4.17(b). The conductance changes to a different value G_d, C_j also changes and r_b remains fairly constant. However, there is an additional capacitance C_d, which results from the storage and motion of minority carriers on either side of the junction. Remember that it takes a finite time, of the order of 1 nanosecond, for the excess charge to recombine and dissipate. High-frequency operation of the diode is therefore limited by this minority carrier storage time. All bipolar devices, such as the bipolar junction transistor, suffer from this limitation. The operation of unipolar devices, such as the metal-semiconductor Schottky diode to be discussed shortly, depends on the transport of majority carriers only and are therefore not limited by minority carrier storage.

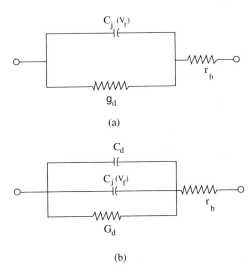

(a)

(b)

Figure 4.17 Equivalent circuits of a p-n junction under (a) reverse bias and (b) forward bias.

4.2.7 Breakdown Phenomena in Junction Diodes

It is generally assumed that the reverse current of a diode remains at a constant, low value for large values of reverse bias. We know that this is not true because of ongoing generation-recombination processes in the depletion region, which increase the current from the saturation value. In addition, there are ohmic leakage currents around the surface of the junction, if it is not properly encapsulated. Such encapsulation is usually done by the deposition of a dielectric such as SiO_x or Si_3N_4 after junction formation. Finally, and most importantly, at large reverse biases junction breakdown takes place. When this happens, the diode current is mostly limited by the resistance of the external circuit. In what follows, the various breakdown phenomena will be briefly described. Such breakdown usually depends on material parameters.

Zener breakdown occurs due to tunneling of carriers across the depletion region and the process is independent of temperature. If the doping densities in the p- and n-regions are high the depletion layer width is reduced and even at moderate reverse biases carriers can tunnel through the depletion barrier and add significantly to the reverse current. The process is shown Fig. 4.18(a). As the bias is increased, the tunneling component increases, and the diode eventually breaks down. In the atomic model it is the ionization of a covalent bond. It is important to remember that for Zener breakdown to occur before carrier avalanching, the doping densities should be high and consequently the depletion region should be narrow. If the diode is designed properly, the current in the diode will be *independent* of voltage after breakdown. Such diodes, usually called *Zener diodes*, are used as voltage regulators.

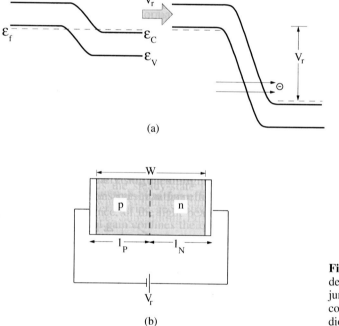

Figure 4.18   (a) Tunneling in a degenerately doped reverse-biased junction and (b) near-punchthrough conditions in a reverse-biased diode.

Avalanche breakdown occurs in diodes with moderate doping density, so that the depletion region is wider. With a large electric field across the depletion region, carriers moving across this region will collide with the lattice. At low fields the momentum and energy of the carriers are redirected, and such collisions are nonionizing. If the energy of the colliding carriers imparted by the field is large enough, then the collision can create an electron-hole pair by breaking a bond. In the band diagram the process is equivalent to raising an electron from the valence band to the conduction band, leaving a hole behind. It is important to realize that for the carriers to gain sufficient energy for ionizing collisions to occur, they must accelerate over some distance, and therefore the depletion region needs to be wide. After a collision event, the original carrier and the pair created by collision move in their respective directions under the field and suffer more ionizing collisions. Thus, a chain reaction sets in, as a result of which the reverse current increases and the diode eventually breaks down. The bias at which this occurs is called the breakdown voltage V_{BR}. The variation of the avalanche breakdown voltage as a function of carrier density in different semiconductors is shown in Fig. 4.19. If the temperature of the diode increases, the lifetime for lattice scattering decreases and that for ionizing collisions increases, resulting in an increase in V_{BR}. The reverse occurs with lowering of temperature. Thus, unlike Zener breakdown, the avalanche breakdown is a temperature-dependent phenomenon. The process just described—*impact ionization*—is characterized by impact ionization coefficients, α_e and α_h, for electrons and holes, respectively. The coefficients define the number of ionizing collisions per unit length. It may be realized that due to this bias-dependent carrier multiplication process, there is a net gain in the number of carriers. For example, if a photon incident on a photodiode produces an electron-hole pair, the gain is unity. But, if this pair creates many more by the avalanche multiplication process, then the external photocurrent is multiplied and a gain is obtained. This is the principle of operation of the avalanche photodiode, which will be discussed in detail in Chapter 8.

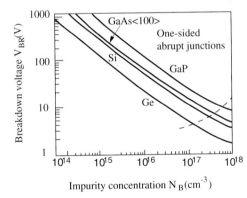

Figure 4.19 Breakdown voltage as a function of doping for one-sided abrupt GaP, GaAs, Si, and Ge p-n junctions (from S. M. Sze and G. Gibbons, *Appl. Phys. Lett.*, **8**, 111, 1966). The tunneling mechanism is dominant for the doping levels higher than corresponding to the dashed line in the low right corner of the figure.

Punchthrough is a breakdown phenomenon dependent primarily on the diode geometry. In a normal reverse-biased p-n junction, as shown in Fig. 4.18(b), holes injected from the contact into the n-region usually recombine in the neutral region before they reach the depletion layer. The same is true for electrons injected into the p-side. Now if the doping is light or if the length of the diode is small, then at some moderate reverse bias, the depletion layer may extend over the entire length. Under these conditions, the minority carriers injected by the contacts cannot recombine and build up in the depletion region and cause large increase of the diode current and eventual breakdown. It is important to remember that this process critically depends on device dimensions, which is not true for the breakdown mechanisms discussed earlier. For a simple analysis, assume an asymmetric p$^+$-n diode in which $N_A \gg N_D$. The width of the depletion region, which then extends almost entirely in the n region, is given by

$$W = \left[\frac{2\epsilon_s\epsilon_o}{qN_D} (V_{bi} + V_r) \right]^{1/2} \tag{4.92}$$

where N_D is the doping density on the n-side. If the length of the n-region is l_N, then for $V_R \gg V_{bi}$ punchthrough occurs when

$$W = l_N = \left[\frac{2\epsilon_s\epsilon_o V_r}{qN_D} \right]^{1/2} \tag{4.93}$$

The punchthrough voltage, V_{pt}, is given by

$$V_{pt} = \frac{qN_D l_N^2}{2\epsilon_s\epsilon_o}. \tag{4.94}$$

A nomogram for depletion layer width, junction capacitance and breakdown voltage for GaAs abrupt junctions is shown in Appendix 5.

4.3 SCHOTTKY BARRIERS AND OHMIC CONTACTS

4.3.1 Introduction

These junctions are special cases of the more general class of metal-semiconductor contacts, whose rectifying properties were first identified as early as 1874. It was realized later that metal-semiconductor contacts prepared in a special way, or in which the work functions of the metal and semiconductor have a required sign of difference, could serve as regions through which connections could be made to external power supplies, circuit elements, etc. A metal-semiconductor rectifying contact, or Schottky diode, was first realized as a point-contact and presently made as evaporated metal films on the semiconductor surface. The current-voltage characteristics of rectifying metal-semiconductor junctions are very similar to that of p-n junctions, although there are important differences in the mechanism of current flow and carrier (type) participation. Both electrons and holes contribute to the current in p-n junctions, whereas Schottky diodes are *unipolar* devices. The current is predominantly due to the thermionic emission of these carriers over the potential barrier that is created at the metal-semiconductor junction due to the difference in work function between them. We are more interested here in ohmic metal-semiconductor contacts, in which equal and large currents flow in both forward and reverse bias directions with small resistances. In other words, the potential drop across the junction should be negligible. The evolution of the I–V characteristics in going from a Schottky to an ohmic contact is shown in Fig. 4.20. The subject of both types of metal-semiconductor contacts has been discussed in detail in several texts. However, a brief discussion, covering the essential principles, is made here.

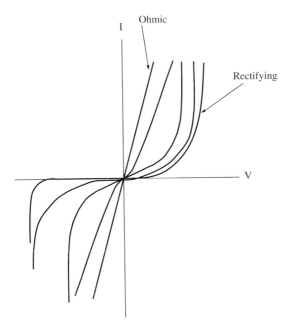

Figure 4.20 Evolution of the current-voltage characteristics of a diode from rectifying to ohmic behavior.

4.3.2 Schottky Barriers

Figure 4.21(a) shows the equilibrium band diagram of a metal of work function ϕ_m and an n-type semiconductor of work function ϕ_s. Note that in this case $\phi_m > \phi_s$. The electron affinity of the semiconductor, measured in energy from the edge of the conduction band to the vacuum level is χ_s. The electron affinity is the energy required to excite an electron from the bottom of the conduction band into vacuum. The work function is the energy needed to excite an electron from the Fermi level into vacuum. When a contact is made between the metal and semiconductor, electrons are transferred from the semiconductor to the metal until the Fermi levels are aligned (i.e., $\mathcal{E}_{Fm} = \mathcal{E}_{Fs}$). Exactly as in a p-n junction, a band-bending or energy barrier and a depletion region are created in the semiconductor, as shown in Fig. 4.21(b). The band-bending or built-in potential V_{bi} is equal to $\phi_m - \phi_s$, and the barrier height ϕ_b is equal to $\phi_m - \chi_s$. To maintain charge neutrality, the positively ionized donors in the depletion layer of the semiconductor are exactly balanced by the charge of a sheet of electrons in the metal at the junction. The barrier height ϕ_b is characteristic of a particular metal-semiconductor combination. Measurement of barrier heights have shown that ϕ_b is often almost independent of the metal species. Surface states resulting from surface imperfections, foreign atoms on the surface, etc., create a double layer on the free surface of a semiconductor, which tends to make the work function independent of the electron concentration in the bulk. If contact is made to a metal, the contact potential difference is compensated largely by a true surface charge rather than by space charge, so that the barrier height is largely independent of the metal. In

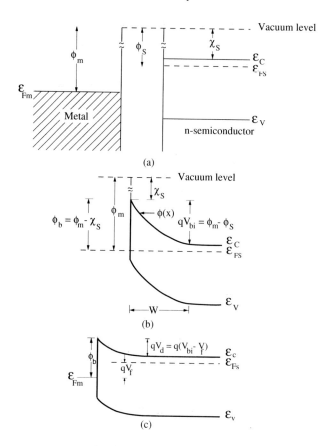

Figure 4.21 Band diagram of metal and n-type semiconductor ($\phi_m > \phi_s$): (a) before contact, (b) at zero applied bias, and (c) under forward bias.

the ideal case, without any surface or interface states, ϕ_b should be a linear function of ϕ_m with a slope less than unity.

Assuming a uniform distribution of ionized impurities in the semiconductor, Poisson's equation yields the one-dimensional parabolic potential energy barrier described by

$$\phi(x) = \frac{qN_D x^2}{2\epsilon_s \epsilon_o} \quad (V) \tag{4.95}$$

for $0 < x < W$ as shown in Fig. 4.21(b). The depletion layer width W is related to the total diffusion potential, V_d, in a forward biased junction [Fig. 4.21(c)] by

$$V_d = \phi_b - \frac{\mathcal{E}_{Fs}}{q} - V_f = \frac{qN_D W^2}{2\epsilon_s \epsilon_o} \tag{4.96}$$

where V_f is the applied forward bias.

The current in a metal-semiconductor contact under bias is determined by the flow of electrons from the semiconductor to the metal and vice versa. In equilibrium, these two components are equal. Under forward bias, electron flow from semiconductor to metal is enhanced due to a reduced band-bending and potential barrier $V_d(V)$. The flow from metal to semiconductor remains unchanged, since ϕ_b remains unchanged. This results in a large net *current* flow from metal to semiconductor. Under reverse bias, due to the increase band bending and barrier, electron flow from semiconductor to metal is almost negligible, and therefore the nearly constant small reverse current is due to the flow of electrons from metal to semiconductor over an unchanged barrier ϕ_b. The reverse current therefore flows from semiconductor to metal.

Current flow and rectification in a metal-semiconductor contact are adequately described by thermionic emission theory, as opposed to drift and diffusion mechanisms. For example, under forward bias, the current results from thermionic emission of electrons over the potential barrier. There are, in general, three modes of current transport: thermionic emission, thermionic-field emission, and field emission. These processes are shown in Fig. 4.22. The latter two involve quantum mechanical tunneling through the barrier at energies below the top of the barrier and are more favored in heavily doped semiconductors and large bias voltages. The barrier changes from a rectifying to an ohmic one as these two modes of current transport dominate. In addition, there is recombination in the depletion region, which becomes important under reverse bias, and some recombination in the neutral semiconductor region. The total current density J for ideal thermionic emission over the barrier is given by

$$J = J_f - J_r = A^{**} T^2 \exp(-q\phi_b/k_B T)(\exp(q V_a/k_B T) - 1) \qquad (4.97)$$

where J_f and J_r are, respectively, the forward and reverse bias current densities, $A^{**}(\cong 8$ A. cm^{-2}. K^{-2} for Au-GaAs) is the effective Richardson constant for the semiconductor, which takes account of the effective mass and some other corrections, and V_a is the applied bias. Departure from the simple thermionic emission model is accounted for by introducing the ideality factor n_f. Thus, under forward bias,

$$J \propto \exp(q V_a/n_f k_B T) \text{ for } V_a \geq 3k_B T/q. \qquad (4.98)$$

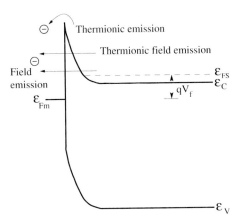

Figure 4.22 Possible conduction processes in a forward-biased Schottky diode with a degenerately doped n-type semiconductor.

In other words,

$$n_f = \frac{q}{k_B T} \left(\frac{dV_a}{d \ln J} \right) \tag{4.99}$$

For a good rectifier $n_f \geq 1$. As shown in Fig. 4.13(b), the linear region of the $\ln J - V$ plot corresponds to $n_f = 1$. At very low biases, the slope of the curve deviates and $n_f \to 2$. Here the current is dominated by generation in the depletion region. Also, for large forward currents, the voltage drop across the series resistance R_s associated with the neutral region of the semiconductor causes the actual voltage drop across the depletion region to be less than the applied bias. Under these conditions the current density is given by

$$J \infty \exp [q(V_a - I R_s)/k_B T]^{-1} \tag{4.100}$$

where I is the current through the diode. Under large forward biases, the voltage difference ΔV for any I gives the drop $I R_s$ across the neutral region. Therefore, a plot of ΔV versus $I R_s$ gives the value of R_s.

The discussion so far was related to a metal-(n) semiconductor junction with $\phi_m > \phi_s$. It is seen that under these conditions a rectifying diode is obtained. Similarly a metal-(p) semiconductor will be rectifying if $\phi_s > \phi_m$ and similar equations are valid.

EXAMPLE 4.5

Objective.   An ideal Schottky barrier diode is formed on an n-type semiconductor with the following parameters at 300°K: $\mathcal{E}_g = 1.1$ eV, $N_C = 2.5 \times 10^{19}$ cm^{-3}, and $N_D = 10^{16}$ cm^{-3}. The metal work function is 4.5 eV and the electron affinity in the semiconductor is 4 eV. To draw the equilibrium band diagrams and show the energy positions of $\mathcal{E}_{FM}, \mathcal{E}_{FS}, \mathcal{E}_C$ and $\mathcal{E}_V$ before and after the diode is formed.

From Eq. 2.78, the energy $(\mathcal{E}_C - \mathcal{E}_F)$ is 0.2 eV. The band diagrams are as shown in Fig. 4.23.

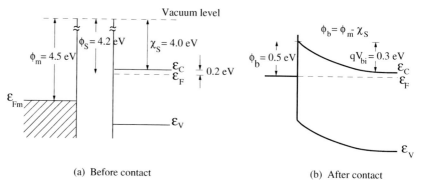

(a) Before contact   (b) After contact

Figure 4.23   Metal-(n) semiconductor junction with $\phi_m > \phi_s$.

At this point it is important to discuss some important differences, including current conduction, between p-n junctions and Schottky barrier diodes. In a Schottky barrier diode made with an n-type semiconductor, the current under forward bias is entirely carried by electrons, even if the doping level is low. In a p-n junction the current is carried predominantly by the majority carrier in the heavily doped side. High-frequency operation of p-n junction diodes is limited by the minority-carrier storage problem. In other words, the minimum time required to dissipate the carriers injected by the forward bias is dictated by the recombination lifetime. In a Schottky diode, electrons are injected from the semiconductor into the metal under forward bias if the semiconductor is n-type. In principle, these "hot" electrons can diffuse back to the semiconductor on the application of a reverse bias. However, they thermalize very rapidly ($\sim 10^{-14}s$) by carrier-carrier collisions, and this time is negligible compared to the minority-carrier recombination lifetime. Finally, Schottky diodes are more suitable for low-voltage applications than p-n junctions. This is because of two reasons. First, the thermionic emission process is much more efficient than the diffusion process and therefore for a given built-in voltage V_{bi}, the saturation current in a Schottky diode is several orders of magnitude higher than in p-n junctions. In addition, the built-in voltage of a Schottky diode is smaller than that of a p-n junction with the same semiconductor. These two factors taken together lead to the fact that to attain the same forward current density, a smaller bias needs to be applied to the Schottky diode than to the p-n junction.

4.3.3 Ohmic Contacts

The band diagram for a metal-(n) semiconductor junction with $\phi_s > \phi_m$ is shown in Fig. 4.24. It is seen that there is no potential barrier, and equal currents will flow under forward and reverse bias. The same is true for a metal-(p) semiconductor with $\phi_m > \phi_s$. Such junctions are termed *ohmic*. Due to the absence of a depletion region, the potential drop across these junctions is negligible at any bias and, therefore, they provide useful regions for contacting a semiconductor device. Unfortunately, metal and semiconductor combinations with the desired relationship between ϕ_m and ϕ_s are hard to come by, and therefore practical contacts cannot be conveniently realized in this way.

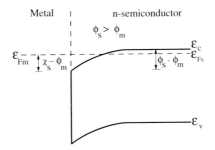

Figure 4.24 Equilibrium band diagram of metal-(n) semiconductor contact with $\phi_s > \phi_m$.

As the doping in the semiconductor is increased, the depletion region becomes narrower. With reference to Fig. 4.22 it is clear that under a reverse bias, when the depletion region becomes very narrow due to band-bending, thermionic field emission and field emission may become dominant. This can result in a rapid increase of current with reverse bias, which is analogous to Zener breakdown in p-n junctions. A similar mechanism of conduction can prevail under large forward biases. These mechanisms are utilized for realizing practical ohmic contacts. The common technique of forming reliable ohmic contacts to semiconductors is to heavily dope the region under the metal contact by diffusion, implantation, or alloying. For this, a metal alloy is first evaporated and then heated above the eutectic temperature. For example, to realize ohmic contacts to n-type GaAs, Au-Ge is evaporated and alloyed at $\sim 400°$C. Ge diffuses into the region under the contact and dopes the semiconductor degenerately n-type. A layer of Ni is usually evaporated on top of the Au-Ge to improve the morphology of the contact.

What is important in an ohmic contact is the differential resistance around zero bias, $R_{CT} = (dV/dI)_{V=0}$. For low doping, in the thermionic emission regime, R_{CT} is independent of N_D, while for highly doped material R_{CT} is proportional to $N_D^{-1/2}$ and conduction is predominantly by field emission. For intermediate values of doping thermionic field emission plays a dominant role. It has been found that in practical contacts $R_{CT} \propto 1/N_D$. This has been explained on the basis of current spreading through nonuniform metal-semiconductor regions that resemble protrusions.

More recently, epitaxial techniques such as MBE and OMVPE have allowed the realization of graded bandgap nonalloyed ohmic contacts. For example, highly doped graded $In_x Ga_{1-x} As$ $(0 \leq x \leq 1)$ is grown as the contact region and a metal layer is evaporated on top. Such a contact does not require annealing, since the highly doped, low bandgap semiconductor reduces the barrier heights sufficiently to allow ohmic conduction. Low-resistance ohmic contacts are absolutely essential for high-performance devices. For example, if the contact resistances of lasers are not very low, the threshold currents are high, which lead to device heating and inferior performance.

The measurement of contact resistance is technologically very important. The total resistance R_T between a small top contact and a very large bottom contact of a uniform piece of semiconductor is given by

$$R_T = r_{spr} + R_{CT} + R_o \qquad (4.101)$$

where r_{spr} is called the spreading resistance under the top contact and accounts for the bulk sample resistance. R_{CT} is the contact resistance of the top contact to be determined and R_o is the resistance of the bottom contact, which is frequently neglected. There are several techniques for measuring contact resistance and a common one is known as the *transfer length method*, originally proposed by Shockley. For a front contact resistance measurement with the geometry shown in Fig. 4.25(a), the total resistance between any two contacts can be shown to be given by

$$R_T = \frac{\rho_s d}{Z} + 2R_{CT} \qquad (4.102)$$

where Z and d are indicated in the figure and ρ_s is the sheet resistance in ohm/$\square$. The measured R_T is plotted as a function of d, as shown schematically in Fig. 4.25(b).

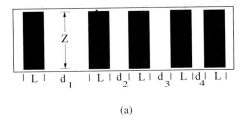

(a)

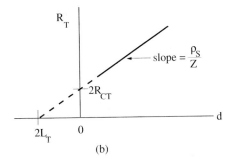

(b)

Figure 4.25 Schematic illustration of the transfer length method (TLM) for measuring contact resistance: (a) test structure with contact geometry and (b) plot of total resistance as a function of contact spacing.

Typically $L \sim 50\mu$m, $Z \sim 100\mu$m and $d \sim 5 - 50\mu$m. The intercept at $d = 0$ is $R_T = 2R_{CT}$, giving the contact resistance. The slope of the plot gives the value of ρ_s. A quantity $L_T = \sqrt{\rho_C/\rho_s}$ is defined as the *transfer length*, where ρ_C is the specific contact resistance in ohm.cm^2, which can also be approximately expressed as $\rho_C = R_{CT}L_T Z$ for $L > 1.5L_T$. From these two equations, ρ_C can be calculated by eliminating L_T and knowing ρ_S. A major limitation of this technique is the uncertainty of the sheet resistance under the contacts.[†]

4.4 SEMICONDUCTOR HETEROJUNCTIONS

4.4.1 Introduction

A heterojunction is formed when two semiconductors with different bandgaps and lattice constants are brought together, usually by epitaxy. Heterojunctions form essential constituents of almost all electronic and optoelectronic devices. For example, in semiconductor lasers heterojunctions provide both optical and carrier confinement. In a bipolar junction transistor, the incorporation of a heterojunction drastically improves the emitter injection efficiency and the current gain of the device. A heterojunction forms a two-dimensional channel of carriers at the interface with superior transport properties, which has been exploited to make high-performance field-effect transistors.

A heterojunction is formed by chemical bonding at the interface. In general, the lattice constants of the two semiconductors are different. If the heterojunction

---

[†]For a more detailed discussion, see D. K. Schroder, *Semiconductor Material and Device Characterization*, Wiley, New York, 1990.

is formed by epitaxy, as is usually the case, then there is a misfit between the two semiconductors, A and B. We are more interested in the perfect heterostructure, in which the lattice constants of semiconductors A and B are perfectly matched. The $Al_xGa_{1-x}As$ alloy system, for the entire composition range, $0 \leq x \leq 1$, is almost perfectly lattice matched to GaAs. It therefore forms a useful and important heterostructure system. Lattice matching can also be achieved by growing the appropriate semiconductors with the right compositions. As we have seen in Chapter 1, the ternary alloys $In_{0.53}Ga_{0.47}As$ ($\mathcal{E}_g = 0.74$ eV) and $In_{0.52}Al_{0.48}As$ ($\mathcal{E}_g = 1.45$ eV) are lattice matched to InP ($\mathcal{E}_g = 1.35$ eV). Therefore, very important heterojunctions, useful for optoelectronic devices working in the 1.3–1.6 μm range can be formed with these semiconductors.

4.4.2 The Ideal Heterojunction

Before analyzing a heterojunction in accordance with the model developed by Anderson, we will first qualitatively examine some interesting features of an isotype n-N heterojunction, formed, for example by n-type GaAs and n-type AlGaAs with a larger bandgap. The band diagrams of the two semiconductors, with their Fermi energies, work functions ϕ_1 and ϕ_2, and electron affinities χ_1 and χ_2, are shown in Fig. 4.26(a). Once the heterojunction is formed, electrons transfer from AlGaAs, because of its smaller electron affinity, to GaAs creating a depletion region in the former and an accumulation region in the latter. The formation of this depletion-accumulation region at the interface region is characteristic of heterojunctions. The band diagram of the heterojunction is shown in Fig. 4.26(b). The band offsets $\Delta\mathcal{E}_C$ and $\Delta\mathcal{E}_V$ in the conduction and valence bands, respectively, will be discussed later. It is important to note that the accumulation region in GaAs, depending on the density of electrons transferred from AlGaAs, forms a triangular potential well with a lateral extent of 100–150 Å. The electrons in this well form a two dimensional electron gas and their energy is quantized, just as in a rectangular quantum well. For a p-type isotype heterojunction, a two-dimensional hole gas will be formed in the well created by the valence band discontinuity.

We will now analyze a heterojunction in accordance with Anderson's model,[†] which assumes an ideal interface free of interfacial defect states. A p-N or anisotype heterojunction is considered, consisting of a small bandgap p-type semiconductor and a large bandgap n-type layer. The band diagrams of the two semiconductors before joining, together with the vacuum level, work function, electron affinity, and Fermi energy, are shown in Fig. 4.27(a). The band diagram after formation, with respect to the vacuum level, is shown in Fig. 4.27(b). The band-bending at the junction is equal to the built-in potential V_{bi}, which is given by

$$qV_{bi} = \phi_1 - \phi_2 = (\mathcal{E}_{g1} + \chi_1 - \Delta\mathcal{E}_{F1}) - (\chi_2 + \Delta\mathcal{E}_{F2})$$

$$= \mathcal{E}_{g1} + \Delta\chi - \mathcal{E}_{F1} - \Delta\mathcal{E}_{F2} \tag{4.103}$$

[†]R. L. Anderson, *Solid State Electronics*, **5**, 341, 1962.

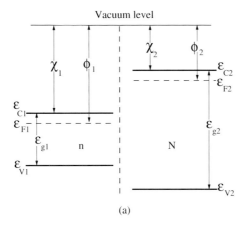

(a)

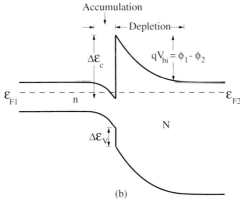

(b)

Figure 4.26   The isototype heterojunction: (a) two n-type semiconductors with bandgaps $\mathcal{E}_{g1}$ and $\mathcal{E}_{g2}$ before junction formation and (b) equilibrium band diagram after formation.

According to the Anderson model, the conduction band offset, $\Delta\mathcal{E}_c$, is given by

$$\Delta\mathcal{E}_c = \chi_1 - \chi_2 = \Delta\chi \tag{4.104}$$

Then the valence band offset is given by

$$\Delta\mathcal{E}_V = (\chi_2 + \mathcal{E}_{g2}) - (\chi_1 + \mathcal{E}_{g1}) = \Delta\mathcal{E}_g - \Delta\chi \tag{4.105}$$

Therefore,

$$\Delta\mathcal{E}_C + \Delta\mathcal{E}_V = \Delta\mathcal{E}_g \tag{4.106}$$

An important and interesting case, applicable to heterojunction bipolar transistors, is as follows. If $\chi_1 = \chi_2$, the offset in the conduction band disappears, and $\Delta\mathcal{E}_V = \Delta\mathcal{E}_g$. In this case

$$qV_{bi} = \mathcal{E}_{g1} - \Delta\mathcal{E}_{F1} - \Delta\mathcal{E}_{F2}$$

$$= \mathcal{E}_{g1} - k_BT \ln\left(\frac{N_C}{n}\right) - k_BT \ln\left(\frac{N_V}{p}\right) \tag{4.107}$$

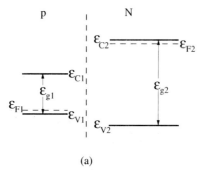

(a)

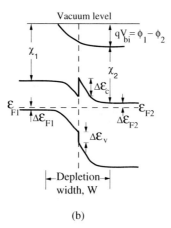

(b)

Figure 4.27 Band diagrams of p-N anisotype heterojunction(a) before and (b) after contact.

where n and p are the free-carrier concentrations in the n- and p-type semiconductors, respectively. Equation 4.107 for the built-in voltage is identical to that for a homojunction. However, there is one important difference in this case. The barrier to hole injection is increased by $\Delta\mathcal{E}_V = \Delta\mathcal{E}_g$, compared to a homojunction. Therefore, the electron injection over the barrier is a much more efficient process than hole injection. This effect is utilized in the design of heterojunction bipolar transistors and phototransistors, to be discussed later.

The total built-in voltage V_{bi} at the junction is divided between the p- and n-regions

$$V_{bi} = V_{bi,p} + V_{bi,n} \tag{4.108}$$

where

$$V_{bi,p} = \frac{\epsilon_2 N_D}{\epsilon_2 N_D + \epsilon_1 N_A} V_{bi}$$

$$V_{bi,n} = \frac{\epsilon_1 N_A}{\epsilon_2 N_D + \epsilon_1 N_A} V_{bi} \tag{4.109}$$

The corresponding depletion layer thicknesses in the p- and n-regions are given by

$$x_{dp} = \left[\frac{2\epsilon_1\epsilon_2\epsilon_0 N_D V_{bi}}{q N_A(\epsilon_2 N_D + \epsilon_1 N_A)} \right]^{1/2}$$

$$x_{dn} = \left[\frac{2\epsilon_1\epsilon_2\epsilon_0 N_A V_{bi}}{q N_D(\epsilon_2 N_D + \epsilon_1 N_A)} \right]^{1/2} \tag{4.110}$$

The capacitance-voltage relationship is

$$C(V) = \left[\frac{q\epsilon_1\epsilon_2\epsilon_0 N_A N_D}{2(\epsilon_1 N_A + \epsilon_2 N_D)(V_{bi} - V)} \right]^{1/2} \text{ (Farad·cm}^{-2}) \tag{4.111}$$

It is assumed here that the n- and p-regions are uncompensated and there is full ionization of the donor and acceptor centers in these regions. Note that these relations reduce to those for a homojunction when the dielectric constants (and bandgaps) of the two semiconductors become identical (i.e., $\epsilon_1 = \epsilon_2$).

The interfacial spikes, produced by the band offsets, may sometimes be undesirable for ideal device operation. Their magnitude can be reduced by compositional grading at the interface, which reduces the electron affinity difference at the interface and therefore the field magnitude. This is shown in Fig. 4.28 for an n-P heterojunction.

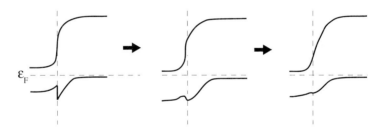

Figure 4.28   Gradual smoothing of the interfacial spike of an n-P heterojunction by compositional grading.

4.4.3 Current-Voltage Characteristics

Under applied forward or reverse bias, the change in the band diagram of a heterojunction at the interface region is very similar to that of a homojunction. Complexities are introduced, however, due to the band offset spikes, and the interfacial defects and traps that are present due to lattice mismatch. There is therefore some uncertainty concerning the current transport mechanism. These include thermionic emission over the barrier, interfacial recombination, and tunneling through the spikes and defect states. Assuming thermionic emission over the heterojunction barrier to be the dominant mechanism, similar to that in a Schottky diode, the current-voltage relation can be expressed as

$$J = \frac{q A^{**} T\, V_{bi}}{k_B} \exp\left[\frac{-q V_{bi}}{k_B T} \right] \left[\exp\left(\frac{q V_a}{k_B T} \right) - 1 \right] \tag{4.112}$$

For a more accurate formulation of the current-voltage characteristics, tunneling, drift, and diffusion effects need to be taken into account.

4.4.4 Real Heterojunction Band Offsets

Anderson's model for band offsets assumes a perfect heterojunction without any interfacial defects. The model states that the conduction band discontinuity is given by the difference of the electron affinities of the two semiconductors. The electron affinities of the common compound semiconductors are listed in Table 4.1. A quick examination reveals that the band offsets measured experimentally are very different from those obtained from the electron affinity rule. The "common anion rule" was proposed to explain experimental data. This rule states that the valence band discontinuity $\Delta \mathcal{E}_V$ at the interface will be "very small" for semiconductors with the same anion arising from the fact that the top of valence band states are predominantly anion states. This rule also failed to explain experimental results. What has been observed, however, is that in heterojunctions made up of two semiconductors with different anions, such as $In_{0.53}Ga_{0.47}As$ and InP (lattice matched) or $Ga_{0.5}In_{0.5}P$ and GaAs (lattice matched), $\Delta \mathcal{E}_V > \Delta \mathcal{E}_C$. The trend is in accordance with the common anion rule.

TABLE 4.1 BANDGAP AND ELECTRON AFFINITY OF SOME IMPORTANT SEMICONDUCTORS.

	$\mathcal{E}_g$(eV)	χ (eV)
GaAs	1.43	4.07
AlAs	2.16	2.62
GaP	2.21	4.3
AlSb	1.65	3.65
GaSb	0.73	4.06
InAs	0.36	4.9
InSb	0.17	4.59
InP	1.35	4.35
ZnTe	2.26	3.5
CdTe	1.44	4.28
ZnSe	2.67	3.9
CdS	2.42	4.87
Ge	0.66	4.13
Si	1.11	4.01

More recently Tersoff[†] has proposed a model, based on analogy with a metal-metal junction, which states that the conduction band discontinuity is given by the difference in the Schottky barrier heights of the two semiconductors forming the heterojunction. In other words,

$$\Delta \mathcal{E}_C = \phi_{b1} - \phi_{b2}. \tag{4.113}$$

[†]J. Tersoff, *Physical Review*, **B30**, 4874, 1984.

In this model, it is postulated that a dipole layer is formed at the interface due to electron tunneling from one material into the energy gap of the other. Reasonable agreement with this model has been observed for lattice-matched GaAs/Al$_x$Ga$_{1-x}$As heterostructures.

Experimentally, early determination of the band offsets was made by Dingle and co-workers in which the band offsets in GaAs/AlGaAs quantum wells was derived from analysis of exciton recombination energies in the luminescence spectra. The fits suggested $\Delta\mathcal{E}_C/\Delta\mathcal{E}_g = 0.85$ for this heterostructure system, and this 85:15 ($\Delta\mathcal{E}_C$: $\Delta\mathcal{E}_V$) rule prevailed for sometime. However, closer examination showed that other fractions may fit the data better and, more significantly, that the fit is quite insensitive to the choice of band offset. A considerably smaller value of $\Delta\mathcal{E}_C/\Delta\mathcal{E}_g \cong 0.6$ was later found by analyzing excitonic energies in photoluminescence excitation spectra from parabolic and square quantum wells. This 60:40 distribution of the discontinuity between the conduction and valence bands has been corroborated by the results of several recent electrical measurements.

Measurements of the conduction band offsets in In$_x$Ga$_{1-x}$As/Al$_{0.3}$Ga$_{0.7}$As quantum wells were made by the author by treating the quantum well as a deep-level trap. The thermal ionization energy of electrons trapped in the well are measured and the band offset is derived from them. Figure 4.29 shows the measured variation of $\Delta\mathcal{E}_C/\Delta\mathcal{E}_g$ with In composition x. It should be cautioned, however, that the quantum wells are biaxially strained and pseudomorphic. The contribution of this strain to the measured data is not known exactly. The offsets are in close agreement to the 60:40 rule.

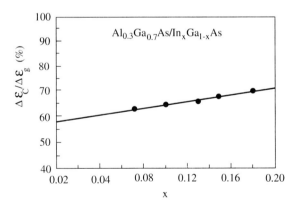

Figure 4.29 Conduction band offsets of In$_x$Ga$_{1-x}$As/Al$_{0.3}$Ga$_{0.7}$As quantum wells measured by the author and co-workers (from N. Debbar et al., *Physical Review*, **B-40**, 1058, 1989).

4.4.5 Application of Heterojunctions to Bipolar Transistors

In Chapter 9 we will be discussing the operating principles of phototransistors. The optical gain of this device is the product of the current gain β_T and the quantum efficiency of the device. It is therefore important that the current gain of the device is reasonably high. In order to put the subject in proper perspective, we will reiterate a few definitions and relations for a p(emitter)- n(base)- p(collector) bipolar transistor. The schematic of a bipolar transistor is shown in Fig. 4.30, where i_E, i_B, and i_C are the emitter, base, and collector currents, respectively. The emitter-base junction is forward biased, while the collector-base junction is reverse biased. Holes are injected from the emitter to the base and after traversing this region they reach the collector where they contribute to the collector current. There are at least three mechanisms that contribute to the base current i_B, but the dominant one is due to the replenishment of electrons in the n-type base region that recombine with the holes injected from the emitter. Also, electrons emitted from the n-type base region to the p-type emitter must be replenished. The terminal currents can be related by several important factors. Neglecting the collector junction reverse saturation current and recombination in the base-collector depletion region, the collector current is made up *almost* entirely of holes injected from the emitter, which do not recombine in the base region. In other words,

$$i_C = B_T i_{E(p)} \tag{4.114}$$

where

$$i_E = i_{E(p)} + i_{E(n)} \tag{4.115}$$

taking into account the electron and hole injection at the forward-biased emitter junction. B_T is the *base transport factor* and the *emitter injection efficiency*, γ_E, is given by

$$\gamma_E = \frac{i_{E(p)}}{i_{E(p)} + i_{E(n)}} \tag{4.116}$$

We also define the *current transfer ratio* as

$$\alpha_T = B_T \gamma_E \tag{4.117}$$

Finally, we define the current gain,

$$\frac{i_C}{i_B} = \beta_T = \frac{\alpha_T}{1 - \alpha_T}. \tag{4.118}$$

Therefore, for high-current gain α_T, B_T, and γ_E should all be as close to unity as possible. Kroemer defined another measure of transistor performance as the *injection deficit* given by

$$\frac{1 - \gamma_E}{\gamma_E} = \frac{i_{E(n)}}{i_{E(p)}} \tag{4.119}$$

In a good transistor the injection deficit should be small. It is possible to lower the injection deficit by use of a heterojunction (i.e., by using an emitter material with a wider

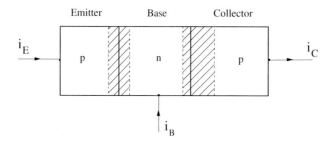

Figure 4.30   Schematic representation of a p-n-p bipolar transistor.

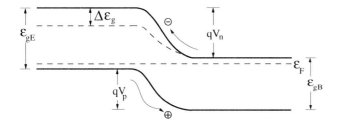

Figure 4.31 (p) emitter- (n) base heterojunction with $\Delta\mathcal{E}_C = \Delta\mathcal{E}_g$. qV_n and qV_p are the potential barriers for electrons and holes, respectively.

bandgap than the base material). The reason becomes clear on considering Fig. 4.31. The potential barrier for electrons flowing from the base to the emitter is larger than the barrier for holes flowing from the emitter to the base. For an applied bias V_a,

$$i_{E(p)} = \frac{qD_h p_o}{L_h}\left[\exp\left(\frac{qV_a}{k_B T}\right) - 1\right]$$

$$i_{E(n)} = \frac{qD_e n_o}{L_e}\left[\exp\left(\frac{qV_a}{k_B T}\right) - 1\right] \tag{4.120}$$

where D_h and L_h are the diffusion constants and diffusion lengths, respectively, for holes in the base and D_e and L_e are the same parameters for electrons in the emitter. n_o and p_o are the equilibrium minority carrier densities in the emitter and base regions. Therefore,

$$\frac{i_{E(n)}}{i_{E(p)}} = \frac{D_e L_h n_o}{D_h L_e p_o} \tag{4.121}$$

We know that

$$n_o = \frac{n_{iE}^2}{N_A}, \quad p_o = \frac{n_{iB}^2}{N_D} \tag{4.122}$$

where n_{iE} and n_{iB} are the intrinsic carrier concentrations in the emitter and base regions, respectively, and N_A and N_D are the majority carrier (doping) concentrations in these regions. Furthermore,

$$\frac{n_{iE}^2}{n_{iB}^2} = \left(\frac{m_{eE}^* m_{hE}^*}{m_{eB}^* m_{hB}^*}\right)^{3/2} \exp\left[-\Delta\mathcal{E}_g / k_B T\right] \tag{4.123}$$

where m_e^* and m_h^* are the effective masses in the two regions and $\Delta\mathcal{E}_g = \mathcal{E}_{gE} - \mathcal{E}_{gB}$. Therefore,

$$\frac{i_{E(n)}}{i_{E(p)}} = \frac{D_e L_h N_D}{D_h L_e N_A}\left(\frac{m_{eE}^* m_{hE}^*}{m_{eB}^* m_{hB}^*}\right)^{3/2} \exp\left[-\Delta\mathcal{E}_g / k_B T\right] \tag{4.124}$$

Remembering that for a homojunction

$$\frac{i_{E(n)}}{I_{E(p)}} = \frac{D_e L_h N_D}{D_h L_e N_A} \tag{4.125}$$

the injection deficit in a heterojunction transistor is largely controlled by the exponential factor. For example, even for $\Delta\mathcal{E}_g = 0.2$ eV, at room temperature, with $k_B T \simeq 0.025$ eV, the injection deficit is *decreased* by a factor of almost 3000.

There are other advantages of having a heterojunction at the emitter-base junction. One can afford to greatly increase the base doping and decrease the emitter doping, without sacrificing the emitting injection efficiency. As a consequence of the higher base doping the base spreading resistance in the direction transverse to current flow is lower. The lower emitter region doping makes the emitter-base junction capacitance small. Both these factors contribute to a higher speed of operation of the transistor.

P R O B L E M S

4.1 **(a)** Discuss in a few words the following:

1. quasi-Fermi levels
2. Debye tailing in a p-i-n junction

(b) Sketch the *hole* quasi-Fermi level in the quasi-neutral n-region of a p^+-n junction at

1. equilibrium
2. reverse bias
3. forward bias

4.2 A semiconductor sample is composed of two regions as shown in Fig. P4.2. Consider regions I and II to be far away from the p-n junction, so that these regions are in equilibrium. At a particular temperature the hole concentration in region I is determined to be $p_1 = 1.1 \times 10^{16}$ cm^{-3}.

(a) Find the intrinsic concentration n_i for the semiconductor at this temperature.

(b) Find the equilibrium electron concentration n in region II at this temperature.

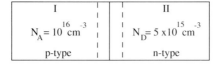

Figure P4.2

4.3 $Al_{0.3}Ga_{0.7}As$ is doped as a $n_1 - n_2$ step junction shown in Fig. P4.3, where $N_{D1} < N_{D2}$. Draw the energy band diagram of such a junction and indicate V_{bi} on this diagram. Also draw the charge density and electric field profile for such a junction.

4.4 A Si p-n junction diode has a rectangular cross-sectional area of 0.5 mm $\times$ 1.0 mm and the thickness of the p- and n-regions are 1.0 mm each. Calculate the forward bias necessary to drive 1 mA and 10 mA through the diode. What is the voltage drop across the bulk p- and n-regions for these

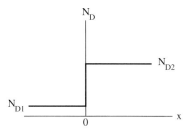

Figure P4.3

forward currents?

[Given: $J_S = 10^{-15} A/cm^2$, $N_A = 10^{16}$ cm^{-3}, $N_D = 10^{17}$ cm^{-3}.]

4.5 Derive Eq. 4.35 under the conditions of Debye tailing in the depletion layer of a junction diode.

4.6 An abrupt p$^+$-n junction is formed with InP having $N_D = 10^{15}$ cm^{-3}. The avalanche breakdown voltage is -80V. What is the minimum thickness of the n-region to ensure avalanche breakdown rather than punchthrough?

4.7 **(a)** Draw the band diagrams of a metal-n semiconductor ($\phi_m > \phi_s$) under a small forward and a large reverse bias. Show clearly the positions of the Fermi level and the different potentials.

(b) A metal and p-type Si ($\phi_s > \phi_m$) are brought together to form a junction. Will the junction be ohmic or rectifying? If the doping in the Si layer is p = 10^{17} cm^{-3}, $\phi_m = 3$ eV, $\chi_s = 2.5$ eV, $\mathcal{E}_g$ (Si) = 1.1 eV, $n_i = 1.5 \times 10^{10}$ cm^{-3}, calculate: (1) V_{bi}, (2) ϕ_b.

4.8 Alternate heavily doped p-type and n-type layers of the same semiconductor, each 100 Å thick, are grown by molecular beam epitaxy. Draw the energy band diagram of this structure and discuss some of its special properties, including electron-hole recombination rates.

4.9 List and explain the differences in the operation of a p-n junction and a metal-n (semiconductor) Schottky diode with $\phi_m > \phi_s$.

4.10 Consider a p-n junction with the following parameters:

$$\text{conductivity } \sigma_p \gg \sigma_n = 1(\Omega - cm)^{-1}$$

$$D_p = 50cm^2/sec$$

$$\mu_n = 1000cm^2/V.s.$$

$$L_h = 1cm$$

$$I = 100mA$$

$$V \gg k_B T/q$$

$$\text{Area A} = 1.25 \times 10^{-2}cm^2$$

You may assume that L_e is of the same order of magnitude as L_h. Is the low injection assumption valid for this device? Explain in detail, including analytical and/or numerical calculations to justify your answer.

4.11 We know that the capacitance of an abrupt p-n junction varies as $V_r^{-1/2}$ for $V_r \gg V_{bi}$ where V_{bi} is the built-in potential and V_r is the reverse bias applied across the junction. The capacitance of a linearly graded junction varies as $V_r^{-1/3}$. In a certain application a capacitance that varies as V_r^{-1} for $V_r \gg V_{bi}$ is needed. Qualitatively, discuss the general form of doping profile needed. Indicate in each of the three cases the variation of the depletion-region width with bias.

4.12 Using the Shockley-Read-Hall theory discussed in Chapter 2, derive an expression for the generation current of a reverse-biased p-n junction diode. What is the temperature dependence of this current?

4.13 In an experiment with a Schottky diode, its current-voltage characteristics are measured as a function of temperature. What parameters can be extracted from such data? Is there any difference between forward-bias and reverse-bias data? If so, explain why.

4.14 Schottky diodes are majority-carrier devices, and therefore can operate at higher speeds compared to p-n junctions. There is still a preference for the p-n junction over the Schottky diode. State what you think may be the reasons.

4.15 The capacitance C of a Schottky diode of 1 cm^2 area is measured as a function of forward and reverse biases. A linear relationship is found that can be expressed as $1/C^2 = A - BV$, where V is the applied bias and A and B are constants. Derive expressions for built-in voltage V_{bi} and the donor density $N_D (cm^{-3})$ in the semiconductor.

4.16 A p-n junction diode has a small alternating voltage superimposed on a steady state forward bias applied to it. Assuming small signal conditions and a low frequency of the modulating bias, derive an expression for the diffusion capacitance C_d.

READING LIST

MILNES, A. G. *Semiconductor Devices and Integrated Electronics*. Van Nostrand Reinhold, New York, 1980.

RHODERICK, E. H. *Metal-Semiconductor Contacts*. Clarendon Press, Oxford, 1978.

SHARMA, B. L., and PUROHIT, R. K. *Semiconductor Heterojunctions*. Pergamon Press, London, 1974.

SHUR, M. *Physics of Semiconductor Devices*. Prentice Hall, Englewood Cliffs, NJ, 1990.

SZE, S. M. *Physics of Semiconductor Devices*, 2nd ed. Wiley, New York, 1981.

TIWARI, S. *Compound Semiconductor Device Physics*. Academic Press, San Diego, 1992.

5

Light-Emitting Diodes

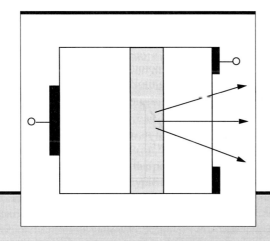

Chapter Contents

5.1 INTRODUCTION

The light-emitting diode (LED) was discovered in 1904. The only semiconductor diode in existence at that time was the silicon carbide (SiC) point contact rectifier. Needless to say, the level of light emission from such devices was very small. Since then, light emission from semiconductor devices has attracted the interest of researchers all over the world. However, it is only recently, with the development of wider bandgap III–V compounds, that visible light emission from a biased p-n junction has been demonstrated. The LED has found wide applications in the form of indicator lamps, display elements, and sensors.

In a junction LED photons of near-bandgap energy are generated by the process of *injection luminescence* or *electroluminescence*, in which a large population of electrons, injected into a normally empty conduction band by forward bias, recombine with holes in the valence band. The device emits light by spontaneous emission, which is different from *stimulated emission* operative in junction lasers. The spontaneous emission process requires a smaller forward bias, and therefore LEDs operate at lower current densities than lasers. An LED also does not require an optical cavity and mirror facets to provide feedback of photons. The emitted photons have random phases, and therefore the LED is an incoherent light source. The linewidth of the spontaneous emission is approximately equal to the photoluminescence linewidth, which is a few times $k_B T$ and is typically 30–50 nm at room temperature. Thus, many optical modes are supported and therefore the LED is a convenient multimode optical source, suitable for use with multimode fibers.

Some of the obvious advantages of the LED as a light source are simpler fabrication procedures, lower cost, and simpler drive circuitry. Light-emitting diodes are, in general, more reliable than lasers because they operate at lower powers and therefore do not suffer catastrophic degradation or even gradual degradation prevalent in lasers. Ideally the LED exhibits a linear light output–current characteristic, unlike that of an injection laser, and is therefore very suitable for analog modulation. The light-current characteristic is also less sensitive to temperature than that in a laser. Some of the obvious disadvantages are the low power output, harmonic distortion due to the multimodal output, and a smaller modulation bandwidth. However, superradiant LEDs are being developed, which have output powers comparable to that of lasers.

5.2 THE ELECTROLUMINESCENT PROCESS

The LED converts input electrical energy into output optical radiation in the visible or infrared portion of the spectrum, depending on the semiconductor material. The energy conversion takes place in two stages: first, the energy of carriers in the semiconductors are raised above their equilibrium value by electrical input energy, and second, most of these carriers after having lived a mean lifetime in the higher energy state, give up their energy as spontaneous emission of photons with energy nearly equal to the bandgap $\mathcal{E}_g$ of the semiconductor. If the semiconductor can be doped p- and n-type, then the energy of the current carriers can be increased by applying a forward bias to a p-n junction. As we have seen in Chapter 4, under forward bias, minority carriers are injected on both sides of the junction and these excess minority carriers diffuse away from the junction, recombining with the majority carriers as they do so. This process is illustrated in Fig. 5.1. The excess hole concentration in the n-side, away from the depletion region, is given by Eq. 4.46 and a similar equation is valid for excess electrons on the p-side. Most of the excess minority carriers on both sides of the junction recombine radiatively with the majority carriers to create photons of frequency v given by

$$h v \cong \mathcal{E}_g \tag{5.1}$$

A small fraction of the excess minority carriers do recombine nonradiatively, and the excess energy of these carriers is dissipated as heat in the lattice. The rate of radiative recombination is normally proportional to the forward bias injection rate, and hence to the diode current under such conditions, given by Eq. 4.55. Most of the recombination occurs close to the junction, although some of the minority carriers diffuse away and hence recombination can occur in regions further away from the junction. The *internal quantum efficiency* of the device is defined as the rate of emission of photons divided by the rate of supply of electrons. As mentioned above, all the injected electrons do not recombine radiatively with holes, and therefore the quantum efficiency may be less than 100%. Under reverse bias applied to the diode, no injection takes place across the depletion region, and consequently no light is emitted.

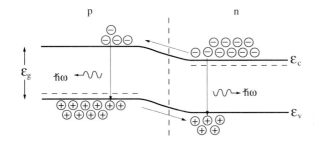

Figure 5.1 Injection of minority carriers in a forward-biased p-n junction leading to spontaneous emission of photons.

In using a semiconductor material for light emission, a wavelength (or energy) of emission close to the bandgap is of importance. As we have seen in Chapter 3, the transitions of interest are band-to-band, impurity-band, donor-to-acceptor, and excitonic transitions. The peak in the density distribution of electrons in the conduction band occurs at an energy $k_B T / 2$ above the band minimum. Therefore, the band-to-band transition energy is slightly larger than $\mathcal{E}_g$, and some of the emitted photons are reabsorbed. This self-absorption does not exist in indirect bandgap semiconductors due to the involvement of phonons. However, because of phonons participating in the radiative transition in indirect semiconductors, their radiative efficiency is low. Fortunately, most radiative transitions involve energy levels within the bandgap for which the emission energy is smaller than the bandgap.

Impurity-induced radiative transitions are usually impurity-band and donor-acceptor transitions, both of which have been discussed in Chapter 3. However, there is an interesting aspect of impurity levels, which is important for the operation of LEDs.

A carrier in an impurity state is "trapped" in the potential of the impurity, and is therefore localized in space. As the impurity level becomes deeper, or its binding energy increases, the localization of the carriers is enhanced. It may be noted that the impurity potential is Coulombic for shallow levels and more like a delta function for deep levels. In either case, as a consequence of the Heisenberg uncertainty principle, a spatial localization of the carriers in the impurity levels causes a spread in their momentum (or k) values. The spread of the impurity states in k-space is sufficient to allow a significant rate of transitions between the impurity states and the band edges *without* the involvement of phonons. This, in effect, increases the radiative efficiency of indirect bandgap semiconductors. An example, which is of tremendous commercial importance, is the presence of O and N increasing the radiative efficiency of GaP. Oxygen produces a deep donor level 0.8 eV below the conduction band, while N produces a shallow isoelectronic trap. The impurity-related transitions are the most important ones for LED operation, since most junction LEDs are made of doped materials. Since the radiation resulting from impurity-related transitions has a lower photon energy than the bandgap, it is not subject to reabsorption to the same extent as band-to-band transitions.

Excitonic transitions have been discussed in Chapter 3. Since the design of LEDs involves doped materials, the radiative transitions of interest are those involving bound excitons. The binding energy of the bound exciton centers depends on the binding energy of the impurity to which the excitons are associated with, and is given by

$$\mathcal{E}_{ex} \cong 0.1\mathcal{E}_i \tag{5.2}$$

where $\mathcal{E}_i$ is the binding energy of the impurity level. Again, the excitons can be bound to neutral or ionized impurities depending on the doping level and the temperature of the sample. Bound excitons may also be sufficiently localized so that electron-hole recombination can take place in indirect bandgap semiconductors via these states, without involving phonons to conserve momentum.

Having described the various near-bandgap transitions via which electrons and holes may recombine radiatively, it is important to answer the question *Which transition is dominant in the operation of a junction LED?* As one may guess at this point, all the transitions described above may be involved, depending primarily on the doping in the sample and also to some extent on the impurity levels and the bias applied. Values of the spontaneous recombination coefficient, B_r, in indirect semiconductors are significantly lower than those in direct bandgap materials. Thus band-to-band transitions are not efficient in indirect semiconductors and usually impurities producing deep levels are incorporated to increase the radiative efficiency. An example, as already mentioned, is O in GaP and GaAsP.

5.3 CHOICE OF LED MATERIALS

The wavelength of light emission required usually dictates the semiconductor materials required, in terms of their bandgap energy. An equally important factor is the ability to heavily dope these materials n- and p-type and thereby fabricate a junction diode.

Lower bandgap materials are required for infrared and far-infrared applications, and larger bandgap materials are needed for a light source in the visible part of the spectrum (around 2 eV). The higher the bandgap, the higher is the melting point of the material, and the lower is their structural stability. Also, the higher bandgap materials have higher resistivity and cannot be easily doped to high levels, as may be desirable for LED fabrication.

The most important III–V compounds for application to LEDs are the binary compounds GaAs and GaP, and their ternary derivative $GaAs_{1-x}P_x$. The properties of these materials were discussed in Chapter 1. GaAs has a direct bandgap of 1.43 eV (λ = 860 nm) at room temperature and therefore emits in the near-infrared region. Light-emitting diodes are usually made by Zn-diffusion into pulled n-GaAs or Si-doping of solution grown GaAs. Si, being amphoteric in nature, can incorporate into the GaAs crystal as a donor or acceptor, depending on growth temperature. In Zn-doped devices, the main radiative transition is a band-to-band process, and consequently there is a large reabsorption of the emitted radiation. The emission wavelength of Si-doped diodes is $\sim$ 940 nm and reabsorption is much less severe. Here the main recombination transition is from an (impurity-related) defect to the conduction band, and the complex defect behaving as an acceptor is believed to have a binding energy of $\sim$ 0.1 eV. Typical external quantum efficiencies of these diodes are of the order of 10%. However, Si-doped GaAs LEDs are more expensive to manufacture than the Zn-doped ones.

GaP is an indirect bandgap semiconductor with $\mathcal{E}_g$ = 2.26 eV (λ = 549 nm), and therefore impurity levels in the bandgap are needed for it to be used as an efficient light emitter. GaP is commonly used for red and green emission. Nitrogen acts as a donor-like isoelectronic trap in GaP with a binding energy of 8 meV, remembering that it belongs to the same group in the periodic table as P. This isoelectronic trap forms a bound exciton, which emits at a near-bandgap wavelength of 550 nm (green). The quantum efficiency is, however, even smaller than 1%. It has been found, however, that higher levels of N doping ($\geq 10^{14}$ cm^{-3}) give rise to defect levels in the bandgap due to the formation of N-N molecular complexes. The corresponding emission is at 590 nm (yellow). Red emission is obtained from GaP simultaneously doped with Zn and O, which occupy, respectively, substitutional Ga and P sites in the lattice. Oxygen as an impurity in GaP gives rise to a deep donor with a binding energy of 0.8 eV. However, when Zn and O atoms occur as nearest neighbors in the lattice, they form an isoelectronic trap with a binding energy of $\sim$ 0.3 eV. The bound exciton associated with this trap level recombines to produce emission at λ = 690 nm.

The ternary $GaAs_{1-x}P_x$ changes from direct to indirect bandgap material for x = 0.45, at which composition $\mathcal{E}_g$ = 2.1 eV. Thus, the direct bandgap and indirect bandgap materials with N doping can be used for the fabrication of diodes to emit red ($GaAs_{0.6}P_{0.4}$), orange ($GaAs_{0.35}P_{0.65}$: N), and yellow ($GaAs_{0.15}P_{0.85}$: N) light.

For longer wavelengths in the near-infrared region of the spectrum, InP-based compounds such as the ternary $In_{0.53}Ga_{0.47}As$ and the quaternaries $In_xGa_{1-x}P_yAs_{1-y}$ are important. These are newer materials, and studies are in progress to fully understand their radiative properties. Devices made of these materials operate in the 1.1–1.6 μm range, which overlaps with the spectral region of low loss and minimum dispersion in optical fibers. They are therefore important for optical-fiber communication. For the far-infrared region of the spectrum, the antimonides such as GaSb, InSb, and their alloys will eventually be important. However, the use of these compounds is hampered by the lack of suitable lattice-matched substrates. Similarly at the other end in the range of bandgaps, the binary gallium nitride (GaN) ($\mathcal{E}_g$ = 3.5 eV) is extremely important for emission of blue light, which would be useful for satellite communication. However, doping the material p-type is a problem. Other promising semiconductor materials that are being developed for making blue LEDs are the IV–IV compound silicon carbide (SiC) and the II–VI compounds. The materials and device technology of SiC are difficult due to its mechanical and chemical properties. The high melting point makes growth of the compound difficult. Doping with Al, B, Be, and Sc give rise to emission in various parts of the spectrum and currently yellow-emitting LEDs are commercially available. Diamond as an LED material also has considerable potential. The II–VI compounds such as ZnS and ZnSe have promise as visible LED materials, but p-type doping of these compounds still eludes success. Similarly compounds such as CdTe, HgTe, ZnTe, and their ternary derivatives with which heterojunctions can also be formed are under intensive study and will be applicable for commercial visible LEDs.

5.4 DEVICE CONFIGURATION AND EFFICIENCY

The schematic of a typical surface-emitting LED is shown in Fig. 5.2. The junction is usually formed by diffusion, or the whole structure is grown by epitaxial techniques. The device is usually designed such that most of the radiative recombination takes place in the side of the junction nearest the surface whereby the probability of reabsorption is greatly reduced. This is ensured by arranging most of the current flowing across the diode to be carried by those carriers (type) that are injected into the surface layer. As shown in the figure, such conditions are achieved in an n^+-p junction where the p-layer is closer to the surface.

The processes occurring in a junction LED can be divided into three stages. The first is the *excitation* or *injection* process, in which the energy of carriers is raised by forward bias injection. Next is the *recombination* process, during which most of these carriers give up their excess energy as photons. Finally, the generated photons must leave the semiconductor and provide the desired optical stimulus to the eye or produce a photocurrent in a detector. This is the *extraction* process. Each of these processes have a characteristic efficiency, and the overall device efficiency, η_o, may then be expressed as

$$\eta_o = \eta_{in}\eta_r\eta_e \tag{5.3}$$

where η_{in}, η_r, and η_e respectively are the injection, radiative recombination, and extraction efficiency. η_o can also be termed the external conversion efficiency.

5.4.1 Injection Efficiency
In an asymmetric n^+-p junction diode, such as the devices shown in Fig. 5.2, electron injection is much more dominant than hole injection. For small carrier diffusion

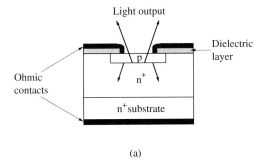

(a)

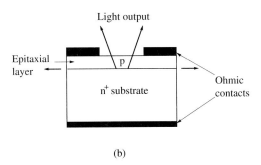

(b)

Figure 5.2   Planar surface-emitting LED structure made by (a) diffusion and (b) epitaxial growth.

lengths, which is usually the case in most III–V compounds, it is almost impossible to determine the injection efficiency experimentally. However, η_{in} can be calculated from junction theory seen in Chapter 4. The fraction of the total diode current that is carried by diffusion of electrons being injected into the p-side of the junction is given by

$$\eta_{in} = \frac{\frac{D_e n_{PO}}{L_e}}{\frac{D_e n_{PO}}{L_e} + \frac{D_h p_{NO}}{L_h}} \tag{5.4}$$

from which we get

$$\eta_{in} = \left(1 + \frac{D_h L_e p_{NO}}{D_e L_h n_{PO}}\right)^{-1} \tag{5.5}$$

where the terms are as defined in Chapter 4.

Using the Einstein relation and the equality $n_o p_o = n_i^2$, Eq. 5.5 becomes

$$\eta_{in} = \left(1 + \frac{\mu_h p_{PO} L_e}{\mu_e n_{NO} L_h}\right)^{-1} \tag{5.6}$$

where, for full ionization $p_{PO} = N_A$, the acceptor doping on the p-side, and $n_{NO} = N_D$, the donor doping on the n-side. In most III–V compounds $\mu_e \gg \mu_h$ and L_e is of the same order as L_h. Under these conditions $\eta_{in} \cong 1$. From Eq. 5.6 it is evident that a high-injection efficiency can be ensured by making $N_D \gg N_A$ (i.e., an asymmetric junction). It should be noted that, in calculating the injection efficiency the recombination in the space charge (depletion) layer of the junction is neglected. From Shockley-Read-Hall theory the current due to space charge recombination is given by

$$I_{rec} \;=\; \frac{qW}{2}s\vartheta_{th}N_T n_i A\left(e^{qV_f/2k_BT} - 1\right) \tag{5.7}$$

where s and N_T are, respectively, the capture cross section and density of the recombination center or traps that are present, $\vartheta_{th} = (3k_BT/m^*)^{1/2}$ is the thermal velocity of the carriers, n_i is the intrinsic carrier concentration, A and W are the area and depletion-layer width of the diode, respectively, and V_f is the applied forward bias. In small bandgap materials the recombination current is negligible due to a lower density of deep-level traps. Note that the product $(s\vartheta_{th}N_T)$ is equal to the inverse of the lifetime of the deep-level recombination center. As the bandgap increases this component of current density increases and can be as high as 10^{-2}A/cm^2 in large bandgap materials. Also, under certain special circumstances, such as in the presence of a large density of interface traps caused by impurities, or dislocations as in a lattice-mismatched heterojunction, space charge recombination may represent a significant shunt path at low forward biases. In general, L_e and L_h are functions of N_A and N_D, respectively. For example, in GaAs, L_e and L_h are fairly constant for N_A, $N_D \leq 10^{18}$ cm^{-3}, but for higher doping, L_e and L_h are inversely proportional to the respective dopings. In GaAs, values of $\eta_{in} \sim 0.8$ can be easily achieved, while that in GaP ranges from 0.6 to 0.8. However, η_{in} is not important in determining η_o, since its value is higher than the other two efficiencies in Eq. 5.3 and a moderate variation in its value does not have a significant effect on the value of η_o.

5.4.2 Recombination Efficiency

The operation of the junction LED relies on the spontaneous emission of photons and the absence of any optical amplification process through stimulated emission, as in a laser, limits the *internal quantum efficiency* (η_i) or radiative efficiency (η_r). The latter is defined as the ratio of the number of photons generated to the number of injected electrons. The radiative efficiency will increase if the possible nonradiative processes can be eliminated. The main nonradiative processes that are operative in an LED structure are Shockley-Read-Hall recombination, band-to-band and trap-assisted Auger recombination, and recombination via surface states.

In direct bandgap materials the radiative recombination efficiency is usually $\sim$ 50% for homojunction devices. In indirect bandgap semiconductors this efficiency is usually very low unless, as we have seen earlier, recombination can take place via an impurity level. In order to achieve a reasonable radiative efficiency in an indirect semiconductor with a substitutional impurity, carriers must be tightly bound to the core of the impurity atom. This requires a deep level in the energy gap and a transition related to such a level will be at a wavelength much larger than that corresponding to the energy gap. For III–V compounds such as GaP or GaAsP, this implies emission in the near-infrared or infrared parts of the spectrum. The best measured values of η_r for GaP: Zn, O, and GaP: N are 30% and 3%, respectively.

Nonradiative surface recombination is more important in direct bandgap materials than in indirect bandgap materials. However, with the use of double heterojunction (DH) structures, as in lasers, the radiative efficiency or internal quantum efficiency can be as high as 60–80%.

5.4.3 Extraction Efficiency and External Conversion Efficiency

It is ultimately the amount of light emerging from the device that determines the overall external efficiency of the device. This efficiency can be low in spite of high values of injection and radiative efficiencies if the light is not extracted efficiently. Several factors determine the extraction efficiency. The absorption coefficient of the semiconductor at the emission wavelength plays an important role. This is dictated, to a large extent, by whether the material has a direct or indirect bandgap. The absorption and luminescence spectra of GaAs and GaP: Zn,O are approximately shown in Figs. 5.3(a) and (b). In GaAs the absorption coefficient at the peak of the output is ~ 10^3 cm^{-1}, which implies that after passing through as little as 2 μm of material about

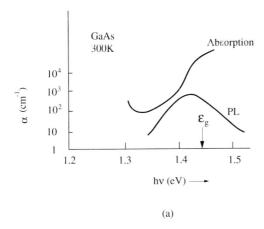

(a)

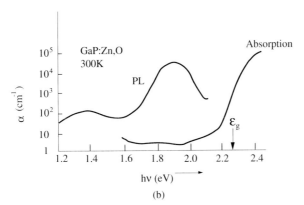

(b)

Figure 5.3 Apsorption and luminescence spectra of (a) GaAs and (b) GaP:Zn,O at room temperature.

half of the emitted radiation is reabsorbed, since $\phi(x) = \phi_o e^{-\alpha x} = \phi_o e^{-0.6} \cong \phi_o/2$ for $\alpha = 3 \times 10^3$ cm^{-1}. This may be contrasted with the case of GaP: Zn,O in which $\alpha \simeq 3$ cm^{-1} at the peak of the emission spectrum and the reabsorption in a few microns of material is negligible. The obvious answer seems to be to place the light-emitting junction very close to the surface in a direct bandgap device. Unfortunately, surface states will reduce the radiative recombination efficiency as the junction approaches the surface. As we shall see in the next section, the use of a heterojunction alleviates this problem to a large extent.

Another cause for the lowering of the external quantum efficiency is the radiation geometry of the LED, which emits through a planar surface. The intensity distribution of the radiation is Lambertian, as shown in Fig. 5.4(a), in which the power radiated from a unit area into a unit solid angle, expressed in units of W sr^{-1} m^{-2} is constant in all directions. However, most of the emitted radiation strikes the semiconductor-air interface at angles greater than the critical angle, θ_c, and so remains trapped by total internal reflection. The high dielectric constant (refractive index) of compound semiconductors makes the critical angle very small. The critical angle is given by

$$\theta_c = \sin^{-1}\left(\frac{n_{r2}}{n_{r1}}\right) \tag{5.8}$$

where n_{r2}(air) = 1 and n_{r1} is the refractive index of the semiconductor. Therefore light originating near the junction will be radiated isotropically, of which only that

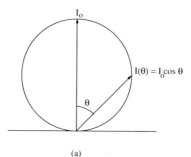

(a)

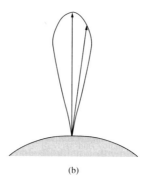

(b)

Figure 5.4   Intensity distribution of radiation from a light-emitting diode: (a) Lambertian and (b) directional.

within a cone of semi-angle θ_c will leave the device surface. It can be shown that the transmission F_T (light radiated to the total light generated) at the semiconductor-air interface is approximately given by (Problem 5.2):

$$F_T = \frac{1}{4} \left(\frac{n_{r2}}{n_{r1}} \right)^2 \left[1 - \left(\frac{n_{r1} - n_{r2}}{n_{r1} + n_{r2}} \right)^2 \right] \tag{5.9}$$

Two techniques that can be used to increase F_T are shown in Fig. 5.5. In the first, shown in Fig. 5.5(a), the semiconductor surface is shaped as a hemisphere. This is, however, expensive and impractical for most applications. The second, and cheaper technique, shown in Fig. 5.5(b), is to use dielectric encapsulation shaped as a dome with a transparent material of high refractive index. For example, with $n_{r1} = 3.6$

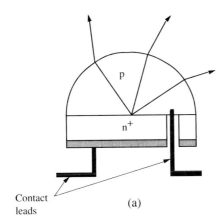

(a)

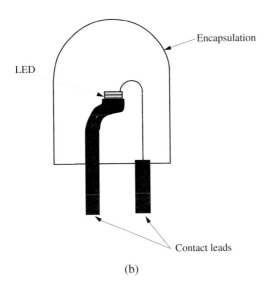

(b)

Figure 5.5 Two types of LED construction to enhance light transmission from device: (a) hemispherically shaped device surface and (b) encapsulated device or dome LED.

and $n_{r2} = 1.6$, a fivefold increase in light output is obtained over the planar semi-conductor/air interface. These are called *dome LEDs*, and are useful for high-volume, low-price applications such as intruder alarms, displays, calculators, etc. Some planar and dome LEDs are also used in optical communication in the near-infrared with fiber bundles. The radiation patterns from these structures are much more directional, as shown in Fig. 5.4(b).

The external power efficiency, or conversion efficiency, η_o, is defined as the ratio of the optical power output P_o to the electrical power input P_e. In other words,

$$\eta_o = \frac{P_o}{P_e} \times 100\%. \tag{5.10}$$

It may be noted that the losses encountered in the extraction process is included in η_o. Typical values for η_o in common LEDs are in the range of 1% to 5%.

Finally, in an LED for daily and common applications it is important to take into account the physiological impact of the device. For this purpose it is more appropriate to express the performance in photometric units, which normalizes the output spectrum of the device with the eye response. Thus, the *luminous efficiency*, or *luminosity*, is expressed as

$$\eta_L \propto \frac{\int_\lambda V(\lambda) P_o(\lambda) d\lambda}{\int_\lambda P_o(\lambda) d\lambda} \tag{5.11}$$

where $P_o(\lambda)$ is the LED emission spectrum and $V(\lambda)$ is the eye response shown in Fig. 5.6. It is apparent that a diode emitting at the peak of the eye response in the green or yellow will appear much brighter than a device emitting an equal amount of energy in the red or blue. As an example, a green-emitting LED ($\lambda = 515$ nm) will appear nearly 30 times brighter than an LED emitting in the red ($\lambda = 630$ nm) but having the same quantum efficiency. Taking the eye response into account

$$\eta_o = \eta_{in} \eta_r \eta_e \eta_L \tag{5.12}$$

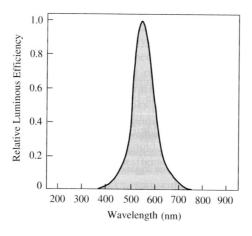

Figure 5.6 Approximate normalized spectral response of the human eye, also known as the photopic visual curve.

5.4.4 Coupling Loss

In most practical devices the light output is coupled into a fiber. This process introduces a further *coupling loss*. Consider the geometry shown in Fig. 5.7. The acceptance angle θ_a of a step index fiber with core index n_{r1} and cladding index n_{r2} ($n_{r1} > n_{r2}$) is given by (Problem 5.3):

$$\theta_a = \sin^{-1}(n_{r1}^2 - n_{r2}^2)^{1/2} = \sin^{-1}(A_n) \tag{5.13}$$

where A_n is the numerical aperture of the fiber. It is assumed that all the light incident on the exposed end of the fiber core within the angle θ_a is coupled by internal reflection. Also, for a Lambertian distribution of the radiated intensity,

$$\Im(\theta) = \Im_o \cos\theta \tag{5.14}$$

Considering a device that is smaller than, and in close proximity to, the fiber core, and assuming a cylindrical symmetry, the *coupling efficiency* η_c can be expressed as

$$\eta_c = \frac{\int_o^{\theta_a} \Im(\theta) \sin\theta d\theta}{\int_o^{\pi/2} \Im(\theta) \sin\theta d\theta} \tag{5.15}$$

which, by substitution of Eq. 5.14 becomes

$$\eta_c = \frac{\int_o^{\theta_a} \Im_o \sin 2\theta d\theta}{\int_o^{\pi/2} \Im_o \sin 2\theta d\theta}$$

$$= \sin^2\theta_a. \tag{5.16}$$

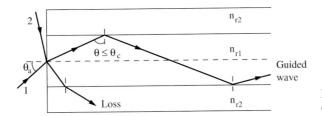

Figure 5.7 Geometry of light coupling into an optical fiber.

EXAMPLE 5.1

Objective.   To determine the coupling characteristics of a silica optical fiber with core and cladding refractive indices of 1.48 and 1.50, respectively.

From Eq. 5.8, the critical angle θ_c is equal to 89.6°. From Eq. 5.13 $A_n = 0.244$. Therefore, the acceptance angle of the fiber, $\theta_a = 15.7°$.

5.5 LIGHT OUTPUT FROM LED

We will now calculate the amount of light emitted by an LED due to forward-bias injection. The photons are emitted, with respect to Fig. 3.2(a), due to the recombination of an electron at energy $\mathcal{E}(\mathbf{k}'') = \mathcal{E}_2$ with a hole at energy $\mathcal{E}(\mathbf{k}') = \mathcal{E}_1$. We assume that $\mathbf{k}'' = \mathbf{k}' = \mathbf{k}$ and $(\mathcal{E}_2 - \mathcal{E}_1) = h\nu = \mathcal{E}$, the energy of the emitted photon. From Eqs. 3.28 and 3.29, we can write

$$k^2 = \frac{2m_r^*}{\hbar^2}(\mathcal{E} - \mathcal{E}_g) \tag{5.17}$$

where m_r^* is the reduced mass given by Eq. 3.36. Therefore, near the bottom of the conduction band

$$\mathcal{E}_2 = \mathcal{E}_g + \frac{m_r^*}{m_e^*}(\mathcal{E} - \mathcal{E}_g) \tag{5.18}$$

Emission of a photon involves the density of states in both the conduction and valence bands. Conservation of energy and momentum forces the photon to interact with electrons and holes of specific energy and momentum. This interaction is described by the *joint density of states*, $N_J(\mathcal{E})$. Equation 5.18 describes a one-to-one correspondence between $\mathcal{E}_2$ and the photon energy $\mathcal{E}$ and allows us to write the incremental relation

$$N(\mathcal{E}_2)d\mathcal{E}_2 = N_J(\mathcal{E})d\mathcal{E} \tag{5.19}$$

from which it follows that

$$N_J(\mathcal{E}) = \frac{m_r^*}{m_e^*}N(\mathcal{E}_2) \tag{5.20}$$

Substitution of Eq. 2.53 for $N(\mathcal{E}_2)$ leads to

$$N_J(\mathcal{E}) = \frac{(2m_r^*)^{3/2}}{2\pi^2\hbar^3}(\mathcal{E} - \mathcal{E}_g)^{1/2} \quad ((eV)^{-1}.cm^{-3}) \tag{5.21}$$

for parabolic electron-hole bands. The spontaneous emission rate $r_{sp}(\mathcal{E})$ and total spontaneous emission rate R_{sp} per unit volume, taking the selection rule into account, are defined by

$$r_{sp}(\mathcal{E}) = P_{em}N_J(\mathcal{E})f_n(\mathcal{E}_2)[1 - f_p(\mathcal{E}_1)] \quad (s^{-1}.(eV)^{-1}.cm^{-3})$$

$$R_{sp} = \int_{\mathcal{E}} r_{sp}(\mathcal{E})d\mathcal{E} \quad (s^{-1}.cm^{-3}) \tag{5.22a}$$

or in an alternate form[†]

$$R_{sp} = \int d\mathcal{E} \frac{n_r q^2 \mathcal{E}}{3\pi\epsilon_o m_o^2 \hbar^2 c^3} \left\{ \int \frac{1}{(2\pi)^3} d^3k\, p_{CV}^2 \delta(\mathcal{E}_2(\mathbf{k}) - \mathcal{E}_1(\mathbf{k}) - \mathcal{E}) \right.$$

$$\left. \times f_n(\mathcal{E}_2)[1 - f_p(\mathcal{E}_1)] \right\} \tag{5.22b}$$

[†]J. Singh, *Physics of Semiconductors and Their Heterostructures*, McGraw-Hill, New York, 1993.

where the integral over d^3k assures that all occupied electron and hole states are included and the δ-function assures that energy is conserved. The integration therefore gives the joint density of states, and Eqs. 5.22(a) and (b) are equivalent. The quantity P_{em} is the emission probability given by

$$P_{em} = \frac{n_r q^2 \mathcal{E} p_{CV}^2 (1 + u_{\mathcal{E}})}{3\pi \epsilon_o m_o^2 \hbar^2 c^3} \tag{5.23}$$

from Eqs. 3.42–3.44. P_{em} also includes the factor $n_{\mathcal{E}}$, which is the photon mode density, as defined in Appendix 4. It should also be noted that the emission probability has a factor $(1 + u_{\mathcal{E}})$ where $u_{\mathcal{E}}$ is the number of photons per mode, or the photon occupation, as defined in Appendix 4. In the case of spontaneous emission, where the photons created by recombination escape, $u_{\mathcal{E}}$ is very small and the factor is unity. For example, in GaAs, where $2p_{CV}^2/m_o = 23$ eV, and assuming $\mathcal{E} = 1.5$ eV, we get $P_{em} = 1.8 \times 10^9 s^{-1}$. Thus, the recombination lifetime $\tau_r = 1/P_{em} = 0.6$ ns. The Fermi functions in Eq. 5.22(a) and (b) ensure that a state in the conduction band is filled and a state in the valence band is empty. The distribution functions $f_n(\mathcal{E}_2)$ and $f_p(\mathcal{E}_1)$ are defined by Eqs. 3.1 and 3.2 and involve the electron and hole quasi-Fermi levels $\mathcal{E}_{fn}$ and $\mathcal{E}_{fp}$, respectively. For example, under thermal equilibrium conditions, the quasi-Fermi levels merge into a single Fermi level within the bandgap. The equilibrium distribution can then be used. Furthermore, if the Boltzmann approximation is valid, then $f_n(\mathcal{E}_2)[1 - f_p(\mathcal{E}_1)] \cong \exp\left(-\frac{\mathcal{E}}{k_B T}\right)$. Under these conditions,

$$r_{sp}(\mathcal{E}) = \frac{(2m_r^*)^{3/2}}{2\pi^2 \hbar^3 \tau_r} (\mathcal{E} - \mathcal{E}_g)^{1/2} e^{-\frac{\mathcal{E}_g}{k_B T}} e^{-\left(\frac{\mathcal{E} - \mathcal{E}_g}{k_B T}\right)} \tag{5.24}$$

Next consider the case of weak injection in an LED, such that the quasi-Fermi levels are still several $k_B T$ away from the bandedges and are within the bandgap. The distribution functions are then given by the exponential approximations of Eqs. 3.3 and 3.4. Under these conditions

$$r_{sp}(\mathcal{E}) = \frac{(2m_r^*)^{3/2}}{2\pi^2 \hbar^3 \tau_r} (\mathcal{E} - \mathcal{E}_g)^{1/2} e^{\left(\frac{\mathcal{E}_{fn} - \mathcal{E}_{fp} - \mathcal{E}_g}{k_B T}\right)} e^{\left(\frac{\mathcal{E} - \mathcal{E}_g}{k_B T}\right)} \tag{5.25}$$

A plot of this function is shown in Fig. 5.8. The total photon flux is obtained by

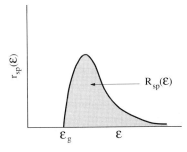

Figure 5.8 Spectral distribution of the injection electroluminescent rate. Also see Problem 5.11.

integration of Eq. 5.25:

$$\Phi_o = V_l \int_o^\infty r_{sp}(\mathcal{E})d\mathcal{E}$$

$$= \frac{V_l}{\sqrt{2}\hbar^3 \tau_r} \left(\frac{m_r^*}{\pi}\right)^{3/2} (k_B T)^{3/2} e^{\left(\frac{\mathcal{E}_{fn}-\mathcal{E}_{fp}-\mathcal{E}_g}{k_B T}\right)} \left(s^{-1}\right) \qquad (5.26)$$

where V_l is the active region volume. In performing this integration P_{em}, a function of photon energy $\mathcal{E}$, has been replaced by $1/\tau_r$. This is because the effect of the value of $\mathcal{E}$ is small compared to the change in $(\mathcal{E} - \mathcal{E}_g)$. When the injection level is increased, the quasi-Fermi levels move into the bands and the probability function $f_n(\mathcal{E}_2)[1 - f_p(\mathcal{E}_1)]$ increases. As a consequence the light output increases.

The responsivity, $\mathcal{R}$, of an LED is defined as the ratio of the emitted optical power $P_o = \Phi_o h\nu$ to the injection current I. In other words,

$$\mathcal{R} = \frac{\Phi_o h\nu}{I} = \eta_o \frac{h\nu}{q}$$

$$= \frac{1.24\eta_o}{\lambda(\mu m)} \ (W/A) \qquad (5.27)$$

EXAMPLE 5.2

Objective.   To calculate the responsivity of an LED with $\eta_o = 3\%$ at $\lambda = 1\mu$m.

$$\mathcal{R} = 0.03 \times 1.24$$

$$= 37.2 \ \mu W/mA$$

According to Eq. 5.27 the output power is proportional to drive current. In this range the responsivity has a constant value. As the injection current is increased, saturation effects take over, and the slope decreases.

5.6 LED STRUCTURES

The structures being discussed here are used in a variety of applications where both low and high radiance are required.

5.6.1 Heterojunction LED

The devices shown in Fig. 5.2 and discussed until now are simple homojunctions. There are two main problems with this structure, which limit the internal quantum efficiency. First, the surface states on the p-layer close to the light-emitting junction give rise to nonradiative recombination. If the surface is far removed from the junction, then the probability of reabsorption is enhanced. Also, the electrons injected from the n^+ region into the p-region as minority carriers diffuse away from the junction and gradually recombine with the majority carriers. Thus, the same number of photons are created in a greater volume of material, thereby reducing the internal quantum efficiency and enhancing reabsorption. Both these problems can be solved in a double heterostructure diode, whose schematic and band diagram are shown in Fig. 5.9. The advantages of this configuration are obvious. The GaAs layer into which electrons are injected from the top n^+ AlGaAs layer can be kept thin (~ 1000–2000 Å), and these electrons are contained within this thickness by the potential barrier of the p-GaAs/p-AlGaAs heterojunction. The top AlGaAs layer serves several important functions. First, the interface state density at this lattice-matched junction is orders of magnitude lower than the free surface state density, thereby drastically reducing nonradiative recombination. Second, being of a larger bandgap, it acts as a *window* layer in which the emitted photons, with energy close to the GaAs bandgap, are not reabsorbed. Finally, the injection efficiency is higher. These factors help to increase both the internal and external quantum efficiencies. Such heterostructure LEDs are usually realized by epitaxial growth and have also been made with InP-based materials such as InGaAsP/InP or InGaAs/InGaAsP/InP, which emit at longer wavelengths.

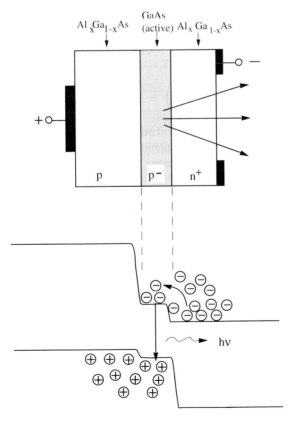

Figure 5.9 Device structure and band diagram of a forward-biased double heterostructure (DH) LED.

5.6.2 Burrus Surface-Emitting LED

This structure was conceived and realized by Burrus and Dawson in order to obtain high radiance and efficient coupling of the emitted light to a fiber. It is essentially a double heterostructure, and is shown schematically in Fig. 5.10. The bottom p$^+$ GaAs and top n$^+$ GaAs layers are included for the realization of low-resistance ohmic contacts. However, to avoid reabsorption of the emitted radiation in the top n-GaAs layer, a deep well is etched to reach the top n-AlGaAs layer. This can be done conveniently by a selective etchant that etches GaAs but not AlGaAs. The well is also used to support the fiber, which is butt-coupled to the device and is held in place with an epoxy resin of appropriate refractive index to enhance the external power efficiency of the device. The thin SiO$_2$ layer at the back isolates the contact layer from the gold heat sink. Photons are generated in the thin p-GaAs region and emission from the top surface is ensured by the heterostructure and reflection from the back crystal face. Thus, the forward radiance of these devices is very high. The top n-GaAs contact layer ensures low contact resistance and thermal resistance, thereby allowing for high current densities and high radiation intensity. The fiber is properly aligned to optimize coupling of the emitted radiation. Nonetheless, there is some loss due to the Lambertian distribution of the radiation intensity and the coupling efficiency is typically 1%–2%.

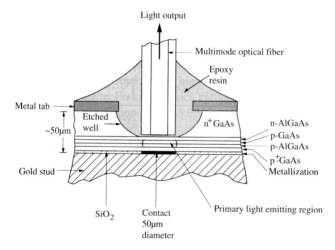

Figure 5.10   The structure of an AlGaAs DH surface-emitting LED (Burrus type) (from C. A. Burrus and B. I. Miller, *Opt. Commun.*, **4**, 307, 1971).

Several other types of lens coupling are used with surface-emitting heterostructures to improve the coupling efficiency. These are schematically shown in Figs. 5.11 (a)–(c). Figure 5.11(a) shows the end of a fiber that is polished into a spherical lens. This collects and collimates the divergent radiation from the LED. The same purpose is served by a glass microlens inserted between the device and the fiber, as shown in Fig. 5.11(b). An integrated lens structure is obtained by etching and polishing the top surface layer of the LED into a spherical lens, as shown in Fig. 5.11(c). This procedure does away with a lot of cumbersome fixing and alignment, but is difficult to implement. It is important to realize that in spite of the problems associated with large

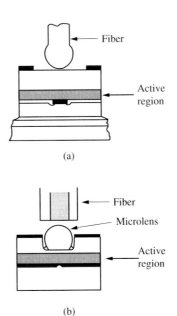

(a)

(b)

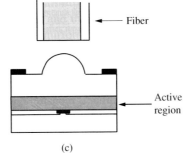

Figure 5.11 Different schemes of lens-coupled LEDs: (a) spherical-ended fiber-coupled device, (b) microlens coupling, and (c) integrated semiconductor lens structure.

(c)

divergence and coupling of the radiant beam, surface-emitting structures will remain important not only for large-volume, low-cost applications, but also for short-distance chip-to-chip communication where surface-emitting sources are essential. For optical computing applications also, surface-emitting sources may play an important role. With the different techniques of lens coupling discussed above, coupling efficiencies of 5%–15% have been achieved.

5.6.3 Guided Wave or Edge-Emitting LED

The surface-emitting LEDs just described are useful for common and less demanding applications. However, in optical communication, where tight coupling of the emitted light to a fiber or waveguide is required, a more collimated light emission is desirable. The basic edge-emitting LED is shown in Fig. 5.12. Generally for superradiant LEDs, a stripe geometry is incorporated, as in a junction laser, to improve the injection efficiency. The active layer is usually lightly doped or undoped,

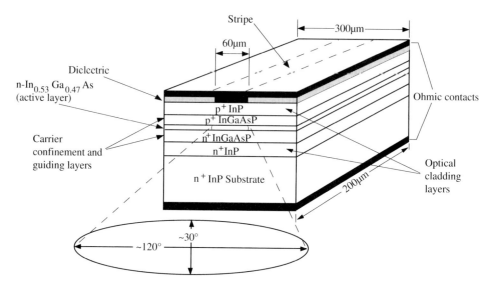

Figure 5.12 Schematic illustration of the structure of a stripe geometry DH InGaAs/InGaAs/InP edge-emitting LED.

and a very large population of carriers for recombination is created in this region by forward-bias injection. The two InGaAsP layers on both sides serve as carrier confinement layers, on the outer sides of which are doped InP layers. These serve as the *cladding* layers, and the region in between forms an optical waveguide. Thus, the wide bandgap layers serve the dual purpose of optical and carrier confinement. The photons are generated in the very thin active region and spread into the guiding layers, without reabsorption, because of their larger bandgaps. It is important to realize that the edge-emitting structure allows a thin active region without sacrificing the radiative recombination efficiency, in comparison with a homojunction surface-emitting LED. The stripe geometry, made by selective metallization on the top surface through a window opened in an SiO$_2$ layer, allows higher carrier injection densities for the same drive current. In other words, the power conversion efficiency is improved.

It is clear that the edge-emitting LED is very similar to a junction laser. However, it does not have a feedback cavity and is still an incoherent light source. Most of the light is made to come out of one edge of the structure. This is ensured by cleaving and putting a reflective coating at the nonemitting end and putting an antireflective coating at the emitting end. The waveguide vastly reduces the divergence of the emitted radiation. In the direction perpendicular to the plane of the layers, the half-power width is approximately 30°. However, in the plane of the layers the output is still Lambertian and the half-power width is approximately 120°. Optical confinement in the transverse direction can be created by selective etching and regrowth of a higher bandgap material. This technique is more common in laser fabrication and will be discussed in Chapter 7. The waveguiding and reduced beam divergence allows more efficient coupling of the radiated beam into fibers. The larger operating current density in a smaller structure can cause heat-sinking problems. However, it is anticipated that edge-emitting LEDs with InP-based materials will play an important role as high-power sources in short-distance optical communication, where dispersion in the fibers will not be too severe.

5.6.4 Drive Circuitry

The LED is operated under forward-bias injection conditions. The amount of forward bias depends on the forward cut-in voltage, which, in turn, depends on the bandgap of the semiconductor. Typical biases are therefore between 1 and 2V. Under operating conditions the current drawn by the LED is $\sim$ 100 mA. An LED can be operated by a direct current (dc) or alternating current (ac) drive circuit. A simple dc drive circuit is shown in Fig. 5.13(a), where the current through the diode provided by the voltage source V is limited by the series resistance R_S. Under operating conditions, the voltage drop across the LED is V_L, the operating voltage of the device. If the operating current is I_L, then the circuit can be described by

$$V - V_L = I_L R_S \tag{5.28}$$

It may often be desirable or necessary to place an LED in an ac circuit. There can be several kinds of circuits, depending on the application, and a common form is shown in Fig. 5.13(b). Here a diode with reversed polarity is placed in parallel with the LED to prevent it from high reverse bias voltages. A capacitor and current-limiting resistor are in series with the two diodes. In this circuit, for current of both polarities the voltage drop across the diode is restricted to a small value. Then the current flowing through the circuit is approximately the supply voltage divided by the impedances due to the resistance and capacitance. The series resistor also prevents the LED from high turn-on transients.

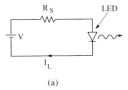

(a)

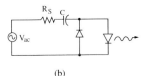

(b)

Figure 5.13 Simple (a) dc and (b) ac drive circuits for light-emitting diodes.

For most communication applications a high-frequency signal is impressed upon the cw output of the light source. This is usually done by internal or external modulation. In external modulation, a modulator is placed in the path of the cw signal from the source, and the signal coming in as an electrical input varies the transmission of the modulator at a high speed. The properties of such optical modulators will be discussed in Chapter 11. Here we will discuss internal modulation, which is usually achieved by varying the drive current of the LED with the signal. Hence it is also called current modulation. In Fig. 5.14 a simple modulation circuit is shown, in which the signal is applied to the base circuit of an n-p-n transistor connected in series with the LED and a current-limiting resistor. It is advantageous to use the transistor because of the gain it provides. The variation of the base current with the signal varies the current flow and thus the forward injection of the LED. A modulated light output is obtained from the diode. The circuit is usually operated such that both the LED and the transistor operate in their linear output regimes. Under these conditions changes in the current flowing through the LED are directly proportional to input signal voltage in the base circuit.

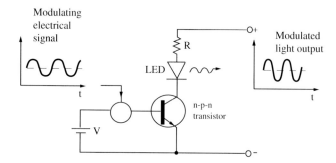

Figure 5.14 Current (internal) modulation circuit for light-emitting diode.

5.7 DEVICE PERFORMANCE CHARACTERISTICS

Once an LED is fabricated, measurements are made to determine its electrical and optical characteristics. For commercial diodes, these characteristics are sometimes provided by the manufacturer. A typical set of performance curves for stripe-geometry AlGaAs and InGaAsP double heterojunction LEDs are shown in Fig. 5.15.

5.7.1 Spectral Response
The spectral response of the AlGaAs LED is shown in Fig. 5.15(a) and is the plot of the relative output intensity against the wavelength in nanometers. The peak intensity is approximately at 835 nm. The spectral linewidth, measured in units of wavelength between the half maximum intensity points—termed the full width at half maximum (FWHM)—is around 40 nm. This is typical of LEDs operating around 0.8 μm. In LEDs emitting at longer wavelengths (1.1–1.6 μm) the material used is typically InGaAsP/InP, and the linewidth of the spectral response is broadened by alloy scattering. The linewidth is also shifted and broadened by heavy doping in the active region. Therefore, in surface-emitting LEDs where the active region is more heavily doped, the emission wavelength moves to longer wavelengths (lower

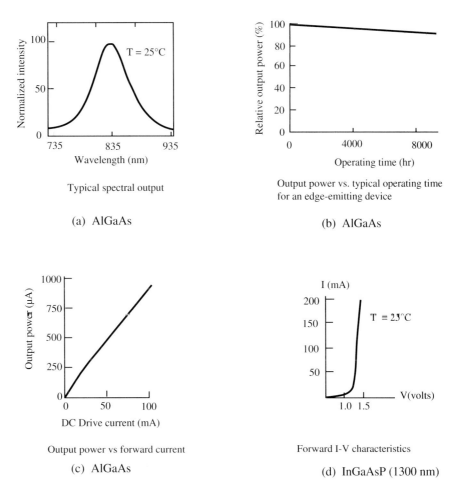

Figure 5.15 Typical performance curves of AlGaAs and InGaAsP light-emitting diodes (courtesy of D. Pooladdej, Laser Diode Inc., New Brunswick, NJ).

energies) and the spectral response curve is broadened. Fluctuations in temperature also affect the response characteristics. The bandgap of most III–V semiconductors decrease at a rate of 5 meV/degree increase of temperature. The linewidth of the spectrum also increases with increase of temperature due to a larger energy spread of the carrier distribution and phonon scattering. Therefore, for applications where the emission wavelength and linewidth need to remain invariant, heat sinking and cooling arrangements are incorporated with the device.

5.7.2 Output Power-Time Characteristics

The lifetime of an electroluminescent device or a source for communication applications is an important parameter. Therefore, curves such as the one shown in

Fig. 5.15(b) are important. This curve, for the AlGaAs LED, shows a power loss of a few percent after operating for 8,000 hours.

5.7.3 Light(Power)-Current Characteristics

These are perhaps the single most important characteristics of an LED. The characteristics of an edge-emitting LED is shown in Fig. 5.15(c). The linearity of the light-current characteristics is important for modulation in analog transmission. Some LEDs can exhibit nonlinearity depending on material properties and device configuration. It is also important to note that edge emitters radiate less optical power into air than surface emitters, although the radiation intensity at the emitting face can be very high in the former. This is because interfacial recombination and reabsorption can play a more dominant role in edge-emitting LEDs. However, because of the higher radiation intensity from them, coupling into fibers with smaller numerical apertures is more efficient. The light-current characteristics shown in Fig. 5.16 illustrate the above points.

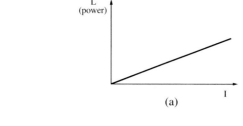

(a)

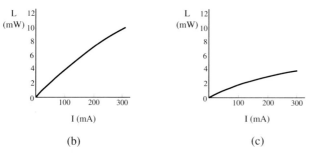

(b)

(c)

Figure 5.16   (a) Schematic illustration of linear light-current characteristics of LED, (b) an AlGaAs surface emitter with a 50μm diameter dot contact, and (c) and AlGaAs edge emitter with a 65 μm wide stripe and 100 μm length. (D. Botez and M. Ettenburg, *IEEE Trans. Electron Devices*, **ED-26(3)**, 1230, ©1979 IEEE).

5.7.4 Diode Current-Voltage Characteristics

These describe the electric operating characteristics of the device. Typical current-voltage characteristics are shown in Fig. 5.15(d). For efficient operation it is important to know the forward threshold voltage, the reverse breakdown voltage and the reverse leakage current at breakdown. It may be noted that the reverse breakdown voltage (5–15V) in LEDs is not very high, since the junction is usually formed of heavily doped materials.

5.8 FREQUENCY RESPONSE AND MODULATION BANDWIDTH

The efficient utilization of a communication system calls for minimizing the transmission time. In other words, the maximum amount of information needs to be transmitted in the least possible time. Rapid transmission of signals can be achieved if the signal changes rapidly with time. A common and convenient measure of signal speed is its bandwidth, which is the width of the signal spectrum in frequency domain. High-speed communication calls for wideband signals to represent the information and wideband systems to accommodate the signals. In optical communication a high-frequency lightwave is modulated by the wideband information signal, thereby reducing the *fractional* bandwidth and simplifying equipment design. For everyday commercial applications and in particular as display devices, the response time of an LED to a drive current pulse is not a limiting factor. The response time of most commercial LEDs is $\leq 1\mu s$ and this is more than adequate. However, for communication applications in which high-frequency internal modulation of the LED as a source is desirable, response times of the order of a nanosecond or less are desirable. Therefore it is important to understand the processes that limit the response time and to explore techniques to reduce it.

There are two main factors, extrinsic and intrinsic, that limit the response time of a LED. The extrinsic factor is the junction capacitance of the diode. This capacitance, together with the resistance in the diode circuit, gives rise to a characteristic RC time constant. The capacitance can be decreased by decreasing the diode area and increasing the reverse bias. The latter is related to the doping. The intrinsic limitation to the response time arises from the charge storage and diffusion capacitance of a p-n junction under forward bias. When the applied bias to the diode is modulated at high speed, the stored charge cannot respond. The excess injected charge during forward bias is removed by diffusion and recombination with minority carriers. Therefore, the recombination lifetime in the junction material plays a major role in determining the modulation bandwidth. As shown in Appendix 6, the frequency response is given by

$$r(\omega) = \frac{1}{[1 + \omega^2 \tau^2]^{1/2}} \tag{5.29}$$

where $r(\omega)$ is the response (power output) at angular frequency ω and τ is the minority carrier lifetime in the injected region. It is evident that for good high-frequency response the recombination lifetime τ should be made small. As we have seen in Chapter 3, the value of τ is also related to doping and injection level. For low-level injection into a p-type material, the value of τ (= τ_r if nonradiative recombination is negligible) is given by Eq. 3.23. Thus, τ_r decreases with increasing doping and the best value that can be obtained is approximately 1 ns. The limit is set by the breakdown and leakage current of the diode and the solubility limit of the dopant specie. An alternate approach is to operate in the bi-molecular recombination regime in lightly doped material. For example, for high-level injection into a thin active region, the value of τ_r is expressed by Eq. 3.22. Under these conditions the injected carrier density Δn into the active region is much greater than the equilibrium value. A simple expression for τ_r as a function of the injection current density J can be derived as follows. If J is entirely a recombination current, then the number of recombinations per unit volume per second in an active region of width d is J/qd. This must be equal to $\Delta n/\tau_r$. Therefore, by using Eq. 3.22,

$$\tau_r = \left(\frac{qd}{JB_r}\right)^{1/2} \tag{5.30}$$

Thus, τ_r may be reduced by reducing d and increasing J. However, a very high value of J may lead to heat-sinking problems and distortion of the modulated signal. It should also be remembered that in conventional surface-emitting LEDs with thin active regions, the recombination lifetime is limited by the surface recombination rate. Edge-emitting LEDs with a thin undoped active region, operating in the bi-molecular recombination region promise the best high-frequency response. It is important to note that all the material and device parameters that are to be adjusted for realizing the best high-frequency response are interrelated.

An alternate way of expressing the response of an LED to a high-frequency current drive is the *modulation bandwidth*. This is defined as the frequency at which the output optical power received at a detector is reduced by 3 dB with respect to power transmitted. As the modulation frequency increases, the LED does not emit the total amount of light, since all the injected carriers cannot recombine. If we assume a linear light(power)-current relationship for both the LED and the detector, then we may define the bandwidth as follows:

$$\Delta f(dB) = 10\log_{10}\frac{P_{out}(f)}{P_{out}(dc)}$$

$$= 10\log_{10}\frac{I_{out}(f)}{I_{out}(dc)} \tag{5.31}$$

where I_{out} is the current at the detector. The 3 dB point occurs when the ratio of the currents is equal to 1/2. The bandwidth is schematically shown in Fig. 5.17. It

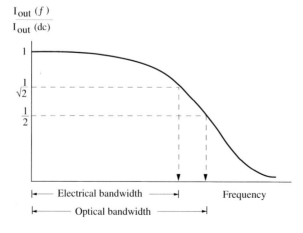

Figure 5.17   Frequency response of a light-emitting diode.

is sometimes also possible to define an electrical 3 dB bandwidth as the ratio of the electrical powers. The 3 dB bandwidth occurs when $[I_{out} (f)/I_{out} (dc)]^2 = 1/2$ or $I_{out} (f)/I_{out} (dc) = 0.707$. Therefore, the electrical bandwidth is smaller than the optical bandwidth.

In conclusion, the modulation bandwidth of an LED depends on the device configuration (surface- or edge-emitting) doping level in the active layer, the lifetime of the injected carriers, and the parasitic capacitance and resistance in the circuit. Finally, it may be remembered that in LEDs the photons are generated by spontaneous recombination, and the radiative lifetime under these conditions is always longer than that under stimulated emission, as in a laser. Therefore, lasers have higher modulation bandwidths. Nevertheless, bandwidths over 1 GHz can now be obtained in practical LEDs. The intensity modulation by variation of the drive current can be applied with both analog and digital signals. The former requires a larger signal-to-noise ratio and is therefore more suited for short-distance applications. Digital modulation has better noise immunity and is therefore more suited for long-distance optical-fiber communication.

5.9 MANUFACTURING PROCESS AND APPLICATIONS

The LED represents the simplest and most robust of the electroluminescent devices. The cross section of a standard red-emitting LED is shown in Fig. 5.18. The device consists of layers of GaAsP grown on a GaAs substrate. Some LEDs (orange-emitting, for example) are grown on GaP substrates. The substrate material is the same as that used for other electronic and optoelectronic devices and is grown in ingots by the Horizontal-Bridgemann or Czochralski techniques. For P-containing compounds, the resulting pressure at the melting point is very high (e.g., 40 atmospheres for GaP). Also, the B_2O_3 used for encapsulation introduces B impurities in the single crystal. For high-quality substrates the defect density varies in the range 10^2–10^5 cm^{-2}.

The epitaxial layers are grown by LPE or VPE. For red LEDs, the n-type layer is doped with Te and for the orange LED the dopant is S. For p-type layers, Zn is the best dopant. NH_3 is used for creating N isoelectronic traps. If Zn-O pairs are needed

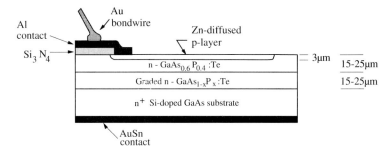

Figure 5.18 Schematic of a red-emitting $GaAs_{0.6}P_{0.4}$ LED with typical dimensions.

to create isoelectronic traps, then Ga_2O_3 is passed through the material, which has already been doped with Zn. GaAsP grown on GaAs substrates may produce misfit dislocations, and therefore graded composition layers are sometimes incorporated between the substrate and the active region. Also, as shown in Fig. 5.18, an n-type substrate and a top p-type Zn-diffused layer are the most common configuration due to the relative ease with which Zn can be incorporated by diffusion. This diffusion is done selectively through an Si_3N_4 mask to form the p-n junction. Contact layers are usually formed of Al and AuSn on the p and n sides, respectively. Alternately, AuBe and AuGe are also used on the p and n sides. Contact to the bias source is made via gold bonds, as shown. To make the light scatter and radiate uniformly from the junction, the metallization on it is made such that the current distribution over the active region is uniform.

The completed LED is encapsulated in a dome, as shown in Fig. 5.5(b), for display and counting applications. For visible LEDs the bottom surface can cause spurious reflections, and therefore a special coating is applied. The most common application of visible LEDs are as lamps and displays. The choice of an LED as a lamp is decided by the color required for the particular application and the requirements of luminous intensity, uniform visibility from all directions, and a contrasting background. LEDs have proved to be very useful for pocket calculators, where red-emitting LEDs are generally used. For this application, several individual chips are required, and each of them should be capable of forming all the numerals. To reduce the size of each chip and yet keep the numerals large, a lens or reflector is used, as shown in Fig. 5.11. In addition, a narrowband filter is placed in the path of the radiated light to improve the contrast ratio. In the fabrication of these devices cost is an important factor.

In applications such as pocket calculators, parallel connections of many small-area LEDs are used to display alphanumerics. Two schemes are generally used. In the first scheme, shown in Fig. 5.19(a), a seven-segment system is used, where each segment can consist of several LEDs. A digital logic control system is used to feed current to the desired segment. The negative terminal of all the LEDs is connected to a common ground terminal. In the second scheme, shown in Fig. 5.19(b), a matrix of LEDs is used, and logic control directs current to specific LEDs to produce a specific display. In both schemes, failure of one or a small number of LEDs will not ruin

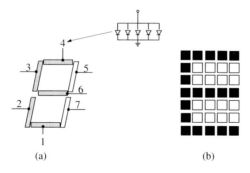

Figure 5.19 (a) Segment architecture and (b) matrix architecture of LEDs for alphanumeric display. In the first, each segment consists of multiple LEDs.

the display. There are competing technologies, such as liquid crystal displays, which have also found wide applications in computing machines and watches.

For sensing and communication applications, infrared LEDs are generally used. For the former GaAs is the most common material. InP-based devices are being used for the latter.

5.10 DEFECTS AND DEVICE RELIABILITY

One of the foremost requirement of LEDs, both for common and specialized applications, is reliability over a long lifetime. Unlike junction lasers, which operate under much larger drive currents, LEDs generally do not suffer catastrophic degradation. But their performance characteristics are known to change with time, and they exhibit gradual degradation. There are two classes of defects that are known to degrade the performance of these devices. The first is dislocations, which may propagate from the substrate during epitaxy into the active region. Currently the dislocation in doped substrates is $\sim 10\text{--}10^2$ cm^{-2} and the density in the active layer can be reduced by small geometry and incorporation of appropriate buffer layers, that act as dislocation filters. The second is deep-level defects in the forbidden energy gap. Deep levels can arise from point defects such as vacancies, interstitial defects, impurity atoms, or complexes formed from combinations of these.

Dislocations in the active area give rise to *dark line defects* and *dark spots*, which cause a more rapid degradation of the device. In essence, the dislocations form absorbing regions. The degradation is dependent on the operating or device temperature, injection current density, and impurity concentration in the active region.

Deep levels, as we have already seen, reduce the radiative efficiency of the LED. In addition, point defects and impurity migration into the active region may be created or enhanced by the recombination process. These give rise to a slow or long-term degradation.[†] The degradation rate is related to the activation energy $\mathcal{E}_d$ of the degradation process by an Arrhenius equation:

$$\gamma_d = C \exp\left(-\mathcal{E}_d / k_B T\right) \tag{5.32}$$

where γ_d is the degradation rate (s^{-1}) and C is a proportionality constant. The activation energy $\mathcal{E}_d$ represents an energy barrier to impurity migration or defect generation. Consequently, the larger is the value of $\mathcal{E}_d$, smaller is the degradation rate. $\mathcal{E}_d$ is related to materials parameters, epitaxial techniques, and the device processing and geometry. The radiant output power as a function of time, $P(t)$, may then be expressed as

$$P(t) = P(0) \exp(-\gamma_d t). \tag{5.33}$$

Based on these equations and values of $\mathcal{E}_d$ in the range of 0.6–1.0 eV, operation lifetimes of $\sim 10^6\text{--}10^9$ hours of continuous wave (cw) operation can be expected,

---

[†]S. Yamakoshi et al., *IEEE J. Quantum Electronics, QE-17(2)*, 167–173, 1981.

depending on the material system. The realization of these lifetimes depends on materials synthesis and device fabrication, and operation with very efficient heat sinking. Without the latter, the junction temperature can quickly become much higher than the ambient temperature.

EXAMPLE 5.3

Objective. A GaAs/AlGaAs LED has an activation barrier of energy $\mathcal{E}_d = 0.6$ eV to long-term degradation. If the value of C is $2 \times 10^2 hr^{-1}$, to find the time after which the radiant output power will fall to half its initial value for room temperature operation.

From Eq. 5.32, $\gamma_d = 7.6 \times 10^{-9} hr^{-1}$. Substituting this value in Eq. 5.33,

$$\frac{P(t)}{P(0)} = 0.5 = \exp(-7.6 \times 10^{-9} t) \tag{5.34}$$

from which $t = 9 \times 10^7 hr$.

It must be remembered, however, that there are other modes of degradation, which can further reduce the operation lifetime.

PROBLEMS

5.1 Discuss the different processes and their efficiencies that combine to produce the external conversion efficiency of an LED. Which one(s) are the most significant? Give reasons for your answer.

5.2 Derive Eq. 5.9.

5.3 Derive Eq. 5.13.

5.4 Without knowing too much about injection lasers at this point, briefly discuss some of the advantages and disadvantages of LEDs in comparison with lasers for long-haul optical-fiber communication.

5.5 A GaAs p-n junction LED has $N_A = N_D = 10^{18}$ cm^{-3}. Light emission predominantly results from electron injection and its recombination in the p-region. Calculate the radiative lifetime. A deep-level recombination center with a lifetime of 5×10^{-9} sec is also present in the active region. Calculate the electron lifetime in the p-region and the radiative recombination efficiency. Assuming that the hole lifetime is equal to the electron lifetime, calculate the electron injection efficiency η_{in} if $D_e = 120$ cm^2/s and $D_h = 0.01 D_e$. What will be the value of η_{in} under a forward bias of 1 V if the recombination current I_{rec} due to the deep level is also taken in account?
[Hint: Taking I_{rec} into account, $\eta_{in} = \frac{I_e}{I_e + I_h + I_{rec}}$. Also, use a value of $B_r = 7.2 \times 10^{-10}$ cm$^3.s^{-1}$.]

5.6   For the LED of Problem 5.5 and under identical operating conditions (neglect I_{rec}) determine the value of the extraction efficiency η_e, if the overall device efficiency is one percent.

5.7   The Zn-O transition is used for light emission at 6900 Å in a GaP LED. Zinc impurities give rise to an acceptor level with $\mathcal{E}_A = 40$ meV and O atoms give rise to a deep donor with $\mathcal{E}_D = 0.8$ eV. Estimate the average spacing between the donor and acceptor atoms. From this result can you estimate the approximate doping levels of Zn and O?

5.8   List the parameters that need to be optimized and the techniques to maximize the external quantum efficiency of an LED.

5.9   Show that the numerical aperture of a step-index fiber is approximately given by the relation

$$A_n = \frac{1}{n_{r0}}\sqrt{2n_{r1}\Delta n_r}$$

where n_{r0} is the index of the medium from which light is coupled, n_{r1} is the index of the fiber core, and Δn_r is the difference of the indices of the fiber core and cladding (n_{r2}). A GaAs LED is coupled through air ($n_r = 1$) to a step-index glass fiber ($n_{r1} = 1.55$). Is the acceptance angle for rays still inside the LED affected by using a dielectric medium ($n_r \simeq 1.55$) instead of air? What is this angle for $\Delta n_r = 0.025$?

5.10   A planar GaAs LED exhibits an external power efficiency of 2% when driven with a current of 40 mA. The voltage developed across its terminals is 2.2 V. Estimate the optical power generated within the device assuming transmission into air.

5.11   Using Eq. 5.25, determine the energy at which the spectral distribution of $r_{sp}(\mathcal{E})$ shown in Fig. 5.8 attains a peak value. Also calculate its full width at half-maximum (FWHM).

5.12   Describe the LED structures commonly used for optical-fiber communication and discuss their merits and disadvantages.

5.13   Light-emitting diodes are made of Si and GaAs. The device dimensions and doping levels are identical. Which one do you expect to exhibit a larger modulation bandwidth and why?

5.14   With reference to Eq. 5.29, explain why edge-emitting LEDs promise high-modulation bandwidths.

5.15   Design a simple dc biasing circuit with a 9V battery for an LED that has a maximum forward current of 2 mA at 2 V.

READING LIST

KRESSEL, H., and BUTLER, J. K. *Semiconductor Lasers and Heterojunction LEDs.* Academic Press, New York, 1977.

SEIPPEL, R. G. *Optoelectronics for Technology and Engineering*, Prentice-Hall, Englewood Cliffs, NJ, 1989.

SZE, S. M. *Physics of Semiconductor Devices*, 2nd ed. Wiley, New York, 1981.

WILLIAMS, E. W., and HALL, R. *Luminescence and Light Emitting Diode*. Pergamon Press, Oxford, 1978.

WILSON, J., and HAWKES, J. F. B. *Optoelectronics: An Introduction*, 2nd ed. Prentice Hall International, United Kingdom, 1983.

6

Lasers: Operating Principles

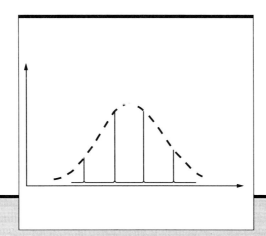

Chapter Contents

6.1 INTRODUCTION

LASER is an acronym for **L**ight **A**mplification by **S**timulated **E**mission of **R**adiation. In 1917, Einstein described the process of stimulated emission. Charles Townes demonstrated stimulated emission for the first time at microwave frequencies, and Theodore Maiman demonstrated it at optical frequencies in a ruby crystal in 1960. The semiconductor laser was invented in 1962. The laser, in principle is an optical waveguide terminated by mirrors or reflecting facets to form a resonant cavity. In that sense, the propagation of optical modes in the structure is similar to the propagation of microwave or millimeter-wave electromagnetic signals through an appropriate guiding structure such as two-wire or coaxial transmission lines and parallel conducting plates, or rectangular waveguides made of conducting material. There is, however, an important difference. In the structures mentioned above, the modes are confined within the guiding regions. In dielectric or semiconductor waveguides, the confinement depends on the index of refraction of the surrounding, or *cladding*, media, and very often there may be leakage of the electromagnetic modes into the cladding region. We will therefore begin with a general discussion of waveguide modes.

6.2 GUIDED WAVES

6.2.1 Waveguide Modes

In a dielectric medium with dielectric constant ϵ_r and permeability μ and containing no charges and conduction currents, the *wave equations* for the electric and magnetic fields **E** and **H** can be derived from Maxwell's equations and can be expressed as

$$\nabla^2 \mathbf{E} + \omega^2 \mu \epsilon_r \mathbf{E} = 0. \tag{6.1}$$

and

$$\nabla^2 \mathbf{H} + \omega^2 \mu \epsilon_r \mathbf{H} = 0 \tag{6.2}$$

We are interested in determining the propagating modes of a dielectric waveguide. These modes must be solutions of the wave equation at a frequency ω. Since we are not dealing with interfaces involving perfect conductors, the tangential components of **E** and **H** must be continuous across the dielectric interfaces, or boundaries. These are the boundary conditions. In a dielectric medium of refractive index $n_r(r)$, Eq. 6.1 can be rewritten as

$$\nabla^2 \mathbf{E}(r) + k_0^2 n_r^2(r) \mathbf{E}(r) = 0 \tag{6.3}$$

where $k_0 = \omega \sqrt{\mu \epsilon_0} = 2\pi/\lambda$. The solution to Eq. 6.3 takes the form of a propagating plane wave

$$\mathbf{E}(r, t) = \mathbf{E}(r) e^{j(\omega t - \beta z)} \tag{6.4}$$

where β is the propagation constant for a wave propagating in the z-direction. Substituting Eq. 6.4 into 6.3, and using cartesian coordinates, Eq. 6.3 can be rewritten as

$$\left(\frac{\partial^2}{\partial x^2} + \frac{\partial^2}{\partial y^2}\right) E(x, y) + \left(k_0^2 n_r^2 - \beta^2\right) E(x, y) = 0 \qquad (6.5)$$

Assuming a semi-infinite slab waveguide, or a planar model, the mathematical solution to Eq. 6.5 becomes simpler, and yet all the insight to the phenomena in a rectangular (channel) waveguide having finite x and y dimensions can be gained. In our geometry, shown in Fig. 6.1, we assume $y \gg x$, so that it may be assumed that $\frac{\partial}{\partial y} = 0$ in Eq. 6.5. The dielectric waveguide is made of three regions with refractive indices n_{r1}, n_{r2}, and n_{r3}. For each of these regions, Eq. 6.5 may be written as

$$\frac{\partial^2}{\partial x^2} E(x, y) + (k_0^2 n_{rl}^2 - \beta^2) E(x, y) = 0 \qquad (6.6)$$

with $l = 1, 2, 3$ for the three regions, I, II, and III. A great deal of insight into the nature of the solutions or propagating modes can be gained even without solving Eq. 6.6 formally.

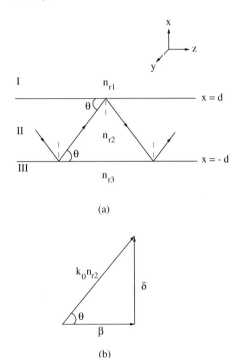

(a)

(b)

Figure 6.1 Three-layer slab-waveguide and the zig-zag ray model.

The usual relationship between the indices in the three regions is $n_{r2} > n_{r3} > n_{r1}$. We consider the nature of the modes as a function of the propagation constant β at a fixed frequency ω. For $\beta > k_0 n_{r2}$, it follows from Eq. 6.6 that $\frac{1}{E} \frac{\partial^2 E}{\partial x^2}$ is positive in all three regions. This leads to exponential solutions in all three regions, with the values of $E(x)$ matched at the two boundaries. The resulting field distribution is shown in Fig. 6.2. The monotonically increasing, unbounded values of the electric field in regions I and II suggest that the solutions do not correspond to real waves.

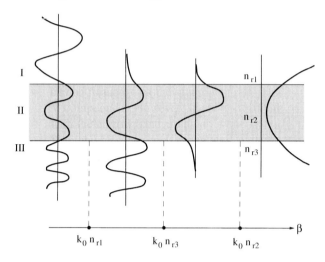

Figure 6.2   Nature of modes in a three-layer slab waveguide as a function of the propagation constant β.

For $k_0 n_{r3} < \beta < k_0 n_{r2}$, $\frac{1}{E} \frac{\partial^2 E}{\partial x^2}$ is positive in regions I and III and negative in region II. Therefore, the solution for the electric field is sinusoidal in region II and exponential in regions I and III, as shown in Fig. 6.2. Under these conditions we have mode confinement and guiding in region II. The exponential solutions in regions I and III do not give propagating modes, and these are essentially evanescent modes. Therefore, it follows that for waveguiding in region II the condition $k_0 n_{r3} < \beta < k_0 n_{r2}$ must be satisfied, and these modes will be confined in region II only if $n_{r2} > n_{r1}, n_{r3}$.

For smaller values of β, such that $k_0 n_{r1} < \beta < k_0 n_{r3}$, Eq. 6.6 gives an exponential solution for $E(x)$ in region I and sinusoidal solutions in regions II and III. The sinusoidal solution in region III represent a loss of power from the guided wave in region II and is therefore undesirable. For $0 < \beta < k_0 n_{r1}$, the solution for $E(x)$ becomes sinusoidal in all three regions and these are called the radiation modes. In this regime, the guide radiates power into both regions I and III.

In the propagation regime, for $k_0 n_{r3} < \beta < k_0 n_{r2}$, the values of β are discrete and these solutions are called the *eigenvalues*. The discrete values of β give rise to the different propagating modes within the guide. On the other hand, the values of β corresponding to the radiation modes are continuous. The number of confined modes depend on the relative values of n_{r1}, n_{r2}, and n_{r3}, the frequency, and the thickness of region II. Remember that we are considering a slab model for the waveguide. Thus, as d increases, the guide first supports mode TE_0, then TE_0 and TE_1, and so on.

EXAMPLE 6.1
Objective.   To establish the same conclusions regarding confinement and guiding in the three-layer waveguide by considering that the modes propagate at some angle to the waveguide axis and undergo a series of total internal reflections at the waveguide boundaries.

The corresponding ray diagram is also depicted in Fig. 6.1, where the ray makes an angle θ with the interface, or the propagation axis of the waveguide. The equation for the propagating wave is still Eq. 6.6 with $n_{rl} = n_{r2}$. Then, for sinusoidal solution within the guide, as discussed, an appropriate solution will be of the form

$$E(x) = C \sin(\delta x + A)e^{-j\beta z} \qquad (6.7)$$

where C and A are constants. Substitution in Eq. 6.6 gives

$$\beta^2 + \delta^2 = k_0^2 n_{r2}^2 \qquad (6.8)$$

This is the equation for a right-angled triangle with sides β, δ, and $k_0 n_{r2}$ and angle θ, as shown in Fig. 6.1(b). The triangle represents plane-wave propagation along the direction of the hypotenuse with a fixed propagation constant equal to $k_0 n_{r2}$. Therefore, as one goes through the regions of decreasing β, θ and δ increase. Thus, for $k_0 n_{r3} < \beta < k_0 n_{r2}$ the values of θ are such that the propagating ray undergoes the desired total internal reflections at the opposite boundaries. For $\beta \leq k_0 n_{r3}$, θ is so large that the modes instead of being reflected are refracted out into regions I and III. Thus, the condition for mode confinement is $\beta > k_0 n_{r3}$, or

$$k_0 n_{r2} \cos \theta > k_0 n_{r3} \qquad (6.9)$$

leading to the condition

$$\theta < \cos^{-1} \left(\frac{n_{r3}}{n_{r2}} \right) = \theta_{cr} \qquad (6.10)$$

where θ_{cr} is the critical angle for total internal reflection at the $n_{r2} - n_{r3}$ boundary. Also, since by our choice $n_{r3} > n_{r1}$, the condition expressed by Eq. 6.10 also guarantees total internal reflection at the $n_{r1} - n_{r2}$ interface. Thus, even with the *zig-zag ray model*, as this is called, the same description of mode confinement and propagation is arrived at.

6.2.2 Propagating Modes in a Symmetric Slab Waveguide

For the design of laser cavities and other waveguiding structures, it is essential to know the mode cut-off conditions for single-mode propagation. The three-layer slab waveguide that we have considered up to this point is asymmetric, since the refractive indices n_{r1} and n_{r3} of the cladding layers are unequal. While a formal solution of the modes in such a waveguide can be calculated, starting from Maxwell's equation, a good insight to the modal properties can be gained by analyzing a symmetric guide, in which $n_{r1} = n_{r3}$ and $n_{r2} > n_{r1}$. As shown in Fig. 6.3, the guiding layer has index n_{r2} and is of thickness 2d.

It can be shown (Appendix 7) that for such a waveguide, the mode cut-off condition is

$$(n_{r2} - n_{r1}) > \frac{m^2 \lambda^2}{16(n_{r1} + n_{r2})d^2}, m = 0, 1, 2, \cdots \qquad (6.11)$$

If $n_{r2} \cong n_{r1}$, the cut-off condition becomes

$$(n_{r2} - n_{r1}) > \frac{m^2 \lambda^2}{32 n_{r2}d^2}, m = 0, 1, 2, \cdots \qquad (6.12)$$

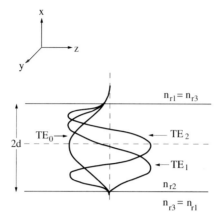

Figure 6.3   Refractive indices and field profiles in a three-layer dielectric slab waveguide.

and, if $n_{r2} \gg n_{r1}$, the condition is

$$(n_{r2} - n_{r1}) > \frac{m^2\lambda^2}{16 n_{r2} d^2}, m = 0, 1, 2. \tag{6.13}$$

The cut-off conditions expressed in Eqs. 6.12 and 6.13 tell us which modes can be supported in a waveguide with a given $\Delta n_r = n_{r2} - n_{r1}$. As mentioned earlier, it can be clearly noted from these equations that the lowest-order mode (m = 0) does not exhibit a cut-off, as do the high-order modes. Thus, in principle, any wavelength could be guided in this mode even for very small Δn_r. However, it may be noted that as Δn_r becomes smaller, or λ/d becomes larger, the confinement of the mode in the guiding region will become very poor and large evanescent tails of the propagating mode will extend in the cladding regions.

6.2.3 Asymmetric and Three-Dimensional Waveguides

The symmetric slab waveguide analyzed above gives a good insight into the nature of the guided modes. However, a more important and practical case of the three-layer planar waveguide is the asymmetric waveguide, in which $n_{r3} \gg n_{r1}$, as shown in Fig. 6.1. A practical example is a thin-film waveguide formed on a substrate of somewhat smaller index and is left either uncoated, or is coated with a metal film. Such structures find use in optical integrated circuits. The analysis for such a guide is very similar to the symmetric case and is outlined in Appendix 8. For an asymmetric guide of thickness $2d$, the mode cut-off condition is

$$\Delta n_r = (n_{r2} - n_{r3}) > \frac{(2m + 1)^2 \lambda^2}{64(n_{r2} + n_{r3})d^2}, m = 0, 1, 2, \cdots \tag{6.14}$$

and assuming $n_{r2} \cong n_{r3}$, the condition becomes

$$\Delta n_r = (n_{r2} - n_{r3}) > \frac{(2m + 1)^2 \lambda^2}{128 n_{r2} d^2}, m = 0, 1, 2, \cdots \tag{6.15}$$

It may be noted that for the asymmetric guide, even the lowest-order, fundamental mode TE_0 has a cut-off. Such a structure can therefore be used as an optical switch or modulator. Equations 6.14 and 6.15 are useful for the design of practical semiconductor waveguides.

Before concluding this section it is important to become aware of other practical guide geometries used in optoelectronics and integrated optics. Two such configurations, the *channel* waveguide and the *strip-loaded* waveguide, are shown schematically in Fig. 6.4. Both belong to the class of three-dimensional waveguides. In the rectangular waveguide, Fig. 6.4(a), the higher-index confining or guiding region is surrounded by media of smaller refractive index. As indicated, a number of materials, with different indices, may be used. Such guides, for example, form the active regions of buried heterostructure lasers, to be described in Chapter 7. The exact solution of the wave-equation for the most general case is very difficult and a slightly approximate solution was provided by Marcatilli.[†] These solutions are outside the scope of this text. However, it is important to note that in this waveguide the fields may have x or y polarization and the order of a mode is expressed as E_{pq}^x or E_{pq}^y (or the H-fields expressed similarly), where the subscripts denote the zeros of the field along the x- and y-directions.

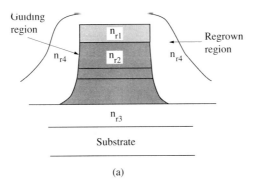

(a)

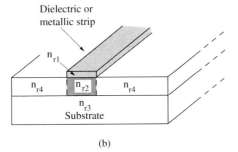

(b)

Figure 6.4   Three-dimensional rectangular waveguide structures used to make laser cavities: (a) buried heterostructure created by epitaxial regrowth and (b) strip-loaded guide. The shaded section is the guiding region.

[†]E. A. J. Marcatilli, *Bell System Technical Journal*, 2071–2102, 1969.

The second important three-dimensional guide is the strip-loaded guide, as shown in Fig. 6.4(b). Here, a dielectric or semiconductor material of smaller refractive index is formed on top of the guiding layer. The presence of the loading strip makes the effective index of the guiding layer underneath it *larger* than the index in the adjacent region of the same layer. As a result, a channel waveguide is formed. Such devices are very useful in integrated optics. The device is also sometimes referred to as an *optical stripline*.

6.3 EMISSION AND ABSORPTION OF RADIATION IN A TWO-LEVEL SYSTEM

Laser action in semiconductors involves the bandstructure and is very different from that in a two-level system. However, it is useful to review the basic principles in a two-level system, in order to elucidate the differences and to know how lasing takes place in a wide variety of materials and atomic systems. The simplest such system is the case of an isolated atom having two energy levels: a ground state with electron energy $\mathcal{E}_1$ and an excited state with energy $\mathcal{E}_2$ as depicted in Fig. 6.5. Under thermal equilibrium conditions most of the atoms are in the ground state and very few atoms have the upper level populated. The distribution of atoms between the two states follows the Boltzmann distribution:

$$\frac{N_2}{N_1} = e^{-h\nu_{12}/k_B T} \tag{6.16}$$

where $h\nu_{12} = \mathcal{E}_2 - \mathcal{E}_1 = \Delta\mathcal{E}$ and N_1 and N_2 are the electron densities in the lower and upper levels, respectively. Consider a photon with energy $\Delta\mathcal{E}$ incident on the hypothetical atomic system. The following two processes can take place: (a) The photon will be absorbed within a path length α^{-1} (where α is the absorption co-efficient) giving rise to an additional electron in the upper level $\mathcal{E}_2$. This electron will relax back to the ground level $\mathcal{E}_1$ after a mean time τ, which is defined as the spontaneous carrier lifetime; and (b) the photon incident on the system can also produce a process which is inverse to the absorption process (i.e., it may induce the downward transition of an electron from the upper to lower level). In this process, called *stimulated emission*, an excited state is kicked into emission by the field of a

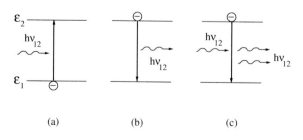

Figure 6.5 (a) Photon absorption, (b) spontaneous emission, and (c) stimulated emission in a two-level atomic system.

passing photon and the induced photon is emitted into the same mode as the incident photon. Stimulated emission generates a photon in a time τ_{st}, called the stimulated emission time, that has the same frequency, direction of propagation, and phase as the stimulating photon. A spontaneous recombination event, which does not require photons to be already present, generates photons propagating in random directions and with random phases, although their frequency is also ν_{12}. These processes can be described in more general terms. Absorption and stimulated emission are both resonant interactions between photons and carriers. When the frequency (or energy) of an incoming photon is equal to or greater than ν_{12}, the photon is absorbed, thereby inducing an upward transition of an electron. At the instant of absorption, before dephasing, the phase of the electron should be equal to that of the photon. On the other hand, a prerequisite for stimulated emission is that energized electrons should exist in the upper level and empty states should exist in the lower level. An incoming photon will resonantly couple with an energized electron of the same frequency and phase, to produce a photon of the same phase as the incoming photon. Remember that the total population of energized electrons, all with the same frequency ν_{12}, but with different phases, will give rise to spontaneous emission. Of the emitted photons, very few are stimulated photons. With the creation of a nonequilibrium population—or population inversion—additional photons are produced in the field of thermal, or random photons, already present and the rate of stimulation grows with the photon density. Normally, under thermal equilibrium conditions the population N_2 is very small. Therefore, *population inversion* is necessary, in which the excited atom population exceeds the ground state population and optical gain becomes possible.

The concepts outlined above are very general and apply to different types of lasing media. There are, however, some important differences. In many types of solid-state lasers and gas lasers radiative transitions take place between discrete energy levels of spatially isolated excited atoms. The spontaneous radiation produced in these cases extends over a very narrow spectral range. In semiconductor materials, on the other hand, the density of atoms, or carriers that contribute to lasing, is 10^{18} cm^{-3} or larger. Therefore, transitions take place between bands of energy and the gain coefficient of the medium under such conditions can be high.

6.4 THE EINSTEIN RELATIONS AND POPULATION INVERSION

It was demonstrated by Einstein in 1917 that the rates of the three processes described in the previous section—absorption, spontaneous emission, and stimulated emission—are related. In what follows, the relationship between the three processes is established. This is the first step toward understanding lasing action in atomic systems.

Consider two energy levels $\mathcal{E}_1$ and $\mathcal{E}_2$ with populations N_1 and N_2 in an atomic system in thermal equilibrium. $\mathcal{E}_2$ is larger than $\mathcal{E}_1$ and normally N_2 is smaller than N_1. In such a system, shown in Fig. 6.6(a), the rate of the upward transitions must equal the rate of the downward transitions. The population of the two levels are

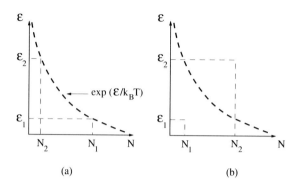

Figure 6.6 (a) Equilibrium (Boltzmann) and (b) nonequilibrium population distributions in a simple two-level lasing medium.

related by Boltzmann statistics according to

$$\frac{N_1}{N_2} = \frac{g_{D1} \exp(-\mathcal{E}_1/k_B T)}{g_{D2} \exp(-\mathcal{E}_2/k_B T)}$$

$$= \frac{g_{D1}}{g_{D2}} \exp\left(\frac{\mathcal{E}_2 - \mathcal{E}_1}{k_B T}\right)$$

$$= \frac{g_{D1}}{g_{D2}} \exp(h\nu_{12}/k_B T) \tag{6.17}$$

where g_{D1} and g_{D2} are the corresponding degeneracies of the levels. The degeneracies are similar to the density of states N_C and N_V in the conduction and valence bands of semiconductors. In other words, the degeneracy parameters indicate the number of subenergy levels within $\mathcal{E}_1$ and $\mathcal{E}_2$. If $\mathcal{E}_1$ and $\mathcal{E}_2$ have single, discrete, energy levels, then the values of g_{D1} and g_{D2} may be taken to be unity.

The absorption rate, or rate of stimulated upward transitions from $\mathcal{E}_1$ to $\mathcal{E}_2$, is proportional to N_1 and the radiation density $\varphi(\nu)$ at the frequency ν_{12}. Therefore, the upward transition rate of an electron from $\mathcal{E}_1$ to $\mathcal{E}_2$ may be expressed as

$$r_{12} = N_1 \varphi(\nu) B_{12} \tag{6.18}$$

where B_{12} is known as the Einstein coefficient for stimulated upward transitions, or absorption. As we have learned earlier, the electrons in level $\mathcal{E}_2$ can come down to $\mathcal{E}_1$ either by spontaneous emission, or stimulated by the incident photons, or radiation field, of energy $h\nu_{12}$. The average lifetime that the carrier spends in the upper level before spontaneous emission takes place is defined as the carrier lifetime τ_{21}, which is the same as the spontaneous recombination lifetime τ. It is important to note that the time for stimulated emission to occur is different and depends on the incident radiation and other factors. The spontaneous emission rate is therefore a product of N_2 and $1/\tau_{21}$, or $N_2 A_{21}$, where $A_{21} = 1/\tau_{21}$ is the Einstein coefficient for spontaneous emission. The rate of stimulated downward transition of electrons from $\mathcal{E}_2$ to $\mathcal{E}_1$ is given by $N_2 \varphi(\nu) B_{21}$, where B_{21} is the Einstein coefficient for stimulated emission.

Therefore, the total transition rate from $\mathcal{E}_2$ to $\mathcal{E}_1$ can be expressed as

$$r_{21} = N_2 A_{21} + N_2 \varphi(\nu) B_{21} \tag{6.19}$$

Under thermal equilibrium conditions, the rates of upward and downward transition must be equal, or $r_{12} = r_{21}$. Therefore,

$$N_1 \varphi(\nu) B_{12} = N_2 A_{21} + N_2 \varphi(\nu) B_{21} \tag{6.20}$$

from which it follows that

$$\varphi(\nu) = \frac{A_{21}/B_{21}}{(B_{12}N_1/B_{21}N_2) - 1} \tag{6.21}$$

Substitution of Eq. 6.17 into Eq. 6.21 leads to

$$\varphi(\nu) = \frac{A_{21}/B_{21}}{[(g_{D1}B_{12}/g_{D2}B_{21})\exp(h\nu_{12}/k_BT)] - 1} \tag{6.22}$$

Under thermal equilibrium conditions we can draw an analogy between the radiation from the two-level system and that from a black-body, for which the radiation density is given in Appendix 4. This comparison leads to

$$B_{12} = \left(\frac{g_{D2}}{g_{D1}}\right) B_{21} \tag{6.23}$$

and

$$\frac{A_{21}}{B_{21}} = \frac{8\pi \nu^3 n_r^3}{c^3} \tag{6.24}$$

If the degeneracies of the levels $\mathcal{E}_1$ and $\mathcal{E}_2$ are equal, or they are both equal to unity, then $B_{12} = B_{21}$. An important parameter for laser action is the ratio of the stimulated emission rate to the spontaneous emission rate. This is given by

$$\frac{B_{21}\varphi(\nu)}{A_{21}} = \frac{1}{\exp(h\nu_{12}/k_BT) - 1} \tag{6.25}$$

For normal two-level atomic systems, which are used for optical sources in the visible range of the spectrum, it can be easily shown that the value of the ratio in Eq. 6.25 is a very small quantity. For $\mathcal{E}_2 - \mathcal{E}_1 \sim k_BT$, $N_2 = 0.37N_1$ and for $\mathcal{E}_2 - \mathcal{E}_1 \sim 2eV$, as in visible radiation, $N_2 = 0$. Such sources are therefore usually incoherent. It can be concluded that to achieve a coherent source in which stimulated emission will dominate, $\varphi(\nu)$ must be increased and N_2 must be made larger than N_1. This is *population inversion*. If $\tau_{21} = \tau_r$, then it is evident that $A_{21} = 1/\tau_r$ and $B_{21} = \lambda^3/8\pi\tau_r$.

EXAMPLE 6.2

Objective.   To calculate the ratio of the spontaneous to stimulated emission rates in a tungsten lamp that radiates at an average frequency of 5×10^{14} Hz at an operating temperature of 1300°K.

From Eq. 6.25

$$\frac{A_{21}}{\varphi(v)B_{21}} \cong e^{hv_{12}/k_B T}$$

$$= \exp\left(\frac{6.6 \times 10^{-34} \times 5 \times 10^{14}}{1.38 \times 10^{-23} \times 1300}\right)$$

$$= e^{18.4}$$

It is evident that the output of such a light source is dominated by spontaneous emission and is therefore incoherent.

To create population inversion, a large amount of energy is needed to excite carriers into $\mathcal{E}_2$. This is achieved by pumping, which creates nonequilibrium conditions in the medium [Fig. 6.6(b)]. In a typical laser pumping is achieved by an external, intense optical beam. In a two-level system, as shown in Fig. 6.6, if $g_{D1} = g_{D2}$, then $B_{12} = B_{21}$, so that once electrons are excited into the upper level, the probabilities of further stimulated upward and downward transitions are equal. Therefore, the best population ratio that can be obtained is unity or $N_1 = N_2$. This is insufficient to produce optical gain and three- and four-level lasers have therefore been proposed and conceived. Such lasers have much lower pumping power requirements.

6.5 GAIN IN A TWO-LEVEL LASING MEDIUM

Before we attempt to understand the concept of gain in a semiconductor, it is easier to consider gain in a single two-level system. Assume a coherent beam of monochromatic radiation of unity cross-sectional area, shown schematically in Fig. 6.7, passing through the lasing medium. In the latter, absorption of the radiation takes place by an electron transition from a lower energy $\mathcal{E}_1$ to an upper level $\mathcal{E}_2$. The intensity, or irradiance, of the light beam changes along its path due to absorption, and this is expressed as

$$\frac{d\Im(x)}{dx} = -\alpha\Im(x) \tag{6.26}$$

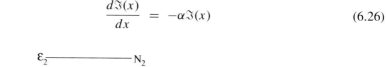
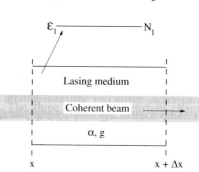

Figure 6.7   Transmission of a coherent beam through a lasing medium.

which, upon integration leads to

$$\Im = \Im_0 e^{-\alpha x} \tag{6.27}$$

Here $\Im_0$ is the incident intensity at $x = 0$ and α is the absorption coefficient of the medium. If N_1 and N_2 are the electron populations in the lower level $\mathcal{E}_1$ and upper level $\mathcal{E}_2$, respectively, then the probabilities of stimulated upward (absorption) and stimulated downward transitions depend on the relative values of N_1 and N_2. The net rate of loss of photon density N_p can be expressed as

$$- h\nu \frac{dN_p}{dt} = N_1 \varphi(\nu) B_{12} - N_2 \varphi(\nu) B_{21} \tag{6.28}$$

which, on substitution of Eq. 6.23 leads to

$$- h\nu \frac{dN_p}{dt} = \left(\frac{g_{D2}}{g_{D1}} N_1 - N_2 \right) \varphi(\nu) B_{21} \tag{6.29}$$

Note that the spontaneous emission process, in which photons are emitted randomly in all directions and phases, is neglected. These photons do not contribute to the coherent beam we have considered. Also by definition, $\varphi(\nu) = N_p h\nu$ where N_p is the number of photons per unit volume having frequency ν.

The intensity, or irradiance, of the photon beam is the energy crossing unit area in unit time. Therefore, for photons of frequency ν,

$$\Im = \varphi(\nu) \frac{c}{n_r}$$

$$= N_p h\nu \frac{c}{n_r} \tag{6.30}$$

The change in photon density in the coherent beam as it passes through the region between x and $x + \Delta x$ is

$$- dN_p(x) = [\Im(x) - \Im(x + \Delta x)] \frac{n_r}{h\nu c} \tag{6.31}$$

and for $\Delta x \to 0$

$$- dN_p(x) = - \frac{d\Im(x)}{dx} \cdot \frac{\Delta x n_r}{h\nu c} \tag{6.32}$$

Now the photons traverse the distance Δx in a time $\frac{\Delta x}{c/n_r}$, and therefore

$$- \frac{dN_p}{dt} = - \frac{d\Im(x)}{dx} \cdot \frac{1}{h\nu}, \tag{6.33}$$

which, on substitution of Eq. 6.26, leads to

$$\frac{dN_p}{dt} = - \frac{\alpha \Im(x)}{h\nu} = -\alpha \varphi(\nu) \frac{c}{n_r} \cdot \frac{1}{h\nu} \tag{6.34}$$

On comparison with Eq. 6.29, and knowing $\nu = \nu_{12}$, gives

$$\alpha \varphi(\nu) \frac{c}{n_r} \frac{1}{h\nu_{12}} = (N_1 - N_2) \frac{1}{h\nu_{12}} \varphi(\nu) B_{21} \tag{6.35}$$

where it is assumed that $g_{D1} = g_{D2}$. From Eq. 6.35 we get

$$\alpha = (N_1 - N_2)\frac{B_{21}n_r}{c} \tag{6.36}$$

As expected, the absorption coefficient depends on the population difference $(N_1 - N_2)$ between the levels $\mathcal{E}_1$ and $\mathcal{E}_2$. In general, for $\mathcal{E}_2 > \mathcal{E}_1$, $N_1 > N_2$, α is positive. This represents a medium with loss. If we can create a population inversion such that $N_2 > N_1$, then α is a negative quantity and the intensity of the beam grows with distance x. We then define a *gain coefficient g* such that

$$\Im(x) = \Im_0 e^{gx} \tag{6.37}$$

and

$$g = -\alpha = (N_2 - N_1)B_{21}\frac{n_r}{c} \tag{6.38}$$

Such gain is essential in a lasing medium.

6.6 LASING CONDITION AND GAIN IN A SEMICONDUCTOR

A semiconductor is not a two- or three-level atomic system, but consists of electrons and holes distributed in energy in their respective bands. Therefore, the concept of population inversion is a little different in a semiconductor. Here laser action and optical gain are produced by creating a nonequilibrium population of the carriers in the bands. To understand the process of stimulated emission in a semiconductor, it is essential to review the absorption and emission processes discussed in Chapter 3. This can be done with the help of the energy band diagrams shown in Figs. 6.8 (a) and (b) for a pure semiconductor at $0°$K. Initially, the valence band is completely filled and the conduction band is empty. We assume that electron-hole pairs are created in the material by incident photons of energy $\mathcal{E} = h\nu$. The electrons occupy the lowermost states of the conduction band, up to the level $\mathcal{E}_{fn}$, and an equal number of holes fill the valence band upto $\mathcal{E}_{fp}$. It may be remembered that while the photoexcitation is incident on the semiconductor, $\mathcal{E}_{fn}$ and $\mathcal{E}_{fp}$ represent the electron and hole quasi-Fermi levels, respectively. Once this situation is created, photons with energy greater than $\mathcal{E}_g$, the bandgap, but less than $\Delta\mathcal{E}_f = \mathcal{E}_{fn} - \mathcal{E}_{fp}$ cannot be absorbed. However, photons with energy in this range can induce stimulated downward transitions from the filled conduction band states to the empty valence band states. With reference to Fig. 6.8 the absorption rate of photons of energy $\mathcal{E}$ can be expressed as

$$r_{abs} = P[1 - f_n(\mathcal{E}_2)]f_p(\mathcal{E}_1)N_p(\mathcal{E}) \tag{6.39}$$

and the stimulated emission rate is

$$r_{st} = Pf_n(\mathcal{E}_2)[1 - f_p(\mathcal{E}_1)]N_p(\mathcal{E}) \tag{6.40}$$

where P is the transition probability and $N_p = n_{\varepsilon}u_{\varepsilon}$ is the density of photons of energy $\mathcal{E}$. Note that both processes are stimulated by a photon. f_n and f_p are Fermi functions defined by Eqs. 3.1 and 3.2. For net stimulated emission or optical gain, the condition $r_{st} > r_{abs}$ must hold. From Eqs. 6.39 and 6.40, the condition becomes

$$f_n(\mathcal{E}_2) > f_p(\mathcal{E}_1) \tag{6.41}$$

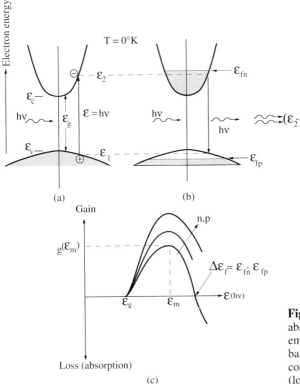

Figure 6.8 Illustration of absorption and stimulated emission processes in a direct bandgap semiconductor and the corresponding dependence of gain (loss) with photon energy.

and under these conditions we have population inversion. Substitution of Eqs. 3.1 and 3.2 leads to

$$\mathcal{E}_{fn} - \mathcal{E}_{fp} > \mathcal{E} \tag{6.42}$$

As the temperature is increased the electron and hole distributions in the respective bands are smeared, but the condition for stimulated emission is still given by Eq. 6.41 or 6.42. These equations also represents a necessary condition for lasing. The process of stimulated downward transition represents gain in a lasing medium. Similarly, stimulated upward transitions, or absorption, represent a loss. The gain and absorption profiles as a function of energy are schematically shown in Fig. 6.8(c) with respect to the population of the bands shown in Fig. 6.8(b). It is seen that the gain becomes positive for $h\nu \cong \mathcal{E}_g$, goes through a maximum, and becomes zero for $h\nu \cong \Delta\mathcal{E}_f = \mathcal{E}_{fn} - \mathcal{E}_{fp}$. At energies much larger than $\mathcal{E}_g$, the occupation of electrons and holes in their respective bands is negligible and the gain curves are given by

$$g(\mathcal{E}) = -\alpha(\mathcal{E}) \tag{6.43}$$

where the absorption process dominates.

The luminescence from a semiconductor represents its spontaneous emission spectrum, which have a certain energy spread, or linewidth. For lasing, it is essential that the gain should be at least equal to the loss at a value of photon energy within this emission spectrum. Usually this occurs at the peak of the spectrum. However, there are many factors that can change this simple picture. Lasers are usually made with highly doped semiconductors and operate under conditions of high forward bias injection of a p-n junction. The high density of injected carriers will alter the density of states at the bandedges, and the requirement for lasing will change. Also, as we have seen in Chapter 3, high doping produces a merging of the impurity levels with the bandedges and the formation of bandtail states. Both factors lead to a change in the lasing wavelength. For example, in a laser in which the active layer is a heavily doped n-type material, the donor level merges with the conduction band, and the radiative transition energy can be larger than the bandgap energy. On the other hand, in lasers where the active region is made of p-type material, for doping levels higher than 10^{18} cm^{-3} the radiative transitions mostly involve shallow acceptor states, which have binding energies between 26 to 32 meV. Therefore, the lasing peak energy is below the bandgap energy.

While in most types of lasers population inversion and subsequent stimulated emission is achieved by optical pumping, in a semiconductor diode lasing is achieved by minority carrier injection across a forward-biased junction. It may be noted, however, that photopumping can be substituted for injection in materials in which junctions cannot be made conveniently. For efficient operation of the laser, the injected carriers need to be confined in the vicinity of the junction. Carrier confinement and optical confinement are achieved in laser diodes with the use of heterojunctions. The electrons and holes recombine in the active region emitting photons.

At this point it is useful to briefly distinguish between the operation of a junction LED and laser. The LED also operates under a forward bias applied to a p-n diode. The injected electrons and holes recombine in the active region to produce photons, which escape. Therefore, the density of photons in any given mode remains insignificantly small and the gain remains less than the total loss. The emission probability discussed in Chapter 5 contains a factor $(u_\mathcal{E} + 1)$ where $u_\mathcal{E}$ is the number of photons per mode. The factor $u_\mathcal{E}$ does not play any role in this regime. This is also the regime of operation of a laser below threshold, where the gain in the device remains less than the loss suffered by the light. However, as we shall soon see, the laser structure is built into a resonant cavity, so that a significant photon density in a mode, or coherent photons, can build up. This is the condition of stimulated emission, and the gain in the device can overcome the losses. Ultimately, as the coherent photon density becomes very high, the laser reaches threshold. At high forward-bias injection the laser operates above threshold.

It is evident that recombination rate and gain are the two most important parameters that determine laser action. The gain is related to the absorption coefficient by Eq. 6.43. The absorption coefficient expressed by Eq. 3.42 was derived under the condition that all the valence band states were occupied and the conduction band states were empty. In a laser operating at room temperature, the occupation of bands needs to be taken into account. The total spontaneous emission rate per unit volume for a transition energy $\mathcal{E}$ is given by Eq. 5.22. The stimulated emission rate is therefore given by

$$R_{st} = \int_{\mathcal{E}} r_{st}(\mathcal{E})d\mathcal{E} = \int_{\mathcal{E}} P_{em}N_J(\mathcal{E})f_n(\mathcal{E}_2)[1 - f_p(\mathcal{E}_1)]d\mathcal{E} \quad (s^{-1}.cm^{-3}) \quad (6.44)$$

where P_{em} is given by Eq. 5.23 and N_J by Eq. 5.21. Note that the expression becomes similar to that for the spontaneous emission rate when $u_\mathcal{E}$ is zero or very small. Similarly, the absorption rate can be written as

$$R_{abs} = \int_{\mathcal{E}} r_{abs}(\mathcal{E})d\mathcal{E} = \int_{\mathcal{E}} P_{abs}N_J(\mathcal{E})f_p(\mathcal{E}_1)[1 - f_n(\mathcal{E}_2)]d\mathcal{E} \quad (s^{-1}.cm^{-3}) \quad (6.45)$$

where P_{abs} and P_{em} are identical and are of the same form as P_{em} in Eq. 5.23 with the occupation factor $(u_\mathcal{E} + 1)$ replaced by $u_\mathcal{E}$. Note that, with reference to Fig. 6.8,

$$\mathcal{E}_2 = \mathcal{E}_C + \frac{m_r^*}{m_e^*}(\mathcal{E} - \mathcal{E}_g) \tag{6.46}$$

and

$$\mathcal{E}_1 = \mathcal{E}_V - \frac{m_r^*}{m_h^*}(\mathcal{E} - \mathcal{E}_g)$$

$$= \mathcal{E}_2 - \mathcal{E} \tag{6.47}$$

The net stimulated emission rate is given by

$$r_{st}^{net}(\mathcal{E}) = r_{st}(\mathcal{E}) - r_{abs}(\mathcal{E})$$

$$= \left(\frac{n_r q^2 \mathcal{E} p_{CV}^2 u_\mathcal{E}}{3\pi \epsilon_o m_o^2 \hbar^2 c^3} \right) N_J(\mathcal{E})[f_n(\mathcal{E}_2) - f_p(\mathcal{E}_1)] \tag{6.48}$$

where the term within the first set of parentheses represents the stimulated (downward) transition probability and its form is given in Eq. 5.23. We can now derive an expression for the gain spectra $g(\mathcal{E})$, which is defined as

$$g(\mathcal{E}) = \frac{\text{power emitted per unit volume}}{\text{power crossing per unit area}} \tag{6.49}$$

Note that the absorption coefficient α can be similarly defined, with the numerator being the power absorbed per unit volume. It can be easily shown that (Problem 6.8) the optical gain is equal to the net stimulated emission rate divided by the photon flux. In other words,

$$g(\mathcal{E}) = \frac{r_{st}^{net}(\mathcal{E})}{\left(\frac{c}{n_r} \right) n_\mathcal{E} u_\mathcal{E}} \quad (cm^{-1}) \tag{6.50}$$

where $n_\mathcal{E}$ is the photon mode density defined in Appendix 4. Substituting for $r_{st}^{net}(\mathcal{E})$, $n_\mathcal{E}$ and $N_J(\mathcal{E})$ (Eq. 5.21) we get

$$g(\mathcal{E}) = \frac{\sqrt{2}(m_r^*)^{3/2}q^2 p_{CV}^2}{3\pi n_r \epsilon_o m_o^2 \hbar^2 c \mathcal{E}}(\mathcal{E} - \mathcal{E}_g)^{1/2}[f_n(\mathcal{E}_2) - f_p(\mathcal{E}_1)] \tag{6.51a}$$

or in the equivalent form

$$g(\mathcal{E}) = \frac{2\pi q^2 \hbar}{3n_r \epsilon_o m_o^2 c \mathcal{E}} \int \frac{d^3 k}{(2\pi)^3} p_{CV}^2 [f_n(\mathcal{E}_2) - f_p(\mathcal{E}_1)]$$

$$\times \ \delta(\mathcal{E}_2(\mathbf{k}) - \mathcal{E}_1(\mathbf{k}) - \mathcal{E}) \tag{6.51b}$$

as in Eq. 5.22b. In these equations $\mathcal{E}_2 - \mathcal{E}_1 = \mathcal{E} = \hbar\omega$, the photon energy and the energies $\mathcal{E}_1$ and $\mathcal{E}_2$ must be such that momentum is conserved. For the gain coefficient to be positive $f_n(\mathcal{E}_2)$ must be larger than $f_p(\mathcal{E}_1)$, which is the population inversion condition stated in Eq. 6.41. Under these conditions an electromagnetic wave traveling through the medium is amplified. When $f_n(\mathcal{E}_2)$ is zero and $f_p(\mathcal{E}_1)$ has a unity value, $g(\mathcal{E}) = -\alpha(\mathcal{E})$ and, in fact, Eq. 6.51 becomes identical to Eq. 3.42. There is a discrepancy of a factor of two-thirds, which accounts for emission into any photon polarization. Under lasing conditions Fermi-Dirac statistics should be used to describe f_n and f_p, since the quasi-Fermi levels are within the respective bands. Equation 6.51 for gain in a semiconductor may be compared with Eq. 6.38 for gain in a two-level atomic system. In a sense $(N_2 - N_1)$ is equivalent to $[f_n(\mathcal{E}_2) - f_p(\mathcal{E}_1)]$. The k-selection rule was assumed to be valid in deriving the equations above. In reality, however, since junction lasers are made with highly doped semiconductors, momentum conservation rules in electron-hole recombination are partly relaxed due to elastic scattering by the impurities.

With reference to the emission rates given in Eqs. 5.22 or 6.44, at low concentrations of injected carriers, Boltzmann statistics is valid and the integral over $f_n(1 - f_p)$ gives a term proportional to $np(= n^2$, if $n = p)$ (Problem 6.10). On the other hand, when the injection is high, the Fermi levels move into the respective bands and one must use Fermi statistics. Under these conditions $R_{sp} \infty n$ and in fact, remembering the discussion of Eq. 5.23,

$$R_{sp} = \frac{n}{\tau_r} \cong \frac{n}{\tau_o} \tag{6.52}$$

where $\tau_o \sim 0.5$ns. It will be seen in Chapter 7 that above threshold the recombination rate or photon output is given by τ_{st}, the stimulated emission lifetime, instead of τ_o. The equations derived in this section are extremely important for understanding and describing the operation of semiconductor junction lasers.

6.7 SELECTIVE AMPLIFICATION AND COHERENCE—NEED FOR LASER CAVITY

It is important to briefly review the ongoing carrier dynamics in a lasing medium, such as a semiconductor. With the provision of an appropriate excitation to create population inversion, such as optical excitation or current injection, both spontaneous and stimulated emission processes are initiated. As the recombination rate increases, the photon density builds up. These photons stimulate further recombination. Therefore, starting with a spontaneous emission spectrum and an inverted population $np > n_i^2$, the initial photon density will *stimulate* further recombinations. Photons at the peak (energy) of the distribution are more numerous, and therefore they stimulate more transitions, compared to the photons at other energies in the spontaneous recombination spectrum. Thus, the output spectrum grows and narrows simultaneously. This is the onset of *superradiance* in the medium, in which the peak usually grows super-linearly in time. This results in spectral narrowing, but the overall radiation remains incoherent. In other words superradiance amplifies the intensity of photons of all phases. Therefore, for lasing two conditions are necessary:

1. The gain must at least be equal to the losses in the medium.

2. The radiation must be coherent.

Coherence is obtained by placing the lasing medium in a Fabry-Perot cavity, where *selective amplification* of one frequency occurs. This frequency is usually at the peak of the spontaneous output or luminescence spectrum. Selective amplification is the result of positive feedback for the frequency at which a standing wave is formed in the cavity and the overall gain is greater or equal to the losses. In most gas or solid-state lasers, such feedback is provided by placing the lasing medium between two plane-parallel or confocal mirrors. In a semiconductor junction laser, the mirrors are usually formed by *cleaving* two ends of the waveguide, as we shall see in the next chapter. The initial light output occurs due to spontaneous transitions between two energy levels, or bands, and this spectrum is selectively amplified as it travels in the cavity between the end mirrors and is "fed back" at the mirrors. Lasing occurs when the gain provided by the medium overcomes the loss in the cavity for a roundtrip.

6.7.1 Threshold Condition for Lasing

We will derive the threshold condition in the context of a semiconductor laser in which a Fabry-Perot cavity is made by cleaving both sides of the waveguide so that the length of the cavity is given by $l = m\lambda/2$. Here λ is a wavelength near the peak of the spontaneous emission spectrum and m is an integer. The cleaved edges, being optically flat to near-perfection, serve as the mirrors of the cavity because of the large refractive index difference between the semiconductor material and air. Nevertheless, a good portion of the optical signal is also transmitted.

For the present analysis, we consider the schematic in Fig. 6.9, where the laser cavity is of length of l and has mirrors of reflectivity R_1 and R_2 at the two ends. Let g and γ be the gain and loss coefficients, respectively. The total loss in the

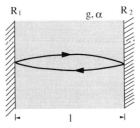

R$_1$ g, α R$_2$

l

Figure 6.9   Schematics of a Fabry-Perot cavity.

system results from a number of different processes. These include: (1) transmission at the mirrors, remembering that transmission at one end is the useful laser output. Therefore, the other end is made highly reflective to minimize transmission loss; (2) absorption, scattering and diffraction losses at the cleaved mirrors; (3) absorption in the cladding regions; and (4) scattering at defects and inhomogeneities in the medium. The loss coefficient γ defined above includes all the losses except the transmission at the ends.

We can calculate the threshold gain by considering the change in intensity of a beam of light undergoing a roundtrip within the cavity. Thus, the light intensity at the center, after traveling the length $2l$ is given by

$$\Im = \Im_0 R_1 R_2 e^{2(g-\gamma)l} \tag{6.53}$$

where $\Im_0$ is the initial intensity. Now if $g > \gamma$, the intensity grows and there is net amplification. Therefore, at threshold, when the roundtrip gain exactly equals the losses

$$\Im = \Im_0 \tag{6.54}$$

and

$$1 = R_1 R_2 e^{2(g_{th}-\gamma)l} \tag{6.55}$$

where g_{th} is the threshold gain. Eq. 6.55 can be expressed in the form

$$g_{th} = \gamma + \frac{1}{2l} \ln\left(\frac{1}{R_1 R_2}\right) \tag{6.56}$$

where the first term on the right represents the losses in the cavity and the second term represents the useful laser output. The latter is seen to depend on the mirror reflectivities. Initially, when lasing starts, the gain may be much larger than the threshold gain. However, with ongoing stimulated emission the degree of population inversion and the gain will decrease. Therefore, after a few oscillations in the value of the roundtrip gain, laser action reaches a steady state and the gain reaches a steady-state value equal to g_{th}. Finally, it should be noted that the efficiency of a laser not only depends on the gain, but also on the pumping efficiency (i.e., the efficiency of creating population inversion). The latter, in turn, depends on transition probabilities and losses in the media.

EXAMPLE 6.3

Objective.   To calculate the mirror reflectances (assumed same) in a laser 10 cm long under threshold conditions given that net gain coefficient of the lasing medium is 0.01 cm^{-1}.

At threshold, with $R_1 = R_2 = R$, from Eq. 6.55,

$$1 = R^2 e^{2(g-\gamma)l}$$

$$= R^2 e^{2 \times 10 \times 0.01}$$

From which we get $R = 0.82$.

It is of interest to compare the value of these parameters with those of a semiconductor Fabry-Perot cavity laser. In a GaAs cleaved cavity $R = 0.36$ and typical cavity lengths are $\sim 100\mu$m. This implies that a GaAs laser has a very large gain coefficient.

We will now discuss the threshold condition in a little more detail. For simplicity we consider a three-layer slab waveguide model, with regions of refractive index n_{r1}, n_{r2}, and n_{r3} (Fig. 6.10) such that the eigenfunctions of the propagating modes in region 2 can be expressed as

$$E_y = \Psi_m(x)\exp(-i\beta_m z) \tag{6.57}$$

where m is an integer and β_m is the propagation constant. The wavefunctions can be normalized to unity such that

$$\int_{-\infty}^{\infty} \Psi_m^*(x)\Psi_n(x)dx = \delta_{mn} \tag{6.58}$$

where δ_{mn} is the Kronecker delta function. In terms of the waveguide geometry, a confinement factor Γ_m may be defined by

$$\Gamma_m = \int_{-d/2}^{d/2} \Psi_m^2(x)dx \tag{6.59}$$

Γ is a measure of the modal confinement, or the fraction of optical intensity in the active region, and is a strong function of the active region thickness d and an asymmetry factor η_a defined by

$$\eta_a = \frac{n_{r2}^2 - n_{r3}^2}{n_{r2}^2 - n_{r1}^2} \tag{6.60}$$

When $d > 0.1\mu$m $\Gamma \to 1$ but can be very small (≤ 0.05) in quantum well lasers where the active region thickness is 50–100 Å. The propagating mode is mainly confined in region 2, but also "spills" over into regions 1 and 3. Therefore the total absorption coefficient can be expressed as

$$\alpha = a_1\alpha_1 + a_2\alpha_2 + a_3\alpha_3 \tag{6.61}$$

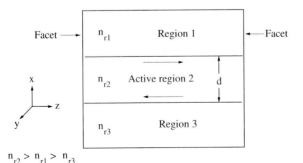

$n_{r2} > n_{r1} > n_{r3}$

Figure 6.10   Three-layer dielectric slab waveguide.

where

$$a_i = \int_i \Psi^2(x)dx \tag{6.62}$$

and $a_2 = \Gamma$, defined in Eq. 6.59, above. We define a quantity Γg, termed *modal gain*, which is a measure of the power transferred from the active region into the propagating mode. At threshold in Eq. 6.61, the net modal $\alpha = 0$ and $\alpha_2 = -g_{th}$. Therefore,

$$\Gamma g_{th} = a_1\alpha_1 + a_3\alpha_3 \tag{6.63}$$

for the special case of a symmetrical waveguide $\alpha_1 = \alpha_3$ and $n_{r1} = n_{r3}$. From the normalization condition

$$a_1 + a_3 + \Gamma = 1 \tag{6.64}$$

Combining Eqs. 6.63 and 6.64,

$$g_{th} = \left(\frac{1-\Gamma}{\Gamma}\right)\alpha_1 \tag{6.65}$$

which can be rewritten as

$$\Gamma g_{th} = (1 - \Gamma)\alpha_1 \tag{6.66}$$

which states that for a propagating mode to remain unchanged in magnitude, the gain must equal the losses. In other words, the power gained by the waveguide mode from the active region is equal to the power drained from the waveguide mode by the passive regions of the guide. Taking the mirror losses and other cavity losses into account, we get

$$\Gamma g_{th} = \gamma + \frac{1}{2l}\ln\left(\frac{1}{R_1 R_2}\right) \tag{6.67}$$

If the reflectivities of the two mirrors are identical,

$$\Gamma g_{th} = \gamma + \frac{1}{l}\ln\left(\frac{1}{R}\right) \tag{6.68}$$

This equation represents the *threshold condition* of the cavity. Remember that g is proportional to the injected carrier density n. The value of n at which $g(n, \mathcal{E}_m)$ (see Fig. 6.8) is equal to $\gamma + \frac{1}{l}\ln\left(\frac{1}{R}\right)$ is the threshold carrier concentration n_{th}. All these carriers can be assumed to recombine spontaneously to yield a threshold current density

$$J_{th} = qdR_{sp}(n_{th}) \tag{6.69}$$

where R_{sp} is the spontaneous emission rate per unit volume and d is the thickness of the active region.

6.8 LINESHAPE FUNCTION AND LINE-BROADENING MECHANISMS

We have seen earlier that for lasing to occur in an optical cavity, the gain must match the total losses at some photon energy within the spontaneous emission spectrum. Looking at it in another way, the field of the initial photons resulting from spontaneous emission induces stimulated emission and gain. If the initial photons are widely distributed in energy, it is harder to achieve coherency. Intuitively, it is clear, that the pumping, either optical excitation or injection current in a diode, will have to be increased. As a matter of fact, the spontaneous emission or photoluminescence spectrum of a semiconductor material does have a finite linewidth, determined by several factors.

Typical emission or photoluminescence spectra, representing band-to-band transitions are shown in Fig. 6.11. For the present analysis the shape of the curve is described by a lineshape or broadening function $S(v)$, such that $S(v)dv$ defines the probability that a given transition between the conduction and valence bands will result in the emission or absorption of a photon with frequency lying between v and $v + dv$. Thus, a photon of frequency v may not stimulate another photon of the same frequency and $S(v)dv$ is the probability that the stimulated photon will have frequency lying between v and $(v + dv)$. It follows that

$$\int_{-\infty}^{\infty} S(v)dv = 1 \tag{6.70}$$

The lineshape $S(v)$ and the linewidth Δv of the spontaneous emission spectrum are eventually determined by various line-broadening mechanisms. These can be classified into *homogeneous* and *inhomogeneous* line-broadening mechanisms. In the former, all parts of the gain medium are affected uniformly, while in the latter, only selected parts are affected. Their cause and nature are briefly reviewed. Homogeneous broadening occurs mainly due to phonon interactions. Both acoustic and optical phonons are involved in the process. The phonon contribution of the linewidth is proportional to the phonon population density. In the case of acoustic phonons this density increases linearly with temperature. On the other hand optical phonons have

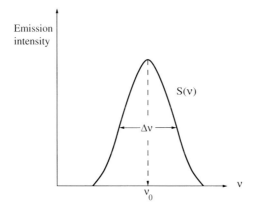

Figure 6.11 Emission spectrum and corresponding lineshape function.

a relatively fixed frequency. The number of phonons thermally excited follows Bose-Einstein statistics. The expression for the total homogeneous linewidth can be written as

$$\Delta \nu_{homo} = AT + \frac{\zeta_{op}}{e^{\frac{\hbar \omega_{LO}}{k_B T}} - 1} \tag{6.71}$$

where the first term represents the acoustic phonon contributions with proportionality constant A and the second term represents the optical phonon contributions. ζ_{op} is optical phonon-broadening constant and ω_{LO} is the longitudinal optical phonon frequency. Calculations have shown that at high temperatures the contribution of the acoustic phonons is always much smaller than that of the optical phonons. This leads to the expression for the homogeneous broadening linewidth

$$\Delta \nu_{homo} = \frac{\zeta_{op}}{e^{\frac{\hbar \omega_{LO}}{k_B T}} - 1} \tag{6.72}$$

Since the phonon interactions are the result of lattice vibrations, homogeneous broadening can be described by a Lorentzian lineshape function.

Inhomogeneous broadening is due to local variations of the electronic properties across the sample. Inhomogeneous contributions arise from localized strain, impurity density variations, alloy clustering, interface roughness in heterostructures, and quantum wells. Of all these factors, only the impurity broadening is temperature dependent. The other contributions to inhomogeneous linewidth are temperature independent. We can thus express the inhomogeneous broadening lineshape as a Gaussian function. In terms of the line-broadening function, the total gain can be expressed as

$$g(\mathcal{E}) = \int g(\mathcal{E}')S(\mathcal{E} - \mathcal{E}')d\mathcal{E}' \tag{6.73}$$

As a simplifying approximation, we may express the lineshape function as

$$S(\nu_0) \cong \frac{\nu_o}{\Delta \nu} \tag{6.74}$$

where ν_o is the frequency corresponding to the peak of the spontaneous emission spectrum.

6.9 LASING THRESHOLD CONDITION IN A TWO-LEVEL SYSTEM

Considering the spontaneous emission spectrum, lasing threshold is usually reached when $\nu = \nu_0$ and $S(\nu_0)$ is a maximum. Therefore, the threshold population inversion can be expressed from Eq. 6.38 as

$$(N_2 - N_1)_{th} = \frac{g_{th}c}{B_{21}n_r S(\nu_0)} \tag{6.75}$$

where the lineshape function is also included. Use is now made of Eq. 6.24, in which A_{21}, the Einstein coefficient for spontaneous transitions is inversely related to the spontaneous radiative lifetime τ_r. Thus, A_{21} can be experimentally determined. Equation 6.75 may therefore be rewritten as

$$N_{th} = (N_2 - N_1)_{th} = \frac{8\pi v_0^3 g_{th} \tau_r n_r^2}{c^2 S(v_0)}. \tag{6.76}$$

With respect to Eq. 6.74, the threshold population inversion is

$$N_{th} = \frac{8\pi v_0^2 g_{th} \tau_r \Delta v n_r^2}{c^2} \tag{6.77}$$

Note that this equation contains terms that are either physical constants or are parameters that can be measured for a real laser.

6.10 AXIAL AND TRANSVERSE LASER MODES

In spite of the feedback provided by the Fabry-Perot cavity, the actual output of a laser consists of a large number of discrete frequency components. These are called the *axial modes* or *longitudinal modes* of the cavity, as shown in Fig. 6.12. The cleaved mirrors at the ends of the resonant cavity will set up standing waves for wavelengths, λ, which satisfy

$$l = \frac{m\lambda}{2n_r} \tag{6.78}$$

where l is the length of the cavity and m is an integer called the mode index or mode number. Each value of m gives an axial mode of the cavity. It may be noted that the distance corresponding to a half wavelength for visible or near-infrared light is so small that the above equation is automatically satisfied in any length of cavity formed by cleaving and many values of m and $\lambda/2$ will satisfy the resonant condition. If m is large, we may write

$$\frac{dm}{d\lambda} = -\frac{2ln_r}{\lambda^2} + \frac{2l}{\lambda}\frac{dn_r}{d\lambda} \tag{6.79}$$

For discrete changes in m and λ,

$$-\delta\lambda = \frac{\lambda^2}{2ln_r}\left(1 - \frac{\lambda}{n_r}\frac{dn_r}{d\lambda}\right)^{-1}\delta m \tag{6.80}$$

and the wavelength change between adjacent modes is

$$\delta\lambda = \frac{\lambda^2}{2ln_r}\left(1 - \frac{\lambda}{n_r}\frac{dn_r}{d\lambda}\right)^{-1} \tag{6.81}$$

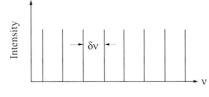

Figure 6.12 Longitudinal modes of a Fabry-Perot cavity.

The different possible frequency components are given by

$$\nu_m = \frac{mc}{2ln_r} \qquad (6.82)$$

The frequency separation between adjacent modes is given by ($\Delta m = 1$)

$$\delta\nu = \frac{c}{2ln_r} \qquad (6.83)$$

which is independent of m. For example, for a GaAs cavity of length 500 μm, the frequency separation is 80 GHz. For all possible values of m, a large number of laser axial modes are possible. However, it is important to realize that only the modes that are within the laser gain curve or spontaneous emission spectrum will grow. Thus, it is important to reduce the linewidth of the laser gain curve, so that only one or two axial modes are produced.

An important parameter is the linewidth of the individual modes. This is determined by phase fluctuations due to noise originating from the spontaneous emission process and is given by the "Schawlow–Townes formula."[†] In a semiconductor laser the spectral linewidth of the individual modes is related to material parameters. The Fabry-Perot cavity, like any other resonant cavity, can be characterized by a Q-factor. This is defined as

$$Q = \frac{\omega_0}{\Delta\omega} \qquad (6.84)$$

where $\Delta\omega$ is the linewidth. The quality factor is a measure of the mirror losses. The *finesse* of the passive cavity is similarly defined as $\frac{2\pi\delta\nu}{\Delta\omega}$. As an example, in a typical junction laser $\delta\nu$ is of the order of tens of GHz, whereas $\Delta\omega/2\pi$ is of the order of 100 KHz.

EXAMPLE 6.4

Objective. The longitudinal modes of a InP injection laser emitting at a wavelength of 0.94 μm are separated by 300 GHz. To determine the length of the optical cavity and the number of longitudinal modes emitted. The refractive index of InP is 3.3.

From Eq. 6.83, the cavity length is

$$l = \frac{c}{2\delta\nu n_r} = \frac{3 \times 10^8}{2 \times 300 \times 10^9 \times 3.3}$$

$$= 151.5 \mu m$$

The mode number m is given by Eq. 6.78 as

$$m = \frac{2n_r l}{\lambda} = \frac{2 \times 3.3 \times 151.5}{0.94}$$

$$\cong 1064$$

[†]See discussions in the following texts: A. E. Siegman, *Lasers*, University Science Books, Mill Valley, CA, 1986 and G. P. Agrawal and N. K. Dutta, *Long Wavelength Semiconductor Lasers*, Van Nostrand Reinhold, New York, 1986.

Of these longitudinal modes, only those within the spontaneous emission spectrum will actually participate in lasing.

As discussed in Sec. 6.2, the laser cavity is formed by a rectangular waveguide, and therefore the propagating modes are TE and TM modes. These modes are usually classified as TE_{pq} and TM_{pq} where p and q give the number of minima in the directions of E and H fields. The lowest-order mode shows a single spot in the output and is called the uniphase mode. Higher-order modes arise from plane waves, which travel slightly off axis due to larger dimensions of the cavity and reinforce themselves after reflection at the mirrors. Higher-order modes can therefore be eliminated by reducing the aperture size of the cavity.

PROBLEMS

6.1 Describe the processes of spontaneous emission, superradiance, and stimulated emission.

6.2 At what wavelength are the rates of spontaneous and stimulated emission equal in a two-level system at room temperature?

6.3 Describe, with suitable diagrams, the basic requirements for lasing and how they are achieved in a semiconductor laser.

6.4 A single mode GaAs laser is to be fabricated with GaAs-$Al_xGa_{1-x}As$ heterojunctions. Using the equations in Sec. 6.2, design a simple heterostructure for an asymmetrical slab waveguide in which the thickness of the active region is 2000 Å. Determine the composition of the ternary layer(s).

6.5 With excitation $h\nu = \mathcal{E}_g$ (bandgap) in a direct bandgap semiconductor laser diode with degenerate n- and p-regions, show that the condition

$$eV > h\nu$$

must be satisfied (where V is the applied bias) for a population inversion and subsequent laser emission. (Assume equal transition probabilities for photon emission and absorption processes.)

6.6 Photons of energy $h\nu_{12} = (\mathcal{E}_2 - \mathcal{E}_1)$ are incident on a two-level system and under thermal equilibrium conditions the rate of upward transitions is equal to the total rate of downward transitions. The Einstein coefficients are B_{12}, B_{21} and A_{21} for stimulated upward, stimulated downward, and spontaneous downward transitions. Show that at high temperatures $B_{12} \cong B_{21}$.

6.7 Under conditions identical to those for Problem 6.6, find an expression for A_{21}/B_{12}, assuming Planck's radiation law.

6.8 Derive Eq. 6.50 where the symbols have meanings outlined in the text.

6.9 Using Eq. 6.51, calculate the gain spectra in InP as a function of injected carrier density.

6.10 From Eq. 5.22, or Eq. 6.44 under low injection conditions and using Boltzmann statistics, show that $R_{sp} \infty\ np$, where n and p are injected carrier densities.

6.11 Discuss some advantages that the laser may have over an LED as a source for optical-fiber communication.

6.12 What steps would you take to improve the (longitudinal) modal purity of a laser? Discuss both material considerations and device design.

6.13 How would you change the geometry of an edge-emitting laser if it was possible to make the reflectance of one of the facets equal to unity? What would be the benefits of such a structure?

READING LIST

ADAMS, M. J. *An Introduction to Optical Waveguides*. Wiley, New York, 1981.

AGRAWAL, G. P., and DUTTA, N. K. *Long Wavelength Semiconductor Lasers*. Van Nostrand Reinhold, New York, 1986.

CASEY, H. C., and PANISH, M. *Heterostructure Lasers: Parts A and B*. Academic Press, New York, 1978.

KRESSEL, H., and BUTLER, J. K. *Semiconductor Lasers and Heterojunction LEDs*. Academic Press, New York, 1977.

SALEH, B. E. A., and TEICH, M. C. *Fundamentals of Photonics*. Wiley, New York, 1991.

SINGH, J. *Physics of Semiconductors and Their Heterostructures*. McGraw-Hill, New York, 1993.

THOMPSON, G. H. B. *Physics of Semiconductor Laser Devices*. Wiley, Chichester, Eng., 1980.

WILSON, J., and HAWKES, J. F. B. *Optoelectronics: An Introduction*, 2nd ed. Prentice Hall International, 1989.

7

Lasers: Structures and Properties

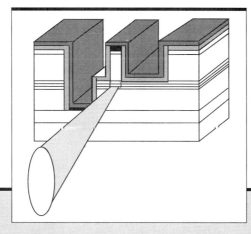

Chapter Contents

7.1 JUNCTION LASER OPERATING PRINCIPLES

The principle of operation of a junction laser is not very different from that of an LED discussed in Chapter 5. We will commence with a discussion of the *homo-junction laser*. In the case of the laser, in addition to the injection provided by the forward-biased p-n junction, there are requirements of stimulated emission and optical feedback. The band diagrams of a simple homojunction are shown in Fig. 7.1. For simplicity, degenerate n- and p-regions are shown, which are almost essential for creating a large density of electrons and holes. A large forward bias is applied to create near-flat-band conditions ($V_a \cong \mathcal{E}_g/q$). Under these conditions, a large density of electrons and holes are injected to the opposite side. In most semiconductor lasers, laser action is made to occur on the p-side by electron injection, since the efficiency of electron injection is higher than that of holes for the same amount of forward bias. Over a diffusion length the injected carrier density is much larger than the equilibrium density, creating nonequilibrium carrier densities and a population inversion in this narrow region, which is also called the *active region*. Over this region electrons and holes recombine radiatively to produce photons and the radiative recombination rate can be high, depending on the injection level. The spontaneously emitted photons can stimulate upward or downward transitions. The former is the absorption process, whereby an electron is raised from the valence band to the conduction band. The latter occurs when the photon stimulates a downward transition of an electron, thereby emitting a photon of the same phase and frequency as the original photon. If the rate of such stimulated emission is made sufficiently high, the medium exhibits gain. One of the conditions for lasing is that in the laser cavity the roundtrip gain must exceed the losses. The losses in active regions of lasers can occur due to free-carrier absorption, scattering at defects and inhomogeneities and other nonradiative transitions.

As stated earlier, external mirrors are not needed in a diode laser and the cleaved planes at the two ends of the waveguide essentially provide the required reflection to

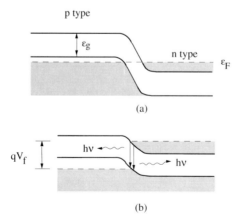

(a)

(b)

Figure 7.1   Degenerate *p-n* junctions: (a) in equilibrium and (b) with forward bias.

form the cavity. The reflectance at the semiconductor/air interface is given by

$$R = \left(\frac{n_{r2} - n_{r1}}{n_{r2} + n_{r1}}\right)^2 \tag{7.1}$$

where n_{r2} and n_{r1} are the refractive indices of the semiconductor and air, respectively. Typically, with $n_{r2} = 3.6$ and $n_{r1} = 1$, $R = 0.32$. The cleavage planes of a crystal are perpendicular to the directions along which the bond strengths are the smallest. For GaAs and other III–V compounds, these are the (110) planes. Therefore, in a typical junction laser the junction coincides with the (100) plane, which is the direction of growth, and the mirrors are formed by cleaving opposite (110) planes perpendicular to the guiding direction.

Within the active region of the junction laser, the gain will eventually exceed the losses with an enhancement in the rate of stimulated emission. Outside the active region the losses will dominate. This is depicted in Fig. 7.2. Population inversion, or gain, occurs over an *active* thickness d, which in a homojunction laser is approximately equal to the diffusion length of the injected carriers. However, the thickness over which the optical mode extends is d_m, which is usually larger than d. The confinement in a homojunction laser is extremely poor, and the active region is defined by the stimulated emission and resulting gain in the region. As the rate of stimulated emission increases in the active region, the roundtrip gain in the cavity overcomes losses and lasing commences. The injection current at which this occurs is called the threshold current, shown schematically in Fig. 7.3. The nature of the spectral output, shown in Fig. 7.4, also changes as incoherent spontaneous emission below threshold is eventually replaced by dominant laser modes above threshold. Below threshold, the spectral output results from the spontaneous recombination in the active region, shown in Fig. 7.4(a). At threshold, the laser modes emerge and are distinctly seen in the output [Fig. 7.4(b)]. These modes can be seen even in an LED due to the boundary conditions imposed by the structure. Beyond threshold, a single narrow peak is observed in the output [Fig. 7.4(c)].

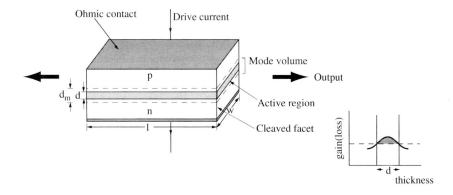

Figure 7.2   Schematics of a broad-area junction laser with cleaved facets.

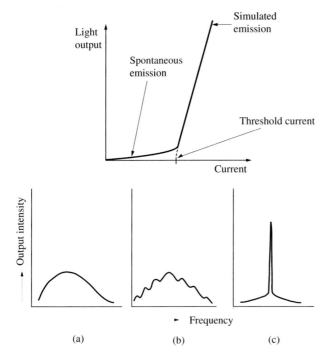

Figure 7.3 Light output-current characteristic of an ideal semiconductor laser.

Figure 7.4 Light output as a function of frequency for a *p-n* junction laser: (a) below threshold, (b) with the laser modes at threshold, and (c) with the dominant laser mode above threshold. The figures are not to scale (from B. G. Streetman, *Solid State Electronic Devices*, 3rd ed., ©1990. Reprinted by permission of Prentice Hall, Englewood Cliffs, NJ).

An interesting analysis of the modal gain in a laser cavity can be done in terms of the peaks and valleys in the output spectrum at the onset of lasing.[†] If after consecutive round trips in the cavity of length l the optical fields interfere constructively, we have an intensity maximum where the expression for field is given by

$$E^+ = E_1 \sum_{p=0}^{\infty} (R_1 R_2)^{p/2} \exp(-p\gamma l) \qquad (7.2)$$

Here E_1 is the incident field, p is an integer, and γ is the net loss coefficient for the waveguide. Equation 7.2 is simply a geometric sum whose constant ratio is less than unity, and therefore will converge to

$$E^+ = \frac{E_1}{1 - (R_1 R_2)^{1/2} \exp(-\gamma l)} \qquad (7.3)$$

Similarly, if the fields interfere destructively, then we have

$$E^- = \sum_{p=0}^{\infty} (R_1 R_2)^{p/2} \exp(-p\gamma l)(-1)^p \qquad (7.4)$$

which converges to

$$E^- = \frac{F_1}{1 + (R_1 R_2)^{1/2} \exp(-\gamma l)} \qquad (7.5)$$

[†]B. W. Hakki and T. L. Paoli, *J. Appl. Phys.*, **44**, 4113, 1973.

Therefore, from Eqs. 7.3 and 7.5,

$$\frac{E^+}{E^-} = \frac{1 + (R_1 R_2)^{1/2} \exp(-\gamma l)}{1 - (R_1 R_2)^{1/2} \exp(-\gamma l)} \tag{7.6}$$

from which, by transposing, we get

$$(R_1 R_2)^{1/2} \exp(-\gamma l) \left(\frac{E^+}{E^-} + 1 \right) = \frac{E^+}{E^-} - 1 \tag{7.7}$$

which leads to

$$\frac{\frac{E^+}{E^-} + 1}{\frac{E^+}{E^-} - 1} = (R_1 R_2)^{-1/2} \exp(\gamma l) \tag{7.8}$$

Taking the natural logarithm of both sides and assuming $R_1 = R_2 = R$, we get

$$\gamma = \frac{1}{l} \ln R + \frac{1}{l} \ln \left(\frac{\frac{E^+}{E^-} + 1}{\frac{E^+}{E^-} - 1} \right) \tag{7.9}$$

It should be noted that it is not common or easy to measure field strengths. Instead, it is usual to measure intensities. Noting that intensity is proportional to the square of the field strength,

$$r = \frac{\Im_P}{\Im_V} = \left| \frac{E^+}{E^-} \right|^2 \tag{7.10}$$

where $\Im_P$ and $\Im_V$ are, respectively, the peak and valley intensities. We now encounter a minor problem. Looking at a typical spectrum, shown in Fig. 7.4(b) we note that peaks and valleys obviously do not occur at the same wavelength, although our intention is to be able to measure gain as a function of wavelength. To circumvent this problem we note that each valley V_i is halfway in wavelength between two adjacent peaks P_i and P_{i+1} and define the ratio of the intensities

$$r_i = \frac{\Im_{Pi} + \Im_{Pi+1}}{2 I_{Vi}} \tag{7.11}$$

The final expression for the net loss in the cavity is then

$$\gamma = \frac{1}{l} \ln R + \frac{1}{l} \ln \left(\frac{r_i^{1/2} + 1}{r_i^{1/2} - 1} \right) \tag{7.12}$$

We can make a few relevant observations from Eq. 7.12. For $r_i \gg 1$, the second term on the rhs goes to zero and

$$\gamma = \frac{1}{l} \ln R \tag{7.13}$$

In other words, the net loss in the cavity (or the guide loss minus gain) exactly balances the mirror losses. Equivalently, the gain is pinned at threshold. This is in agreement with what we had said earlier (i.e., the appearance of the multiple peaks in the spectral output is the attainment of threshold and onset of lasing). For $r_i \cong 1$, the loss becomes very large. In other words, the pumping or injection is insufficient.

7.2 THRESHOLD CURRENT

7.2.1 Threshold Current Density of a Semiconductor Laser Treated as a Two-Level System

The threshold population inversion in a two-level lasing medium is given by Eq. 6.77. However, in a junction laser, where population inversion is strictly not the quantity $(N_2 - N_1)$, but is caused by an injection current, a threshold current density is more meaningful. Remember also that in a semiconductor laser, electrons and holes are distributed in bands. Nonetheless, it is interesting to try to calculate the threshold current of a junction laser assuming the semiconductor to be a two-level system. In other words, the carriers are assumed to occupy single discrete levels in the bands and N_{th} in Eq. 6.77 is assumed to be equivalent to the threshold density of injected carriers n_{th}. In regions where the mode extends beyond the active region, the losses may be larger than the gain resulting from stimulated emission. To offset this, the threshold injection or pumping needs to be increased. Therefore,

$$n_{th} = \frac{d_m}{d} \left(\frac{8\pi v_0^2 g_{th} \tau_r \Delta v n_r^2}{c^2} \right) \tag{7.14}$$

where d_m and d are shown in Fig. 7.2. Let the current density through the laser diode be J. Therefore, the number of electrons injected per second per unit volume of the active region is J/qd. This number of carriers is removed, in the steady state, by recombination with a lifetime τ, in accordance with the rate equation

$$\frac{dn}{dt} = \frac{J}{qd} - \frac{n}{\tau} = 0 \tag{7.15}$$

Also,

$$\frac{1}{\tau_r} + \frac{1}{\tau_{nr}} = \frac{1}{\tau} \tag{7.16}$$

where τ_{nr} is the nonradiative recombination lifetime and τ_r is the spontaneous radiative recombination lifetime. At threshold,

$$\frac{n_{th}}{\tau} = \frac{J_{th}}{qd} \tag{7.17}$$

Therefore, from Eq. 7.14,

$$J_{th} = \frac{qd_m}{\tau} \left(\frac{8\pi v_0^2 g_{th} \tau_r \Delta v n_r^2}{c^2} \right) \tag{7.18}$$

Substitution for g_{th} from Eq. 6.56 leads to

$$J_{th} = \frac{8\pi v_0^2 q d_m \tau_r n_r^2 \Delta v}{\tau c^2} \left[\gamma + \frac{1}{2l} \ln \left(\frac{1}{R_1 R_2} \right) \right] \tag{7.19}$$

A close examination of this equation, which gives a form of the threshold current in a junction laser, reveals several serious flaws. First, J_{th} does not depend on d, the active layer thickness. Instead, it is proportional to d_m. If the active region is wide enough, or if the mode is sufficiently confined, then $d \cong d_m$ and the discrepancy is removed. A more serious problem is the lack of any temperature dependence. This comes about because we had assumed that carriers occupy a single level in the band. Therefore, in reality, Eq. 7.19 is not a valid expression for the threshold current of a junction laser.

7.2.2 Threshold Current Density from the Spontaneous Emission Rate

The current flowing in a junction laser is simply given by

$$J = q d R_{sp} \quad (A.cm^{-2}) \tag{7.20}$$

where R_{sp} $(cm^{-3}.s^{-1})$ is the total spontaneous recombination rate given by Eq. 5.22 and d is the width of the active region. The quasi-Fermi levels are calculated for a range of injected carrier concentrations using Fermi statistics. For an undoped active region it is usual to assume $n = p$. The values of $R_{sp}(\mathcal{E}, n)$ and $g(\mathcal{E}, n)$ can then be calculated using Eqs. 5.22 and 6.51, respectively. Calculated values of $g(\mathcal{E}, n)$ are shown in Fig. 7.5. The peak gain $g(\mathcal{E}_m, n)$ is then plotted against injected carrier density, as schematically shown in Fig. 7.6. For most lasers this relationship is superlinear but shown linearity in some region. The value of n at which the peak gain matches the computed cavity loss for known values of γ, Γ, l, R_1, and R_2 is the threshold concentration n_{th}. The spontaneous recombination rate $R_{sp}(n_{th})$ is then

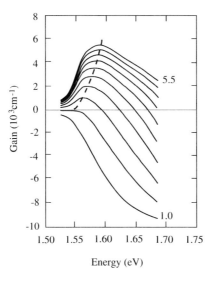

Figure 7.5 Calculated TE mode gain coefficient in a 50 Å GaAs/Al$_{0.3}$Ga$_{0.7}$As lattice-matched quantum well laser at 300°K for various carrier injections (10^{12} carriers/cm^2) in steps of 0.5 $\times 10^{12}$/cm^2. The dashed line is the photon energy ($\mathcal{E}_m$) at which the gain coefficient is maximum (courtesy of J. Singh and J. P. Loehr, University of Michigan).

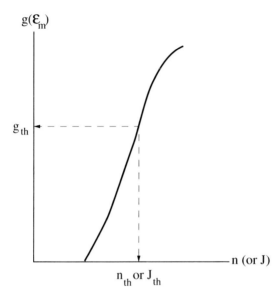

Figure 7.6 Schematic plot of peak gain versus injection current (or carrier density). J_{th} usually corresponds to a point in the linear region of the plot.

known and the threshold current density is given by

$$J_{th} = \frac{qd}{\eta_i} R_{sp}(n_{th}) \tag{7.21}$$

At high values of injection current (or injected carrier density) the gain tends to saturate, and the slope of the gain curve in Fig. 7.6 decreases. Gain saturation occurs due to the limitation imposed by the carrier density of states. The latter dictates the number of carriers that can be injected to produce photons of a specified energy $\mathcal{E}$.

We will now discuss the transparency condition in a laser. A related parameter is the *transparency current* J_{th}^o. From the rate equation (7.15) we get, under steady-state conditions,

$$n = \frac{\tau}{qd} J \tag{7.22}$$

If we assume that the peak modal gain $\Gamma g(\mathcal{E}_m, n)$ is linearly related to n near the point of operation, it can be expressed as

$$\Gamma g(\mathcal{E}_m) \cong \alpha \left(\frac{n}{n_{nom}} - 1 \right) \tag{7.23}$$

where α is the absorption coefficient of the material. n_{nom} is the injected carrier density that makes the medium transparent. In other words, emission and absorption are balanced and $g(\mathcal{E}_m) = 0$. A typical value of n_{nom} in high-quality InP-based quantum well lasers is of the same order (but $\sim$ a factor of 2 lower) as n_{th}. In heterostructure lasers the difference is larger. From the last two equations, we can write

$$\Gamma g(\mathcal{E}_m) = \alpha \left(\frac{J}{J_{th}^o} - 1 \right) \tag{7.24}$$

where the transparency current density J_{th}^o is given by

$$J_{th}^o = \frac{qd}{\tau} n_{nom}$$

$$= \frac{qd}{\eta_i \tau_r} n_{nom} \tag{7.25}$$

where $\eta_i = \tau/\tau_r$ from Eq. 3.13. At threshold $\Gamma g(\mathcal{E}_m) = \Gamma g_{th}$ is equal to the total losses $(\gamma + \frac{1}{l} \ln \frac{1}{R})$. It follows from Eq. 7.24,

$$J_{th} = \frac{\Gamma g_{th} + \alpha}{\alpha} J_{th}^o \tag{7.26}$$

Therefore, the threshold current density is larger than the transparency current unless α is much larger than the total cavity losses including the mirror losses. Substitution of Eqs. 7.23 and 7.25 into Eq. 7.26 leads to

$$J_{th} = J_{th}^o + \frac{qd}{\Gamma \eta_i \tau_r \frac{\partial g}{\partial n}} \left[\gamma + \frac{1}{l} \ln \frac{1}{R} \right] \tag{7.27}$$

where $\frac{\partial g}{\partial n} = \frac{g(\mathcal{E}_m)}{n} = \frac{\alpha}{\Gamma n_{nom}}$ is defined as the *differential gain* of the medium. Equation 7.27 enables the experimental determination of the transparency current, provided the value of the differential gain is known from an independent measurement, to be described later. When the cavity length becomes very large, or $\frac{1}{l} \cong 0$, the mirror losses are negligible and

$$J_{th(\frac{1}{l}=0)} = J_{th}^o + \frac{qd\gamma}{\Gamma \eta_i \tau_r \frac{\partial g}{\partial n}} \tag{7.28}$$

Measurements are made on lasers of different cavity lengths l. A plot of J_{th} versus $1/l$, as shown schematically in Fig. 7.7(a), yields the intercept on the J_{th} axis, and from it the transparency current J_{th}^o, provided γ, τ_r, Γ, and η_i are also known. Figure 7.7(b) shows schematically the plot of I_{th} versus cavity length l. Phenomenologically, it can be seen that the overall curve is a combination of the two dashed curves. The linear curve represents the fact that the threshold current increases with cavity length (J_{th} remaining constant). The hyperbolic curve indicates that as the cavity length decreases, the mirror losses become dominant.

One of the challenges in laser research and development is the minimization of the threshold current J_{th}. From Eqs. 7.25 and 7.27 it is apparent that J_{th} can be reduced by reducing d, increasing η_i, increasing the differential gain and reducing the cavity and mirror losses. With the other parameters remaining constant and $\Gamma \cong 1$, J_{th} decreases with decrease of d. However, with further reduction of d, Γ becomes less than unity as the optical mode extends beyond the active region. As a result the threshold current increases again. This behavior is schematically shown in Fig. 7.7(c). In a heterojunction laser, which we will learn about in the next section, Γ remains nearly equal to unity due to confinement of the mode within the well-defined active region. Much lower threshold current densities are therefore achieved.

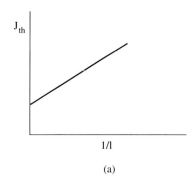

(a)

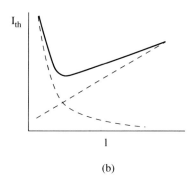

(b)

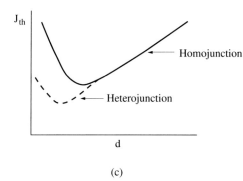

(c)

Figure 7.7 Schematic representation of (a) variation of J_{th} with inverse cavity length; (b) variation of I_{th} with cavity length; and (c) variation of J_{th} with active region thickness.

EXAMPLE 7.1

Objective.   To estimate the behavior of the light-current characteristics of a junction laser.

A simple insight of the light-current characteristics of a laser below and above threshold can be obtained as follows. Assume a laser cavity that has $l = 200\mu$m, $W = 10\mu$m, and $d = 0.1\mu$m. The active volume V_{ca} is therefore $2 \times 10^{-10} cm^3$. The photon flux can be assumed equal to

$$R_{sp} = \frac{n}{\tau_r} = \frac{I}{qV_{ca}}$$

where I is the injection current. The photon flux therefore increases linearly with current. Remember that $\tau_r \cong \tau_o \sim 0.6$ns (e.g., in GaAs). The threshold condition is gradually approached and assume that $I_{th} = 100$ mA, which is typical in a heterostructure laser. The corresponding $n_{th} = 1.9 \times 10^{18} cm^{-3}$. As the current is increased above threshold, the carrier concentration does not increase much above n_{th} since the photon output increases many times. This is expressed by

$$\frac{I}{qV_{ca}} = \theta \frac{n_{th}}{\tau_{st}}$$

where τ_{st} is the lifetime for stimulated emission and θ represents the fraction of the spontaneous emission, couples into the cavity mode satisfying the requirements of a *coherent* photon. It has a value ranging from 10^{-1} to 10^{-5} depending on the cavity volume and operating wavelength. If the laser is biased above threshold at 200 mA and $\theta = 10^{-1}$, $\tau_{st} = 30$ps. Therefore, the output in the stimulated emission regime is nearly 20 times that in the spontaneous emission region. A break, as shown in Fig. 7.3, is observed in the light-current characteristics. This aspect will be discussed in a later section.

7.2.3 Power Output

As the injection current exceeds the threshold current, laser oscillations build up with a large rate of stimulated emission. However, the latter reduces the population inversion, and this fluctuation continues for a few cycles until the current is clamped at the threshold value. We will later study these transient effects. With reference to the light-current output of a laser (Fig. 7.3), the optical power output into the modal volume due to a current $J(> J_{th})$ can be expressed as

$$P = A(J - J_{th})\frac{\eta_i h\nu}{q} \tag{7.29}$$

where A is the junction area and η_i is the internal quantum efficiency of the semiconductor. Equation 7.29 expresses the fact that the number of photons generated in the modal volume per second is equal to the number of electrons injected per second times the internal quantum efficiency. Now a fraction of this power is coupled out through the cleaved facets as useful laser output and the rest is used to overcome the losses, γ, within the cavity. Therefore, the output power of the laser is

$$P_0 = A(J - J_{th})\left(\frac{\eta_i h\nu}{q}\right)\frac{\frac{1}{2l}\ln\left(\frac{1}{R_1 R_2}\right)}{\left[\gamma + \frac{1}{2l}\ln\left(\frac{1}{R_1 R_2}\right)\right]} \tag{7.30}$$

The differential quantum efficiency, η_d, is defined as the increase in light output due to an increase in the drive current. Thus,

$$\eta_d = \frac{d(P_0/h\nu)}{d\left(\frac{A}{q}|J - J_{th}|\right)} \tag{7.31}$$

which, on substitution of Eq. 7.30, is expressed as

$$\eta_d = \eta_i \left[\frac{\ln \left(\frac{1}{R} \right)}{\gamma l + \ln \left(\frac{1}{R} \right)} \right] \tag{7.32}$$

where $R_1 = R_2 = R$. From this equation it is evident that η_i and γ can be estimated from a measured dependence of η_d on cavity length l. An important measure of the overall efficiency of the laser is the power conversion efficiency, η_P, defined as

$$\eta_P = \frac{P_0}{V_f A J} \tag{7.33}$$

where V_f is the applied forward bias. Substitution of Eq. 7.30 leads to

$$\eta_p = \frac{\eta_i h\nu}{q V_f} \left(\frac{J - J_{th}}{J} \right) \frac{\ln \left(\frac{1}{R} \right)}{\gamma l + \ln \left(\frac{1}{R} \right)} \tag{7.34}$$

For high-injection (flat-band) conditions $q V_f \cong \mathcal{E}_g \cong h\nu$ and $J \gg J_{th}$. Also, by optimum coupling of light out of the cavity $\frac{1}{l} \ln (\frac{1}{R})$ can be made much larger than γ, the cavity losses. Under these conditions $\eta_p \cong \eta_i$. Thus, if η_i is large, then the external power efficiency of the laser can also be made large.

EXAMPLE 7.2

Objective.   To calculate the efficiency of a GaAs laser that operates with a drive current much larger than J_{th}. The cavity length is 250 μm, the cavity loss coefficient is 10 cm^{-1}, and the internal quantum efficiency is unity.

$R = 0.32$ for GaAs cleaved facets. In Eq. 7.34, $J \gg J_{th}$ and $V_f \cong \frac{h\nu}{q}$ leads to

$$\eta_p \cong 1.0 \times \frac{\ln \left(\frac{1}{0.32} \right)}{(250 \times 10^{-4} \times 10) + \ln \left(\frac{1}{0.32} \right)}$$

$$= 0.82$$

Therefore, η_p has a value close to η_i (1.0).

7.2.4 Temperature Dependence of Threshold Current

In spite of very efficient heat sinking, the temperature of a laser diode may increase with time because of the large injection current density. It is desirable to have the lasing characteristics and, in particular, the threshold current, independent of device or ambient temperature. Unfortunately, all junction lasers exhibit a temperature dependent threshold current, as shown in Fig. 7.8, which can be described by the empirical equation:

$$J_{th} = J_{th0} \exp \left(\frac{T}{T_0} \right) \tag{7.35}$$

where T is the device temperature and T_0 is the threshold temperature coefficient. As seen in Fig. 7.8, the threshold current increases with increasing temperature. The

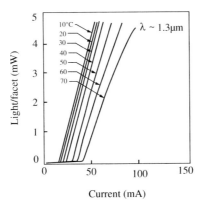

Figure 7.8 Light-current characteristics of an etched mesa buried heterostructure laser at different temperatures (from G. P. Agrawal and N. K. Dutta, *Long Wavelength Semiconductor Lasers*, Van Nostrand Reinhold, New York, ©1986. Reprinted by permission of AT&T Bell Laboratories).

degree of temperature dependence depends on the value of T_0, which in turn, is related to the material used and the type of device structure. In general, T_0 is higher in wider bandgap materials, signifying a weaker temperature dependence of J_{th}. For example, in AlGaAs lasers, T_0 ranges from 100° to 200°K. The increase in threshold current with temperature can be qualitatively accounted for by considering the increasing spread in energy of carriers in the conduction and valence bands. However, a much stronger temperature dependence has been observed in InGaAsP lasers used for optical-fiber communication and there must be an additional factor to account for this. As the bandgap is reduced, the efficiency of Auger recombination, which is a nonradiative process, increases. It is believed that this is mainly responsible for low values of T_0 (40°–80°K) in devices made with this material. Carrier leakage from the active region is also cited as a cause for low T_0.

EXAMPLE 7.3

Objective.   To compare the ratios of J_{th} at 20°C and 80°C for a AlGaAs injection laser with $T_0 = 160°$K and a InGaAsP laser with $T_0 = 55°$K.

For the AlGaAs laser $\frac{J_{th}(20°C)}{J_{th}(80°C)} = 0.68$

For the InGaAsP laser $\frac{J_{th}(20°C)}{J_{th}(80°C)} = 0.36$ It is obvious that there is a greater variation in the threshold current of the InGaAsP laser with temperature.

7.3 HETEROJUNCTION LASERS

The device described until now is a homojunction laser, since the p- and n-regions are made of the same semiconductor. The drawbacks of such a laser are twofold. First, the active region d is ill defined by the diffusion length of the carriers. Second, the thickness d_m defining the mode volume is also ill defined, and can far exceed d. The cavity losses will therefore increase. From Eq. 7.27 it is evident that large values of d and γ will result in a large value of J_{th}. Therefore, such homojunction lasers cannot be operated continuously at room temperature (continuous wave, or cw operation), which is a major drawback. To reduce the drive current both the carriers and the optical modes need to be confined, and these can be achieved with the use of heterojunctions.

The schematics, band diagram, and radiation field distribution of single- and double-heterojunction lasers are shown in Fig. 7.9. Carrier confinement is effectively achieved in a single-heterojunction (SH) laser, while optical confinement on *both* sides of the active region is obtained with a double-heterojunction (DH) structure. To understand carrier confinement, consider the single-heterojunction structure shown in Fig. 7.9(a). The electrons injected from the n^+ GaAs layer to the p GaAs active layer is blocked from diffusing over a large distance by the potential barrier provided by the higher-bandgap p-AlGaAs layer. Thus, the active region thickness d is the thickness of the p-GaAs layer, and if d is made small then the same carrier injection level can be obtained with a smaller drive current. Therefore, the power efficiency of the laser improves. The higher bandgap of the AlGaAs layer not only confines carriers, but also the optical modes, since the refractive index of the higher bandgap layer is smaller than that of the GaAs active region. It is easy to see that the double-heterojunction laser shown in Fig. 7.9(b) confines the modes on both sides of the active region and reduces the cavity loss γ. The threshold current will there-

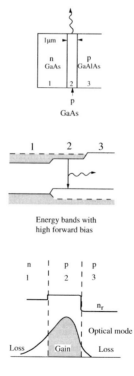

Figure 7.9 Layer structure, band diagram, refractive index profile, and optical confinement in (a) single-heterostructure and (b) double-heterostructure lasers.

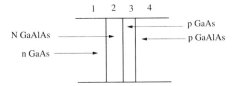

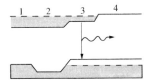

Energy bands with high forward bias

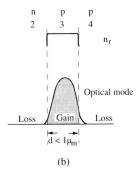

(b)   **Figure 7.9** (*continued*)

fore be lower in a DH laser. Typical values of d in DH lasers range from 0.1 to
0.3 μm.

Further improvements in device performance can be effected by reducing the
transverse dimension of the active region to 2–10 μm. Transverse stripe geometry
can be formed in three different ways, schematically shown in Fig. 7.10. In the
first, a well-defined ridge is formed by mesa-etching, and the entire ridge is exposed
to air, or covered by a dielectric layer. In the second, the transverse dimension is
defined by proton bombardment, and in the third, mesa-etching is followed by the
regrowth of a wider bandgap layer for better index matching and carrier confinement.
Since the active layer is buried, these lasers are also called buried stripe geometry lasers. With the reduction of the transverse dimension of the active region, the
current to achieve the same carrier injection level is greatly reduced. For example,
the operating current of homojunction lasers can be several amperes, while that of
DH transverse stripe geometry lasers are approaching 100 mA with output powers of
$\sim$ 10mW. The stripe geometry limits the lateral spread of injected carriers inside the

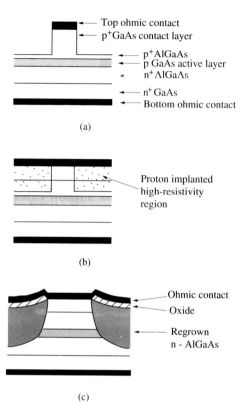

(a)

(b)

(c)

Figure 7.10 Different schemes of forming transverse stripe-geometry lasers: (a) mesa-etching; (b) proton implantation; and (c) mesa-etching and regrowth. The heterostructure layers in (b) and (c) are the same as in (a).

active layer. Stripe geometry heterostructure lasers in which the lateral variation of the optical gain confines the optical mode under the stripe are also called *gain-guided lasers*. On the other hand, heterostructure lasers in which mode confinement occurs mainly through lateral variations of the refractive index are termed *index-guided lasers*.

Stripe geometry lasers provide a few other important advantages. The output is usually of single mode, and because of the small dimensions of the active region, the output light can be easily coupled to fibers or an integrated waveguide. This becomes very important for practical applications. Also, the smaller geometry of the active region removes the "kinks" usually observed in the light-current characteristics of broad-area lasers. The anomaly is caused by the active region being divided into narrow filaments due to the interaction of the optical mode with the carrier distribution arising from the refractive index profile. The output appears to originate from several waveguides adjacent to each other, rather than uniformly from one guide. The nonlinearities and kinks represent power switching between the dominant lateral modes. As the transverse geometry of the active region is reduced, the kinks are gradually eliminated, as seen in Fig. 7.11.

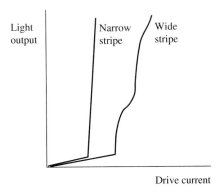

Figure 7.11   Light-current characteristics of a junction laser as a function of stripe width. "Kinks" are observed for wide stripes, as discussed in the text.

We will now analyze[†] the carrier confinement properties of a single-heterojunction laser, shown schematically in Fig. 7.12(a), where it is assumed that the GaAs active region is p-type, and the doping in this region is generally different from the p-doping in the wide-bandgap AlGaAs confining layer alongside. It is also assumed

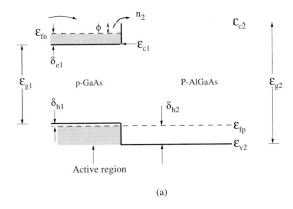

(a)

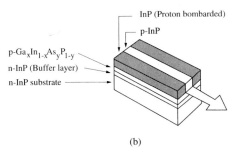

(b)

Figure 7.12   (a) Electron injection and emission in a forward-biased single heterojunction (from H. Kressel and J. K. Butler, *Semiconductor Lasers and Heterojunction LEDs*, Academic Press, New York, 1977); and (b) Schematics of a InGaAsP/InP DH laser.

[†]See H. Kressel and J. K. Butler, *Semiconductor Lasers and Heterojunction LEDs*, Academic Press, New York, 1977.

that the injection level is high, so that the quasi-Fermi level is deep in the bands. This reduces the effective confining barrier ϕ. Interfacial spikes at the p-P heterojunction are neglected. We denote parameters in the active region with a subscript 1 and those in the confining layer with subscript 2. The quasi-Fermi levels in the active region are $\mathcal{E}_{fn}$ and $\mathcal{E}_{fp}$. In terms of the energies denoted in Fig. 7.12(a), the effective energy barrier confining electrons is

$$\phi = \Delta\mathcal{E}_g - \delta h_1 - \delta e_1 - \delta h_2 \tag{7.36}$$

where

$$\Delta\mathcal{E}_g = \mathcal{E}_{g2} - \mathcal{E}_{g1} \tag{7.37}$$

We are interested in deriving an expression for the threshold current considering leakage of injected electrons from the active region of thickness d by diffusion. Remember that hole injection from the active layer to the adjacent n-layer can occur too, but this effect is small and is being neglected. Then the true, or observed, threshold current density is the sum of the injected current density J_{th}^i and the diffusion current J_{diff}. In other words

$$J_{th} = J_{th}^i + J_{diff} = \frac{q n_{th} d}{\tau} + \frac{q D_{e2} n_2}{L_{e2}} \tag{7.38}$$

where n_{th} is the injected carrier density and n_2 is the minority carrier (electron) density in the p - AlGaAs layer at the p - GaAs/P-AlGaAs heterointerface. n_{th} and n_2 can be evaluated by considering the appropriate density of states function and Fermi-Dirac statistics. Thus, the excess injected electron density in the active region, n_{th}, is given by

$$n_{th} = \frac{4\pi}{h^3} (2m_{e1}^*)^{3/2} \int_{\mathcal{E}_{c1}}^{\infty} \frac{(\mathcal{E} - \mathcal{E}_{c1})^{1/2}}{1 + \exp[(\mathcal{E} - \mathcal{E}_{fn})/k_B T]} d\mathcal{E} \tag{7.39}$$

Knowing n_{th}, the position of the quasi-Fermi level at the hetero-interface, $\mathcal{E}_{fn}$, can be determined. Then the electron density n_2 at the edge of wide-bandgap AlGaAs layer can be determined from

$$n_2 = \frac{4\pi}{h^3} (2m_{e2}^*)^{3/2} \int_{\mathcal{E}_{c2}}^{\infty} \frac{(\mathcal{E} - \mathcal{E}_{c2})^{1/2}}{1 + \exp[(\mathcal{E} - \mathcal{E}_{fn})/k_B T]} d\mathcal{E} \tag{7.40}$$

If $(\mathcal{E}_{c2} - \mathcal{E}_{fn}) \geq 3k_B T$, Boltzmann statistics is valid, and

$$n_2 = N_{c2} \exp[-(\mathcal{E}_{c2} - \mathcal{E}_{fn})/k_B T] \tag{7.41}$$

where N_{c2} is the effective density of states in the higher-bandgap layer. The diffusion current can then be expressed as

$$J_{diff} = \frac{q D_{e2} N_{c2}}{L_{e2}} \exp\left(-\frac{\phi}{k_B T}\right) \tag{7.42}$$

Under conditions of high-injection $\delta h_1 \cong 0$. Also, in the AlGaAs layer

$$\delta h_2 \cong k_B T \ln\left(\frac{p_2}{N_{V2}}\right) \tag{7.43}$$

using the Boltzmann approximation, where p_2 is the hole concentration. Therefore, Eq. 7.42 becomes

$$J_{diff} = \frac{q D_{e2} N_{c2}}{L_{e2}} \left(\frac{N_{V2}}{p_2} \right) \exp \left[-\frac{(\Delta \mathcal{E}_g - \delta e_1)}{k_B T} \right] \qquad (7.44)$$

and the total threshold current becomes

$$J_{th} = \frac{q n_{th} d}{\tau} + \frac{q D_{e2} N_{c2} N_{V2}}{L_{e2} p_2} \exp \left[-\frac{(\Delta \mathcal{E}_g - \delta e_1)}{k_B T} \right] \qquad (7.45)$$

A similar analysis can be done for a n-N heterojunction. Some relevant remarks can be made at this point. For high (p-type) doping in the confining AlGaAs layer, it is reasonable to assume $p_2 \cong N_{V2}$. The leakage current can then be expressed as

$$J_{diff} = \frac{q D_{e2} N_{c2}}{L_{e2}} \exp \left[-\frac{(\Delta \mathcal{E}_g - \delta e_1)}{k_B T} \right] \qquad (7.46)$$

and it can be shown that the value of J_{diff} is greatly reduced. It is evident from Eq. 7.46 that the leakage current will increase if N_{c2} is high. This can happen if the AlGaAs layer is indirect bandgap, in which case the effective masses and density of states in the L or X minima are much higher than those in the Γ minimum.

Another important factor that will increase the threshold current is interfacial recombination at the heterointerface. This region may contain interface states and traps and the quality of epitaxy becomes important in realizing perfect heterostructures. In mismatched heterostructures, there is the added complexity due to possible dislocations.

7.3.1 Losses in Heterostructure Lasers

The oscillation or threshold condition is given by Eq. 6.68. The end or mirror losses depend on the cavity length and the facet reflectivities. All the other losses are included in γ. The two main components of γ are the free-carrier loss α_{fc} and the scattering loss. The latter results from irregularities and defects in the active layer and at the heterostructure interfaces. Free-carrier absorption is an unavoidable loss mechanism in lasers. It results from the scattering of injected free carriers due to the various mechanisms that limit carrier mobility. Therefore, in indirect bandgap semiconductors, in which the scattering rate is high, laser oscillations are never achieved. By fitting data, the free-carrier absorption rate near the band edge of GaAs can be expressed by the empirical relation[†]

$$\alpha_{fc}(cm^{-1}) = 3 \times 10^{-18} n + 7 \times 10^{-18} p. \qquad (7.47)$$

The scattering loss is due to scattering of the propagating modes out of the active guiding region by dielectric or heterostructure imperfections. A related loss mechanism is the coupling loss, which results from the optical modes spreading beyond the wider energy gap cladding layers.

[†]H. C. Casey and M. Panish, *Heterostructure Lasers, Parts A and B*, Academic Press. New York, 1978.

7.3.2 Heterostructure Laser Materials

The first heterostructure system that has been used very successfully for the fabrication of lasers is the GaAs/Al$_x$Ga$_{1-x}$As system. The lattice mismatch between GaAs and AlAs is only 0.06%. A consequence of this is that the hetero-interfaces are near perfect and free from interfacial defects and dislocations, which may give rise to nonradiative recombination. Lasers have been grown with Al$_x$Ga$_{1-x}$As active regions, since the alloys remain direct bandgap until $x = 0.4$. This provides tunability in the emission wavelength. GaAs/AlGaAs heterostructure lasers can be easily grown by all the epitaxial techniques discussed in Chapter 1. In addition, with the use of meltback and regrowth techniques, transverse stripe geometry lasers are realized. From the point of view of cost and materials purity, liquid phase epitaxy remains the most common technique for realizing heterostructure lasers. The most serious limitation of GaAs/AlGaAs lasers is that the emission wavelength is around 0.7–0.8 μm, whereas with present optical fibers, it is desirable to have lasers emitting in the wavelength region of 1.1 to 1.6 μm. Therefore, for optical-fiber communication, another heterostructure system is needed. Nonetheless, GaAs/AlGaAs lasers will remain useful for a host of common applications and for short-distance communication and local-area networks, where fiber loss and dispersion do not become very critical.

Two heterostructure systems have emerged as contenders for use in optical-fiber communication. With reference to Fig. 1.2 these are the In$_{1-x}$Ga$_x$As$_y$P$_{1-y}$/InP and In$_x$Ga$_y$Al$_{(1-x-y)}$As/In$_{0.52}$Al$_{0.48}$As systems, both lattice matched to InP. In the first, the active region is made of the quaternary compound In$_{1-x}$Ga$_x$As$_y$P$_{1-y}$, which spans the bandgap energy range from InP ($\mathcal{E}_g = 1.35\,eV$; $x = 0$, $y = 0$) to In$_{0.53}$Ga$_{0.47}$As($\mathcal{E}_g = 0.74\,eV$; $x = 0.47$, $y = 1.0$). This heterostructure system is more easily grown by LPE, OMVPE and gas-source MBE. In the second system, the active region is made of the lattice-matched quaternary In$_x$Ga$_y$Al$_{1-x-y}$As alloy, which spans the same energy (wavelength) range as the P-containing compounds. In this alloy, there is mixing in only the group III sublattice, whereas in the former, there is mixing in both group III and group V sublattices. From the point of view of epitaxial techniques used, the Al-containing compounds are easier to grow. However, the quantum efficiency of the P-containing quaternaries have generally been higher. It must be remembered that only a single composition is lattice matched to InP for a particular bandgap energy. This places stringent limitations on growth parameters. Nonetheless, InP-based heterostructure lasers, such as one shown in Fig. 7.12(b), have become important as optical sources in fiber communication. The bandstructure parameters of InGaAsP are listed in Table 7.1.

There is growing interest in developing both visible lasers and far-infrared sources in the 2–10 μm range. For the former, the III–V compounds of interest are InGaAlP and the nitrides. Other contenders are the II–VI compounds CdZnS and CdSSe. For infrared applications, InAs, InSb and their ternary derivatives are important. An important contender is the II–VI HgCdTe system, which is also being grown very successfully by a variety of epitaxial techniques. Some of these materials may require low-temperature operation. Similarly for emission wavelengths in the range of 10–30 μm, IV–VI alloys such as PbSnTe are used. These light sources find applications in

TABLE 7.1 BAND-STRUCTURE PARAMETERS OF $In_{1-x}Ga_xAs_yP_{1-y}$ LATTICE MATCHED TO InP ($y = 2.16x$) (from G. P. Agrawal and N. K. Dutta, *Long Wavelength Semiconductor Lasers*, Van Nostrand Reinhold, New York, 1986. Reprinted by perimission of copyright ©AT&T Bell Laboratories).

Parameter	Dependence on the Mole Fractions x and y
Energy gap at zero doping	$\mathcal{E}_g$ (in eV) $= 1.35 - 0.72y + 0.12y^2$
Heavy-hole mass	$m^*_{hh}/m_0 = (1-y)[0.79x + 0.45(1-x)] + y[0.45x + 0.4(1-x)]$
Light-hole mass	$m^*_{lh}/m_0 = (1-y)[0.14x + 0.12(1-x)] + y[0.082x + 0.026(1-x)]$
Dielectric constant	$\epsilon = (1-y)[8.4x + 9.6(1-x)] + y[13.1x + 12.2(1-x)]$
Spin-orbit splitting	$\Delta(eV) = 0.11 + 0.31y - 0.09y^2$
Conduction-band mass	$m_{ce}/m_0 = 0.080 - 0.039y$

the automotive industry. Various compound semiconductors used for lasers are shown in Fig. 7.13.

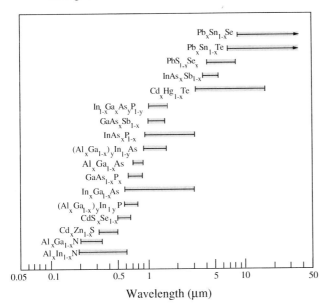

Figure 7.13 Compound semiconductors for lasers. The narrow bandgap devices require cooling for operation (adapted from B. E. A. Saleh and M. C. Teich, *Fundamentals of Photonics*, ©1991. Reprinted by permission of Wiley, New York).

7.4 DISTRIBUTED FEEDBACK LASERS

7.4.1 Introduction

In the junction lasers discussed so far, feedback necessary for lasing action is obtained by the cavity facets formed by cleaving. Spectral emission of higher modal purity can be obtained in a very ingenious way by using a corrugated structure or grating. In such a structure, a periodic variation of the refractive index within the cavity along the direction of wave propagation is produced. Feedback in such a cavity occurs due to the energy of the wave propagating in the forward direction being continuously fed back in the opposite direction by Bragg diffraction at the corrugation, or grating. Lasers using such corrugations as feedback elements are called *distributed-feedback* (DFB) or *distributed-Bragg reflector* (DBR) lasers. The first DFB lasers utilized thin film dyed gelatin on a glass substrate and were investigated and analyzed by Kogelnik and Shank.[†] At least three approaches have been used to analyze laser action in these periodic structures, which differ slightly in their formalism. The coupled-wave analysis of Kogelnik and Shank assumes plane-wave propagation in the periodic structure, from which the threshold conditions can be obtained.

7.4.2 Coupled-Mode Theory

Consider a waveguide with a permittivity profile $\epsilon(x, y)$ through which light propagates in the z-direction, as shown in Fig. 7.14(a). The electric field of the electromagnetic wave is comprised of the guided modes

$$\mathbf{E} = \sum_m a_m \mathbf{E}_m(x, y) e^{j(\beta_m z - \omega t)} \tag{7.48}$$

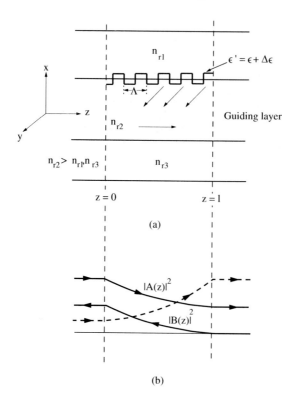

Figure 7.14 Schematic representation of the principle of distributed feedback (DFB) laser: (a) corrugated index profile in guiding region; (b) incident and reflected intensities in corrugated section; and (c) output spectrum (from G. Keiser, *Optical Fiber Communications*, 2nd ed., McGraw-Hill, New York, 1991).

[†]H. Kogelnik and C. V. Shank, *Appl. Phys. Lett.*, **18**, 152, 1971; *Jour. Appl. Physics*, **43**, 2327, 1972.

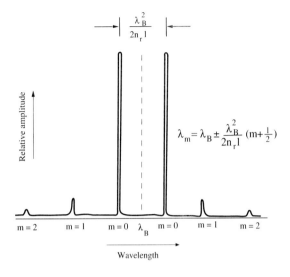

(c)

Figure 7.14 (*continued*)

where a_m is not a function of z. The corrugation, or waveguide perturbation, is described by

$$\epsilon'(x, y, z) = \epsilon(x, y) + \Delta\epsilon(x, y, z) \tag{7.49}$$

where

$$\Delta\epsilon(x, y, z) = \epsilon_0 \Delta n_r^2(x, y, z) \tag{7.50}$$

The propagating modes must still satisfy the wave equation

$$[\nabla^2 + w^2 \mu \epsilon']\mathbf{E} = 0 \tag{7.51}$$

The solutions are expressed as

$$\mathbf{E} = \sum_m A_m(z)\mathbf{E}_m(x, y)e^{j(\beta_m z - \omega t)} \tag{7.52}$$

where the modal amplitudes depend on the z-coordinate.

Now assume that only two modes in the periodic structure are strongly coupled. The contribution of the other modes to the points at which Bragg diffraction occur is minimal. We identify them with coefficients A and B and set

$$A = a(z)e^{-j\Delta\beta z/2},$$

$$B = b(z)e^{j\Delta\beta z/2} \tag{7.53}$$

The quantity $\Delta\beta$ needs some explanation. We assume the period of the perturbation $\Delta n_r^2(x, y, z)$ to be Λ, such that $p\pi/\Lambda = \beta$ for some integer p. In other words, coupling is a maximum for wavelengths close to the Bragg wavelength $\lambda_B = \frac{2\pi n_r}{\beta}$, where n_r is the effective refractive index of the mode. Thus,

$$\Delta n_r^2(x, y, z) = \Delta n_r^2(x, y) \sum_{-\infty}^{\infty} a_p e^{j(2p\pi/\Lambda)z} \tag{7.54}$$

and

$$\Delta\beta = \beta_A - \beta_B + \frac{p\pi}{\Lambda} \tag{7.55}$$

Note that first-order gratings (p = 1) provide the strongest coupling. We can now write the coupled-mode equations as

$$\frac{da}{dz} = \frac{j\Delta\beta}{2}a + j\kappa_{AB}b \tag{7.56}$$

$$\frac{db}{dz} = -j\frac{\Delta\beta b}{2} - j\kappa_{AB}^* a$$

where $\kappa_{AB}(cm^{-1})$ is termed the coupling constant. These equations can be written in matrix form as

$$\frac{d}{dz}\begin{bmatrix} a \\ b \end{bmatrix} = \begin{bmatrix} j\frac{\Delta\beta}{2} & j\kappa_{AB} \\ -j\kappa_{AB}^* & -j\frac{\Delta\beta}{2} \end{bmatrix}\begin{bmatrix} a \\ b \end{bmatrix} \tag{7.57}$$

which is an eigenvalue equation. Let us assume a dependence of the form

$$\begin{bmatrix} a \\ b \end{bmatrix} = \begin{bmatrix} A_0 \\ B_0 \end{bmatrix}e^{\delta z} \tag{7.58}$$

Substitution of this into Eq. 7.57 leads to

$$\begin{bmatrix} j\frac{\Delta\beta}{2} - \delta & j\kappa_{AB} \\ -j\kappa_{AB}^* & -j\frac{\Delta\beta}{2} - \delta \end{bmatrix}\begin{bmatrix} a \\ b \end{bmatrix} = 0 \tag{7.59}$$

The determinant of the coefficient matrix is zero. Thus,

$$\delta^2 + \frac{(\Delta\beta)^2}{4} - |\kappa_{AB}|^2 = 0 \tag{7.60}$$

The general solution, which accounts for both forward and backward traveling waves, can be expressed as

$$\begin{bmatrix} a \\ b \end{bmatrix} = \begin{bmatrix} a_+ \\ b_+ \end{bmatrix}e^{\delta z} + \begin{bmatrix} a_- \\ b_- \end{bmatrix}e^{-\delta z} \tag{7.61}$$

where

$$\delta = \sqrt{|\kappa_{AB}|^2 - (\Delta\beta/2)^2} \tag{7.62}$$

For the forward traveling wave, we have from Eq. 7.59

$$\left(j\frac{\Delta\beta}{2} - \delta\right)a_+ = -j\kappa_{AB}b_+ \tag{7.63}$$

or

$$\frac{b_+}{a_+} = \frac{\delta - j\frac{\Delta\beta}{2}}{j\kappa_{AB}} \tag{7.64}$$

Similarly, for the backward traveling wave

$$\frac{b_-}{a_-} = \frac{-\delta - j\frac{\Delta\beta}{2}}{j\kappa_{AB}} \tag{7.65}$$

We consider the three-layer slab waveguide shown in Fig. 7.14(a), in which there is a corrugated interface between the media with indices n_{r1} and n_{r2} in a section from z = 0 to z = l. The periodicity of the square corrugation, Λ, causes a coupling between the forward and backward waves with amplitudes A and B, respectively, and this coupling is used in the distributed feedback laser for mode selectivity. A wave with an amplitude $A(0)$ is incident from the left on the corrugated section. The backward wave, caused by periodic reflections at the corrugation is assumed to have an amplitude $B(z)$, with the boundary condition B(l) = 0. The amplitudes of the incident and reflected waves are shown in Fig. 7.14(b). From Eqs. 7.61, 7.64, and 7.65, we can write

$$\begin{bmatrix} a \\ b \end{bmatrix} = a_+ \begin{bmatrix} 1 \\ \frac{\delta - j\frac{\Delta\beta}{2}}{j\kappa_{AB}} \end{bmatrix} e^{\delta z} + a_- \begin{bmatrix} 1 \\ \frac{-\delta - i\frac{\Delta\beta}{2}}{j\kappa_{AB}} \end{bmatrix} e^{-\delta z} \tag{7.66}$$

By definition, and from the boundary condition,

$$a_+ + a_- = A(0) \tag{7.67}$$

and

$$a_+ \begin{bmatrix} \frac{\delta - j\frac{\Delta\beta}{2}}{j\kappa_{AB}} \end{bmatrix} e^{\delta l} - a_- \begin{bmatrix} \frac{\delta + j\frac{\Delta\beta}{2}}{j\kappa_{AB}} \end{bmatrix} e^{-\delta l} = 0 \tag{7.68}$$

From Eqs. 7.67 and 7.68,

$$a_+ \begin{bmatrix} \frac{\delta - j\frac{\Delta\beta}{2}}{j\kappa_{AB}} \end{bmatrix} e^{\delta l} - [A(0) - a_+] \begin{bmatrix} \frac{\delta + j\frac{\Delta\beta}{2}}{j\kappa_{AB}} \end{bmatrix} e^{-\delta l} = 0 \tag{7.69}$$

which, by transposing, leads to

$$a_+ \begin{bmatrix} \frac{\delta - j\frac{\Delta\beta}{2}}{j\kappa_{AB}} e^{\delta l} + \frac{\delta + j\frac{\Delta\beta}{2}}{j\kappa_{AB}} e^{-\delta l} \end{bmatrix} = A(0) \frac{\delta + j\frac{\Delta\beta}{2}}{j\kappa_{AB}} e^{-\delta l} \tag{7.70}$$

or

$$a_+ \begin{bmatrix} \frac{2\delta}{j\kappa_{AB}} \cosh \delta l - \frac{\Delta\beta}{\kappa_{AB}} \sinh \delta l \end{bmatrix} = A(0) \frac{\delta + j\frac{\Delta\beta}{2}}{j\kappa_{AB}} e^{-\delta l} \tag{7.71}$$

which can be expressed as

$$a_+ [2\delta \cosh \delta l - j\Delta\beta \sinh \delta l] = A(0) \begin{bmatrix} \delta + j\frac{\Delta\beta}{2} \end{bmatrix} e^{-\delta l} \tag{7.72}$$

Therefore

$$a_+ = \frac{A(0)\left[\delta + j\frac{\Delta\beta}{2}\right]e^{-\delta l}}{2\delta \cosh \delta l - j\Delta\beta \sinh \delta l} \tag{7.73}$$

Similarly, starting from Eq. 7.68 it can be shown that

$$a_- = \frac{A(0)\left[\delta - j\frac{\Delta\beta}{2}\right]e^{\delta l}}{2\delta \cosh \delta l - j\Delta\beta \sinh \delta l} \tag{7.74}$$

By combining Eqs. 7.64 and 7.65 with Eq. 7.61, we can also get

$$b_+ = \frac{A(0)\left[\delta^2 + \left(\frac{\Delta\beta}{2}\right)^2\right]e^{-\delta l}}{j\kappa_{AB}(2\delta \cosh \delta l - j\Delta\beta \sinh \delta l)}$$

$$b_- = \frac{-A(0)\left[\delta^2 + \left(\frac{\Delta\beta}{2}\right)^2\right]e^{\delta l}}{j\kappa_{AB}(2\delta \cosh \delta l - j\Delta\beta \sinh \delta l)} \tag{7.75}$$

Thus, the desired solutions are

$$a(z) = \frac{A(0)\left[\delta \cosh \delta(l - z) - j\frac{\Delta\beta}{2} \sinh \delta(l - z)\right]}{\delta \cosh \delta l - j\frac{\Delta\beta}{2} \sinh \delta l} \tag{7.76}$$

$$b(z) = \frac{jA(0)\kappa_{AB} \sinh \delta(l - z)}{\delta \cosh \delta l - j\frac{\Delta\beta}{2} \sinh \delta l} \tag{7.77}$$

We can then define a reflectance Θ_R for the corrugation at $z = 0$ as

$$\Theta_R = \left|\frac{b(0)}{a(0)}\right|^2 = \frac{\kappa_{AB}^2 \sinh^2 \delta l}{\delta^2 \cosh^2 \delta l + \left(\frac{\Delta\beta}{2}\right)^2 \sinh^2 \delta l} \tag{7.78}$$

Similarly, a transmission coefficient can be defined by $T = \left|\frac{a(l)}{a(0)}\right|^2$.

Under phase-matching conditions, $\Delta\beta = 0$, and from Eqs. 7.76 and 7.77, we get

$$a(z) = A(0)\frac{\cosh \delta(l - z)}{\cosh \delta l} \tag{7.79}$$

and

$$b(z) = A(0)\frac{\kappa_{AB}}{\delta}\frac{\sinh \delta(l - z)}{\cosh \delta l}. \tag{7.80}$$

The mode powers, given by $|a(z)|^2$ and $|b(z)|^2$ are plotted in Fig. 7.14(b). It is seen that the incident power drops off in the corrugated region. For sufficiently large arguments of the cosh and sinh functions, this drop is exponential. It is important to note that the drop in power does not represent absorption, but the reflection of power into the reflected wave $b(z)$. If the corrugated guiding medium is not passive, but has gain, the incident power grows with distance (shown by the dashed line in Fig 7.14(b)) and lasing action is obtained. The gain coefficient is incorporated in the exponentials of Eq. 7.53. The output spectrum of an idealized DFB laser is shown in Fig. 7.14(c). In an ideal DFB laser structure with perfect antireflection coated ends, the two zero-order modes on either side of λ_B will have the same low threshold gains and would lase simultaneously. In real structures the imperfection of the facets and grating and randomness of the cleaving process lift the degeneracy in threshold gain and make the device single mode. The slightest noise or other instability can do the same thing. The first-order mode (m = 1) is usually more than 30dB down from the zero-order mode.

The schematic of a practical GaAs/AlGaAs DFB laser ⬤ shown in Fig. 7.15(a). The active layer is p-type GaAs and has a thickness of 0.2 μm. It is confined between a n-Al$_{0.3}$Ga$_{0.7}$As layer and a p-Al$_{0.17}$Ga$_{0.83}$As layer. Thus, optical confinement and gain are provided within the p-GaAs layer. The feedback is provided by the corrugation at the interface between the p-Al$_{0.07}$Ga$_{0.93}$As and the p-Al$_{0.3}$Ga$_{0.7}$As layer where the main index discontinuity for optical guiding occurs. Such corrugation is formed by ion-milling or reactive ion beam etching (RIBE) techniques. Thus, the role of the thin ($\sim 0.1\ \mu m$) p-Al$_{0.17}$Ga$_{0.83}$As layer is twofold— it helps to confine the guided modes in the p-GaAs layer, and thus avoids nonradiative recombination at defects formed by corrugation; at the same time, it ensures leakage of part of the optical field into the corrugated layer. Since a small fraction of the confined mode power reaches the corrugated layer, the distributed feedback coefficient is very weak. In practical lasers, the corrugation period $\Lambda - m\lambda/2 = 0.35\ \mu$m (for m = 3), and the corrugation length

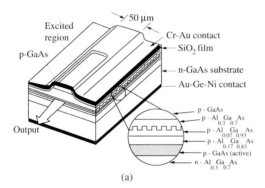

(a)

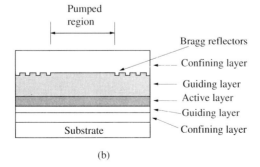

(b)

Figure 7.15 (a) A GaAs/AlGaAs double-heterostructure distributed feedback laser and (b) schematics of a distributed Bragg reflector (DBR) laser (from K. Aiki, *IEEE Journal of Quantum Electronics*, **QE-12**, 601, ©1976 IEEE).

is ~ 700 μm. In realizing the complete DFB heterostructures, the epitaxy is done in two steps. In the first step, the heterostructure upto the p-$Al_{0.07}Ga_{0.93}As$ layer is grown. The corrugations are then fabricated and the cladding and contact layers are regrown. The laser is usually made as a mesa stripe-geometry diode. The DBR laser is schematically shown in Fig. 7.15(b). Here, the end mirrors are replaced by gratings, with an active waveguide region in between.

7.5 THE CLEAVED-COUPLED-CAVITY (C^3) LASER: A TECHNIQUE FOR OBTAINING NARROW SPECTRAL LINEWIDTH

An important application of junction lasers is as sources for optical-fiber communication. Optical fibers, as we have seen, have their lowest loss and dispersion in the wavelength range 1.3 to 1.6 μm. In particular, single-mode silica fibers have their lowest loss at 1.55 μm. However, at this wavelength, the dispersion is not a minimum and, hence, different frequency components will be transmitted at different speeds within the fiber. The output of a conventional semiconductor laser has a linewidth, or wavelength, spread of $\sim$ 50–100 Å. An alternative is to work at 1.3 μm, where the fiber dispersion is minimal, but the losses are higher than that at 1.55 μm. Therefore, in a long-haul system more reamplification is necessary. Also, lattice-matched materials emitting at 1.3 μm are not available and strained or mismatched materials are needed. The dislocation density in such materials is usually very high. It would therefore be very useful to have a truly monochromatic junction laser source. The cleaved-coupled-cavity, or C^3, laser[†] comes close to achieving this objective in a simple and elegant manner.

The operation of a C^3 laser is based on the principle of a common vernier scale. It uses the technique of coupled-cavity resonance to filter out all but a few of the Fabry-Perot modes, so that the spectral output becomes very narrow. To realize the C^3 laser, one starts with a laser having a typical cavity length of say, 250 μm. This is then cleaved close to the center in a direction parallel to the original, cleaved facets. The cleaving is done on-chip with the substrate intact by techniques such as reactive-ion etching or reactive-ion beam etching. In effect, two shorter laser diodes of slightly different lengths and having their cavities aligned to each other, are produced. For example, the lengths of the two cavities may be 115 and 140 μm. The actual difference between the two lengths is not important. The schematic of a C^3 laser is shown in Fig. 7.16(a) and a fabricated and bonded device is shown in Fig. 7.16(b). If the two cavities are driven by the same forward current, then, by virtue of the different lengths, the lasing modes of the two cavities will not overlap. The *coupled* cavity system will support only the common resonant modes. For example, if the lengths of the two diode cavities differ by 10%, only every tenth axial mode will be reinforced by both cavities. In effect, the modes of the coupled lasers is spaced an order of magnitude further apart than those of a single laser. Therefore, the number

---

[†]W. T. Tsang, N. A. Olsson and R. A. Logan, *Appl. Phys. Lett.*, **42**, 650, 1983; *Electronics Letters*, **19**, 438, 1983.

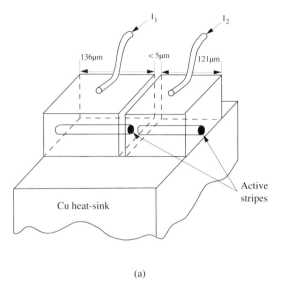

(a)

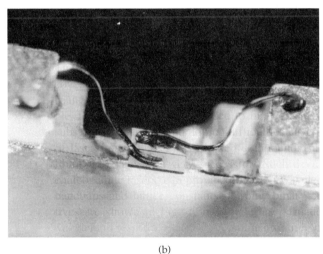

(b)

Figure 7.16 (a) Schematic representation and (b) photograph of a cleaved-coupled-cavity (C^3) laser (courtesy of W. T. Tsang, AT&T Bell Laboratories, Murray Hill, NJ).

of modes within the frequency spread of the gain curve is drastically reduced and the output becomes highly monochromatic. If the current in one of the cavities is changed, then due to the change in injected carrier density, the refractive index of the active region of that laser changes. This will change the wavelength of the stimulated emission and cause a shift of the Fabry-Perot modes. This causes a *large* shift in the frequency of the common resonant mode of the coupled cavities. Thus, frequency tuning can be achieved by change in drive current of one of the diodes. The change in current required for the frequency tuning is extremely small. For example, a change of 1mA will produce a 20 Å shift in wavelength.

7.6 QUANTUM WELL LASERS

In double-heterostructure lasers, the active layer thickness is typically of the order of 0.1 to 0.3 μm and this thickness is decided from considerations of gain and mode confinement. If d is made smaller, as in a quantum well, J_{th} is expected to decrease further. In a single quantum well laser $d = L_z \sim 50 - 100 \text{Å}$. However, the overlap of the electron and hole wavefunctions may start to decrease in a very small well. Mode confinement will also be very poor. There are other causes for the reduction of J_{th} in quantum well lasers. The density of states of electrons and holes in bulk material varies as $\mathcal{E}^{1/2}$, and from Fermi-Dirac statistics we know that the probability of occupation of the levels in the bands decreases rapidly as $\mathcal{E}$ increases. The carriers are distributed over a wide energy range in the bands with a relatively small density at the bandedges. As a consequence, population inversion is harder to achieve. It is desirable to have a large density of states of carriers in both bands at energies close to the bandedge. This is provided in quantum well lasers. If fact, some of the lowest threshold currents have been measured in quantum well lasers.

The density-of-states function for a quantum well, as derived in Sec. 2.5, is shown in Fig. 7.17(a). This applies for both electrons and holes in the respective bands. The probability of occupation of the states, continuous or discrete, is given by the Fermi-

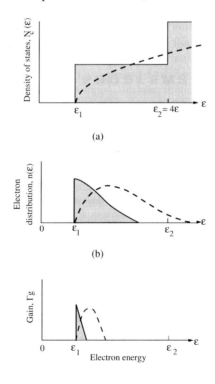

(a)

(b)

(c)

Figure 7.17   (a) Two-dimensional density of states, (b) electron distribution, and (c) gain spectrum in a quantum well. The dashed curves show the equivalent three-dimensional parameters taking the quantum well size into account.

Dirac function, which decreases rapidly as $\mathcal{E}$ increases. The corresponding electron distribution, given by Eq. 2.69, is illustrated in Fig. 7.17(b) and the modal gain is illustrated in Fig. 7.17(c). The three-dimensional density of states in the quantum well multiplied by the well size is also shown by the dotted line in Fig. 7.17(a), with the origin of the energy axis chosen at $\mathcal{E}_1$, for ease of comparison. The corresponding electron distribution and gain profiles are shown by the dotted lines in Figs. 7.17(b) and (c), respectively. In the three-dimensional case the electrons are spread over a relatively wide energy range with a small density at the edge. In sharp contrast, for the two-dimensional case, due to the abrupt increase of the density of states at the subband energy, the electrons are spread over a smaller energy range with a relatively high density at the subband edge. This situation implies that population inversion is achieved with a lower injected carrier density in a quantum well laser than in a conventional diode laser. As a consequence, the modal gain curve is more peaked, as shown in Fig. 7.17(c). It is therefore expected that the threshold current will be low in quantum well lasers. In lower-dimensional quantum-confined systems, such as *quantum wires* (1-D) and *quantum boxes* (0-D) the density of states of electrons and holes at the bandedges are peaked and the joint density of states (and gain) are expected to be even higher. Research and development of lasers made with these novel artificial materials are well under way.

The lowering of threshold current due to quantum-sized effects can be largely offset by the small width of the gain region in a single quantum well laser, which causes the optical confinement to be poor. The problem can be solved by using MQW structures for the gain region and a separate confinement heterostructure (SCH) to enhance the optical confinement factor. In particular, Graded Refractive Index–SCH (GRIN-SCH) structures, shown in Figs. 7.18(a) and (b) have resulted in the lowest threshold current densities.

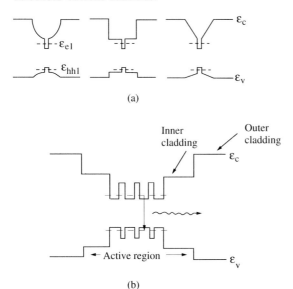

(a)

(b)

Figure 7.18 Band diagrams of the active region of (a) single quantum well and (b) multi-quantum well separate-confinement heterostructure (SCH) and graded refractive index-SCH (GRIN-SCH) lasers.

EXAMPLE 7.4
Objective.   To calculate the joint density of states in a 100 Å GaAs quantum well at the subband energy.

The joint density of states in a 2-D system between two subbands of the same quantum number is given by

$$N_J^{2D} = \frac{m_r^*}{\pi \hbar^2 L_z} \quad ((eV)^{-1} cm^{-3})$$

In GaAs $m_r^* \cong 0.05 m_o$. Therefore $N_J^{2D} \cong 2 \times 10^{19} (eV)^{-1}.cm^{-3}$, which is of the same order as that in bulk GaAs. Also, note that N_J^{2D} has a staircase distribution and this amount is contributed by each pair of allowed subbands.

7.6.1 Strained Quantum Well Lasers

The properties of strained semiconductors and heterostructures were discussed in Chapter 1. Because of the effects of strain on the bandstructure, deformation potential calculations have shown that there are dramatic changes of the optical properties. Most of these changes may be used to our advantage in designing better lasers. It may be remembered that in a quantum well the degeneracy between the light and heavy-hole bands is lifted at $\mathbf{k} = 0$ due to the large difference in effective masses, which determine the subband energies. As an example, in a 100 Å GaAs-Al$_{0.3}$Ga$_{0.7}$As quantum well the separation between the light-hole and heavy-hole subbands is approximately 10 meV.

In the conduction band of III–V semiconductors biaxial strain shifts the position of the bandedge and has a small effect on the effective mass. A more dramatic effect can be observed in the valence bandstructure. In addition to a change in the bandgap, the energy difference between the light- and heavy-hole bands also changes. Since, as mentioned above, there is already a split between the light- and heavy-hole bands, it is clear that strain will either serve to push the two bands closer to degeneracy, as is the case with tensile strain, or will serve to move them further apart, as is the case with compressive strain (Fig. 1.18). Finally, due to the changes in the coupling between the light and heavy-hole bands, the hole effective mass is reduced, almost by a factor of 3 or more. If the density of states in the valence band is reduced with reduction in effective mass, and it becomes comparable to that in the conduction band, then population inversion becomes more efficient. As a result, the differential gain $\frac{\partial g}{\partial n}$ increases and single-mode lasing occurs at lower injection currents. This is the case of compressively strained wells. In the case of tensile strain, the opposite happens when the HH and LH states merge and $N_V(\mathcal{E})$ becomes large. For higher values of strain, the bands decouple again with the LH bands being higher in energy. In that case $\frac{\partial g}{\partial n}$ increases again. Calculated and measured gain and differential gain of strained quantum well lasers are shown in Figs. 7.19(a) and (b).

In addition, the Auger rates are expected to change as the amount of strain changes for two reasons. First, strain will change the separation between the bandedge and the split-off band. Second, since the threshold current density will change, the carrier

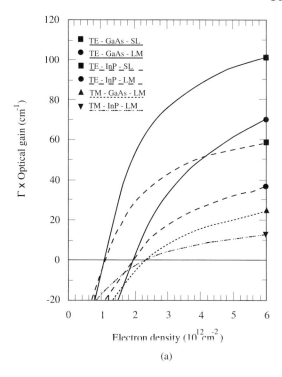

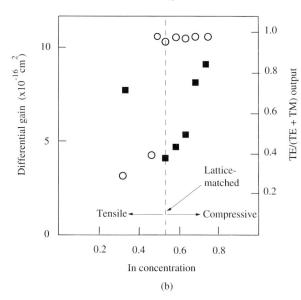

Figure 7.19 (a) Calculated optical gain versus carrier concentration at the lasing wavelength for GaAs- and InP-based QW lasers. Here SL ≡ compressively strained quantum well and LM ≡ lattice-matched quantum well. The TM gain for strained material is not shown, since it liesvery close to the zero gain line (courtesy J. Singh and Y. Lam, University of Michigan); and (b) differential gain and output polarization as a function of $In_xGa_{1-x}As$ well composition in pseudomorphic $In_{1-x}Ga_xAs/InGaAsP/InP$ MQW lasers measured by the author and co-workers (D. Nichols and P. Bhattacharya, *Applied Physics Letters*, **61**, 2129, 1992).

densities present in the device will change and, since the Auger rate is proportional to the cube of the carrier concentration, it will change too.

There is another important and profound effect of strain in the output of quantum well lasers. The heavy-hole bands are coupled to the TE mode, while the light-hole bands are coupled to the TM modes. In a lattice-matched quantum well laser, one expects the output to be a mixture of TE and TM modes. With the application of compressive strain the modes are more TE-like, and with the application of tensile strain the modes are more TM-like. At a very small value of tensile strain, when the light-hole and heavy-hole bands merge, the output is a pure mixture of TE and TM modes. These features are seen in the calculated gain curves of Fig. 7.19(a) and the measured data of strained $In_xGa_{1-x}As/InP$ quantum well lasers in Fig. 7.19(b). This polarization dependence of the output could be extremely important for many applications.

In GaAs-based strained quantum well lasers the wells consist of compressively strained $In_xGa_{1-x}As$. The barriers are usually GaAs or $Al_xGa_{1-x}As$. In InP-based strained quantum well lasers the wells are of $In_{0.53\pm x}Ga_{0.47\mp x}As$ with InGaAlAs or InGaAsP barriers and in this case, depending on the sign of x, the well regions can be under compressive or tensile strain—the former for $x > 0.53$ and the latter for $x < 0.53$. Measured threshold current densities in broad area lasers with 800 μm cavity length as a function of tensile and compressive strain are shown in Fig. 7.20. The behavior is exactly as predicted theoretically. The slight increase in threshold current for high compressive strain values is probably due to the onset of dislocation generation and/or higher Auger recombination rates.

A natural question that comes to mind is the number of quantum wells one should use. From theory and experimental data it is clear that a single-quantum well (SQW) laser can demonstrate very low threshold currents and threshold current densities. This

0.2 µm	$In_{0.53}Ga_{0.47}As$	p $(2.0 \cdot 10^{18})$
1.7 µm	InP	p $(2.0 \cdot 10^{17})$
400 Å	InGaAsP ($\lambda = 1.1$ µm)	undoped
1600 Å	InGaAsP ($\lambda = 1.2$ µm)	undoped
500 Å	$In_xGa_{1-x}As$ - 50 Å (4 Wells) In GaAsP ($\lambda = 1.2$ µm) - 100 Å	undoped
1600 Å	InGaAsP ($\lambda = 1.2$ µm)	undoped
400 Å	InGaAsP ($\lambda = 1.1$ µm)	undoped
1.3 µm	InP	n $(8.0 \cdot 10^{17})$
	n^+ InP Substrate	

(a)

Figure 7.20 (a) Typical InP-based strained quantum well laser heterostructures grown by chemical beam epitaxy in the author's laboratory and (b) measured threshold current density as a function of the alloy composition in the quantum well (D. Nichols et al., *IEEE Journal of Quantum Electronics*, **28**, 1239, ©1992 IEEE).

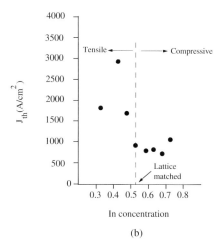

(b)

Figure 7.20 (*continued*)

is because the laser can operate with a higher differential gain, compared to a bulk laser, for the same threshold gain. A number of quantum wells in multiquantum well (MQW) laser results in higher gain, which can overcome the relatively high mirror losses for short cavity length devices. However, if too many wells are used, problems arise from non-uniform injection and, in strained quantum well lasers, from generation of dislocations. The same arguments can be made on a more quantitative basis. We expect that the threshold gain will be independent of the number of wells and is given by Eq. 6.68. As the number of wells goes from 1 to m, the amount of gain required from each well decreases by a factor of m. Now, since $J \sim mn^2$,[†] in order to benefit by using m wells, the threshold carrier density n in a well must decrease by a factor of $\sqrt{m}$ as the required gain per well drops. From Fig. 7.6 it is clear that the differential gain is a monotonically decreasing function of carrier density. Therefore, if one operates devices with very low threshold gain, the differential gain will be relatively high and a large reduction in the gain will mean only a small reduction in carrier density and thus only a small reduction, if any, in the threshold current density. However, the opposite will be true if one operates at a higher loss line (i.e. with high guide losses or with very short cavities). Therefore, it seems that MQW lasers are more favored for high-current densities.

Several other considerations have to be taken into account in the design of high-performance quantum well and double-heterostructure lasers. In addition to choosing the appropriate thickness of the active region or the number of quantum wells to ensure good carrier confinement, the waveguiding geometry should ensure single transverse modal confinement. The modes should not extend too far out into the ohmic contact layers. This will degrade the optical confinement factor and increase the free-carrier absorption.

[†]J. P. Loehr and J. Singh, *IEEE J. of Quan. Electron*, **27**, 708, 1991.

7.7 SURFACE-EMITTING LASERS

For many applications where there is a need for chip-to-chip communication and large arrays of sources, surface-emitting lasers may be very useful. These devices have, in addition, the potential for generation of high-power, broad-area emission with a narrow beam width, and the possibility of optical pumping of solid-state lasers. In addition, wafer-level testing becomes feasible.

In principle, a surface-emitting laser requires a vertical cavity with top and bottom mirrors. Therefore, in this structure the electric field is in the same direction as the mode propagation. If the dimension in the radial direction is large enough, one can assume plane-wave propagation. The approximation breaks down, however, if the radial dimension is shrunk. There are inherent advantages and disadvantages in using a vertical cavity configuration. The cavity length is much smaller, and therefore the frequency separation between the modes is increased. This ensures better modal purity than a conventional cleaved cavity laser, in which the cavity length is of the order of 100 μm. However, there is a price to be paid for this. Since the cavity length is greatly reduced, the roundtrip gain is less. Thus, the threshold current densities are very high. For example, in the early devices the threshold current densities were $\sim 25 KA/cm^2$ and the devices had difficulties operating at room temperature. This problem can be minimized to some extent by having very high reflectivity mirrors.

The schematic of a simple surface-emitting structure is as shown in Fig. 7.21(a). The top and bottom reflectors are multiquantum well quarter-wave Bragg reflectors realized by epitaxy. There are currently several schemes that are being investigated

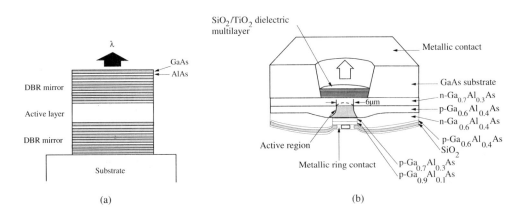

Figure 7.21 (a) Conceptual design of vertical cavity surface-emitting lasers with distributed Bragg reflectors and (b) structure of AlGaAs surface-emitting laser with dielectric reflectors (from S. Kinoshita and K. Iga, *IEEE J. Quantum Electronics*, **QE-23**, 882–888, ©1987, IEEE).

for the realization of reliable surface-emitting lasers. One such scheme is shown in Fig. 7.21(b). Here the top and bottom mirrors are created by artificial dielectrics. In spite of some of the technological barriers, two-dimensional arrays of surface-emitting lasers with a cw output power of ~ 750 mW and a power density in excess of 25 W/cm^2 have been made. With the advent of quantum well lasers, with higher material and modal gains, the performance characteristics of surface-emitting lasers are expected to improve.

Some of the problems that exist in the realization of efficient surface-emitting lasers are formation of low-resistance contacts and efficient current injection. For ultralow-threshold lasers it is also necessary to have a small gain volume and a single optical mode. All these features can potentially be realized in *microcavity lasers*, where the diameter of the surface-emitting lasers is shrunk to the order of a few microns. Several forms of such lasers are being studied, including microdisk lasers, in which advantage is taken of the large dielectric discontinuity between an optically thin semiconductor layer and a surrounding low index medium.

It is appropriate at this point to say a few words about semiconductor laser *arrays*. There is a limit to the output power of a single laser set by current leakage, that increases with increase in drive current, and facet degradation. For many applications, higher power levels are desirable. One way to achieve this is to design an array With edge-emitting lasers, a linear array can be made, in which each laser can either be gain guided or index guided. Such an array is fabricated by epitaxial regrowth and is shown schematically in Fig. 7.22(a). With surface-emitting lasers it is possible to realize two-dimensional arrays, shown in Fig. 7.22(b). In an array, which emits in a narrow coherent beam, there is a definite phase relationship between neighboring elements. The beam can be electronically steered by changing this phase relationship. To understand the modal characteristics of laser arrays, a coupled mode analysis is usually done. The normal modes of a laser array are usually referred to as array modes or *supermodes*.

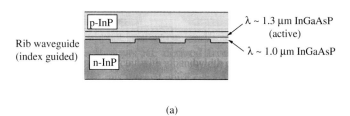

(a)

Figure 7.22   (a) Edge-emitting array of index-guided lasers made by regrowth and (b) SEM micrograph of ion-beam etched microlaser array. The devices have been treated chemically, etching the AlAs layers of the Bragg reflector, therefore revealing the structure (courtesy of J. L. Jewell, Photonics Research, Inc., Broomfield, Colorado).

Figure 7.22 (*continued*)

7.8 RARE-EARTH DOPED LASERS

There is an ever-increasing need for luminescent sources that are stable, have narrow spectral width, high gain, a low threshold current, and that show a minimum shift of the lasing wavelength with change in temperature. Quantum well lasers, as we have seen, meet some of these requirements. Rare-earth ions in insulating crystals have been studied extensively as the gain media for laser operations. Rare-earth ions have also been incorporated into III–V semiconductors and silicon by doping (during growth) and ion-implantation. Sharply structured luminescence bands, which arise from intracenter $4f$–$4f$ transitions between crystal field split spin-orbit levels of the trivalent rare-earth ions, were observed [Fig. 7.23(a)]. The wavelength of these emissions (1–1.5 μm) practically do not depend on the bandgap energy of the host semiconductors but rather on the particular rare-earth ion incorporated only.

There are several potential advantages of rare-earth doped lasers. Since the rare-earth transition is atomic level in nature, a reproducible scheme of obtaining precisely the same lasing wavelength can be achieved from wafer to wafer and from device to device, provided that the overall gain at the rare-earth transition energy is highest. Furthermore, the wavelength of this line should not shift with heat-sink temperature due to material dispersion and bandgap variation with temperature of the host semiconductor. Other important consequences include the possible linewidth narrowing due to atomic-level transitions and immunity to external optical reflections, etc. These emissions, excited by nonequilibrium carriers are, therefore, very useful for the fabri-

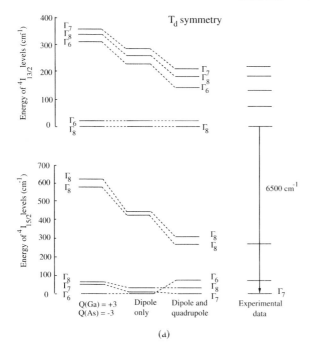

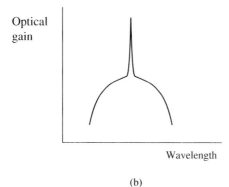

Figure 7.23 (a) Energy level diagram for E_r^{3+} in GaAs (from S. B. Stevens and C. A. Morrison, U. S. Army Laboratory Command, Harry Diamond Laboratories, Report #HDL-TM-91-16); and (b) overall gain profile of the combined gains due to rare-earth ion transitions and bandedge transitions in a semiconductor.

cation injection lasers if the emission line is placed (in energy) on top of the broad gain peak of the bandedge transition of the host semiconductor [Fig. 7.23(b)].

It is obvious from early studies that the gain of the rare-earth doped active region needs to be increased and the effect of such doping on the material properties need to be carefully characterized. There are several other important issues regarding rare-earth doping that need more investigation. The understanding of the mechanisms that lead to the observed transitions are quite unclear. At present, it is unclear as to how the rare-earth ions are being pumped, either through carrier transfer from the semiconductor carrier continuum to the discrete levels of the rare-earth ions or through optical pumping by the photons generated in the semiconductor through direct bandedge emission, or yet other unknown possible mechanisms.

7.9 ALTERNATE PUMPING TECHNIQUES

So far, we have learned that the semiconductor laser is pumped by an injection current across a forward-biased p-n junction to produce a small region in which a large rate of stimulated emission is achieved. It is tacitly assumed that a near-ideal p-n junction can be formed with the materials used. This is mostly true for the wider bandgap III–V compounds. However, it is very difficult to form heavily doped p-doped material with the narrow bandgap binaries such as InAs and InSb, and their ternary derivatives. Similar problems are encountered with II–VI and IV–VI compounds such as ZnS, ZnSe, and CdSe. For these materials, other means of pumping have to be considered. Some of the alternate techniques that have been used are optical pumping, electron-beam pumping, and avalanche breakdown induced pumping. The first one has been used most successfully, but before discussing it, we will briefly discuss the other two techniques.

In electron-beam pumping, the lasing semiconductor is excited by a suitable beam of energetic electrons. The electron beam penetrates the laser material to a depth of 5–10 μm and creates a degenerate population of electrons and holes. Just as in a junction laser, spontaneous emission first builds up, followed by stimulated emission and lasing. Usually a pulse electron beam is used to avoid overheating. The efficiency of such lasers is usually low due to the large density of hot carriers produced by the electron beam.

Avalanche breakdown induced lasing has been demonstrated in p-π-p GaAs structures, where avalanche breakdown is initiated at the high-field regions of the p-π boundaries. A degenerate population of electrons and holes is formed by avalanche multiplication, which eventually leads to lasing.

In optical pumping, the laser material is usually excited by another junction laser, as schematically shown in Fig. 7.24. It is advantageous to have the energy of the

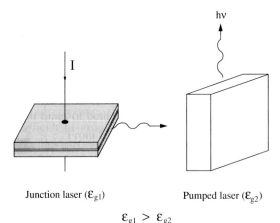

Junction laser (ε_{g1})   Pumped laser (ε_{g2})

$$\varepsilon_{g1} > \varepsilon_{g2}$$

Figure 7.24 Schematic illustration of optical pumping. Both lasers are mounted on heat sinks during actual operation.

pumping light to be higher and very close to the energy corresponding to the bandgap of the lasing material. Otherwise, surface absorption becomes dominant, and surface recombination effects reduce the optical efficiency. An inherent advantage in optical pumping is the large volume of material that can be excited. However, as a concluding remark it is fair to state that current injection is still the most efficient technique to drive a semiconductor laser.

7.10 DEVICE FABRICATION

The steps involved in device fabrication, depend, to a certain extent, on the junction formation technique. In GaAs lasers made by diffusion, the p-n junction is formed by diffusing Zn into a n-type GaAs substrate. The junction depth, varying between 1 and 100 μm, depends on the time and temperature of diffusion. The diffusion is usually done in an inert atmosphere using a 2% solution of Zn in Ga saturated with GaAs. This is the simplest form of a homojunction laser. More complicated homojunction and heterojunction laser structures are formed by epitaxial and fabrication techniques described in Chapter 1.

Once the laser diode wafer is made, broad area or stripe geometry lasers of the conventional Fabry-Perot type, as shown in Fig. 7.25, are made. Ridge lasers

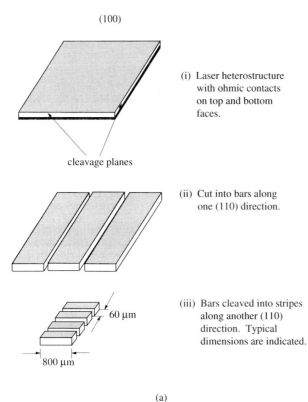

(100)

cleavage planes

(i) Laser heterostructure with ohmic contacts on top and bottom faces.

(ii) Cut into bars along one (110) direction.

(iii) Bars cleaved into stripes along another (110) direction. Typical dimensions are indicated.

60 μm

800 μm

(a)

Figure 7.25 (a) Formation of broad-area lasers by cleaving and (b) schematic of a single-mode ridge laser made by photolithography and wet- and dry-etching.

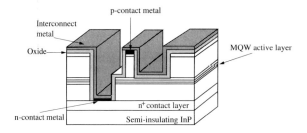

(b)

Figure 7.25 (*continued*)

are defined by photolithography and wet chemical or dry etching. For testing and operation the lasers are mounted on a diamond heat sink, which itself is mounted on a tin-plated copper block. Drastic reduction of threshold current densities with the use of quantum well heterostructures and efficient mounting and heat sinking ensures continuous operation of several thousand hours.

The linearity of the light-current characteristics in lasers and high-power LEDs is extremely important for systems and communication applications in which the output is subjected to high-frequency modulation. A typical direct analog modulation scheme is illustrated in Fig. 7.26. A high-frequency signal s(t) with amplitude ΔI is superimposed on the dc bias current I_{dc} to give a modulated output power

$$P(t) = P_0[1 + ms(t)] \tag{7.81}$$

where P_0 is the dc or cw optical power output when no high-frequency signal is

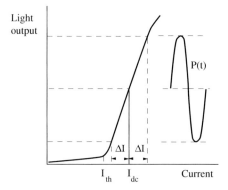

Figure 7.26 Illustration of direct analog modulation scheme showing the dc bias point and modulating (current) input and modulated (light) output.

applied. m is the modulation index defined by

$$m = \frac{\Delta I}{I_m} \tag{7.82}$$

where $I_m = I_{dc} - I_{th}$. To avoid distortion of the output signal, the total swing $2\Delta I$ of the modulating signal should be confined to the linear region of the light-current characteristics. Typical values of m vary between 0.2 and 0.5. There are two types of distortion caused by nonlinearities in the output characteristics. In general, such nonlinearities create higher-order harmonics in the output signal. For example, if the modulating signal is described by $s(t) = A \sin \omega t$, the output signal might be of the form

$$P(t) = B_0 + B_1 \sin \omega t + B_2 \sin 2\omega t + \cdots + B_n \sin n\omega t \tag{7.83}$$

which contains higher-order harmonics of the input signal. This type of distortion is called *harmonic distortion*. The magnitude of the distortion is defined by

$$(\text{distortion})_{n^{th}\text{order}} = 20 \log_{10} \frac{B_n}{B_1} \tag{7.84}$$

Another type of distortion is *intermodulation distortion* in which the modulating signal is of the form $s(t) = A_1 \sin \omega_1 t + A_2 \sin \omega_2 t$. The output signal will be of the form

$$P(t) = \sum_{p,q} B_{pq} \sin(p\omega_1 + q\omega_2), \, p, q = 0, \pm 1, \pm 2 \cdots \tag{7.85}$$

Thus, the signal includes all the higher-order harmonics of ω_1 and ω_2 plus the sum and difference frequencies. The latter are responsible for intermodulation distortion. It is therefore clear that the light-current characteristics should be as linear as possible.

7.11 MEASUREMENT OF LASER CHARACTERISTICS

The basic measurements that are made to characterize a laser are diode current-voltage characteristics, light-current characteristics and lifetime, or output power-time charac-teristics. The measurement techniques are very similar to those for the LED. However, some important differences should be kept in mind. Junction lasers operate at much higher power levels, and therefore heat dissipation becomes a critical issue. Mounting of lasers for these measurements and for normal operation has to be done with special care. Usually, the fabricated laser is mounted on an efficient heat sink. For high-performance operations, a feedback circuit and thermoelectric cooling are provided to maintain the temperature of the device constant. The whole assembly is depicted in Fig. 7.27. In spite of all these, cw operation of a laser may not be possible unless the threshold currents are very low. It is therefore common to determine the laser char-acteristics under pulsed biasing conditions. For example, to measure the light-current characteristics, a pulsed current bias is applied to the diode and the output power is measured with a calibrated Si photodiode.

The differential gain of a laser can be measured following the analysis and tech-nique of Hakki and Paoli, outlined in Sec. 7.1. The gain is calculated from the

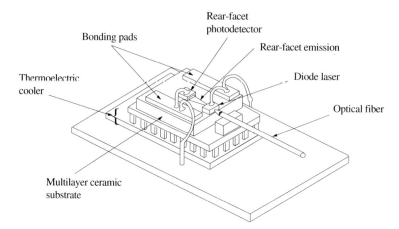

Figure 7.27   Mounting of a laser diode transmitter using a thermoelectric cooler for temperature stabilization (from G. Keiser, *Optical Fiber Communications*, 2nd ed., McGraw-Hill, New York, 1991).

emission spectra of the laser diodes. The gain is measured from the peaks and valleys of the output spectrum in accordance with Eq. 7.12. By varying the input current, one can measure the gain for a variety of values of the injection carrier density n. The differential gain $\frac{dg}{dn}$ can then be calculated assuming a linear variation of gain with injected carriers over a reasonably broad range of carrier densities.

The modal properties and angular dependence of the power output (beam shape) are given by the near- and far-field pattern, respectively, of the laser diode. To measure the near-field pattern, the laser with its heat sink is mounted on a precision translation stage. The diode is biased above threshold under cw or pulsed conditions and a collimating lens is used to focus the image of the output at infinity. The collimated beam is detected by an IR camera and viewed in a monitor. After correcting for the magnification of the collimating lens, the near-field intensity profile as a function of x and y (that define the cavity) is determined. For the determination of the far-field pattern, the laser diode is mounted on a rotating stage. The angular (θ, ϕ) dependence of the radiation is suitably detected and recorded after amplification. These measurements are schematically described in Fig. 7.28.

7.12 LASER MOUNTING AND FIBER COUPLING

Like LEDs, laser diodes have to be properly packaged and mounted before they can be used. To add to the complexity, since lasers are operated with much higher drive currents, heat dissipation and heat sinking become critical issues. Proper mounting and packaging should provide proper positioning of the laser chip on the mount, hermetically sealed leads, a coupled (or lens-coupled) "pigtail" fiber, thermoelectric cooling for operation at constant temperatures, and photodetector to monitor and control the

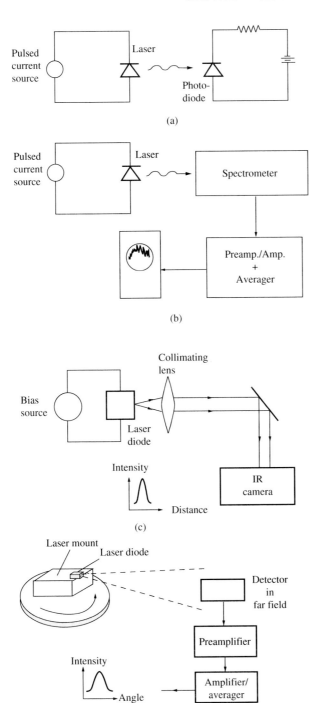

Figure 7.28 Schematics for the measurements of (a) light-current characteristics, (b) differential gain, (c) near-field pattern, and (d) far-field pattern.

output power. Typical laser packaging is illustrated in Fig. 7.29. The completed mount with electrical enclosure and a multipin connector assembly is shown in Fig. 7.30. Both GaAs/AlGaAs and InGaAs/InP lasers are available with such packaging.

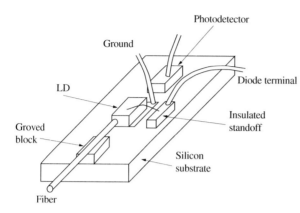

Figure 7.29   Mounted laser diode with monitoring photodetector and fiber pigtail (from J. C. Palais, *Fiber Optic Communications*, 2nd ed., ©1988. Reprinted by permission of Prentice Hall, Englewood Cliffs, NJ).

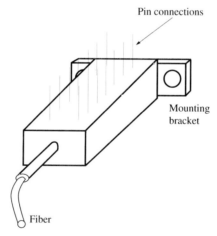

Figure 7.30   Packaged laser with fiber output, and multi-pin connector (from J. C. Palais, *Fiber Optic Communications*, 2nd ed., ©1988. Reprinted by permission of Prentice Hall, Englewood Cliffs, NJ).

7.13 MODULATION OF LASERS: RATE EQUATIONS

One of the most important aspects of laser operation is its transient or temporal behavior. In other words, when the drive current is modulated at high frequency, we would like to know how the laser reacts to it and whether the light output faithfully reproduces the current drive pulse or not. This behavior determines the modulation bandwidth of the laser. In order to understand the temporal behavior of the laser we must formulate the rate equations for carriers and photons. It is useful to remember that the carriers are supplied by injection, while the photons are supplied by stimulated emission.

We will begin with the assumption that the background impurity concentration in the active region is low, and therefore the density of injected carriers is equal to the electron or hole concentration in the band. These carriers are supplied by the injection current and removed by nonradiative and spontaneous recombination. Because of the small dimension of a laser, we ignore any spatial variation of carrier and photon density along the z-direction, and assume that the overlap function Γ between the electromagnetic mode and the carrier distribution is constant. The net rate of change of electron population n with time can then be written as

$$\frac{dn}{dt} = \frac{J}{qd} - \frac{n}{\tau} - R_{st} \tag{7.86}$$

where the first term on the right-hand side is the carrier production term and the second and third terms represent carrier losses via spontaneous and stimulated emission, respectively. R_{st} represents the carrier-photon interaction. Also,

$$\tau = \frac{\tau_r \tau_{nr}}{\tau_r + \tau_{nr}} \tag{7.87}$$

where τ_r and τ_{nr} are the radiative and nonradiative carrier lifetimes, respectively. Now, the gain coefficient $g = \frac{1}{N_p}\frac{dN_p}{dx}$, where N_p is the number of photons per unit energy interval in the cavity mode. Thus, assuming $n_r = 1$,

$$R_{st} = h\nu\frac{dN_p}{dt} = cgN_p h\nu \tag{7.88}$$

Note that this equation is identical to Eq. 6.50 with r_{st} replaced by R_{st} and $n_\varepsilon u_\varepsilon$ by $N_p h\nu = \varphi(\mathcal{E})$. Assuming a linear dependence of g on n, as expressed by Eq. 7.23, we can write,

$$R_{st} = \Omega(n - n_{nom})N_p h\nu \tag{7.89}$$

where Ω is a factor equal to $c\frac{\partial g}{\partial n}$. n_{nom} is the injected carrier density required to make the cavity transparent. Substituting Eq. 7.89 in Eq. 7.86, the rate equation for the inverted carrier density becomes

$$\frac{dn}{dt} = \frac{J}{qd} - \frac{n}{\tau} - \Omega(n - n_{nom})N_p h\nu \tag{7.90}$$

To see how the carriers interact with photons, we have to write the rate equation for the photons, which are largely supplied through stimulated emission and to a small extent by spontaneous emission. The photons are depleted through cavity losses. Thus, the rate equation for photons becomes

$$h\nu\frac{dN_p}{dt} = \Omega(n - n_{nom})N_p h\nu + \frac{\theta n}{\tau_r} - \frac{N_p h\nu}{\tau_p} \tag{7.91}$$

The first term on the right-hand side represents the rate of increase due to stimulated photon emission. The second term represents the spontaneous emission rate into the mode. θ represents the fraction of the spontaneous emission, which couples into the cavity mode. The last term represents the cavity and coupling losses, and τ_p is the photon lifetime whose value is ~ 1 ps. It is the average time that the photon remains in the cavity before it is absorbed or emitted through the facets. Equations 7.90 and 7.91 are the coupled rate equations, which provide useful insights to laser characteristics. We will now derive the steady-state and transient behavior of the laser from solutions of these equations.

7.13.1 Steady-State Solution or Static Characteristics

When a laser is turned on, there are initially transient effects, and oscillations in the photon density, as we shall see later. After a while, fluctuations in carrier and photon densities subside and a steady state is reached. Under these conditions $\frac{dn}{dt} = 0$ and $\frac{dN_p}{dt} = 0$. We can get an insight into the behavior of the light (power)-current characteristics from the steady-state solution. In the region below threshold, where spontaneous emission dominates, the stimulated emission term in Eqs. 7.90 and 7.91 can be neglected. With $\frac{dn}{dt} = \frac{dN_p}{dt} = 0$, we get

$$n = \frac{\tau}{qd} J \tag{7.92}$$

and

$$h\nu N_p = \frac{\theta \tau \tau_p}{\tau_r qd} J \tag{7.93}$$

With θ being a small number, it is clear from Eq. 7.93 that N_p increases slowly with increasing J. Above threshold, where stimulated emission dominates, the spontaneous emission term in Eq. 7.91, $\theta n/\tau_r$, can be neglected and the steady-state solution is

$$n = \overline{n} = n_{nom} + \frac{1}{\Omega \tau_p} \tag{7.94}$$

where $\overline{n}$ is the steady-state injected carrier population for $J > J_{th}$ and $\overline{n}$ remains constant when stimulated emission occurs and is independent of the photon density in the mode. Substituting Eq. 7.94 in Eq. 7.90 and setting $\frac{dn}{dt} = 0$, we get

$$\overline{\varphi} = h\nu \overline{N}_p = \frac{\tau_p}{qd}(J - J_{th}) \tag{7.95}$$

where

$$J_{th} = \frac{\overline{n}qd}{\tau} \tag{7.96}$$

and $\overline{N}_p$ and $\overline{\varphi}$ are the steady-state coherent photon densities. Thus, the light-current characteristic in the stimulated emission regime is linear with a much larger slope, by a factor $\tau_r/\theta\tau$, than that in the spontaneous emission regime. A sharp "break" is therefore produced in the light-current characteristic. Also, the intercept on the current axis defines the threshold current density J_{th}. These features are observed in Fig. 7.3. The slope of the light-current characteristics in the stimulated emission regime is, from Eq. 7.95,

$$h\nu \frac{d\overline{N}_p}{dJ} = \frac{\tau_p}{qd} \tag{7.97}$$

Each injected carrier produces a photon by stimulated emission; therefore, the photon emission rate (N_p/τ_p) and hence the optical power emitted are linear functions of the current above threshold. It is also evident that the total *spontaneous* emission rate $\bar{n}/\tau$ is constant since $\bar{n}$ is constant. Since the gain coefficient is a function of the injected carrier concentration, it will also saturate at threshold.

Under steady-state conditions, Eq. 7.91 can be rewritten as

$$\Omega(n - n_{nom}) = \frac{1}{\tau_p} - \frac{\theta n}{\tau_r N_p h\nu} \tag{7.98}$$

It is assumed that $n > n_{nom}$. Equation 7.98 is a mathematical statement of the fact that the sum of the stimulated and spontaneous transition rates is equal to the total rate of depletion of photons. As the output power N_p increases, the spontaneous emission processes described by the second term on the right becomes smaller and smaller. As a consequence, the stimulated emission, or gain rate, is clamped at threshold value and is equal to the photon loss rate.

The total steady-state power is obtained by adding Eqs. 7.93 and 7.95,

$$h\nu N_p = \frac{\tau_p}{qd}\left[J - \frac{\bar{n}qd}{\tau}\right] + \frac{\theta\bar{n}\tau_p}{\tau_r} \tag{7.99}$$

or

$$h\nu N_p = \frac{\tau_p}{qd}[J - J_{th}] + N_{p(spont)}h\nu \tag{7.100}$$

in which the second term on the right is constant, since $\bar{n}$ remains constant. A simple way to understand the light-current characteristics is to consider the photon modes. Below threshold, the photons emitted by spontaneous emission goes into a very large number (see Example 6.4) of modes. Above threshold, the photons produced by the injection go into a very few modes, which represents coherent light. Thus, the intensity of emission is greatly enhanced.

In deriving the equations above, it is assumed that the laser cavity supports one dominant cavity mode. In reality, since the dimensions of the laser cavity, and in particular the longitudinal dimension, can be much larger than the wavelength, many cavity modes may be excited. However, it can be shown that the photon density in the main mode increases indefinitely with current, while that in the secondary modes tend to saturate. Therefore, the predictions from the equations above are quite valid.

The simple theory presented here predicts a linearly increasing power with current. This is not strictly true because of the influence of the optical field on the gain coefficient. The stimulated recombination rate of photons into the dominant cavity mode is inversely proportional to the photon density in that mode. With increase of injection, the stimulated recombination rate increases and the stimulated emission lifetime can become *comparable* to the average carrier lifetime in the bands ($\sim 10^{-12}s$). Thus, the population of the bands providing carriers for stimulated emission is decreased, thereby decreasing the inversion and the gain coefficient. A few thousand photons in a mode is sufficient to reduce the gain coefficient.

7.13.2 Transient Phenomena and Frequency Response

We will now analyze the temporal behavior of the injection laser using the coupled rate equations. The objective is to be able to estimate the modulation bandwidth of the laser. Several things happen when a laser is turned on and the injection current is a high-frequency signal. Usually in a current modulation scheme, the laser is biased with a dc current above J_{th} and a small-signal high-frequency current is superimposed on it. Transient phenomena occur because of the time required for the electron and photon populations to come into equilibrium.

7.13.2.1 Response to a Current Density Step Excitation.

When a step current density J, as shown in Fig. 7.31, is applied, there is a delay between the application of this current pulse and the initiation of the laser optical pulse. This is because with the application of the current pulse the carrier density increases from an initial value n_i to a final value n_f in a time interval t_d, which is obtained from the integration of Eq. 7.86 as

$$t_d = \tau \ln \left(\frac{J - \frac{qdn_i}{\tau}}{J - \frac{qdn_f}{\tau}} \right) \tag{7.101}$$

When the injected carrier density exceeds $\bar{n}$ given by Eq. 7.94, the photon population begins to build up with a time constant $\sim \tau_p$. Since $\tau(\sim 10^{-9}\, \text{sec})$ is much longer than $\tau_p(\sim 10^{-12}\, \text{sec})$, it is fairly accurate to assume $n_i = 0$ and $n_f = \bar{n}$. In this case,

$$t_d = \tau \ln \frac{J}{J - J_{th}} \tag{7.102}$$

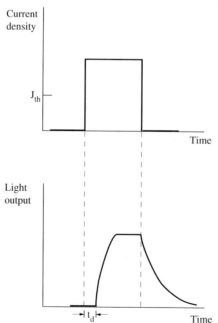

Figure 7.31 Large-signal current applied to a pulse and corresponding light output.

7.13.2.2 Relaxation Oscillations and Oscillatory Optical Output.   In
the previous section, we have seen that the photon population builds up rapidly after
injection. At the same time, the carrier density is also depleted, until it falls below $\bar{n}$.
At this point, $\frac{dN_p}{dt}$ becomes negative and the value of R_{st} becomes very small. The
carrier population starts to build up again, this time from a higher initial value and so
does the photon population. A damped oscillatory optical output is observed, where
the initial swing in the photon density is quite large (Fig. 7.32). The perturbation in
the carrier density is much smaller. The system, in essence, behaves as a tuned circuit
and a resonant condition can be achieved in the system transfer function, which is the
ratio of the light output to the injection current density, at some characteristic oscillation frequency. These *relaxation oscillations* set an upper limit to the direct current
modulation frequency. Figure 7.32 gives an idea of the order of magnitude of n and
N_p in practical lasers. In what follows, we will derive an expression for the oscillation
frequency.[†] It should be remembered that the coupled rate equations are nonlinear and
numerical methods are generally needed for their solution. We will assume, however,
that the perturbations in the carrier and photon populations are not very large, compared to their steady-state values. This assumption allows us to linearize the coupled
equations. It is assumed that the laser diode is turned on by a step function current
density at some value above threshold and is held constant at this value. In other
words, there is no high-frequency, small-signal, modulation of the current itself.

In terms of steady-state and time-varying components, the carrier and photon

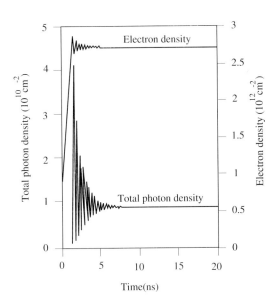

Figure 7.32 Calculated transient
response of the electron and
photon densities in a InP-based
lattice-matched In$_{0.53}$Ga$_{0.47}$As
QW laser showing the relaxation
oscillations. The switching is from
0 to 1000 A/cm^2 (courtesy of J.
Singh, University of Michigan).

[†] See H. Kressel and J. K. Butler, *Semiconductor Lasers and Heterojunction LEDs*, Academic Press,
New York, 1977.

densities can be expressed as

$$n = \bar{n} + \Delta n$$

$$N_p = \overline{N}_p + \Delta N_p \tag{7.103}$$

where it is assumed that Δn and ΔN_p are small quantities. Substitution of these equations into the rate equation for electrons leads to the following equation containing time-varying quantities, where it is assumed that $\bar{n}$ and $\overline{N}_p$ remain constant and the product $(\Delta n \Delta N_p)$ can be neglected:

$$\frac{d(\Delta n)}{dt} = \left\{ \frac{J}{qd} - \Omega(\bar{n} - n_{nom})\overline{N}_p h\nu - \frac{\bar{n}}{\tau} \right\}$$

$$- \left\{ \Omega\overline{N}_p h\nu \Delta n + \Omega(\bar{n} - n_{nom})\Delta N_p h\nu + \frac{\Delta n}{\tau} \right\} \tag{7.104}$$

Similarly, for the photons, we obtain by substituting Eqs. 7.103 into Eq. 7.91 and neglecting the spontaneous emission term,

$$\frac{d(\Delta N_p)}{dt} = \left\{ \Omega(\bar{n} - n_{nom}) - 1/\tau_p \right\} \Delta N_p + \Omega\overline{N}_p \Delta n$$

$$+ \left\{ \Omega(\bar{n} - n_{nom}) - 1/\tau_p \right\} \overline{N}_p \tag{7.105}$$

Since $\frac{d\bar{n}}{dt} = 0$ and $\frac{d\overline{N}_p}{dt} = 0$, it follows from the steady-state rate equations that the quantity within the first brace of Eq. 7.104 and the quantities within the braces of Eq. 7.105 have values of zero. With these simplifications we get

$$\frac{d(\Delta n)}{dt} = -(\Omega\overline{N}_p h\nu + \frac{1}{\tau})\Delta n - \Omega(\bar{n} - n_{nom})\Delta N_p h\nu \tag{7.106}$$

and

$$\frac{d(\Delta N_p)}{dt} = \Omega\overline{N}_p \Delta n \tag{7.107}$$

With use of Eq. 7.94, the equations above can be combined to give the identical equations

$$\frac{d^2(\Delta n)}{dt^2} + \left(\Omega\overline{N}_p h\nu + \frac{1}{\tau} \right) \frac{d(\Delta n)}{dt} + \frac{\Omega\overline{N}_p h\nu}{\tau_p} \Delta n = 0 \tag{7.108}$$

$$\frac{d^2(\Delta N_p)}{dt^2} + \left(\Omega\overline{N}_p h\nu + \frac{1}{\tau} \right) \frac{d(\Delta N_p)}{dt} + \frac{\Omega\overline{N}_p h\nu}{\tau_p} \Delta N_p = 0$$

Solutions to these differential equations are of the forms

$$\Delta n = \Delta n_0 \exp[-(Q - j\omega_r)t] \tag{7.109}$$

and

$$\Delta N_p = \Delta N_{p0} \exp[-(Q - j\omega_r)t] \qquad (7.110)$$

where

$$Q = \frac{1}{2}\left(\Omega \overline{N}_p h\nu + \frac{1}{\tau}\right) \qquad (7.111)$$

and

$$\omega_r = \left(\frac{\Omega \overline{N}_p h\nu}{\tau_p} - \frac{\Omega^2 \overline{N}_p^2 (h\nu)^2}{4} - \frac{\Omega \overline{N}_p h\nu}{2\tau} - \frac{1}{4\tau^2}\right)^{1/2} \qquad (7.112)$$

From Eqs. 7.94 and 7.95, for $J > J_{th}$,

$$\Omega \overline{N}_p h\nu \cong \frac{J}{qd(\overline{n} - n_{nom})} \qquad (7.113)$$

and from Eq. 7.96,

$$\frac{1}{\tau} = \frac{J_{th}}{qd\overline{n}} \qquad (7.114)$$

Therefore, if the laser is biased well above threshold, such that $\overline{n} \gg n_{nom}$,

$$Q \cong \frac{1}{2\tau}\left(\frac{J}{J_{th}} + 1\right) \qquad (7.115)$$

and

$$\omega_r \cong \left[\frac{1}{\tau\tau_p}\left(\frac{J}{J_{th}} - 1\right) - Q^2\right]^{1/2} \qquad (7.116)$$

Now since τ is $\sim 10^{-9}s$ and τ_p is $\sim 10^{-12}s$, the first term within the parentheses is larger than the second. Thus,

$$\omega_r = 2\pi f_r \cong \left[\frac{1}{\tau\tau_p}\left(\frac{J}{J_{th}} - 1\right)\right]^{1/2} \qquad (7.117)$$

It is seen that the resonance frequency f_r increases with decreasing spontaneous carrier lifetime and photon lifetime and with increasing value of the ratio J/J_{th}. Also, decay time of the oscillation, or its envelope, is proportional to $1/Q$. This decay time is of the order of the carrier lifetime. The decay time is also reduced by increasing the value of the ratio J/J_{th}. The implications of these results are the following. If high-frequency small-signal current modulation (i.e., $J = J_0 e^{j\omega t}$) is applied to the diode, then serious distortion may occur when ω approaches ω_r. Also, the modulation index is a maximum at $\omega = \omega_r$, above which frequency it drops rapidly. The effect of injecting photons at the stimulated emission energy from another source is equivalent to increasing J, thereby increasing ω_r. The power needed to do this is estimated to be much less than the emitted power. Values of $\omega_r \sim$ 2-30 GHz can be obtained in high-performance lasers.

7.13.2.3 High-Frequency Modulation of Laser Diodes.   In this case, we wish to determine the steady-state response of a laser to a high-frequency sinusoidal current drive. In other words, the laser is biased by a dc drive current, on top of which is superimposed a small-signal high-frequency current of the form $J = J_0 e^{j\omega t}$. The modulating frequency is expected to be > 1 GHz. It is important to determine if the laser output pulse fully replicates the current pulse, without any distortion and loss of modulation depth. We will also calculate the transfer function, which is the ratio of the photon output to the injected current density. In addition to the time-varying components of the carrier and photon densities, as in the last case, we also assume that the diode is biased above threshold by a dc current $\bar{J} > J_{th}$ and an additional small-signal AC component $\Delta J(t)$ is added. Hence,

$$J = \bar{J} + \Delta J \tag{7.118}$$

Following similar mathematical steps as in the last case, and retaining the term accounting for the small number of photons generated by spontaneous emission in Eq. 7.91, Eqs. 7.106 and 7.107 can be rewritten as,

$$\frac{d(\Delta n)}{dt} = \frac{\Delta J(t)}{qd} - \left(\Omega \bar{N}_p h\nu + \frac{1}{\tau} \right) \Delta n - \frac{\Delta N_p h\nu}{\tau_p}$$

$$\frac{d(\Delta N_p)}{dt} = \left(\Omega \bar{N}_p h\nu + \frac{\theta}{\tau} \right) \Delta n \tag{7.119}$$

where the substitution $\Omega(\bar{n} - n_{nom}) = 1/\tau_p$ is made. Combining the two equations gives

$$\frac{d^2(\Delta n)}{dt^2} + \left(\Omega \bar{N}_p h\nu + \frac{1}{\tau} \right) \frac{d(\Delta n)}{dt} + \frac{1}{\tau_p} \left(\Omega \bar{N}_p h\nu + \frac{\theta}{\tau} \right) \Delta n = \frac{1}{qd} \frac{d(\Delta J)}{dt} \tag{7.120}$$

and similarly,

$$\frac{d^2(\Delta N_p)}{dt^2} + \left(\Omega \bar{N}_p \, h\nu + \frac{1}{\tau} \right) \frac{d(\Delta N_p)}{dt} + \frac{1}{\tau_p} \left(\Omega \bar{N}_p h\nu + \frac{\theta}{\tau} \right) \Delta N_p$$

$$= \frac{\Delta J}{qdh\nu} \left(\Omega \bar{N}_p h\nu + \frac{\theta}{\tau} \right) \tag{7.121}$$

For a sinusoidal modulation and response, we have

$$\Delta J = \Delta J_0 \, e^{j\omega t}$$

$$\Delta N_p = \Delta N_{p0} \, e^{j\omega t} \tag{7.122}$$

$$\Delta n = \Delta n_0 \, e^{j\omega t}$$

Substitution into Eq. 7.121 for the photon density leads to

$$(j\omega)^2 \Delta N_{p0} + (j\omega) \left(\Omega \, \bar{N}_p h\nu + \frac{1}{\tau} \right) \Delta N_{p0} + \frac{1}{\tau_p} \left(\Omega \bar{N}_p h\nu + \frac{\theta}{\tau} \right) \Delta N_{p0}$$

$$= \frac{\Delta J_0}{qdh\nu} \left(\Omega \bar{N}_p h\nu + \frac{\theta}{\tau} \right), \tag{7.123}$$

from which, the desired transfer function is

$$h\nu \frac{\Delta N_{p0}}{\Delta J_0} = \frac{\frac{1}{qd}\left(\Omega\overline{\varphi} + \frac{\theta}{\tau}\right)}{\left[\frac{1}{\tau_p}\left(\Omega\overline{\varphi} + \frac{\theta}{\tau}\right) - \omega^2\right] + j\omega\left[\Omega\overline{\varphi} + \frac{1}{\tau}\right]} \tag{7.124}$$

Where $\overline{\varphi}$ is the average value of the spectral radiation density or total photon density. If we bias the laser well above threshold, so that the spontaneous emission term can be neglected, Eq. 7.124 can be simplified to

$$h\nu \frac{\Delta N_{p0}}{\Delta J_0} = \frac{\tau_p/qd}{\left\{1 - \frac{(\omega\tau_p)^2}{\tau_p\Omega\overline{\varphi}}\right\} + j\omega\tau_p} \tag{7.125}$$

The resonance condition gives

$$\omega_r^2 = \frac{\Omega\overline{N}_p h\nu}{\tau_p} \tag{7.126}$$

Thus, the resonant frequency increases as the photon lifetime is shortened or when the dc power (drive current) is increased. Frequency response of InGaAs/AlGaAs strained quantum well lasers is shown in Fig. 7.33. It is apparent that higher-power operation might lead to higher-modulation bandwidth; however, after a while this process becomes self-destructive, since facet degradation sets in. Higher bandwidths are also obtained by reducing the cavity length, maintaining the drive current the same. It is clear that modulation bandwidths of several gigahertz are achievable with these lasers.

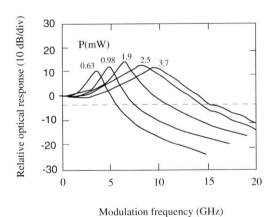

Figure 7.33 Microwave modulation response of a $10 \times 200\mu$m In$_{0.35}$Ga$_{0.65}$As MQW laser. The device has a -3 dB bandwidth of 15.5 GHz at the highest power (from S. D. Offsey et al., *IEEE J. Quantum Electron.*, **27**, 1455–1462, ©1991, IEEE).

Now, from Eq. 7.125, the amplitude of the transfer function is given by

$$\left|\frac{h\nu\Delta N_{p0}}{\Delta J_0}\right| = \frac{\tau_p/qd}{\left[\left(1 - \frac{(\omega\tau_p)^2}{\tau_p\Omega\overline{\varphi}}\right)^2 + (\omega\tau_p)^2\right]^{1/2}} \tag{7.127}$$

As $\omega \to 0$, $\left| \dfrac{h\nu \Delta N_{p0}}{\Delta J_0} \right| \to \dfrac{\tau_p}{qd}$. Therefore, in the limit

$$\Delta N_{p0} \cong \frac{\tau_p}{qd} \Delta J_0 \tag{7.128}$$

which is the steady-state solution derived earlier. Here $\Delta J_0 = J - J_{th}$. As $\omega \to \infty$,

$$\left| \frac{h\nu N_{p0}}{\Delta J_0} \right| \cong \frac{\tau_p/qd}{\omega^4 \tau_p^4 / \tau_p^2 \Omega^2 \overline{\varphi}^2} \to 0. \tag{7.129}$$

Thus, as the modulation frequency becomes very high, the photons cannot react fast enough and the light output does not change with time.

Finally, before concluding this section, we will consider some of the equations in more exact form. Remember that we had assumed the overlap integral $\Gamma = 1$. This may not necessarily be true, in which case, from Eq. 7.88,

$$R_{st} = \frac{\Gamma cg N_p h\nu}{n_r} \tag{7.130}$$

where n_r is the refractive index of the active region. The rate equations can then be written as

$$\frac{dn}{dt} = \frac{J}{qd} - R_{sp} - \Gamma g \frac{c}{n_r} N_p h\nu$$

$$h\nu \frac{dN_p}{dt} = \Gamma g \frac{c}{n_r} N_p h\nu + \theta R_{sp} - \frac{N_p h\nu}{\tau_p} \tag{7.131}$$

where R_{sp} is the spontaneous recombination rate. For small-signal modulation, Taylor's expansion around median values neglecting higher-order terms leads to

$$R_{st} = \overline{R}_{st} + \frac{\partial \overline{R}_{st}}{\partial n} \Delta n + \frac{\partial \overline{R}_{st}}{\partial N_p} \Delta N_p$$

$$R_{sp} = \overline{R}_{sp} + \frac{\partial \overline{R}_{sp}}{\partial n} \Delta n$$

$$g = \overline{g} + \frac{\partial \overline{g}}{\partial n} \Delta n \tag{7.132}$$

If the device is biased above threshold,

$$\overline{g} \cong g_{th} \tag{7.133}$$

and

$$\Gamma g \frac{c}{n_r} = \frac{c}{n_r} \left(\gamma + \frac{l}{l} \ln \frac{1}{R} \right) \cong \frac{1}{\tau_p} \tag{7.134}$$

An alternate form of Eq. 7.124 for the transfer function is

$$h\nu \frac{\Delta N_{po}}{\Delta J_0} = \frac{\frac{1}{qd} \left(\theta \frac{\partial \mathbf{R}_{sp}}{\partial n} + \Gamma \frac{c}{n_r} \frac{\partial g}{\partial n} \overline{\varphi} \right)}{\left[\frac{1}{\tau_p} \left(\Gamma \frac{\partial \mathbf{R}_{sp}}{\partial n} + \Gamma \frac{c}{n_r} \frac{\partial g}{\partial n} \overline{\varphi} \right) - \omega^2 \right] + j\omega \left(\frac{\partial \mathbf{R}_{sp}}{\partial n} + \Gamma \frac{c}{n_r} \frac{\partial g}{\partial n} \overline{\varphi} \right)} \tag{7.135}$$

and the expression corresponding to Eq. 7.125 for a laser biased well above threshold is

$$hv\frac{\Delta N_{po}}{\Delta J_0} = \frac{\tau_p/qd}{\left\{1 - \frac{(\omega\tau_p)^2}{\tau_p\Gamma\frac{c}{n_r}\frac{\partial g}{\partial n}\overline{\varphi}}\right\} + j\omega\tau_p}$$

(7.136)

Equations 7.135 and 7.136 are written in terms of known and measurable parameters of lasers, and therefore it is easier to analyze laser performance in terms of these equations.

EXAMPLE 7.5

Objective.   To derive an expression for the resonance frequency in terms of the differential gain $\partial g/\partial n$.

Substituting Eqs. 7.95, 7.96, and 7.134 into Eq. 7.117, we get

$$\omega_r = \left(\frac{\overline{\varphi}c\frac{\Gamma g}{n}}{n_r\tau_p}\right)^{1/2}$$

(7.137)

Under small signal conditions (i.e., small changes in $\overline{n}$ and g), $\frac{g}{n} \cong \frac{\partial g}{\partial n}$. Therefore,

$$\omega_r = \left(\frac{\overline{\varphi}\frac{c}{n_r}\Gamma\frac{\partial g}{\partial n}}{\tau_p}\right)^{1/2}$$

(7.138)

Note that this equation does not contain the lifetime τ.

7.13.2.4 Auger Recombination Rates.

We had derived an expression for the turn-on delay (Eq. 7.102) in Sec. 7.13.2.1. Auger recombination rates can be determined from this turn-on delay. Auger recombination plays an important role in determining the total threshold current, particularly in long-wavelength lasers. In other words,

$$J_{th} = J_{th(r)} + J_{th(nr)}$$

(7.139)

Neglecting the stimulated recombination term, Eq. 7.90 can be rewritten as

$$\frac{dn}{dt} = \frac{I}{qV_{ca}} - R(n)$$

(7.140)

where V_{ca} is the active volume of the laser and the recombination rate $R(n) = \frac{n}{\tau}$ is given by Eq. 3.91, which contains the Auger recombination coefficient C. The turn-on delay can therefore be expressed as

$$t_d = qV_{ca}\int_0^{n_{th}}\frac{1}{I - qV_{ca}R(n)}dn$$

(7.141)

The turn-on delay can be measured and the Auger recombination coefficient can be estimated from Eq. 3.91 and 7.141. Measured delay times and Auger recombination coefficients in InP-based strained-quantum well lasers are shown in Fig. 7.34. It should be pointed out that the n^3 dependence of the Auger rates, given in Eq. 3.91 holds in the limit where Boltzmann distribution is valid. As a laser is driven toward threshold and beyond, the Fermi levels are pushed into the bands and the Auger recombination ultimately becomes proportional to n. Therefore, in the strictest sense n^3 needs to be replaced by an appropriate polynomial in n to derive the correct Auger rates.

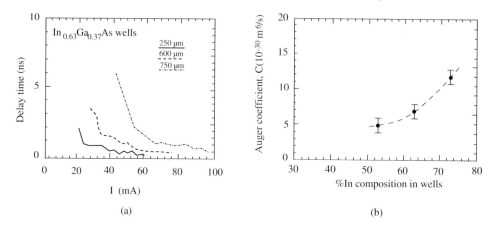

(a)

(b)

Figure 7.34 (a) Turn-on delay times in InP-based compressively strained $In_{0.63}Ga_{0.37}As$ QW lasers of various lengths and (b) Auger coefficients for various well compositions measured by the author and co-workers.

7.14 MODE LOCKING OF SEMICONDUCTOR LASERS

Mode locking is a technique of obtaining intense narrow pulses. A laser oscillating on an inhomogeneously broadened transition will have many longitudinal modes separated by an intermodal spacing of $\delta v = c/2ln_r$, where l is the cavity or resonator length. In general, these modes can oscillate with different phases. In mode locking, the phases of these components are locked together by external means. The modes then constitute components of a Fourier series expansion of a periodic function of period $t_M = \frac{1}{\delta v} = 2ln_r/c$ and therefore form a periodic pulse train. Mode locking is achieved by incorporating a switch or an absorbing region within the cavity. This switch is such that it blocks light at all times except when the light pulse crosses it. The different longitudinal modes have, in general, different phases, and groups of them can have the same phase at different regions of the cavity and different instants of time. If all the modes are equiphase, even by accident, they will form a giant narrow pulse that will be transmitted by the switch. Once the phase-locked oscillations start, they will continue to be locked. Any other combination of phases will add to a much smaller amplitude and will be blocked. In other words, they will add to the loss of the system. The action of the switch should therefore be intensity dependent. In a semiconductor laser system, such a switch, also called a *saturable absorber*, can be formed with quantum wells. The excitonic absorption peak of a quantum will is strong at low intensity. At very high intensities, on the other hand, the excitonic absorption is bleached due to carrier screening and bandgap renormalization effects. A semiconductor laser with a saturable absorber is schematically shown in Fig. 7.35. In this configurable it is essentially a coupled cavity system, with separate gain (active) and absorbing (passive) segments. It can be shown[†] that the peak power is the average power in a longitudinal mode times the number of modes locked together, m. In other words,

$$P_{peak} = m P_{avg} \tag{7.142}$$

The pulse-width Δt is given by the approximate relationship

$$P_{peak} \Delta t \cong P_{avg} \tau_p \tag{7.143}$$

If it is assumed that $\tau_p = \frac{2 l n_r}{c}$, the roundtrip time for a mode and photon loss due to absorption is neglected, then

$$\Delta t \cong \frac{2 l n_r}{cm} \tag{7.144}$$

Now the number of (locked) oscillating modes is approximately equal to the linewidth $\Delta \nu$ of the spontaneous emission divided by the mode spacing $c/2 l n_r$. With this substitution in Eq. 7.144 we get

$$\Delta t \sim \frac{1}{\Delta \nu} \tag{7.145}$$

which is an interesting result. As an example, a laser emitting 1 eV (1.24 μm) light oscillates on a transition of linewidth 10 meV. This corresponds to $\Delta \lambda = 124$Å and $\Delta \nu = 2.4 \times 10^{12}$ Hz. The linewidth of the mode-locked laser is 0.4ps. This is the ideal case, when all the modes are locked.

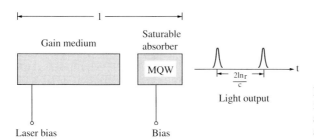

Gain medium

Saturable absorber

MQW

Light output

Figure 7.35   Mode locking in a semiconductor laser with a multiquantum well saturable absorber.

Laser bias

Bias

7.15 ANOMALOUS BEHAVIOR AND DEVICE RELIABILITY

In the simple treatment outlined in this chapter, it has been assumed that laser power and frequency remain constant in time once the device attains steady state. However, the output of a laser suffers from intensity and phase fluctuations and frequency shifts. The origin of these fluctuations is in the quantum nature of the lasing process. A proper

---

[†]See, for example, J. T. Verdeyen, *Laser Electronics*, *2nd* ed., Prentice Hall, Englewood Cliffs, NJ, 1989.

description of laser operation therefore requires a quantum mechanical formulation of the rate equations, which is outside the scope of this text. It is useful, however, to be aware of the anomalous phenomena.

It was seen that internal modulation of a laser is achieved by high-frequency, small-signal modulation of the drive current. The latter, however, leads to a periodic shift of the steady-state mode frequency because of changes in the refractive index of the emitting region arising from modulation of the carrier concentration. This phenomenon is known as *frequency chirping*, or wavelength chirping. In quantum well lasers, the active volume is small and the carrier density to be modulated is therefore small. This leads to lower chirping.

If adequate heat sinking and cooling is not provided, then the temperature of the laser can increase as it is driven beyond threshold. This can lead to a lowering of the lasing energy and the *lasing output mode* of a single-mode laser can shift to a higher-wavelength longitudinal mode. Such a shift occurs over a very small range (~ 1 mA) of drive current and is quite random. This phenomenon, known as *mode hopping*, can lead to kinks in the light-current characteristics of single mode lasers.

We have discussed the phenomenon of relaxation oscillations when a laser is turned on. Similar oscillations or *self-pulsations*, usually at a lower frequency (~ 1 GHz) can be observed in a laser under cw operation. The exact origin of these pulsations are not very clear, and they have been attributed to defects and traps, dark line defects and other absorbing regions in the active region, filamentation, and quantum noise effects. The phenomenon can be understood very simply as follows: An absorbing region, such as a dark line defect, will enhance the absorption of the photons created in the active region, thereby creating more carrier pairs, more recombinations, and more gain. The enhanced stimulated emission will reduce the gain. This will increase the absorption again, and the whole cycle is repeated. In reality, the process may be much more complicated.

The subject of noise in lasers is a very wide one. There are several sources of noise. Quantum noise is inherent in the laser operation itself and results from random spontaneous and stimulated transitions, which cause fluctuations in the output intensity. Anomalous features such as kinks and self-pulsations can lead to instability in operation and noise. In multimode devices, lack of mode stability can lead to partition noise.

Defects in the active region and imperfect mirrors lead to laser degradation. Catastrophic degradation usually results from imperfectly cleaved or mechanically damaged mirrors and can lead to complete device failure. Gradual degradation of laser performance is caused by the energy released in the active region by nonradiative recombination at defects and traps and the subsequent defect migration into the active region aided by this energy.

In conclusion, it is fair to state that reliability is an important aspect of laser operation, especially as the heterostructures and devices are assuming a more complex nature. Cross-talk and heat dissipation can give rise to nonoptimal behavior in laser arrays. Needless to say, even the phenomena outlined in this section needs more detailed understanding.

EXAMPLE 7.6

Objective.   To estimate the relative modulation of the output wavelength of a DFB laser due to a refractive index change of one part in a million.

For p = 1, the Bragg condition in a DFB laser is

$$\frac{\pi}{\Lambda} = \beta = \frac{2\pi}{\lambda_0} n_r \tag{7.146}$$

where Λ is the period of the grating. From Eq. 7.146,

$$\lambda_0 = 2\Lambda n_r \tag{7.147}$$

Therefore,

$$\frac{\delta\lambda_0}{\lambda_0} = \frac{\delta n_r}{n_r} = 10^{-6}$$

which is a small wavelength change. However, the corresponding frequency change can be large.

7.16 LONG-WAVELENGTH SEMICONDUCTOR LASERS

The materials that we have mostly discussed so far are GaAs- or InP-based, which cover the emission wavelengths of 0.7–1.6 μm. For molecular spectroscopy in terrestrial and extraterrestrial regions longer emission wavelengths are required. In principle, the antimonides can be used to make lasers emitting in the 2–10 μm range, but substrate availability and doping difficulties have impeded progress. The development of lasers made with IV–VI mixed compounds such as PbSnTe, PbSSe, PbSeTe, and PbSnSeTe have been very successful. Such devices, usually called lead-salt or lead-chalcogenide lasers, can provide emission wavelengths in the range of 3–30 μm. The bandgaps and lattice constants of some binary compounds and their ternary derivatives are shown in Fig. 7.36. Diode lasers made with these materials usually have

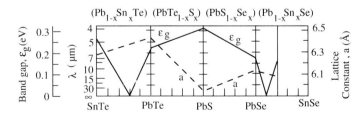

Figure 7.36 Bandgap (solid line) and lattice constant (dashed line) of $Pb_{1-x}Sn_xTe$, $PbTe_{1-x}S_x$, $PbS_{1-x}Se_x$, and $Pb_{1-x}Sn_xSe$ at 77°K as a function of composition x. The lines are drawn assuming linear extrapolation between the endpoints (from H. Preier, *Applied Physics*, **20**, 189, 1979).

to be operated at temperatures far below $300°$K and threshold current densities in the range of 100–1000 A/cm^2 have been measured at operating temperatures below $77°$K. Quantum well lasers have been realized with these materials, and alloying with rare-earth species, such as Eu or Yb, has been demonstrated.

Long wavelength lasers can, in principle, be realized by making use of intersub-band transitions in a quantum well. It is necessary to invert the population between the bands. Calculations have shown[†] that such transitions are allowed. The process is schematically shown in Fig. 7.37. Population inversion can be achieved by using resonant tunneling. The intersubband energy difference and consequently the emission wavelength can be tuned by changing the well and barrier parameters, and such devices are in the early stages of their development.

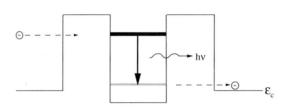

Figure 7.37   The conduction band diagram and energy subbands in a resonant tunneling heterostructure. The dark arrow indicates the long-wavelength lasing transition, while the dashed arrows indicate paths for electron population and depopulation.

PROBLEMS

7.1   What is the difference between a super-radiant light-emitting diode and a laser? Give your answer with illustrations.

7.2   Describe, with suitable diagrams, the principle of operation of a heterojunction laser.

7.3   In what applications would you use a broad-area laser, and in what applications would you use a quantum well or buried stripe geometry laser?

7.4   Could a quantum well be used to fabricate a very far-infrared (long wavelength) laser? If so, what would be the principle of operation?

7.5   Design a single-mode double-heterostructure laser for an emission wavelength of 1.50 μm. Design the heterostructure with particular attention to carrier and mode confinement. Use InP-based lattice-matched materials for your design.

7.6   Calculate the threshold current of a GaAs laser with an undoped active region of width $d = 0.2\mu m$ starting from a calculation of $R_{sp}(\mathcal{E})$ and $g(\mathcal{E})$.

[†]J. P. Loehr, J. Singh, R. K. Mains, and G.I. Haddad, *Applied Physics Letter*, **59**, 2070, 1991.

Assume the following values of the relevant parameters:

$$l = 400 \mu m$$

$$R_1 = R_2 = 0.9$$

$$\gamma = 10^3 cm^{-1}$$

$$\Gamma = 0.95$$

$$\eta_i = 0.9$$

7.7 The figure below shows the observed temperature dependence of the spectral distribution of an InP laser diode. The injection current is maintained constant. Explain the observed data.

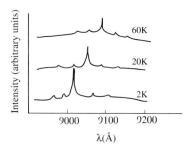

Figure P7.7

7.8 The differential quantum efficiency of an InP injection laser with negligible cavity losses is 30%. The voltage applied to the device is 2.5 V. Calculate the external power efficiency of the device.

7.9 The following equations hold for a junction laser:

$$\frac{dn}{dt} = AI - \frac{n}{\tau} - \frac{Bn\varphi}{\tau}$$

$$\frac{d\varphi}{dt} = \frac{Bn\varphi}{\tau} - C\varphi$$

Here n is the excess carrier population, φ the photon density, $\frac{Bn\varphi}{\tau}$ the stimulated emission, $C\varphi$ the loss of photons through the end walls, n/τ a relaxation term, I the current, and AI the injection rate. $A, B,$ and C are appropriate parameters. Find

(a) The steady-state solution

(b) The threshold current for laser action

7.10 Distinguish between homogeneous and inhomogeneous line broadening. What is the dominant excitonic line-broadening mechanism in a GaAs/AlGaAs MQW laser grown by MBE? Will such broadening affect the operation of a quantum well laser? List the major advantages of using a quantum well laser.

7.11 The density of states function in a material is as shown below. Are there any advantages in using this material in the active region of a laser? Sketch the spectral gain and comment on the threshold current.

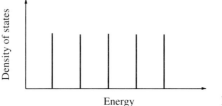

Figure P7.11

7.12 Describe briefly the principle of operation of a distributed feedback laser. What is its principle advantage? Sketch the transmission and reflection characteristics of a corrugated section of length l as a function of detuning $\Delta\beta l \cong [(\omega - \omega_0)\frac{l}{c}]n_r$. Assume $\kappa_{AB}l = 1.84$.

7.13 Describe (with figures and equations, if necessary) what happens as a function of time when (a) a laser is driven from zero current bias to a current $I > I_{th}$ by a step impulse; (b) a high-frequency small signal is superimposed on a dc bias $I > I_{th}$ and the frequency of the modulating signal is gradually increased.

7.14 A GaAs double-heterojunction laser has $f_r = 3$ GHz. Calculate the differential gain given $\overline{\varphi} = 10^{18}$ cm^{-3}, $\Gamma = 0.9$, and $\tau_p = 2$ ps.

READING LIST

AGRAWAL, G. P., and DUTTA, N. K. *Long Wavelength Semiconductor Lasers.* Van Nostrand Reinhold, New York, 1986.

CASEY, H. C., and PANISH, M. *Heterostructure Lasers: Parts A and B.* Academic Press, New York, 1978.

KEISER, G. *Optical Fiber Communications.* McGraw-Hill, New York, 1991.

KRESSEL, H., and BUTLER, J. K. *Semiconductor Lasers and Heterojunction LEDs.* Academic Press, New York, 1977.

THOMPSON, G. H. B. *Physics of Semiconductor Laser Devices.* John Wiley, Chichester, Eng., 1980.

VERDEYEN, J. T. *Laser Electronics,* 2nd ed. Prentice Hall, Englewood Cliffs, NJ, 1989.

WANG, S. *Fundamentals of Semiconductor Theory and Device Physics.* Prentice Hall, Englewood Cliffs, NJ, 1989.

8

Photodetectors

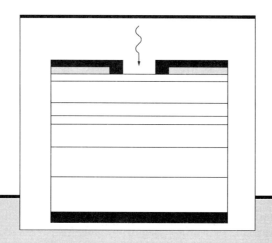

Chapter Contents

8.1 INTRODUCTION

A photodetector is an optoelectronic device that absorbs optical energy and converts it to electrical energy, which usually manifests as a photocurrent. There are generally three steps involved in the photodetection process: (1) absorption of optical energy and generation of carriers, (2) transportation of the photogenerated carriers across the absorption and/or transit region, with or without gain, and (3) carrier collection and generation of a photocurrent, which flows through external circuitry. The process of detection is sometimes associated with *demodulation*, when a high-frequency optical signal is converted into a time-varying electrical signal and further processed and rectified. Photodetectors are used to detect optical signals ranging over a very wide range of the optical spectrum.

Photodetectors are widely used in optical communication systems. In this application, detectors receive the transmitted optical pulses and convert them, with as little loss as possible, into electronic pulses that can be used by a telephone, a computer, or other terminal at the receiving end. The performance requirements from the detector are high sensitivity, low noise, wide bandwidth, high reliability, and low cost. Another common application of photodetectors is the monitoring of laser transmitters. A large-area detector is placed close to one facet of a transmitting laser source. The photocurrent generated in the detector is used in a circuit to maintain the laser output at a near-constant level in spite of temperature fluctuations. For communication applications, there is usually a need for high-speed detectors. For several other applications, high gain is necessary. Therefore, bandwidth and gain are fundamental physical trade-offs and the final application decides which photodetector is the most suitable.

The three main types of detectors are photoconductors, PIN diodes, and avalanche photodiodes. The first and the third types have internal gain. PIN photodiodes have no internal gain but can have very large bandwidths. In the sections to follow, we will learn about the physical principles of these devices. We will also learn about more specialized photodetectors such as the phototransistor, the modulated barrier photodiode, and the metal-semiconductor-metal (MSM) photodiode in the next chapter.

Photodetectors are also classified into *intrinsic* and *extrinsic* types. An intrinsic photodetector usually detects light of wavelength close to the bandgap of the semiconductor. Photoexcitation creates electron-hole pairs, which contribute to the photocurrent. An extrinsic photodetector detects light of energy smaller than the bandgap energy. In these devices the transition corresponding to the absorption of photons involves deep impurity and defect levels within the bandgap. Thus, absorption of a photon usually raises an electron from a deep level to the conduction band, or an electron from the valence band to the deep level, leaving behind a hole in the valence band. The electron or hole in the respective bands contributes to the photocurrent. There is another type of extrinsic photodetector, which involves transitions between subband energies in a quantum well. Since the energy difference between these subbands is usually small and of the order of 100 meV, such devices are used for far-infrared detection. There are, however, some drawbacks and limitations regarding the polarization of light that can be absorbed. The different intrinsic and extrinsic processes are illustrated in Fig. 8.1.

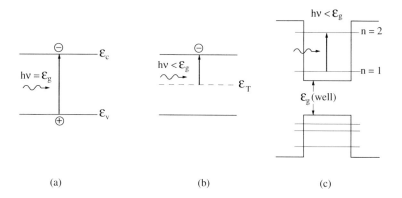

(a)   (b)   (c)

Figure 8.1   Different mechanisms of photodetection: (a) for intrinsic light $\left(h\nu \geq \mathcal{E}_g\right)$; (b) for extrinsic light utilizing a deep level; and (c) for extrinsic light utilizing intersubband transitions in a quantum well.

The quantum efficiency η of a photodetector is given by

$$\eta = \frac{I_{ph}/q}{P_{inc}/h\nu} = \frac{I_{ph}}{q} \cdot \frac{h\nu}{P_{inc}} \tag{8.1}$$

which is the number of carriers (electron-hole pairs) collected to produce the photocurrent I_{ph} divided by the number of incident photons. P_{inc} is the incident optical power. The quantity η in Eq. 8.1 is also defined as the *external* quantum efficiency of the device, η_{ext}. The *internal* quantum efficiency η_i is the number of pairs created divided by the number of photons absorbed and is usually very high, if not unity, in pure, defect-free materials. The external quantum efficiency depends on the absorption coefficient of the material and the thickness of the absorbing region:

$$\eta_{ext} \quad \text{or} \quad \eta \infty (1 - e^{-\alpha a}) \tag{8.2}$$

where α is the absorption coefficient of the semiconductor and a is the thickness of the active region. It is obvious, by virtue of Eq. 8.2, that the quantum efficiency is a function of the photon wavelength. The *responsivity* of a detector, $\mathcal{R}$, is defined as

$$\mathcal{R} = \frac{I_{ph}}{P_{inc}}$$

$$= \frac{\eta q}{h\nu}$$

$$= \frac{\eta \lambda (\mu m)}{1.24} \quad (A/W) \tag{8.3}$$

So ideally, for a fixed value of η, $\mathcal{R}$ should increase linearly with λ as schematically shown in Fig. 8.2. In reality, however, η is dependent on the absorption coefficient α, which in turn depends on the incident wavelength λ. Therefore, for practical detectors the long-wavelength cut-off of the spectral response is determined by the absorption edge, or bandgap, of the semiconductor. A short-wavelength cut-off is also observed in the spectral response characteristics because at short wavelengths the value of α is very large in most semiconductors and all the incident optical energy is absorbed near the surface. A real spectral response characteristic is also shown in Fig. 8.2.

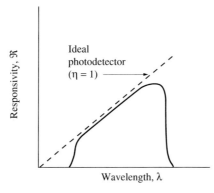

Figure 8.2 Schematic illustration of the responsivity of ideal and real photodetectors.

EXAMPLE 8.1
Objective.   To determine the long-wavelength cut-off of an InP photodetector.

To be absorbed by this photodetector, the photon energy must be greater than the bandgap $\mathcal{E}_g$. In other words, $\frac{hc}{\lambda} \geq \mathcal{E}_g$. The long-wavelength cut-off is therefore defined as $\lambda = \frac{hc}{\mathcal{E}_g}$. InP has a bandgap of 1.35 eV at 300°K. Therefore,

$$\lambda_C = \frac{6.626 \times 10^{-34} \times 3 \times 10^8}{1.35 \times 1.6 \times 10^{-19}}$$

$$= 0.92 \mu m.$$

8.2 PHOTOCONDUCTORS

The photoconductor is perhaps the simplest optical detector that exhibits an internal gain mechanism and clearly demonstrates the gain-bandwidth limitation. Its operation is based on the increase in conductivity of a specific region with photoexcitation. The photogenerated electrons and holes are collected by opposite contacts and result in a photocurrent. The schematic of a photoconductor in its simplest form, with two top contacts, is shown in Fig. 8.3. The active layer is formed epitaxially, or sometimes by ion implantation, on a high-resistivity substrate, and suitable contacts are delineated on top by evaporation of metals and subsequent alloying. The thickness of the active layer should be large enough so that it can absorb a significant fraction of the incident light but at the same time small enough so as to minimize the noise current resulting

Incident
photon

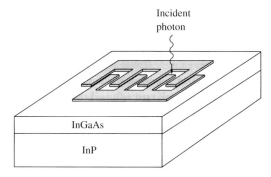

InGaAs

InP

Figure 8.3 Schematic of an
InGaAs/InP photoconductor with
interdigitated top contacts.

from a low resistance of the semiconductor layer. As we shall see later, the separation
between the contact pads, either in linear or interdigitated form, is also an important
parameter in the operation of the device. A suitable bias is applied across the contacts
to collect the photogenerated carriers. In InGaAs photoconductors, $\eta_{ext} \sim 80\%$–90%
with an absorption region thickness of 2 μm. To increase the quantum efficiency, an
antireflective coating, or a wider bandgap *window* layer needs to be formed on the
surface of the absorbing region.

Another important consideration in the performance of the device is noise generation. The noise is principally generated by the large dark current of the device and
is known as *Johnson* or *thermal noise*, that has its origins in the random motion of
carriers that contribute to the current. The resulting noise current, i_J, was formulated
by Nyquist from thermodynamic arguments and is expressed as

$$\bar{i}_J^2 = \frac{4k_B T B}{R_C} \tag{8.4}$$

where B is the bandwidth of the device and R_C is the resistance of the photoconducting channel. Current always flows through the photoconductor, and therefore the
signal-to-noise ratio, S/N, can be approximately expressed as

$$\frac{S}{N} \simeq \frac{(\text{conductivity})_{\text{light}}}{(\text{conductivity})_{\text{dark}}} \tag{8.5}$$

Hence, it is obvious that the dark conductivity has to be reduced to reduce noise.

EXAMPLE 8.2

Objective.   A photoreceiver circuit with a bandwidth of 1 GHz is connected to a load resistance of 1 KΩ. To calculate the rms thermal noise current of this resistor at room temperature.
From Eq. 8.4, the mean square noise current is

$$\bar{i}_J^2 = \frac{4 \times 1.38 \times 10^{-23} \times 300 \times 1 \times 10^9}{1 \times 10^3}$$

$$= 1.66 \times 10^{-14} A^2$$

and the rms thermal noise current is $1.29 \times 10^{-7} A$. This noise level is much higher than the
usual level of shot noise in a detector.

The rate and density of data transmission depend on the response speed of the device. The photogenerated electrons and holes move in opposite directions in the active region under the applied bias. The resulting photocurrent will persist until both carriers are collected at the electrodes, or until they recombine in the bulk of the semiconductor *before* reaching the respective contacts. We are really concerned with the recombination of minority carriers and therefore the parameter of importance is the minority-carrier recombination time τ. The time for detection of the photogenerated current is limited by the transit time between electrodes of the faster carrier—usually the electrons. Therefore, the shortest response time (maximum bandwidth) can be obtained by minimizing the distance between the contacts. It is assumed that the bias is such that the carriers travel at their saturation velocity. Note that the persistence of the slower hole in an n-type channel after the electron is collected will increase the response time, and therefore

$$\text{Bandwidth} \propto \frac{1}{\tau} \qquad (8.6)$$

where $\tau = \tau_h$. However, the continued persistence of the hole in the channel will draw more electrons to maintain charge neutrality. This constitutes a photocurrent gain, Γ_G, which is defined as

$$\Gamma_G = \frac{\tau}{t_{tr}} \qquad (8.7)$$

where t_{tr} is the transit time of the electrons. Therefore, phenomenologically, it is possible to write

$$\text{Gain} \times \text{Bandwidth} = K \qquad (8.8)$$

where the constant K is determined by the electron transit time.

We will now study the principle of operation of the photoconductor in more detail. Reference is made to Fig. 8.4, which shows the schematic of an ideal photoconductor in slab form with contacts at the two opposite edges. The photoconductor has length L, thickness a, and width b, and a voltage V is applied to the opposite contacts,

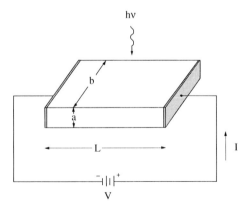

hv

Figure 8.4 Schematic of an ideal photoconducting slab for analysis. The device has side contacts.

leading to a current I flowing through the photoconductor circuit. This current can be expressed as

$$I = abq(n\mu_e + p\mu_h)\frac{V}{L}$$

$$= I_D + I_{ph} \tag{8.9}$$

where I_D is the dark current and I_{ph} is the photocurrent produced by incident light. n and p are the concentrations of free carriers and μ_e and μ_h are their respective mobilities. The number of photons incident on the semiconductor per second is equal to $P_{inc}/h\nu$ where P_{inc} is the incident optical power. If the corresponding generation rate is G and all the incident photons are absorbed, then the internal quantum efficiency, η_i, of the photoconductor is given by

$$\eta_i = \frac{G}{R_{abs}} = \frac{GV_v}{P_{inc}/h\nu} \tag{8.10}$$

where V_v is the volume. An optical power P_{inc} is incident on the top surface, and if the thickness a is small enough, then some light will be transmitted after absorption in the material. The light power leaving the lower surface is given by

$$P(a) = P_{inc}\,e^{-\alpha a} \tag{8.11}$$

where α is the absorption coefficient of the photoconductor material. Also,

$$GV_v = \eta_i \frac{P_{inc} - P(a)}{h\nu}$$

$$= \eta_i \frac{P_{inc}}{h\nu}(1 - e^{-\alpha a}) \tag{8.12}$$

and the quantum efficiency can be expressed as

$$\eta = \eta_i(1 - e^{-\alpha a}) \tag{8.13}$$

The spectral dependence of the absorption coefficient α of common semiconductors was discussed in Chapter 3. From that the spectral dependence of the generation rate can be derived, and this is schematically shown in Fig. 8.5. This dependence ultimately determines the spectral response of the photoconductor.

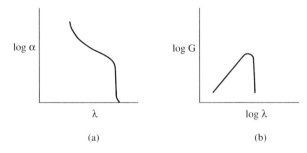

Figure 8.5 Schematic illustrations of (a) variation of absorption coefficient with λ and (b) corresponding spectral dependence of generation rate G.

We will next examine the process of current gain in photoconductors. In order to calculate the increase of carrier concentration by the absorption of light, one needs to find out how long the excess carriers exist in the photoconductor before they recombine or are swept out. Also, the injection of carriers through the contacts has to be considered. One of two processes can occur: successive reinjection of the faster carrier or loss by recombination. If all the carriers are swept out before recombining and if there is no replenishment by reinjection,

$$I_{ph} = qGV_v \tag{8.14}$$

For the general case, a gain Γ_G has to be included, and

$$I_{ph} = qGV_v\Gamma_G$$

$$= q P_{inc} \frac{\eta}{h\nu} \Gamma_G \tag{8.15}$$

The value of $\Gamma_G < 1$ if carriers recombine before sweeping out and $\Gamma > 1$ if reinjection can occur. The transit times of carriers through the device are given by

$$t_{tr}^e = \frac{L}{\vartheta_e} = \frac{L^2}{\mu_e V} \tag{8.16}$$

for electrons and

$$t_{tr}^h = \frac{L}{\vartheta_h} = \frac{L^2}{\mu_h V} \tag{8.17}$$

for holes. Here ϑ_e and ϑ_h are the respective saturation velocities.

8.2.1 DC Photoconductor

We will now examine the operation and gain of a photoconductor with dc photoexcitation (unmodulated light) incident on it. Before recombination occurs, the gain Γ_G can be defined as

$$\Gamma_G = \frac{\text{(path traveled by electron)} + \text{(path traveled by hole)}}{L} \tag{8.18}$$

We will assume first that both contacts of the photoconductor are ohmic and calculate the gain of the device in three bias voltage regimes.

When V is small, the drift velocity of the photogenerated carriers is small. Therefore there is no sweep-out of the carriers and recombination dominates. The average transit times of both electrons and holes are larger than the recombination lifetime, or $\frac{t_{tr}^e}{2}, \frac{t_{tr}^h}{2} > \tau$. From Eqs. 8.16, 8.17, and 8.18, the gain is given by

$$\Gamma_G = \frac{\mu_e \frac{V}{L}\tau + \mu_h \frac{V}{L}\tau}{L}$$

$$= \frac{\mu_e + \mu_h}{\mu_e} \frac{\tau}{t_{tr}^e} \tag{8.19}$$

and for $\mu_h \ll \mu_e$, which is generally the case,

$$\Gamma_G = \frac{\tau}{t_{tr}^e} < 1. \tag{8.20}$$

When V is made larger, since the electrons travel faster than holes, the situation may be described by the conditions $t_{tr}^e < \tau, t_{tr}^h > \tau$. In this case, there is sweep-out and reinjection of electrons only. The gain Γ_G is again given by Eq. 8.18, but now

$$\Gamma_G = \frac{\tau}{t_{tr}^e} > 1 \tag{8.21}$$

Finally, when the bias voltage V is made very large, $t_{tr}^e < \tau$ and $t_{tr}^h < \tau$. In this case, sweep-out and reinjection of both types of carriers lead to space-charge limited currents. The gain Γ_G is also reduced to unity. The variation of gain with bias is depicted in Fig. 8.6(a). This analysis can be extended to the case where one contact is ohmic while the other is rectifying, or blocking. For simplicity of discussion, we assume that the contact through which holes can be reinjected is blocking. In the first two regimes of bias voltage discussed above, the gain behavior remains the same. However, in the case of very large bias, since holes cannot be reinjected, the gain Γ remains greater than unity. The voltage dependent gain for this type of photoconductor is also depicted in Fig. 8.6(b).

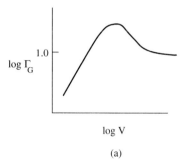

(a)

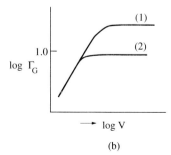

(b)

Figure 8.6 Variation of gain with bias in a photoconductor: (a) both contacts ohmic; and (b) (1) one contact ohmic, one blocking, (2) both contacts blocking.

While the case of two ohmic contacts covers the normal photoconductor, it is instructive to contrast it with the other extreme, namely, the case of two blocking contacts. In this case, the reinjection of both types of carriers is prevented. When V is small $t_{tr}^e > \tau$ and $t_{tr}^h > \tau$. As before, in this case $\Gamma_G < 1$. For larger bias values $t_{tr}^e < \tau$ and $t_{tr}^h > \tau$. Because there is no reinjection, the electron path is restricted to an average of $L/2$. The gain is therefore expressed by

$$\Gamma_G = \frac{\frac{L}{2} + \mu_h \frac{V}{L}\tau}{L}$$

$$= \left(\frac{1}{2} + \frac{\tau}{t_{tr}^h}\right) < 1 \tag{8.22}$$

Finally, for large V, $t_{tr}^e < \tau, t_{tr}^h < \tau$. Because there is no possibility of reinjection, the path lengths of both carriers are restricted and

$$\Gamma_G = \frac{\frac{L}{2} + \frac{L}{2}}{L} = 1 \tag{8.23}$$

This is the regime of normal operation of a p-i-n diode, which we will study in the next section.

8.2.2 AC Photoconductor

We will now treat the case of an amplitude-modulated light signal incident on the photoconductor. The power incident on the detector is described by

$$P_{inc} = P_0 + P_1 e^{j\omega t} \tag{8.24}$$

which leads to a generation rate

$$\hat{G} = G_0 + \hat{G}_1 e^{j\omega t} \tag{8.25}$$

The time-dependent electron concentration is expressed by the rate equation

$$\frac{dn}{dt} = \hat{G} - \frac{n}{\tau} \tag{8.26}$$

A solution for the photogenerated electron concentration Δn is obtained from Eq. 8.26 as

$$\Delta n = G_0\tau(1 - e^{-t/\tau}) + \frac{\hat{G}_1\tau}{1 + j\omega\tau}\left(e^{j\omega t} - e^{-t/\tau}\right) \tag{8.27}$$

Under steady-state conditions, when $t \gg \tau$,

$$\Delta n = G_0\tau + \frac{\hat{G}_1\tau}{1 + j\omega\tau}e^{j\omega t} \tag{8.28}$$

and, from Eqs. 8.7 and 8.15,

$$i_{ph} = qP_0\frac{\eta}{h\nu}\frac{\tau}{t_{tr}^e} + qP_1\frac{\eta}{h\nu}\frac{\tau}{t_{tr}^e}\frac{1}{1 + j\omega\tau}e^{j\omega t}$$

$$= I_0 + i_1 \tag{8.29}$$

It is evident that the dc component of the photocurrent, I_0, is the same as expressed earlier in Eq. 8.15. The ac component of photocurrent, i_1, has in addition the factor $(1 + j\omega\tau)^{-1}$. This indicates that for large response at low frequencies τ should be large. At high frequencies, $j\omega\tau \gg 1$ and therefore i_1 is independent of τ and inversely proportional to $t_{tr}^e = L^2/\mu_e V$. The latter calls for a small contact-to-contact distance L, and large applied bias V. However, the space-charge limited regime, discussed earlier, should be avoided and for this it is necessary to ensure that the transit time t_{tr}^e is larger than the dielectric relaxation time $\tau_d = \rho\epsilon_r\epsilon_0$ where ρ is the resistivity and ϵ_r is the relative dielectric constant of the semiconductor.

8.2.3 Gain and Bandwidth

As we have seen earlier, the photoconductive gain is given by Eq. 8.7. It should be remembered that the recombination process is probabilistic and is governed by the equation

$$\zeta(t) = \frac{1}{\tau} e^{-t/\tau} \tag{8.30}$$

where $\zeta(t)$ is the probability that a carrier will recombine after time t. To test this definition, we can express the average recombination time τ as

$$\tau = \int_0^\infty t\zeta(t)dt = \int_0^\infty \frac{t}{\tau}e^{-t/\tau}dt$$

$$= \tau \tag{8.31}$$

If an impulse of radiation is incident on the photoconductor, the resultant current will be of the form

$$i(t) = i_0 e^{-t/\tau} \tag{8.32}$$

In frequency domain the response can be expressed as

$$i(\omega) = \int_0^\infty i_0 e^{-t/\tau} e^{-j\omega t} dt$$

$$= \frac{i_0\tau}{1 + j\omega\tau} \tag{8.33}$$

and the power (frequency) response is given by

$$|i(\omega)|^2 = \frac{i_0^2\tau^2}{1 + \omega^2\tau^2} \tag{8.34}$$

The cut-off frequency, f_c, is defined as the frequency at which $\omega\tau = 1$. In other words,

$$f_c = \frac{1}{2\pi\tau} \tag{8.35}$$

and the gain-bandwidth product of the photoconductive detector is defined as

$$\Gamma_G f_c = \frac{\tau}{t_{tr}} \cdot \frac{1}{2\pi\tau}$$

$$= \frac{1}{2\pi t_{tr}} \tag{8.36}$$

Therefore, by reducing the spacing between the electrodes, or contacts, the gain and the gain-bandwidth product can be increased.

8.2.4 Noise in Photoconductors

One of the big drawbacks of a photoconductive detector is the large dark current and the associated noise. Resistance, thermal, or Johnson noise results from the random motion of carriers with average energy $k_B T$ contributing to the dark current of the device. The noise current due to Johnson noise is given by Eq. 8.4. The corresponding random voltage produced across the open-circuited terminals has a Gaussian distribution with a mean value of zero. The spectral density of Johnson noise is constant and is given by

$$\rho_J(f) \; = \; 2R_C k_B T \;\; (V^2/Hz) \tag{8.37}$$

Johnson noise is the dominant source of noise at high frequencies. In the equivalent circuit model it is usual to replace the noisy photoconductor with a noiseless device in parallel with a fictitious noiseless resistance equal to that of the photoconductor and an ideal noise current or voltage generator.

Fluctuations in the generation and recombination processes within the photoconductor will cause fluctuations in the carrier concentrations, and hence fluctuations in the conductivity of the semiconductor. If a current flows through the sample, a fluctuating voltage or noise is observed across its terminals. Generation noise becomes more important in small bandgap semiconductors in which, unfortunately, the dark current and Johnson noise are also high. The generation-recombination noise current is given by[†]

$$\overline{i_{GR}^2} \; = \; \frac{4q\Gamma_G I_0 B}{1 + \omega^2 \tau^2} \tag{8.38}$$

where I_0 is the steady-state light-induced output current. Generation-recombination noise is also called *shot noise*. Since the probability of thermal generation of electron-hole pairs is proportional to $\exp(-\mathcal{E}_g/2k_B T)$, a reduction of the operating temperature will reduce the generation-recombination noise. A rule of thumb is that T, the operating temperature should be less than $\mathcal{E}_g/25k_B$. It is seen from Eq. 8.38 that i_{GR} is almost independent of frequency for $f \ll 1/2\pi\tau$ and for $f > 1/2\pi\tau$ the noise current declines with increasing frequency with a $1/f^2$ dependence. Thus, at high frequencies the dominant noise mechanism will be Johnson noise.

At very low frequencies, less than 1 KHz, *flicker noise* becomes important. Flicker noise arises from surface and interface defects and traps in the bulk of the semiconductor. The spectral distribution of the noise current is given by

$$\overline{i_f^2} \; \infty \; \frac{1}{f} \tag{8.39}$$

[†]A. van der Ziel, *Fluctuation Phenomena in Semiconductors*, Butterworths Scientific Publication, London, 1959; T. E. Jenkins, *Optical Sensing Techniques and Signal Processing*, Prentice Hall International, Englewood Cliffs, NJ, 1987.

and hence Flicker noise is also known as 1/f noise. Summarizing what has just been discussed, the spectral distribution of noise current in a photoconductor is as shown in Fig. 8.7. At practical operating frequencies, thermal noise and shot noise determine the noise performance of the photoconductor. The RF noise equivalent circuit of a photoconductor is illustrated in Fig. 8.8 where i is the total current and G_C is the conductance resulting from the dark and photocurrents.

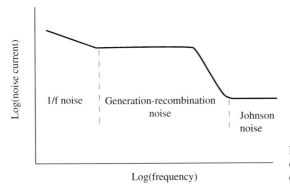

Figure 8.7 Schematic illustration of spectral distribution of noise current in a photoconductor.

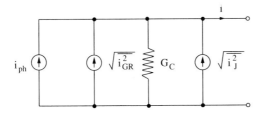

Figure 8.8 Noise equivalent circuit of a photoconductor.

Important parameters which describe the noise performance of a photoconductor are the *noise equivalent power* (NEP) and detectivity D^*. To estimate these, it is first essential to know the signal-to-noise ratio expressed as

$$\frac{S}{N} = \frac{i_{ph}^2}{i_J^2 + i_{GR}^2} \tag{8.40}$$

where the Flicker noise contribution is neglected. From Eq. 8.29, for a modulated optical power $P_1 e^{j\omega t}$ incident on the photoconductor, the rms optical power is $P_1/\sqrt{2}$ and the rms photocurrent can be expressed as

$$i_{ph} = \frac{q P_1 \eta}{\sqrt{2} h\nu} \frac{\tau}{t_{tr}} \frac{1}{(1 + \omega^2 \tau^2)^{1/2}} \tag{8.41}$$

Combining Eqs. 8.4, 8.38, 8.40, and 8.41, the signal-to-noise (power) ratio can be expressed as

$$\frac{S}{N} = \frac{\eta P_1}{8 B h\nu} \left[1 + \frac{k_B T}{\Gamma_G q} (1 + \omega^2 \tau^2) \frac{G_C}{I_0} \right]^{-1} \tag{8.42}$$

It is important to note that the S/N ratio is inversely proportional to the bandwidth of the photoconductor. The noise equivalent power is defined as the incident rms optical power required to produce a S/N of unity in a bandwidth of 1 Hz and is a measure of the minimum detectable signal. This can be easily deduced from Eq. 8.42. The detectivity D^* is defined as

$$D^* = \frac{A^{1/2}B^{1/2}}{(NEP)} \qquad (cm.Hz^{1/2}.W^{-1}) \tag{8.43}$$

where A is the area of the photoconductor on which light is incident. The reference bandwidth is usually taken as 1 Hz. Thus D^* is usually expressed as $D^*(\lambda, f, 1)$ where λ is the wavelength and f is the frequency of modulation.

8.3 JUNCTION PHOTODIODES

8.3.1 Introduction

The junction photodiode is the most common photodetection device that is used in a multitude of ordinary and specialized applications. They can be made with any semiconductor material in which a p-n junction can be formed by any of the techniques outlined in Sec. 4.1. Most photodiodes are of the intrinsic type where the energy of photons absorbed is equal to the bandgap energy. Photodiodes useful for the spectral range from visible to far infrared can be made with elemental and compound semiconductors. The basic principle of operation of a junction photodiode is very simple. The diode is operated under reverse-biased conditions, when the depletion layer is quite wide and a small dark current—the reverse saturation current—flows through the device. Upon photoexcitation on a p-n homojunction diode, photons are absorbed mainly in the depletion region and also in the neutral regions, particularly on the top, where the light is incident. The absorbed photons create electron-hole pairs. The photogenerated carriers in the depletion region are accelerated in opposite directions by the reverse bias and give rise to a photocurrent, the magnitude of which depends on the quantum efficiency. In general, if one photon produces one e^-–h^+ pair, the internal quantum efficiency is unity. The carriers that are produced within a diffusion length of the depletion layer, on either side of it, also reach the depletion region by diffusion and are then accelerated by the bias, thereby contributing to the photocurrent. Photoexcitation is therefore detected as an increase in the reverse-biased current of a junction photodiode. The process is schematically shown in Fig. 8.9. This type of a device is commonly called a depletion layer photodiode. It is obviously desirable to have a large depletion layer width, so that more photons can be absorbed, giving rise to a larger number of electron-hole pairs. However, in this case the transit time of carriers across the depletion region will increase, consequently increasing the response time of the diode and degrading its high-speed performance. In practical diodes, a compromise is made between the responsivity of the device, which is a measure of the photocurrent produced, and the bandwidth, which is related to the transit time of carriers, the junction capacitance, and the resistance of the diode and the circuit. It is also usual to have an asymmetric p-n junction in which the depletion

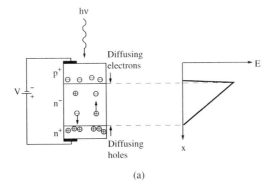

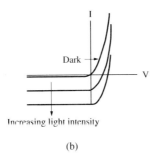

Figure 8.9 (a) Junction photodiode showing carrier drift (depletion) and diffusion regions and (b) current-voltage characteristics in the dark and under illumination.

region extends almost entirely in the lightly doped side and to make the top layer of a higher-bandgap material so that all the light is absorbed only in the depletion region of the lower-bandgap material. In such heterojunction photodiodes diffusion effects are eliminated and the response speed is improved. Also, since all the photons are absorbed in the depletion layer, the photoresponse is optimized. To achieve higher levels of photocurrent, or responsivity, or for the detection of very low levels of light, it is customary to use *avalanche photodiodes*, in which the photocurrent is enhanced by avalanche multiplication. However, although the gain increases, the bandwidth is reduced, since it takes time for the avalanche process to build up. We will discuss the characteristics of these devices in detail in Sec. 8.4.

With this brief introduction to junction photodiodes, we will now describe and discuss the characteristics of p-i-n and avalanche photodiodes in the following sections.

8.3.2 p-i-n (PIN) Photodiodes

The p-i-n (or PIN) photodiode is a junction diode in which an undoped i-region (p^- or n^-, depending on the method of junction formation) is inserted between p^+ and n^+ regions. Because of the very low density of free carriers in the i-region and its high resistivity, any applied bias drops almost entirely across the i-layer, which is fully depleted at zero or a very low value of reverse bias. Care must also be taken to see that all photogenerated carriers (electron-hole pairs) are collected. Therefore, the PIN diode has a "controlled" depletion layer width, which can be tailored to meet the requirements of photoresponse and bandwidth. For high response speed, the depletion layer width should be small and for high quantum efficiency, or responsivity, the width should be large. Therefore, a tradeoff is necessary. The absorption and carrier generation processes in a PIN photodiode are shown in Fig. 8.10. For practical applications, photoexcitation is provided either through an etched opening in the top contact, or an etched hole in the substrate, as schematically shown in Fig. 8.11(a). The latter reduces the active area of the diode to the size of the incident light beam.

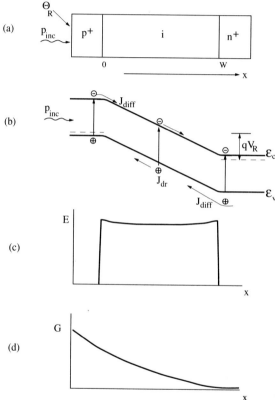

Figure 8.10 Absorption and carrier generation in a reverse-biased p-i-n diode.

Since there is no internal optical gain associated with the operation of a PIN diode (unity gain device), the maximum internal quantum efficiency η_i is 100% and the gain-bandwidth product is equal to the bandwidth. By careful choice of material parameters and device design, very large bandwidths can be attained. The response speed and bandwidth are ultimately limited either by transit time effects or by circuit parameters. The transit time of carriers across the depletion or i-layer depends on its width and the carrier velocity. Even for moderate reverse biases it can be assumed that carriers drift across the i-layer (of width $\sim 1\mu$m) with saturation velocity. The transit time can be reduced by reducing the thickness of the i-layer. There is the effect of diffusion of carriers created outside the i-layer, which can lower the response speed.

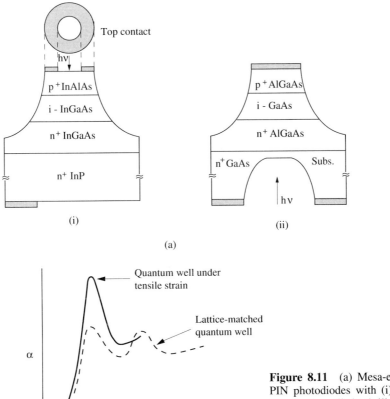

(a)

(b)

Figure 8.11 (a) Mesa-etched PIN photodiodes with (i) top illumination, (ii) back illumination; and (b) schematic illustration of the enhancement in excitonic absorption coefficient in a quantum well under slight tensile strain.

This can be minimized by fabricating the junction close to the illuminated surface. Finally, the device capacitance can be reduced by using a small-area device. This, of course, can limit the number of photons that can be absorbed. For optimal design it is customary to have

$$t_{tr} = \frac{1}{2}(\text{modulation period}) \tag{8.44}$$

and

$$t_{tr} = \tau_{RC} \tag{8.45}$$

the RC time constant. To summarize, the key to achieving a high-performance PIN diode (high quantum efficiency and large bandwidth) is to illuminate the diode through the substrate, ensure total depletion of the *i*-layer, and operate the device at a low reverse bias. The latter is important for digital operation and for low-noise performance. The principle source of noise in a PIN diode is generation-recombination noise, which is 1–2 orders of magnitude smaller than the Johnson noise, since the dark current in a reverse-biased junction is very small. The PIN diode is therefore a more sensitive device than a photoconductor.

8.3.2.1 Quantum Efficiency and Frequency Response.

We will now analyze the operation of a p-i-n photodiode shown schematically in Fig. 8.10 along with its energy band diagram and optical power profile. Light absorbed in the different regions of the device produce electron-hole pairs. Carriers produced within the depletion region or within a diffusion length of it will be separated by the electric field and their motion leads to current flow in the external circuit. Photogenerated carriers move by drift and diffusion and therefore the total current density through the reverse-biased depletion layer is

$$J = J_{dr} + J_{diff} \tag{8.46}$$

We will derive an expression for J under the assumptions that generation within the depletion region and absorption in the top p^+ layer are negligible. The second condition implies that the thickness of the top layer is much smaller than α^{-1}, where α is the absorption coefficient of the incident photoexcitation. The origin of the distance coordinate is assumed to coincide with the p^+–i junction. The pair generation rate, assuming $\eta_i = 1$, is given by

$$\begin{aligned} G(x) &= \phi_0 \, \alpha \, e^{-\alpha x} \\ &= \frac{P_{inc}(1 - \Theta_R)}{Ah\nu} \alpha \, e^{-\alpha x} \end{aligned} \tag{8.47}$$

where ϕ_0 is the incident photon flux (photons/sec/cm^2), Θ_R is the reflectivity of the top surface, and A is the area of the diode, usually equal to the junction area. In practical diodes, the value of Θ_R is made as small as possible by depositing an antireflection coating on the top surface. The drift current can then be expressed as

$$\begin{aligned} J_{dr} &= -q \int_0^W G(x)dx \\ &= -q\phi_0(1 - e^{-\alpha W}) \end{aligned} \tag{8.48}$$

where W is the width of the i-layer. We will next consider the diffusion component of the photocurrent, remembering that there is negligible absorption and production of carriers in the top p^+ layer. We are therefore mainly concerned with the diffusion of minority holes produced in the n^+ layer beyond the i-layer. The continuity equation for holes in this region at steady state, including generation and recombination, is given by

$$D_h \frac{\partial^2 p_N}{\partial x^2} - \frac{p_N - p_{NO}}{\tau_h} + G(x) = 0 \tag{8.49}$$

This equation can be solved with the boundary conditions

$$p_N = p_{NO} \quad \text{at } x = \infty$$

$$p_N = 0 \quad \text{at } x = W \tag{8.50}$$

The second condition comes as a surprise, but is easy to reason, since the minority carrier density goes to zero at the edge of the depletion region in a reverse-biased diode (refer to Fig. 4.15). The solution to Eq. 8.49 is then of the form

$$p_N = p_{NO} - (p_{NO} + Ce^{-\alpha W})e^{(W-x)/L_h} + Ce^{-\alpha x} \qquad (8.51)$$

where

$$L_h = \sqrt{D_h \tau_h} \qquad (8.52)$$

and

$$C = \frac{\phi_0 \alpha L_h^2}{D_h(1 - \alpha^2 L_h^2)} \qquad (8.53)$$

The diffusion current is then given by

$$J_{diff} = -qD_h \left(\frac{\partial p_N}{\partial x}\right)_{x=W}$$

$$= -q\phi_0 \frac{\alpha L_h}{1 + \alpha L_h}e^{-\alpha W} - qp_{NO}\frac{D_h}{L_h} \qquad (8.54)$$

Adding Eqs. 8.48 and 8.54, the total current is given by

$$J = -q\phi_0 \left(1 - \frac{e^{-\alpha W}}{1 + \alpha L_h}\right) - qp_{NO}\frac{D_h}{L_h} \qquad (8.55)$$

Under normal operating conditions the second term in Eq. 8.55 is small as the value of p_{NO} is normally very small. Under these conditions the total current is proportional to the incident flux ϕ_0, which is the expected result. The external quantum efficiency of the diode can be expressed as

$$\eta_{ext} = \frac{|J/q|}{P_{inc}/Ah\upsilon}$$

$$= (1 - \Theta_R)\left(1 - \frac{e^{-\alpha W}}{1 + \alpha L_h}\right) \qquad (8.56)$$

It is clear that to attain a high device quantum efficiency, a low reflection coefficient at the top surface is desirable. As mentioned before, this is usually realized in practice by special dielectric coatings, which have $\Theta_R \cong 0$. Also, for high η_{ext} a necessary condition is

$$\alpha W \gg 1 \qquad (8.57)$$

which implies that α or W should be large. This condition ensures that all or most of the photons incident on the device are absorbed. However, if W is made too large, the transit time of carriers across the depletion region becomes large and the device response speed is reduced. It is therefore desirable that the absorption coefficient α is high. Unfortunately, in most bulk semiconductors the value of α at the bandedge is determined by the bandstructure and therefore cannot be changed. Finally, if we assume that the diffusion coefficient L_h is very small, and $\eta_i < 1$, then

$$\eta_{ext} = \eta_i (1 - \Theta_R)(1 - e^{-\alpha W}) \tag{8.58}$$

A novel scheme of artificially enhancing the bandedge absorption coefficient can be employed in strained quantum well materials. In the p-i-n diode the i-region is a multiquantum well in which the well regions are under biaxial tensile strain. Referring to the discussion in Sec. 1.5, for a very small amount of tensile strain, the light- and heavy-hole resonances in the excitonic spectra can be made to *merge*. This increases the effective mass and the density of states, thereby enhancing the oscillator strength of the transition from hole subband to electron subband and the absorption coefficient. The effect is illustrated schematically in Fig. 8.11(b). In fact, the absorption coefficient is almost doubled at the heavy-hole exciton wavelength. As a consequence, diodes with smaller areas and smaller thickness of the i-region can be designed. Another technique of enhancing the absorption is to use microcavity photodiodes, to be described in the next chapter.

If the photodiode is designed to be sufficiently small, and still has the desired responsivity, then the frequency response, or bandwidth, is determined by transit time effects. This is because the photogenerated carriers require a finite time to traverse the depletion layer. If the incident photoexcitation is modulated at high frequency, then a phase difference between the photon flux and the photocurrent will be created. To analyze the frequency response of the diode, we consider the photogenerated carriers traverse the fully depleted i-layer of width W (Fig. 8.10) with saturation velocity under the influence of the applied reverse bias V_R. Use is made of the continuity equation for both electrons and holes

$$\frac{\partial p}{\partial t} = \frac{-(p - p_0)}{\tau} + G - \frac{1}{q} \nabla \cdot \mathbf{J_h} \tag{8.59}$$

$$\frac{\partial n}{\partial t} = \frac{-(n - n_0)}{\tau} + G + \frac{1}{q} \nabla \cdot \mathbf{J_e} \tag{8.60}$$

and the current density equations

$$J_h = q\mu_h p E - q D_h \frac{\partial p}{\partial x} \tag{8.61}$$

$$J_e = q\mu_e n E + q D_e \frac{\partial n}{\partial x} \tag{8.62}$$

$$J = J_h + J_e + \epsilon_s \frac{\partial E}{\partial t} \tag{8.63}$$

where p_0 and n_0 are the equilibrium hole and electron densities, τ is the recombination time, E is the electric field, and μ_h and μ_e are the hole and electron mobilities, respectively.

The following assumptions are usually valid: (a) The transit time of carriers is much shorter than the bulk recombination lifetime, so that the recombination term can be neglected from the continuity equations, and (b) the electric field is large, so that the diffusion current can be neglected. The incident photon flux is assumed to be of the form $\phi_{inc} = \phi_0 + \phi_1 e^{j\omega t}$ and it is assumed that each photon produces an electron-hole pair. Then the generation rate and carrier densities are given by

$$\hat{G}(x, t) = G_0(x) + \hat{G}_1(x)e^{j\omega t} \tag{8.64}$$

$$p(x, t) = p_{dc}(x) + p_1(x)e^{j\omega t} \tag{8.65}$$

$$n(x, t) = n_{dc}(x) + n_1(x)e^{j\omega t} \tag{8.66}$$

where the subscript 1 denotes the incremental ac quantities and the subscript dc denotes dc quantities. With these assumptions and rearranging the terms, we can write the continuity equations for ac quantities as

$$\frac{\partial J_{e1}}{\partial x} - j\omega \frac{J_{e1}}{v_e} = -q\,\hat{G}_1 \tag{8.67}$$

$$\frac{\partial J_{h1}}{\partial x} + j\frac{\omega J_{h1}}{v_h} = q\,\hat{G}_1 \tag{8.68}$$

where v_h and v_e are the hole and electron saturation velocities and $\hat{G}_1 = \alpha\phi_0 e^{-\alpha x}$. Solving for $J_{e1}(x)$ and $J_{h1}(x)$ and using the boundary condition $J_{e1}(0) = 0$ and $J_{h1}(W) = 0$, we get

$$J_{h1}(x) = -\alpha q\phi_0 \left[\frac{e^{-\alpha x} - e^{-\alpha W + \frac{j\omega(W-x)}{v_h}}}{(\alpha - j\omega/v_h)} \right] \tag{8.69}$$

and

$$J_{e1}(x) = -\alpha q\phi_0 \left[\frac{e^{\frac{j\omega x}{v_e}} - e^{-\alpha x}}{(\alpha + j\omega/v_e)} \right] \tag{8.70}$$

Maxwell's equation expressing Ampere's circuital law is of the form:

$$\nabla \times \mathbf{H} = \mathbf{J_c} + \frac{\partial \mathbf{D}}{\partial t}$$

$$= \mathbf{J} \tag{8.71}$$

where the second term on the right is the displacement current and $\mathbf{D} = \epsilon_s \mathbf{E}$. Since $\nabla \cdot (\nabla \times \mathbf{H}) = 0$, it follows that

$$\nabla \cdot \mathbf{J} = 0 \tag{8.72}$$

and

$$J = \frac{1}{W} \int_0^W \left(J_c + \epsilon_s \frac{\partial E}{\partial t} \right) dx \tag{8.73}$$

Solving for the short-circuit current, wherein terms containing the applied bias V are equal to zero, one obtains

$$J(\omega) = \frac{1}{W} \int_0^W (J_{e1}(x) + J_{h1}(x))dx \tag{8.74}$$

which by substitution of Eqs. 8.69 and 8.70 gives

$$J(\omega) = q\phi_0\alpha W \left[\frac{e^{-\alpha W} - 1}{\alpha W(\alpha W - j\omega t_{tr}^h)} + \frac{e^{-\alpha W}(e^{j\omega t_{tr}^h} - 1)}{j\omega t_{tr}^h(\alpha W - j\omega t_{tr}^h)} \right]$$

$$+ q\phi_0\alpha W \left[\frac{1 - e^{j\omega t_{tr}^e}}{j\omega t_{tr}^e(\alpha W + j\omega t_{tr}^e)} + \frac{1 - e^{-\alpha W}}{\alpha W(\alpha W + j\omega t_{tr}^e)} \right] \qquad (8.75)$$

where t_{tr}^e and t_{tr}^h are the electron and hole transit times, respectively. The current expressed in Eq. 8.75 is just the current source in the equivalent circuit of the photodiode shown in Fig. 8.12, and gives the intrinsic frequency response of the diode. If the diode capacitance and other extrinsic circuit parameters shown in Fig. 8.12 are taken into account, the frequency response of these can be included in a transfer function $H(\omega)$ and the output current can be expressed as

$$J_0(\omega) = J(\omega)H(\omega) \qquad (8.76)$$

where $J_0(\omega)$ is the output current across the load and $H(\omega)$ is the transfer function of the equivalent circuit which can be found from simple circuit analysis as

$$H(\omega) = \frac{R_D}{a + j\omega(b - c\omega^2) - d\omega^2} \qquad (8.77)$$

with

$$a = R_S + R_L + R_D$$

$$b = R_S R_L C_p + L_S + (R_S + R_L)R_D C_j + R_L R_D C_p$$

$$c = R_S L_S C_p R_D C_j$$

$$d = (R_S R_L C_p + L_S)R_D C_j + R_S L_S C_p + R_D C_p L_S \qquad (8.78)$$

In the equivalent circuit of Fig. 8.12 C_j is the junction capacitance originating from the depletion region. C_p is termed the parasitic capacitance, which is external to the wafer. It is a sum of the stray capacitance shunted across the wafer from the leads, the so-called *fringing* capacitance, and the package or encapsulation capacitance. L_S

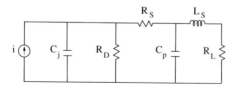

R_D= 100 MΩ	C_j = 80 fF
R_L= 50 Ω	C_p= 15 fF
R_S= 10 Ω	L_S= 60 pH

Figure 8.12 Equivalent circuit of a PIN photodiode. Typical values of the circuit elements of a reverse-biased ($\sim 10V$) InGaAs PIN diode also are listed.

is the total series inductance, mostly originating in the leads. R_D is the diode shunt resistance, or junction resistance, which is caused by carriers within the depletion region. It has a large value at zero and reverse biases and a very small value under forward bias. R_S is termed the series resistance of the diode, which is the sum of the contact resistances (spreading resistance) and the resistance of the undepleted regions of the diode. R_L is the load resistance. Typical values of these elements for a reverse-biased diode are listed in Fig. 8.12. Neglected in this equivalent circuit is the capacitance of the undepleted region of the diode, which is assumed negligible under reverse-bias conditions. Equation 8.76 expresses the output photocurrent in terms of the intrinsic and extrinsic parameters of the diodes. The calculated output photocurrents are shown in Fig. 8.13 for different values of the *i*-region thickness W.

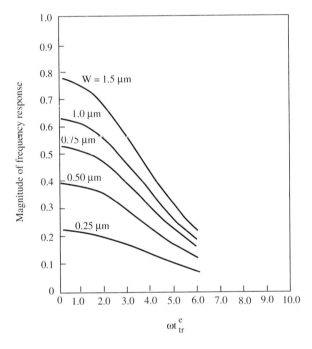

Figure 8.13 Calculated frequency response of InGaAs PIN photodiode for different *i*-region widths. The equivalent circuit element values listed in Fig. 8.12 have been used. The value of t_{tr}^e is constant for a fixed value of W.

Another important parameter for high-speed photodiodes is their impulse response, which is a measure of the speed of the diode. To get the impulse response, the discrete Fourier transform of the photocurrent in the frequency domain is calculated. This is shown in Fig. 8.14(a). From the frequency-dependent and time-dependent photocurrents the relationship between the full-width-at-half-maximum (FWHM) of the impulse response, the rise time of the impulse response (t_r), and the f_{3dB} frequency bandwidth can be determined. From this analysis the 3-dB bandwidth, f_{3dB}, can be expressed as

$$f_{3dB} = \frac{\Gamma_d}{2\pi t_r} \tag{8.79}$$

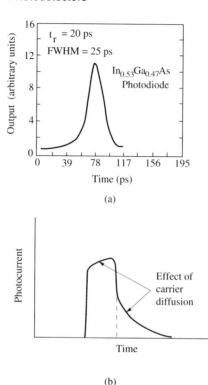

Figure 8.14 (a) Calculated temporal response of small (20 μm diameter) PIN diode with $W = 0.7\mu$m and (b) schematic illustration of the effects of carrier diffusion on the temporal response of a photodiode to a step optical impulse.

where Γ_d has a value of 2.8. The 3dB bandwidth is the frequency at which the response of the diode drops to half its value at low frequencies. The relation between FWHM and f_{3dB} can be expressed as

$$f_{3dB} = \frac{1}{2(FWHM)} \tag{8.80}$$

It is useful to reiterate the effects of diffusion on the temporal response. It has to be ensured that as few carriers as possible are generated *outside* the depletion region. For this to happen, the condition $W = (x_n + x_p) \geq [\alpha(\lambda)]^{-1}$ must approximately hold. Here W is the depletion layer width and λ is the wavelength of operation. For carriers traveling at saturation velocities, the drift time across the depletion region of a high-speed diode is ~ 1 ps. The diffusion time, on the other hand, is $\sim 1 - 10$ ns. The effect of diffusion on the temporal response of a high-speed photodiode is shown in Fig. 8.14(b). The most obvious effect is the long tail in the impulse response. The design of high-speed junction diodes is discussed in Sec. 8.3.3. We have largely discussed the p-i-n diode in which the i-region is a very lightly doped p-type (π) layer or n-type (ν) layer. A related device is the p-n photodiode. In reality it is configured as a p^+–n–n^+ junction in which the n-region, where the photons are absorbed, may not be fully depleted even at large bias. Therefore p-n photodiodes tend to have a lower response speed than the p-i-n diodes because of a larger diffusion current in the former.

8.3.2.2 Noise Performance.   In general, the p-i-n diode or any junction photodiode is operated under reverse bias conditions, and the dark current is very small.
Therefore, the shot noise, caused in this case only by generation within the depletion
region, is very small compared to that in photoconductors. The resistance or Johnson
noise dominates in these devices and even this can be minimized by optimizing the
circuit parameters. In what follows, we will derive equations that describe the noise
performance of p-i-n photodiodes. We consider modulated optical power, expressed by

$$P(\omega) = P_{inc}(1 + e^{j\omega t}) \tag{8.81}$$

incident on the photodiode, where it is assumed that the modulation index is unity.
The rms optical power is $P_{inc}/\sqrt{2}$ and the ac or rms photocurrent is given by

$$i_{ph} = \frac{q\eta P_{inc}}{\sqrt{2}h\nu} \tag{8.82}$$

In addition, there is usually a background radiation, albeit of very low intensity, in
the ambience in which the detector is placed, which gives rise to a photocurrent
I_B. Finally, a reverse current flows through the device due to thermally generated
electron-hole pairs in the depletion region. This dark current I_D contributes to the
total system noise and gives rise to random fluctuations of the output current. Because
of the randomness of the generation of these currents, they give rise to shot noise in
the output, and the equivalent noise current can be expressed, from Eq. 8.38, as

$$\overline{i_S^2} = 2q(I_{ph} + I_B + I_D)B \tag{8.83}$$

where B is the bandwidth of the device. Note that the rhs of Eq. 8.83 does not have
a frequency term, and is obtained from Eq. 8.38 under the condition $\omega\tau = 1$. The
corresponding frequency $f_c = 1/2\pi\tau$ is called the *noise equivalent bandwidth*.

    In addition to shot noise, there is resistance or Johnson noise at the output caused
by the various resistances in the diode circuit of Fig. 8.12. These resistances are the
diode shunt resistance R_D, the series resistance R_S, and the load resistance R_L. If the
diode is followed by a preamplifier in a receiver circuit, then the input resistance R_i
of the latter is in parallel to R_L. The mean-square noise current is given by

$$\overline{i_J^2} = \frac{4k_B T B}{R_{eq}} \tag{8.84}$$

where

$$R_{eq} = \left(\frac{1}{R_D} + \frac{1}{R_L} + \frac{1}{R_i} \right)^{-1} \tag{8.85}$$

The series resistance R_S is usually very small, a few ohms, and is neglected. The noise
performance of a photodiode with its associated circuit is schematically represented
in Fig. 8.15. We can now calculate the signal-to-noise ratio of the photodiode. The
signal power is given by $i_{ph}^2 R_{eq}$ and the total noise power at the output is $[\overline{i_S^2} + \overline{i_J^2}]R_{eq}$.

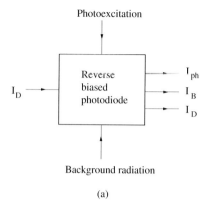

Photoexcitation

Background radiation

(a)

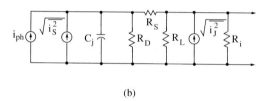

(b)

Figure 8.15   (a) Different current components in a photodiode and (b) noise-equivalent circuit.

Therefore, the signal-to-noise power ratio can be expressed as

$$\left(\frac{S}{N}\right)_{power} = \frac{\frac{1}{2}\left(\frac{q\eta P_{inc}}{h\nu}\right)^2}{2q(I_{ph} + I_B + I_D)B + \frac{4k_BTB}{R_{eq}}} \tag{8.86}$$

The total noise current of the device can then be expressed as

$$\overline{i_N^2} = \overline{i_S^2} + \overline{i_J^2} \tag{8.87}$$

The sensitivity of a photodiode is an important characteristic and is defined as the minimum optical input power needed to achieve a signal-to-noise ratio greater than a given value. A convenient measure of sensitivity is the NEP. Thus, when the rms signal power is equal to NEP, the photocurrent is equal to the noise current and the signal-to-noise ratio is unity. Also, NEP is calculated for bandwidth B = 1Hz. From Eq. 8.86, for S/N = 1,

$$\frac{q\eta P_{inc}}{\sqrt{2}h\nu} = \sqrt{\overline{i_N^2}}$$

$$= \left[2q(I_{ph} + I_B + I_D) + \frac{4k_BT}{R_{eq}}\right]^{1/2} B^{1/2} \tag{8.88}$$

from which we get

$$\text{NEP} = \frac{h\nu}{q\eta}\left[2q(I_{ph} + I_B + I_D) + \frac{4k_BT}{R_{eq}}\right]^{1/2} \; (W) \qquad (8.89)$$

It is clear from Eq. 8.89 that to increase the sensitivity of the photodiode, η and R_{eq} should be as large as possible, and the unwanted currents I_B and I_D should be as small as possible. Thus, the NEP is a measure of the minimum detectable power $(P_{inc})_{min}$.

EXAMPLE 8.3

Objective.   To calculate the value of NEP limited by background radiation for a $In_{0.53}Ga_{0.47}As$ photodiode with $I_B = 10^{-7}A$ and $R_L = 10^7\Omega$ used for detection of 1.3 μm radiation with $\eta = 0.7$. It may be assumed that the dark current is less than 1 nA.

At $\lambda = 1.3\mu$m, $h\nu = 0.95$ eV and $h\nu/q = 0.95$ V. From Eq. 8.89,

$$\text{NEP} = 1.414 \times \frac{0.95}{0.7} \times 1.3 \times 10^{-13}$$

$$= 2.5 \times 10^{-13}\,W$$

It is assumed that both R_i and R_D are much larger than the load resistance R_L. Thus a large value of R_{eq} is required for NEP to be limited by I_B of I_D.

8.3.3 Heterojunction Diodes

A junction photodiode is always operated under reverse-biased conditions, and it is imperative that the reverse leakage current is reduced to the absolute minimum. In photodiodes made with large bandgap materials, this is fairly easily achieved, since the reverse breakdown voltage is high enough and the diode can be operated at a bias that is far below the breakdown voltage. For optical communication applications, photodiodes are made of InP-based materials such as $In_{0.53}Ga_{0.47}As$ ($\mathcal{E}_g = 0.74$ eV) or InGaAsP ($\mathcal{E}_g = 0.74 - 1.1$ eV). Homojunctions made of these materials have a low breakdown voltage and a large reverse leakage current. Another aspect of photodiode design is the required bandwidth for high-speed operation. We have seen that for fast temporal response the area of the diode should be small and the thickness of the absorption layer should be small, while maintaining an acceptable value of responsivity. A rule of thumb for optimal design is that the transit time of carriers in the absorption or i-region should be approximately equal to the RC time constant of the diode and associated equivalent circuit. The active area of the diode can be kept small by illuminating the diode with a well-collimated beam through a hole etched in the substrate. The breakdown voltage can be increased by using a heterojunction and placing the p-n junction either at the heterojunction, or in the large bandgap material. These schemes are shown in Fig. 8.16(a) and (b). Illuminating the diode from the substrate side also ensures that the light is only absorbed in the n^- layer and the wider bandgap layers act as windows. In Fig. 8.16(a) the top p^+ layer is typically created by Zn-diffusion. In Fig. 8.16(b), the diode structure is realized completely by epitaxy. Mesa-shaped diodes are made by a combination

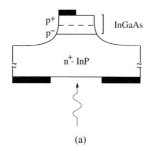

(a)

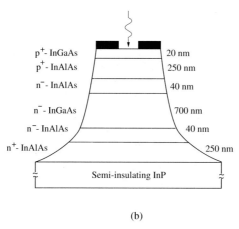

(b)

Figure 8.16 Two types of heterojunction photodiodes: (a) InGaAs/InP diode made by epitaxy and diffusion and (b) InGaAs/InAlAs/InP diode made by epitaxy.

of metal-deposition for contacts and subsequent alloying, etching, and photolithography. A silicon dioxide or silicon nitride layer is usually deposited on the surface for passivation and reduction of surface leakage current. A high-speed diode has a typical diameter of 10–50 μm, and the i-region (n^- or p^-) thickness is 0.5–1.0 μm.

The top contacts shown in Figs. 8.16(a) and (b) are usually connected monolithically to a large contact pad placed on the dielectric layer. The large pad facilitates bonding to the external circuit and power supply. This scheme usually introduces stray capacitance due to the conducting substrate. In other words, an additional capacitor is formed by the top contact pad, the dielectric layer and the conducting substrate. For high-speed photodiodes it is therefore common to grow the structure on a high-resistivity substrate, and place both n-type and p-type contacts and contact pads on the substrate by selective etching and metallization. This is schematically shown in Fig. 1.36. The photomicrograph of an InGaAs/InAlAs/InP photodiode with its bonding pads is shown in Fig. 8.17. Some performance characteristics of such diodes are shown in Fig. 8.18. The use of heterojunctions greatly helps the design and fabrication of near- and mid-infrared photodiodes in which the absorption region is a small bandgap semiconductor.

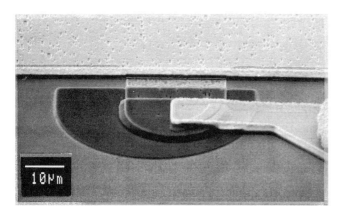

Figure 8.17 SEM micrograph of a high-speed $p^+ - n^- - n^+$ $In_{0.53}Ga_{0.47}As/In_{0.52}Al_{0.48}As$ photodiode grown by molecular beam epitaxy and fabricated by the author and co-workers. The top of the picture shows the bottom n-contact pad. Connection from the top p-contact to the p-contact pad on the right is made by an air bridge. The thickness of the n^- InGaAs absorption region is 0.7 μm.

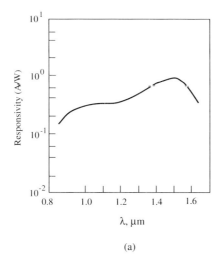

(a)

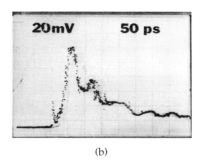

(b)

Figure 8.18 Performance characteristics of $In_{0.53}Ga_{0.47}As/In_{0.52}Al_{0.48}As$ heterojunction photodiodes grown by MBE and fabricated in the author's laboratory: (a) responsivity of 100 μm diodes with a 2 μm InGaAs absorption region and (b) temporal response of a diode of area $(20 \times 25)\mu m^2$ and 0.75 μm InGaAs absorption region to a 1 ps optical pulse. The deconvolved risetime and FWHM are 22 and 27 ps, respectively (Y. Zebda et al., *IEEE Electron Device Letters*, **EDL-8**, 579, ©1987 IEEE).

8.4 AVALANCHE PHOTODIODES

8.4.1 Introduction

For many applications, where very low levels of light are to be detected, it is desirable to use a detector with a large sensitivity. This translates to a requirement for large optical gain. We have seen that a photoconductor exhibits internal optical gain. However, the magnitude of this gain may not be sufficient. Large gains can be obtained in an *avalanche photodiode* (APD). The device is essentially a reverse-biased p-n junction that is operated at voltages close to the breakdown voltage. Photogenerated carriers in the depletion region travel at their saturation velocities, and if they acquire enough energy from the field during such transit, an *ionizing* collision with the lattice can occur. The field necessary to produce an ionizing collision is in the range of 10^4 to 10^5 V/cm, depending on the material. *Secondary* electron-hole pairs are produced in the process, which again drift in opposite directions, together with the primary carrier, and all or some of them may produce new carriers. The process is known as *impact ionization*, which leads to carrier multiplication and gain. Depending on the semiconductor material and device design, very large avalanche gains (~ 200 or more) can be achieved, and the avalanche photodiode therefore exhibits very high sensitivity.

The overall noise performance of an avalanche photodiode is determined by shot noise arising from the unmultiplied dark current, and noise from the probabilistic avalanche process itself. There are random fluctuations in the actual distance between successive ionizing collisions. These fluctuations give rise to variations in the total number of secondary carriers generated for each primary carrier injected into the gain region. This leads to randomness or noise in the total signal current, and the magnitude of the noise depends on the mean avalanche gain.

The avalanche process is itself asymmetric (i.e., the probability for initiating an avalanche is usually greater for one type of carrier than for the other). For example, in silicon, ionizing collisions are 30–50 times more frequent with electrons than with holes. This asymmetry is characterized by the ratio α_e/α_h, where α_e and α_h are the impact ionization coefficients for electrons and holes, respectively. They are discussed in more detail in the next section. Avalanche multiplication noise is lowest in devices in which the avalanche process is initiated by the carrier with the highest ionization rate. Also, as it may seem obvious by now, the value of α_e/α_h should be as large or as small as possible. In these cases the recycling of carriers through the depletion region is minimized, and consequently the noise is reduced because there is less fluctuation in the ionization process compared to the case when both carriers participate.

The largest bandwidth of an avalanche photodiode, which is obtained under unity gain conditions, is transit time limited. Under these conditions the device operates as a p-i-n diode. When the device exhibits gain, the bandwidth is reduced, since the avalanche process takes time to build up. In other words, it takes a finite amount of time for an ionizing collision to occur. Also, if $\alpha_e/\alpha_h \cong 1$, both types of carriers continuously recycle and persist in the depletion region, leading to a long response time, and consequently a small bandwidth. The best case is obtained for $\alpha_e/\alpha_h \to 0$ or ∞, in which case the response is limited only by the duration of a single transit.

With this general description of the avalanche photodiode, we will now study its operation and characteristics in more detail, starting with an analysis of the avalanche multiplication process itself. We will also learn about avalanche multiplication in multiquantum wells and staircase superlattices and the application of these structures to the design of low-noise avalanche photodiodes. Finally, practical photodiode structures and their characteristics will be described and discussed.

8.4.2 Avalanche Multiplication: Ionization Threshold Energies

In an avalanche photodiode, the process of impact ionization occurs in the high-field depletion region. As shown in Fig. 8.19, impact ionization is a three-carrier collision process where a high-energy primary carrier produces an additional electron-hole pair. The final carriers are left with finite kinetic energy and momentum. In general, the energy of the impact-ionizing primary carrier is greater than the bandgap energy $\mathcal{E}_g$. It can be shown from energy minimization considerations that the final carriers must have the same group velocity. The actual energy lost by the primary carrier is the *threshold ionization energy*. To study the avalanche process, the concepts of *ionization* and *ionization coefficient* must be understood. Impact ionization coefficients for electrons and holes, α_e and α_h, which characterize the avalanche process are fundamental material parameters. The coefficients are the reciprocal of the average distance traveled by electrons and holes in the direction of the electric field to create the electron-hole pairs. α_e and α_h can also be defined as the number of ionizing collisions per unit length (i.e., the number of secondaries). In Si, $\alpha_e > \alpha_h$, as shown in Fig. 8.20(a). However, in most III–V compounds $\alpha_e \cong \alpha_h$, which contributes to increasing the avalanche noise in photodiodes made of these materials. The measured impact ionization coefficients in InP are shown in Fig. 8.20(b). No analytical expression for the dependence of the coefficients on electric field exists, but from the measurement of these parameters in a variety of semiconductors it is found that their

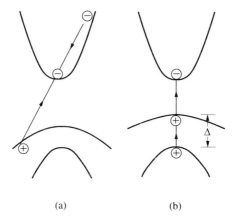

(a)   (b)

Figure 8.19 Impact ionization processes: (a) electron-induced, in which the conservation of energy and momentum may require higher energy of the primary electron since states near $\mathbf{k} = 0$ may not be available; and (b) hole-induced, in semiconductors where $\mathcal{E}_g \cong \Delta$, the split-off energy gap. At sufficiently high electric fields the holes can easily populate the split-off band.

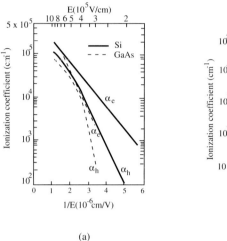
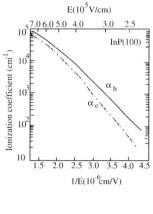

(a)   (b)

Figure 8.20   Measured impact ionization coefficients for electrons and holes in (a) Si and GaAs (from M. Zambuto, *Semiconductor Devices*, McGraw-Hill, Inc., 1989; adapted from R. A. Logan and S. M. Sze, *J. Phys. Soc. Jpn. Suppl.*, **21**, 434, 1966, and S. L. Miller, *Physical Review*, **105**, 1246, 1957), and (b) InP (from L. W. Cook, et al., *Appl. Phys. Lett.*, **40**, 589, 1982).

relation to electric field is roughly given by

$$\alpha_{e,h}(E) \;=\; \alpha_\infty \exp\!\left[-\left(\frac{b}{E}\right)^m\right] \tag{8.90}$$

in units of inverse length. Here α_∞ is the value of $\alpha_{e,h}$ for $E \to \infty$ and b and m are arbitrary coefficients. For example, in GaAs, $\alpha_\infty(cm^{-1}) = 1.3 \times 10^6$, $b(V/cm) = 2 \times 10^6$, and $m = 2$. In GaAs $\alpha_h \cong \alpha_e$, but in Si $\alpha_e \cong 30\alpha_h$, and in Ge $\alpha_e \cong 5\alpha_h$.

The electric field required to create impact ionization depends on the bandgap energy. At room temperature this field ranges from $\sim 10^4$ V/cm in InAs to more than 10^5 V/cm in GaP. The threshold ionization energy $\mathcal{E}_i$ is defined as the minimum energy needed for impact ionization. The value of $\mathcal{E}_i$ influences the ionization rates of electrons and holes. For the simplest case of *parabolic* conduction and valence bands, described by $\hbar^2 k^2/2m^*$, it can be shown that[†]

$$\mathcal{E}_{ie} \;=\; \mathcal{E}_g\!\left(1 + \frac{m_e^*}{2m_h^*}\right) \tag{8.91}$$

for electrons, and

$$\mathcal{E}_{ih} \;=\; \mathcal{E}_g\!\left(1 + \frac{m_h^*}{2m_e^*}\right) \tag{8.92}$$

[†]See, for example, B. K. Ridley, *Quantum Processes in Semiconductors*, Oxford University Press, New York, 1982.

for holes. Note that the conductivity effective masses are used here. These equations also depend on the crystal orientation. If $m_e^* = m_h^*$,

$$\mathcal{E}_{ie} = \mathcal{E}_{ih} = \frac{3}{2}\mathcal{E}_g \tag{8.93}$$

This model assigns the lowest ionization threshold to the carrier with the smallest effective mass. It also follows from Eq. 8.93 that

$$\mathcal{E}_{ie} + \mathcal{E}_{ih} = 3\mathcal{E}_g \tag{8.94}$$

independent of the effective mass ratio. In the case of nonparabolic bands, as is usually the case, the threshold ionization energies can no longer be defined by the simple equations 8.91 and 8.92. For impact ionization to occur, the carrier energy must exceed the threshold ionization energy. As mentioned earlier, the impact ionization process is probabilistic and can also be characterized by an impact ionization cross section σ_i. When the carriers have threshold energy $\mathcal{E}_i$ the value of σ_i is zero. Above threshold, the value of σ_i is a strongly increasing function of the electron energy. In other words, the probability of impact ionization per unit time for one type of carrier is proportional to the ionization cross section for that carrier.

8.4.3 Multiplication and Ionization Coefficients in p-i-n and p-n Junction Diodes

The avalanche process is initiated by carrier multiplication in a high-field region. The latter is defined by *multiplication coefficients*. Imagine a piece of semiconductor material in which a number of electrons corresponding to an electron current J_{ei} is injected at one end. By the avalanche multiplication process these electrons will multiply in number and at the output, a larger electron current J_{eo} is obtained. The hole current through the sample will proportionately adjust so that the total current remains constant. The electron multiplication coefficient is then defined as

$$M_e(E) = \frac{J_{eo}}{J_{ei}} \tag{8.95}$$

Similarly, for holes,

$$M_h(E) = \frac{J_{ho}}{J_{hi}} \tag{8.96}$$

It is important to know how M_e and M_h are related to α_e and α_h. We will consider two cases, a p-i-n diode and a $p^+ - n - n^+$ diode.

In the p-i-n diode, it is assumed that the electric field due to an applied bias is uniform across the i-region of width W, as shown in Fig. 8.10. Electron-hole pairs are assumed to be created in the i-region by photoexcitation, and they will drift in opposite directions. Since the electric field is assumed constant across the i-region, $\alpha_e(E)$ and $\alpha_h(E)$ can be assumed to be constant. Also, for simplicity, assume $\alpha_e = \alpha_h$. The carriers accelerated in opposite directions will suffer ionizing collisions, creating more pairs. The multiplication coefficient can be expressed as

$$M = 1 + \alpha_e W + (\alpha_e W)^2 + (\alpha_e W)^3 + \cdots \cdots (\alpha_e W)^n + \cdots \cdots$$

$$= \frac{1}{1 - \alpha_e W} \text{(for } \alpha_e W < 1) \tag{8.97}$$

Avalanching occurs when the current at breakdown goes to ∞ (i.e., $M \to \infty$). The condition for avalanching is therefore

$$\alpha_e W = 1 \tag{8.98}$$

In other words, breakdown occurs when only one carrier is produced at the end of the transit.

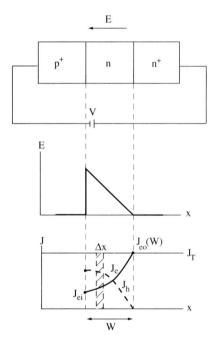

Figure 8.21 Electric field and electron and hole current densities in an avalanche photodiode.

We will next analyze the case of the $p^+ - n - n^+$ diode, shown in Fig. 8.21 together with the electric field distribution and the various current components in the n-region. We consider the transport and collision of carriers in this region. The electrons drifting toward the n^+-layer will produce a current J_e and holes drifting toward the p^+ region will produce a current J_h. The total current J_T is given by

$$J_T = J_e + J_h \tag{8.99}$$

and, for electrons,

$$M_e(E) = \frac{J_{eo}(W)}{J_{ei}} \tag{8.100}$$

where W is the thickness of the n-region. We next take an incremental slice of the n-region of thickness Δx and analyze the current components in more detail, assuming that the avalanche process is initiated by electrons. At the input end, the current density is J_e (or J_{ei} if the slice is taken at the edge of the n region). Then at the output of the incremental slice, the additional components of current are due to

1. Electrons generated by electrons
2. Electrons generated by holes

Also, as shown in Fig. 8.21, $J_T = J_{eo}(W)$ neglecting the reverse saturation current. The *increase* in current can therefore be expressed as

$$\Delta J_e = J_e \alpha_e \Delta x + J_h \alpha_h \Delta x \qquad (8.101)$$

from which

$$\frac{dJ_e}{dx} = J_e \alpha_e + J_h \alpha_h \qquad (8.102)$$

and, from Eq. 8.99,

$$\frac{dJ_e}{dx} = J_e \alpha_e + (J_T - J_e)\alpha_h \qquad (8.103)$$

rearranging which, we get the first order differential equation

$$\frac{dJ_e}{dx} + (\alpha_h - \alpha_e)J_e = \alpha_h J_T \qquad (8.104)$$

Remember that α_e and α_h are functions of electric field, which itself is not uniform across the incremental slice. It is easy to show that Eq. 8.104 is valid across the entire multiplication, or n region. Since the impact ionization coefficients are of the form $\alpha[E(x)]$, Eq. 8.104 is of the general form

$$\frac{dy}{dx} + P(x)y(x) = Q(x) \qquad (8.105)$$

Case I: $\alpha_e(E) = \alpha_h(E)$
From Eq. 8.104 it follows that

$$\frac{dJ_e}{dx} = \alpha_e J_T \qquad (8.106)$$

the solution of which is of the form

$$J_e(W) = J_T \int_0^W \alpha_e dx + J_{ei} \qquad (8.107)$$

and since $J_{eo}(W) = J_e(W) = J_T$, it follows that

$$J_{eo}(W) = J_{eo}(W) \int_0^W \alpha_e dx + J_{ei} \qquad (8.108)$$

and invoking the definition of the electron multiplication coefficient, it follows that

$$M_e = M_e \int_0^W \alpha_e dx + 1 \tag{8.109}$$

or

$$M_e = \frac{1}{1 - \int_0^W \alpha_e dx} \tag{8.110}$$

Breakdown in this case corresponds to the condition

$$\int_0^W \alpha_e dx = 1. \tag{8.111}$$

Case II: $\alpha_e(E) \# \alpha_e(E)$

We assume, for this general case that the avalanche process is initiated by electrons. We then have to solve Eq. 8.104, which is of the general form expressed by Eq. 8.105. The solution to the latter is of the standard form

$$y = \left[\int_0^x Q e^{\int_0^x P dx'} dx + C \right] \Big/ e^{\int_0^x P dx'} \tag{8.112}$$

where C is the constant of integration. Therefore, the solution of Eq. 8.104 with the boundary condition $J_T = J_e(W) = M_e J_{ei}$ is given by

$$J_e(x) = \frac{J_T \left\{ \frac{1}{M_e} + \int_0^x \alpha_e \exp\left[\int_0^x (\alpha_h - \alpha_e) dx' \right] dx \right\}}{\exp\left[\int_0^x (\alpha_h - \alpha_e) dx' \right]} \tag{8.113}$$

and since $J_e(W) = J_T$, it follows that

$$1 - \frac{1}{M_e} = \int_0^W \alpha_e \exp\left[\int_0^x (\alpha_h - \alpha_e) dx' \right] dx \tag{8.114}$$

and consequently

$$M_e = \frac{1}{1 - \int_0^W \alpha_e \exp\left[\int_0^x (\alpha_h - \alpha_e) dx' \right] dx} \tag{8.115}$$

and

$$M_h = M_e \exp\left[\int_0^x (\alpha_h - \alpha_e) dx' \right] \tag{8.116}$$

Breakdown occurs when $M_e \to \infty$, or

$$\int_0^W \alpha_e \exp\left[\int_0^x (\alpha_h - \alpha_e) dx' \right] dx = 1 \tag{8.117}$$

Similarly, if the avalanche process is initiated by holes, the breakdown condition is

$$\int_0^W \alpha_h \exp\left[\int_0^x (\alpha_e - \alpha_h) dx' \right] dx = 1 \tag{8.118}$$

Equations 8.117 and 8.118 are equivalent. Therefore, the breakdown and avalanching process depends on the carrier population and collision rate in the depletion region and not on the carrier type that initiates the avalanche process. When the avalanche breakdown condition is satisfied in a particular type of diode by Eq. 8.98, 8.111, 8.117, or 8.118, the breakdown voltage can be calculated by Poisson's equation. Breakdown occurs whenever E_M, the maximum field in the depletion region is equal to or larger than a *critical field* E_{CR}. The value of E_{CR} depends on the impurity concentration and is shown in Fig. 8.22. The calculated breakdown voltages in different semiconductors for one-sided abrupt junctions were shown in Fig. 4.19. It may be noted that the impact ionization process is dependent on the crystal orientation, since the scattering rates are orientation dependent.

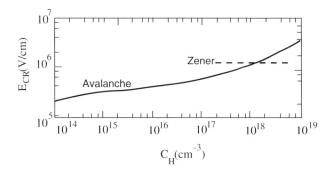

Figure 8.22   Breakdown electric field vs. impurity concentration for one-sided junction in Si (from M. Zambuto, *Semiconductor Devices*, McGraw-Hill, 1989; adapted from S. L. Miller, *Physical Review*, **105**, 1246, 1957).

The multiplication coefficient or factor, M, is also called the *avalanche gain*, which is therefore related to the impact ionization coefficients. Very high values of avalanche gain, $\sim 10^3$, can be obtained in photodiodes biased near the breakdown voltage. However, a high avalanche gain implies that the avalanche process takes longer to build up and persists longer. In detecting high-frequency modulated light, the bandwidth is reduced. This is the fundamental gain-bandwidth limitation of avalanche photodiodes.

8.4.4 Measurement of Multiplication Factors and Impact Ionization Coefficients

The measurement of the impact ionization coefficients is not a direct one, and the value of the ionization coefficients are derived from measurement of multiplication coefficients that will be described shortly. Very stringent requirements regarding diode structure and experimental conditions need to be met before reliable results are obtained. First, pure electron and hole injection should be obtained in the same diode, rather than in complementary devices. The best technique of ensuring this is by photogeneration of the carriers to be injected. It is important to be able to measure the photocurrent *without* avalanche gain very accurately. Only then can reliable values of M_e and M_h be obtained. The electric field profile in the multiplication region should be accurately known and the variation of the electric field with distance should be very gradual. The latter ensures that α_e and α_h are functions of the electric field only.

Finally, it is important to ensure that the avalanche gain is uniform across the active area of the device. All of these requirements are satisfied in a p-i-n diode in which the electric field is uniform across the i-region. In a p-i-n diode, the following equations have been derived to relate α_e and α_h with M_e and M_h:[†]

$$\alpha_e(E) = \frac{1}{W}\left[\frac{M_e(V) - 1}{M_e(V) - M_h(V)}\right]\ln\left[\frac{M_e(V)}{M_h(V)}\right]$$

$$\alpha_h(E) = \frac{1}{W}\left[\frac{M_h(V) - 1}{M_h(V) - M_e(V)}\right]\ln\left[\frac{M_h(V)}{M_e(V)}\right] \tag{8.119}$$

where $V = EW$ is the applied reverse bias and W is the thickness of the i-region. Similar equations have been derived for abrupt p-n junctions.

A typical device structure for the measurement of impact ionization coefficients is shown in Fig. 8.23(a). 100–300 μm diameter mesa diodes are made by photolithogra-

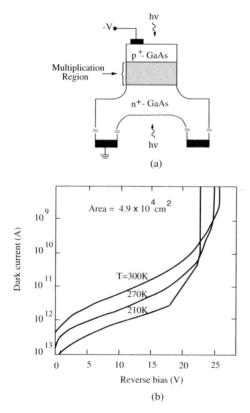

Figure 8.23 (a) Mesa-etched photodiode for the measurement of carrier multiplication coefficients and (b) typical temperature-dependent current-voltage characteristics of such a diode (F. Y. Juang, *MBE Growth and Characterization of Superlattice Avalanche Photodiodes*, Ph.D. Thesis, University of Michigan, 1987).

[†]G. E. Stillman and C. M. Wolfe, "Avalanche Photodiodes," in *Semiconductors and Semimetals*, Vol. 12, Chapter 5, ed. R. K. Willardson and A. C. Beer, Academic Press, New York, 1977.

phy and etching. Grooves are made on the backside by selective etching to facilitate photoexcitation from this side. Large-area backside and annular top ohmic contacts are formed by evaporation and alloying. Note that the top p^+ layer and the n^+ layer on the substrate must be thick enough so that the photoexcitation incident from either side is absorbed completely in these layers and does not reach the i-layer. Only then can pure carrier injection be ensured. In other words, when light is absorbed in the top p^+ layer, excess minority electrons are produced, and these are injected into the avalanching i-layer by the reverse bias. Similarly the bottom n^+ layer acts as an injector for holes. The i-region is typically 2 μm thick.

Before proceeding with the measurement of the multiplication coefficients it is important to ensure that avalanche multiplication is the dominant breakdown mechanism in the diode. This can be done by measuring the reverse current-voltage characteristics of the diode as a function of temperature. If avalanche multiplication is the dominant breakdown process, the reverse breakdown voltage will be higher at higher temperatures. This occurs because as the temperature increases, the probability of nonionizing collisions increases, and consequently the probability of ionizing collisions decreases. A competing process for breakdown is tunneling, either across the bandgap or via a trap level in the gap, but these processes are largely independent of temperature. Measured temperature dependent I–V characteristics for GaAs-AlGaAs p-i-n diodes in which the i-region is a MQW are shown in Fig. 8.23(b). It is clear that in this case avalanching is the dominant breakdown mechanism.

An arrangement for the measurement of the multiplication coefficients is shown in Fig. 8.24(a). A beam splitter and chopper are introduced between the excitation source and sample to allow phase-sensitive detection. The diode photocurrent is amplified by a current sensitive preamplifier and lock-in amplifier and is fed to the y-axis of an x-y recorder. The reverse bias applied to the diode is fed to the x-axis. The bias is swept from zero to near-breakdown and the corresponding photocurrent for both electron and hole injection is recorded. To obtain the electron and hole multiplication

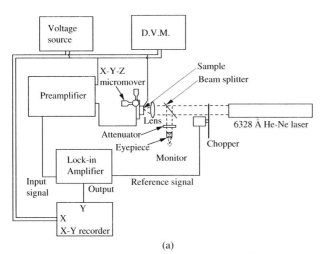

(a)

Figure 8.24 (a) Schematics for the measurement of carrier multiplication coefficients and (b) data obtained from GaAs/AlGaAs avalanche photodiodes (F. Y. Juang, *MBE Growth and Characterization of Superlattice Avalanche Photodiodes*, Ph.D. Thesis, University of Michigan, 1987).

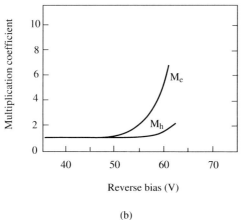

(b)   **Figure 8.24** (*continued*)

factors, appropriate stable unmultiplied values of the photocurrent at low bias are used as reference. Measured values of M_e and M_h in a p-i-n diode with a GaAs/AlGaAs MQW *i*-region are shown in Fig. 8.24(b) as a function of reverse bias. From these values of M_e and M_h, α_e and α_h are calculated using Eq. 8.119.

8.4.5 Noise Performance of Avalanche Photodiodes

Avalanche photodiodes—like p-i-n diodes—are limited by the shot noise associated with the primary, unmultiplied bulk signal, dark, and background currents. However, the principal source of noise in these devices is the avalanche process that gives gain. As we have learned before, the avalanche process is probabilistic. As in any statistical process, there are random fluctuations in the actual distance between successive ionizing collisions. These variations, in turn, give rise to fluctuations in the total number of secondary carriers generated for each primary carrier injected into the gain region. This fluctuation will cause excess noise in the total signal current. To minimize avalanche noise, the electron-to-hole ionization coefficient ratio, $k_r = \alpha_e/\alpha_h$, should be kept as large, or as small, as possible, by making the avalanche process asymmetrical. The reason is that in materials in which the probability of ionization is nearly identical for both types of carriers—as is typical of GaAs, InP, and Ge—a large density of carriers of both types is generated during avalanche; secondary holes, as well as secondary electrons can generate further carriers. For electron injection into the high-field region, the excess noise factor for electrons is given by[†]

$$F_e = M_e \left\{ 1 - (1 - k_r)[(M_e - 1)/M_e]^2 \right\} \qquad (8.120)$$

For hole injection, the excess noise factor for holes is given by

$$F_h = M_h \left\{ 1 - \left[1 - \left(\frac{1}{k_r} \right) \right] [(M_h - 1)/M_h]^2 \right\} \qquad (8.121)$$

[†]R. J. McIntyre, *IEEE Trans. Electron Devices*, **ED-13**, 164, 1966.

Since avalanche noise is smaller for materials with larger ionization coefficient ratios, such devices can be used at higher gain than detectors made with materials having ionization coefficient ratios close to unity. At some high value of gain, the avalanche noise from the APD dominates all other sources including the receiver amplifier, and a further increase in gain only reduces the signal-to-noise ratio of the receiver. This tradeoff point occurs when the noise from the APD equals the noise from the receiver. The optimum gain at which this occurs depends on how rapidly the avalanche noise increases with gain.

The response time of an APD also depends on the ionization coefficient ratio, k_r. In the best case, $k_r \rightarrow \infty$, the response time is limited only by the duration of a single transit across the depletion region. In the worst case, $k_r \rightarrow 1$, the secondary carriers continuously recycle through the depletion region, which induces a very long response time, along with high gain. The gain-bandwidth product of an APD is greatest in devices and materials where the ionization coefficient ratio is maximized.

The expression for the signal-to-noise ratio of an avalanche photodiode is very similar to that of a p-i-n diode given by Eq. 8.86. The avalanche gain of the diode enhances the dark current, the current produced by the optical signal and the current produced by any background light. Therefore, from Eq. 8.82,

$$i_{ph} = \frac{q\eta P_{inc}\Gamma_G}{\sqrt{2}h\nu}$$ (8.122)

Similarly, the mean square shot noise is given by

$$\overline{i_S^2} = 2q(I_{ph} + I_B + I_D)\Gamma_G^2 FB$$ (8.123)

where F is the excess noise factor. The thermal noise is the same as that for a p-i-n diode, given by Eq. 8.84. Therefore, for an input optical signal having unity modulation index, the signal-to-noise power ratio is given by

$$\left(\frac{S}{N}\right)_{power} = \frac{\frac{1}{2}\left(\frac{q\eta P_{inc}\Gamma_G}{h\nu}\right)^2}{2q(I_{ph} + I_B + I_D)F\Gamma_G^2 B + \frac{4k_B TB}{R_{eq}}}$$ (8.124)

The NEP for the avalanche photodiode is given by an expression, almost identical to Eq. 8.89, in which Γ_G appears in the denominator. It is clear that the avalanche gain can reduce the NEP of the device and thereby enhance its sensitivity compared to the p-i-n photodiode. Among photodetectors, avalanche photodiodes have demonstrated the best sensitivity. In a digital system the *bit error rate* (BER) depends on the S/N ratio according to the equation

$$\text{BER} = \frac{1}{2}\text{erfc}\left(\frac{S}{2\sqrt{2}N}\right)$$ (8.125)

8.4.6 Practical Avalanche Photodiodes

Two of the most important objectives in APD design are reduction of dark current and enhancement of device speed. In order to obtain the best performance from an APD, several structural and materials requirements must be met. First and foremost it is important to ensure uniformity of carrier multiplication over the entire photosensitive area of the diode. The material in which avalanching occurs must be defect free, and great care must be taken in device fabrication. Defects in the absorption and multiplication regions can give rise to *microplasmas*, which are local areas in which breakdown occurs at voltages lower than the rest of the junction. Such defects arise from bulk or edge dislocations propagating from the substrate or are initiated in the active region during growth and processing. The defect density can be reduced by having a small active area, typically between 10 and 100 μm in diameter.

The dark current in an APD, which depends largely on the material used, arises from generation-recombination in the depletion region, leakage current, and tunneling current. A common problem in APDs is the excessive leakage current at the junction edges. In Si diodes the common technique used to alleviate this problem is to incorporate a guard ring, which is an n-p junction created by selective-area diffusion around the periphery of the diode. In addition, a SiO_x film is deposited to minimize surface leakage currents, or edge leakage currents in mesa-diodes. The incorporation of guard rings is more common in Si APDs, an example of which is shown in Fig. 8.25.

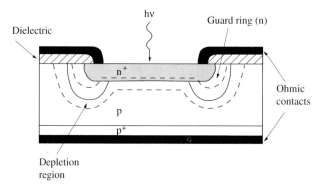

Figure 8.25 Guard ring created in Si APD by selective diffusion.

The choice of materials for an APD is obviously dictated by the application. For applications where the system is designed to operate in the near-infrared region, GaAs/AlGaAs diodes are used. For fiber-optic communication InGaAsP ($0.74 \leq \mathcal{E}_g \leq 1.35$ eV) is the material of choice. For example, $In_{0.53}Ga_{0.47}As$ has a bandgap of 0.74 eV, which corresponds to a wavelength of 1.55 μm. Homojunction diodes made of these low bandgap materials have a high reverse leakage current at high voltages due to band-to-band tunneling. The tunneling probability across the bandgap of a semiconductor is exponentially proportional to $\mathcal{E}_g^{-3/2}$. Therefore, the tunneling-induced reverse leakage current can be reduced by incorporating the high-field avalanching region of the diode in high-bandgap material. These considerations have led to the design and realization of the *separate absorption and multiplication* (SAM) APD, or SAM-APD.[†] These structures combine low leakage, due to the junction being placed in the high-bandgap material (e.g., InP), with sensitivity at long wavelengths provided

[†]K. Nishida, K. Taguchi, and Y. Matsumoto, *Appl. Phys. Lett.*, **35**, 251, 1979.

by the low-bandgap absorption region (e.g., InGaAs). Dark currents of the order of pA-nA can be obtained with gains ~ 10. However, there is a potential problem with the operation of a SAM-APD. Holes can accumulate at the valence band discontinuity at the InGaAs/InP heterojunction and thereby increase the response time. A graded bandgap quaternary InGaAsP layer can be inserted between the InP and InGaAs layer to alleviate this problem. The resulting structure is known as a *separate-absorption-graded-multiplication* (SAGM)–APD or SAGM-APD,[†] an example of which is shown in Fig. 8.26.

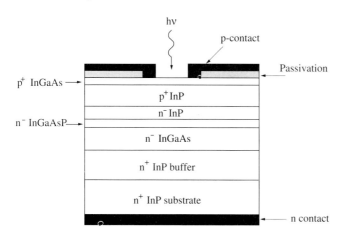

Figure 8.26 Schematic of an InP-based SAGM-APD.

To improve the performance of these devices a thin delta-doped region can be inserted between the absorption and avalanche regions. This layer ensures a low field in the absorption region, which is desirable. Gain-bandwidth products over 100 GHz have been measured in delta-doped SAGM-APDs. In the design of these devices it is important to ensure that all the absorption takes place in the absorption region, and of the photogenerated electrons and holes, the carrier type, which has a higher impact ionization coefficient in the multiplication region, is injected into it. This will ensure low-noise operation. In practical devices, the 1–2 μm thick absorption region is undoped. The graded layer (0.1–0.3 μm) and the avalanching layer (1–2 μm) are doped to $\sim 1 \times 10^{16} cm^{-3}$. The p^+ layer can be thin and doped to 10^{17}–$10^{18} cm^{-3}$.

For high-speed operation of these devices it is desirable to deplete all or most of the absorption region. Otherwise, carrier transit is dominated by the slow diffusion of holes. The temporal response of well-designed APDs usually shows a fast rise time (100–200 ps), which is determined by the transit and collection of the faster electrons and the time for the avalanche process to build up. The fall-time, on the other hand, can be very large ~ 10–100 ns, and this is believed to be dominated by the trapping and recombination of holes. SAGM-APD devices are currently the most suitable for optical communication systems.

[†] Y. Matsushima, K. Sakai, and Y. Noda, *IEEE Electron. Dev. Lett.*, **EDL-2**, 179, 1981.

It is perhaps fair to mention some of the drawbacks of APDs. They have more complicated structures compared to photoconductors and PIN diodes and can therefore involve higher manufacturing costs. The biasing voltage is, in general, quite high (20–50 V). Finally, the avalanche gain is extremely sensitive to bias and temperature. This makes their use and operation more expensive, since elaborate temperature compensation and supply voltage stabilization is needed to ensure stable operation of the device.

EXAMPLE 8.4

Objective.   An avalanche photodiode detects 1.5 μm photoexcitation with a responsivity of 0.7 A/W when operated with a gain of 30. 10^{11} photons are incident on the device per sec. To calculate the quantum efficiency and the rms photocurrent of the device.

From Eq. 8.3,

$$\eta = \frac{1.24 \times 0.7}{1.5}$$

$$= 0.58$$

From Eq. 8.122,

$$i_{ph} = \frac{\Gamma_G}{\sqrt{2}} \eta \left(\frac{P_{inc}}{h\nu} \right) q$$

where the quantity within parentheses is the number of incident photons. Therefore,

$$i_{ph} = (30 \times 0.58 \times 10^{11} \times 1.6 \times 10^{-19})/\sqrt{2}$$

$$= 20 \mu A$$

8.4.7 Superlattice Avalanche Photodiodes

The best performance of an avalanche photodiode is obtained when $\alpha_e/\alpha_h \rightarrow 0$ or ∞. Unfortunately, in most compound semiconductor materials $\alpha_e/\alpha_h \cong 1$. Different ways and means to enhance or reduce the value of α_e/α_h in artificially structured materials have therefore been explored. A viable multilayered structure, in which enhancement of α_e/α_h can be obtained, is shown in Fig. 8.27(a). It is called a *staircase superlattice* and in a practical photodiode this material would form the avalanching region. It should be mentioned that the staircase superlattice is difficult to synthesize, since it involves the growth of precisely controlled quaternary graded layers in each period. In order to understand the process of impact ionization in a staircase superlattice, it is necessary to look at the band profile under an applied bias, as shown in Fig. 8.27(b). The electrons, which drift from left to right, gain in energy at each step and undergo impact ionization. The process is repeated at each step, resulting in a large multiplication of the number of electrons. The holes, on the other hand, do not encounter such large potential steps, and their multiplication remains as in bulk semiconductors. The structure therefore behaves as a solid state *photomultiplier* of electrons and a large value of α_e/α_h can be obtained.

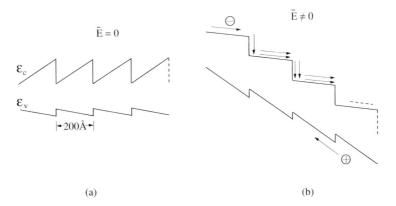

Figure 8.27 A "staircase" superlattice (a) under zero bias and (b) with an applied transverse bias.

Similar enhancement of α_e/α_h can be obtained in a multiquantum well structure with large well and barrier thicknesses (200–500 Å) and/or deep wells in the conduction band. Similarly, α_h/α_e will be enhanced if the band offset is higher in the valence band, resulting in deep-hole quantum wells. The process of preferential multiplication of one type of carrier is similar to that in the staircase superlattice. Data obtained from such a GaAs/AlGaAs MQW structure are shown in Fig. 8.28. When well and barrier sizes become smaller, and approach 100Å or less, enhancement in the multiplication factors and in the value of α_e/α_h can still be obtained. However, the mechanism for such enhancement, if the conduction band offset is not very large, is different. For example, for GaAs/Al$_{0.3}$Ga$_{0.7}$As MQW, where $\Delta\mathcal{E}_C : \Delta\mathcal{E}_V$ is 57 : 43, enhancement

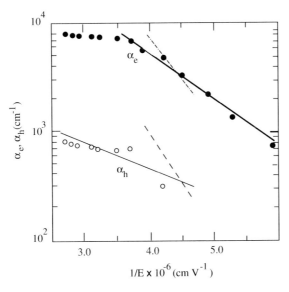

Figure 8.28 Measured electron and hole ionization coefficients in a 420 Å GaAs/570 Å Al$_{0.4}$Ga$_{0.7}$As multilayered heterostructure (from P. Bhattacharya, *Intl. Journal of Optoelectronics*, Taylor & Francis, **5**, 51, 1990). The dashed lines represent measurements made by F. Capasso et al.(*Applied Physics Letters*, **40**, 38, 1982).

of α_e/α_h can still be obtained. In these structures, the electrons are not truly confined in the wells because of their small effective mass. The electron behavior is similar to that in a bulk semiconductor. The holes, on the other hand, are more confined and experience more scattering in the wells, and therefore their ionization rate can be reduced. The overall effect is an enhancement of α_e/α_h.

Superlattice avalanche photodiodes, in addition to promising low noise performance and gain, can also provide tunability of the spectral response. In such diodes, the absorption and multiplication region of the diode is made of multiquantum wells with varying sizes of wells and barriers. The measured spectral response and avalanche gain of GaAs/AlGaAs superlattice photodiodes, typically with 20–30 periods, are shown in Figs. 8.29(a) and (b), respectively. The data of Fig. 8.29(b) indicate

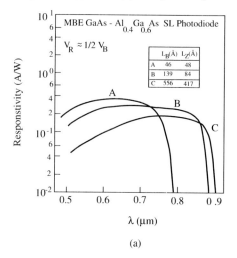

(a)

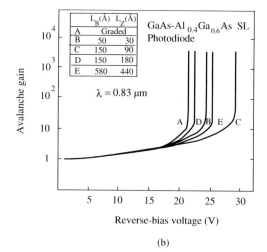

(b)

Figure 8.29 (a) Spectral response and (b) avalanche gain of GaAs/AlGaAs superlattice avalanche photodiodes (from P. Bhattacharya, *Intl. Journal of Optoelectronics*, Taylor & Francis, 5, 51, 1990).

that if stabilized bias supplies are used, very large values of gain can be realized with avalanche photodiodes. The measured temporal response of a 80ÅGaAs/90 ÅAl$_{0.1}$Ga$_{0.9}$As superlattice APD of 30 μm diameter to 10 ps excitation pulses is shown in Fig. 8.30. The intrinsic speed of these devices is set by the emission and collection of carriers created in the well region, which in turn is a sum of the emission or tunneling times from the wells, and the carrier transit time across the entire absorption region. The latter, with large-enough bias applied across the diode, is very small. Therefore, field-assisted thermionic emission and tunneling times, which can be in the range of 60–100 ps, sets the intrinsic limit. The extrinsic limit is set by the diode area, capacitance, and resistance. With proper design, therefore, these APDs can be made to have very large bandwidths. Combined with their potential for low-noise operation, SL APDs can prove to be very important devices.

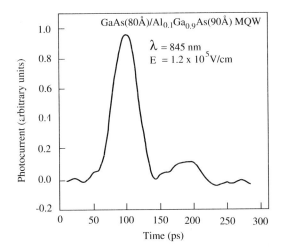

Figure 8.30 Measured impulse response of GaAs/Al$_{0.1}$Ga$_{0.9}$As MQW (0.75 μm) photodiode to 10 ps optical pulse (S. Goswami, *A Quantum Well Phototransistor for Switching Applications*, Ph. D. Thesis, University of Michigan, 1992).

A superlattice or MQW with a high α_e/α_h or α_h/α_e can be inserted as the multiplication region in a SAGM-APD. Efficient, low-noise operation is then ensured if the carriers with the higher multiplication factor are injected in this region from the absorption region.

8.5 HIGH-SPEED MEASUREMENTS

We have learned about the design and operation of high-speed photodetectors. Some techniques that are commonly used to measure the high-speed response of these diodes are now briefly described.

8.5.1 Impulse Response Measurements

An optical source, usually producing pulses shorter than the response time of the device under test, is required. Depending on the spectral response of the device and the materials with which the device is made, different types of lasers can be used.

(a)

(b)

Figure 8.31   (a) Coplanar waveguide carrier with detector mounted and (b) high-frequency measurement test fixture with biasing circuit.

These can be mode-locked dye lasers, a Ti-sapphire laser, or a color center laser. A cw laser can also be driven by a comb generator to produce very short pulses. The device is mounted on a coplanar waveguide carrier (Fig. 8.31) and is biased with a high-frequency biasing circuit. The time-domain impulse response of the detector is usually measured on a sampling oscilloscope with a high-speed sampling head. The overall response obtained in this way is a convolution of the device response including circuit and parasitics, the measurement system including the sampling gate width, and the laser pulse width. Alternately, the output of the photodiode can be directly fed to a microwave spectrum analyzer from which the transfer function and frequency response of the complete detector circuit are obtained.

8.5.2 Optical Heterodyning

Photodetectors are square law devices, which imply that if they are excited by laser light having an electric field E, the photocurrent I_{ph} is proportional to E^2. This is the basis of the optical heterodyning technique. Light from two near-identical lasers with slightly differing frequencies, obtained by temperature tuning one of them, is simultaneously incident on the test device. The product of the two laser fields at different frequencies produce a beat frequency, which is detected by the photodetector, with its output fed to a microwave spectrum analyzer. The limitation of this technique is the bandwidth of the transmission line on which the device is mounted and that of the spectrum analyzer.

8.5.3 Electro-optic Measurement Technique

For measuring the high-speed response of passive or active devices, directly in the time-domain, electro-optic sampling is emerging as the most useful technique. This technique exploits the Pockel's effect, in which an external electric field applied to a nonlinear crystal changes the birefringent properties of the crystal, and therefore the polarization of the light passing through it. Hence, if the crystal is placed between crossed polarizers, the transmitted light intensity changes as a function of the applied field. Many different variations of electro-optic sampling can be implemented. The most popular and convenient arrangement is called the external "finger-probe" scheme. Since most circuits have an open electrode structure, there are fringing electric fields above the surface of the circuit. Therefore, a small electro-optic crystal dipped into this fringing field will experience a change in its birefringent properties. This method has the advantage of needing no hardwired connection to the device, since only a capacitive coupling is used in the measurement process.

A schematic of the actual system is depicted in Fig. 8.32. A short pulse femtosecond laser source, running at a high repetition rate, is divided into two beams. The

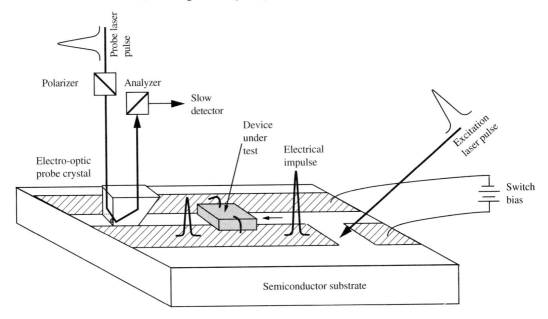

Figure 8.32   Schematic for measuring high-speed device response by electro-optic sampling.

excitation beam is focused on a photoconductive gap switch, embedded in a broadband transmission line structure, here a coplanar strip. The excitation laser pulse generates carriers in the photoconductor gap, which are swept by the applied bias to the other electrode, so as to generate a current (voltage) pulse traveling down the transmission line, as shown. This electrical impulse reaches the device under test, which can either be wire-bonded or can be integrated in the test structure. Depending on the device response, an output voltage waveform is transmitted, whose electric field is monitored by the electro-optic transducer. The other beam, which is the probe beam, is passed through a variable optical delay line, and then through a combination of the polarizer, the electro-optic transducer, and the analyzer, successively. By varying the optical delay of the probing beam, the sampling probe pulses have the opportunity to probe the amplitude at many points in a voltage waveform, as it passes the transducer. This is manifested by a change in the transmitted intensity of the probe beam in a polarizer-analyzer assembly as shown, which is then detected by a slow photodetector using lock-in techniques. The temporal response is limited to about 300 fs, and excellent sensitivity can be achieved at kilohertz modulation frequencies.

8.5.4 Fiber Coupling

In a lot of applications where detectors are used, the light to be detected is focused onto the active area of the device, or is incident through a back hole etched in the substrate. In optical communication systems, where the transmission medium is invariably a low-loss fiber, the end of the fiber is usually butt-coupled to the detector, which may form a part of the front-end photoreceiver circuit [Fig. 8.33(a)]. The end of the fiber may

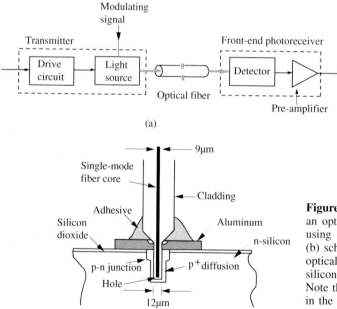

Figure 8.33 (a) Schematic of an optical transmission system using an optical fiber and (b) schematic cross section of an optical fiber interconnected to a silicon p-n junction photodiode. Note that the photodiode is formed in the cylindrically etched hole by diffusion (from P. Prucnal et al., *Optics Letters*, **11**, 109, 1986).

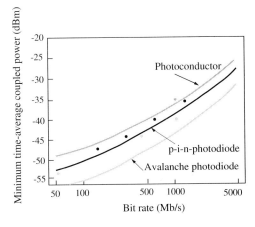

Figure 8.34 Calculated and measured minimum detectable power as a function of bit rates for three types of detectors. The calculations are for detectors operating at 1.3 μm with a bit error rate of 10^{-9} (from S. R. Forrest, *IEEE Spectrum*, **23**, 81, ©1986 IEEE).

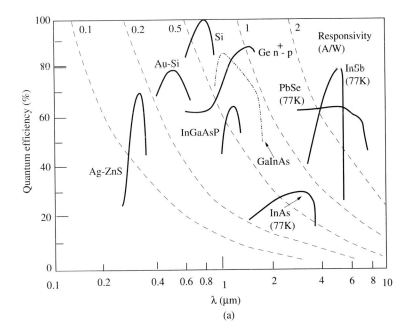

(a)

Figure 8.35   (a) Quantum efficiency and responsivity and (b) detectivity for various photoconductors and photodiodes (from S. M. Sze, *Physics of Semiconductor Devices*, 2nd ed., Wiley, New York, 1981). The data are adapted from H. Melchior, *Physics Today*, 32, November 1977; H. Melchior, "Demodulation and Photodetection Techniques," in F. T. Arecchi and E. O. Schulz-Dubois, eds., *Laser Handbook*, Vol. 1, North-Holland, Amsterdam, 1972, 725–835; and P. W. Kruse et al., *Elements of Infrared Technology*, Wiley, New York, 1962.

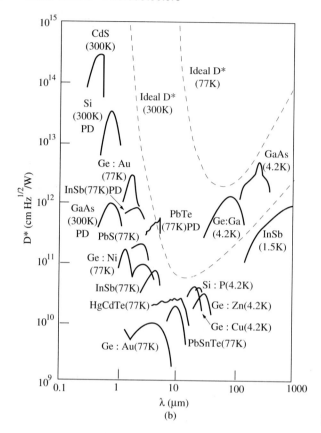

(b)

Figure 8.35 (*continued*)

be shaped into a lens to focus the light, or the fiber can be directly coupled, as shown in Fig. 8.33(b). For receiving low levels of signal power, these coupling techniques become important.

8.6 COMPARISON OF DIFFERENT DETECTORS

The application usually decides the detector of choice. A useful figure of merit is the minimum detectable time-averaged signal power. As we have seen, a photoconductor is an attractive device, considering its simplicity of fabrication and ease of operation. However, its high dark current and the associated Johnson noise make it unsuitable for high-performance communication applications. For such applications, an important performance index is the bit error rate defined by Eq. 8.125. The BER in digital communication is a measure of how often a transmitted 1 is mistakenly identified by the detector as a 0, and vice versa. A BER of 10^{-9} (one mistake per 10^9 transmitted bits), which is considered acceptable, can be achieved if the time-averaged signal power is six times the RMS noise power.

It is apparent that the best combination of bandwidth and sensitivity is obtained by the p-i-n diode and the APD. With proper design, the response speed of an APD should approach that of a p-i-n diode, but more work is needed to make this possible. At the present time the SAGM-APD exhibits the best performance in practical communication systems, in spite of its higher bias requirements, which can be incompatible with integrated circuit design. Figure 8.34 shows a comparison of the measured and calculated minimum detectable optical power in photoconductors, p-i-n diodes and APDs. The solid curves were calculated assuming total capacitance of 0.5 and 1.0 pF and a dark current of 50 nA. Based on this figure, the APD is expected to outperform the other two types of photodetectors. It is appropriate to conclude with a comparison of the performance characteristics of different devices, made with various material systems, as shown in Figs. 8.35(a) and (b).

For low-voltage applications and for very high sensitivity, special detection schemes may be necessary. These can involve individual devices such as phototransistors and modulated barrier photodiodes, or integrated photoreceivers. The special detection devices will be described in the next chapter. Integrated devices, such as a PIN-FET photoreceiver, have emerged as important and useful circuits because of the ability to combine attractive features of both devices, ease of fabrication, and low-voltage operation. Integrated photoreceivers will be discussed in Chapter 12.

PROBLEMS

8.1 What type of photodetector would you recommend using for the following applications:
 (a) low-noise
 (b) high gain
 (c) large bandwidth

8.2 Derive expressions for the gain of a photoconductor with dc excitation at different levels of increasing applied bias if the device has one ohmic contact (for electron flow) and the other blocking (to holes).

8.3 Does the output of a photoconductive detector (with two ohmic contacts) increase when the applied voltage is increased? If so, can we increase the voltage indefinitely?

8.4 Describe the disadvantages of a homojunction photodiode, and how they can be overcome by using a heterojunction. Draw the schematic of a heterojunction photodiode suitable for detecting 1.5 μm light.

8.5 Compare the performance characteristics of a p-i-n photodiode and a p-n photodiode.

8.6 Distinguish between *internal* and *external* quantum efficiency.

8.7 A photodiode ceases to operate when light of wavelength greater than 1.1 μm is incident upon it. What is the material of the absorption region of the diode?

8.8   The parasitic elements of a p-i-n photodiode are represented by the junction capacitance C_j and series resistance R_S. The area of the diode is A. If carriers travel with saturation velocity ϑ_s across the i-region, show that the minimum response time of the diode is given by

$$\tau = [AR_S\epsilon_s\epsilon_0/\vartheta_s]^{1/2}$$

8.9   Calculate the frequency response limitation of a photodiode that has an area of 1 mm^2 and the absorption region is made of material having $N_D = 10^{15}$ cm^{-3} and $\epsilon_r = 11.7$. The bias applied to the device is 10 V and assume that the photogenerated carriers travel at saturation velocity at this bias. The thickness of the light-absorbing region is 1.0 μm (assume $R_L = 50$ ohm).

8.10   A silicon p-i-n photodiode is illuminated by 100 nW of light with $\lambda = 1.0$ μm. The quantum efficiency of the device is 55% and its dark current at the reverse bias that it is operated is negligible. Calculate the photocurrent and the rms shot noise current if $B = 5$ MHz.

8.11   The quantum efficiency of an InGaAsP/InP avalanche photodiode is 80% when detecting 1.3 μm radiation. With an incident optical power of 1.0 μW, the output current of the device is 20 μA. Calculate the avalanche gain or current multiplication factor of the device.

8.12   For a one-sided abrupt $p^+ - n$ junction with doping level $n = N_D$, show that the critical field E_{CR} is related to the breakdown voltage V_{BR} by the approximate relation

$$V_{BR} = \frac{\epsilon_s\epsilon_0}{2qN_D}E_{CR}^2$$

8.13   Explain the operation of a SAGM-APD in which a thin (5–10 Å) delta-doped region is incorporated between the absorption and multiplication regions. Show schematically the heterostructure for a diode to absorb 1.5 μm radiation and the doping and electric field profiles.

8.14   A PIN photodiode has a responsivity of 0.6 A/W at 0.8 μm and the dark current of the device is 1 nA. Calculate the NEP at 300°K if $R_{eq} \cong R_L = 100$ ohms. Also calculate the minimum detectable power if the bandwidth of the device is 10 MHz.

READING LIST

Forrest, S. R., "Optical Detectors: Three Contenders." *Spectrum*, **23**, 76–84, May 1986.

Green, P. E. *Fiber Optic Networks*. Prentice Hall, Englewood Cliffs, NJ, 1993.

Articles in *Semiconductors and Semimetals, vol. 22, Part D: Photodetectors*, ed. R. K. Willardson and A. C. Beer, Academic Press, Orlando, 1985.

SALEH, B. E. A., and TEICH, M. C. *Fundamentals of Photonics*. Wiley, New York, 1991.

STILLMAN, G. E. and WOLFE, C. M. "Avalanche Photodiodes," *In Semiconductors and Semimetals; vol. 12; Infrared Detectors II*, ed. R. K. Willardson and A. C. Beer, Chapter 5, Academic Press, New York, 1977.

SZE, S. M., *Physics of Semiconductor Devices*, 2nd ed. Wiley, New York, 1981.

Special Detection Schemes

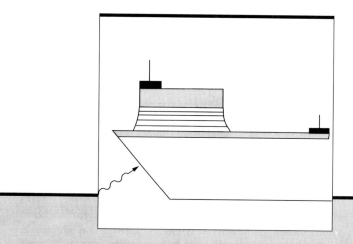

Chapter Contents

9.1 INTRODUCTION

In the previous chapter, we learned about the design, principles, and operating characteristics of the three basic types of photodetectors, namely, the photoconductor, the p-i-n photodiode, and the avalanche photodiode (APD). These devices satisfy most of the detection needs in optical communication and other everyday applications. Nonetheless, because of certain drawbacks in these devices, and for special detection needs, more advanced devices and schemes have been conceived. For example, the APD is, in many ways, an ideal detector with internal gain. However, the large bias values required, together with the requirement of bias stability, and the noise associated with the avalanche process have renewed interest in other devices such as the phototransistor and the modulated barrier photodiode. For simplicity of operation and ease of fabrication metal-semiconductor (Schottky) diodes and metal-semiconductor-metal (MSM) diodes are being used. For communication applications it is necessary to use *coherent* detection, or realize true wavelength selective detection (i.e. detect a preselected wavelength carrying information from a number of simultaneously transmitted ones). Finally, special schemes are used to design detectors for use at long wavelengths, where normal junction diodes and photoconductors have very high dark currents due to the small bandgap of the materials used. In this chapter, we will learn about these devices and detection schemes.

9.2 PHOTOTRANSISTOR

The phototransistor is an optical detection device that provides internal gain and is therefore attractive for many applications. As the name suggests, it is a bipolar transistor that is excited by an optical signal, and the latter is converted into a photocurrent, which is the collector current of the device. More often than not, the base contact is not made and there is no base current. This is called the *floating-base* mode of operation. In a sense, the optical signal acts as the base current in this device. The device is also analogous to a junction photodiode followed by an amplifier, such as in a photoreceiver, which we will study in Chapter 12. In the phototransistor, the reverse-biased base-collector junction, at which most of the light is absorbed, is the junction diode and the current that is generated is amplified by transistor action. This analogy should not be extended too far, however, since in a phototransistor the two processes are integrated.

A detailed analysis of a phototransistor and a calculation of its optical gain involves a solution of the current transport equations, as in a bipolar transistor. This analysis is described in detail in other texts and is beyond the scope of this book. However, the principles of operation, the major assumptions, and the mathematical form of the optical gain are described here.

The band diagram of an n-p-n heterojunction phototransistor is shown in Fig. 9.1(a). The wide-bandgap emitter serves two purposes. First, it acts as a window layer, so that the optical signal is absorbed only in the base and the base-collector depletion regions. Second, as seen in Chapter 4, it enhances the current gain of the transistor. The

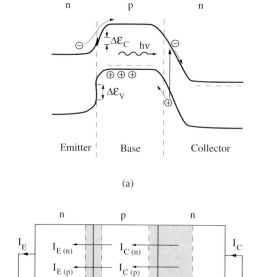

(a)

(b)

Figure 9.1 (a) Energy band diagram of a heterojunction bipolar phototransistor under operating bias conditions and (b) one-dimensional representation showing the major current components.

device with its major current components is shown in Fig. 9.1(b). The base-collector junction operates as a reverse-biased photodiode and therefore the phototransistor, as shown in Fig. 9.1(b), operates in a common-collector configuration.

In an n-p-n heterojunction bipolar transistor the emitter injection efficiency is given, by analogy with Eq. 4.124, as

$$\gamma_E = \left[1 + \left(\frac{D_e L_h N_D}{D_h L_e N_A}\right)^{-1} \left(\frac{m_{eB}^* m_{hB}^*}{m_{eE}^* m_{hE}^*}\right)^{-3/2} \exp\left(-\frac{\Delta\mathcal{E}_g}{k_B T}\right) \tanh\left(\frac{W_b}{L_e}\right)\right]^{-1} \quad (9.1)$$

The base transport factor can be shown to be

$$B_T = \frac{I_C}{I_{E(n)}} = \frac{1}{\cosh\left(\frac{W_b}{L_e}\right)} \quad (9.2)$$

where W_b is the thickness of the neutral base region. The current transfer ratio α_T and the common-emitter current gain β_T are then given by Eqs. 4.117 and 4.118.

The optical gain Γ_G of the phototransistor is the ratio of the number of photo-generated carriers creating the photocurrent to the number of incident photons. Thus,

$$\Gamma_G = \frac{h\nu}{q} \frac{I_{ph}}{P_{inc}} \quad (9.3)$$

where I_{ph} is the optically generated component of the collector current and P_{inc} is the incident optical power. To calculate I_{ph} the following assumptions are usually made. It is assumed that the base-emitter and base-collector depletion layer widths are much smaller than the thicknesses of these regions and that outside the depletion region the base remains neutral. The latter requires that the concentration of minority photogenerated electrons in the base is much less than that of majority holes. Steady-state conditions are assumed to exist throughout the device. The optical excitation produces additional components of collector current due to absorption of photons in the base, the base-collector depletion region, and the neutral collector region. In addition, some photogenerated holes move toward the emitter and modify the emitter potential and therefore the emitter injection. In this respect, the optical excitation acts as a base current and can lower the gain of the transistor. If $\Delta \mathcal{E}_g$ is large enough, then γ_E is also large. If the base is very thin, then $W_b/L_e \ll 1$. Finally, it may also be assumed that the absorption of photons in the collector region is only limited to the depletion layer and not the neutral collector. Under these conditions it can be shown that[†]

$$\Gamma_G = \eta \beta_T \left[1 + e^{-\alpha W_b} \left(1 - e^{-\alpha(x_c - W_b)} \right) \right] \tag{9.4}$$

where η is the quantum efficiency of the phototransistor and x_c denotes the edge of the depletion layer in the collector region. Referring to Fig. 9.1(b), if it is assumed that W_b is small, and the depletion layer width is large enough so that $x_c > \frac{1}{\alpha}$, where α is the absorption coefficient, then

$$\Gamma_G \cong \eta \beta_T \tag{9.5}$$

Therefore, under the approximations mentioned above, the dc optical gain of a phototransistor is the product of the quantum efficiency and the current gain. In other words, the current gain of a phototransistor is given by

$$\beta_T = \frac{h\nu}{q\eta} \frac{I_{ph}}{P_{inc}} \tag{9.6}$$

Unlike the avalanche photodiode, the gain of a phototransistor can be uniform with bias. Floating-base devices are usually fabricated as mesa-diodes with large areas. Γ_G can be $\sim$ 100–500 in these devices. However, due to the storage of the photogenerated minority carriers in the base, the response speed is rather poor. The modulation bandwidth of such phototransistors is usually less than 1 GHz. To obtain a better temporal response, or frequency bandwidth, device dimensions need to be reduced considerably and a base terminal may be added to remove the excess carriers. At the same time, the transit times and circuit parameters need to be optimized as in a bipolar transistor. The gain of the device is, however, greatly reduced. The layout of a high-speed GaAs/AlGaAs phototransistor and its temporal response are shown in Fig. 9.2.

The uniformity of the gain with bias makes the phototransistor an attractive device for optical communication applications, in which a InP/InGaAs heterojunction is used.

[†]See J. C. Campbell, "Phototransistors for Lightwave Communications," in *Semiconductors and Semimetals*, Vol. 22, Part D, eds. R. K. Willardson and A. C. Beer, Academic Press, New York, 1985.

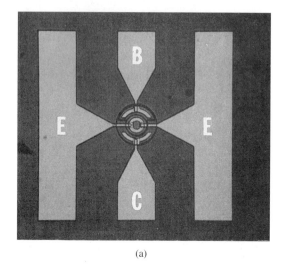

(a)

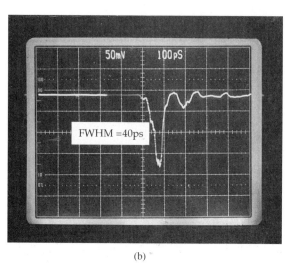

(b)

Figure 9.2 High-speed phototransistor fabricated and measured by the author and co-workers: (a) microphotograph of n-$Al_{0.3}Ga_{0.7}As$ (emitter) - p^+-GaAs (base) - n-$In_{0.1}Ga_{0.9}As/Al_{0.2}Ga_{0.8}As$ MQW (collector) heterojunction phototransistor. The device, grown by MBE, is a top (emitter) to bottom (collector) mesa-etched structure on semi-insulating GaAs substrate in which the emitter optical window diameter is $4\mu m$ and all contact pads are on the top surface; and (b) measured temporal response of the phototransistor to 10 ps pulses at 850 nm from a mode-locked dye laser. $V_{CE} = 5V$, $I_B = 10\mu A$, and the current gain is 10.

A typical long-wavelength device is shown schematically in Fig. 9.3. The dark current of the device can be quite large, and this is an obvious disadvantage in terms of noise performance.

9.3 MODULATED BARRIER PHOTODIODE

The Schottky diode is a majority carrier device that can be used as a high-speed photodetector. The dominant current transport mechanism in the device is thermionic emission over the potential barrier formed at the metal-semiconductor interface. The barrier height cannot be varied by any means, but is intrinsically determined by the metal-semiconductor work function difference. In actual practice, it is pinned by

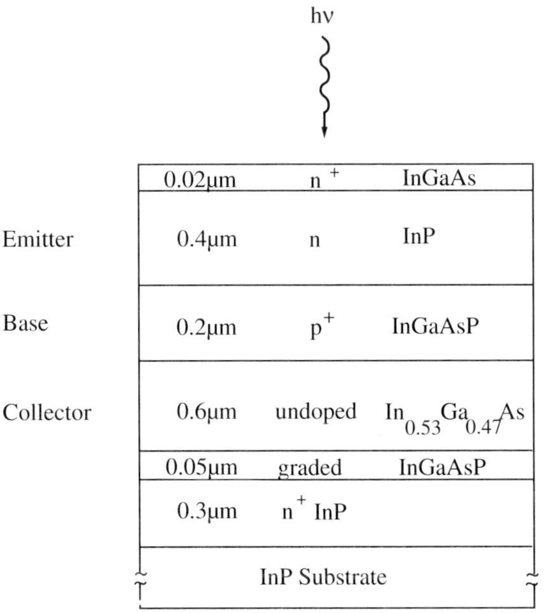

hν
0.02μm n+ InGaAs
Emitter
0.4μm n InP
Base
0.2μm p+ InGaAsP
Collector
0.6μm undoped In0.53Ga0.47As
0.05μm graded InGaAsP
0.3μm n+ InP
InP Substrate

Figure 9.3 Schematic of a InGaAs/InGaAsP/InP heterojunction bipolar phototransistor that can be easily grown by the OMVPE technique.

interface states at this junction. A novel variation of the Schottky diode is shown schematically in Fig. 9.4(a) together with the band diagram and charge distribution. The thickness and doping of the p^+ layer is so chosen that it is fully depleted at zero bias. It can be seen that the barrier height can be modified by varying the

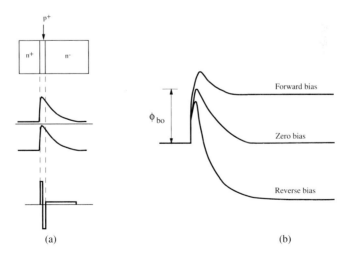

p+
n+
n-
φbo
Forward bias
Zero bias
Reverse bias
(a)
(b)

Figure 9.4 (a) Layer schematic, band diagram and charge distribution of modulated barrier photodiode and (b) variation of band diagram with applied bias.

doping of the p^+ layer. The variation of the band diagram with bias is shown in Fig. 9.4(b). There is significant barrier lowering with reverse bias. The dominant conduction mechanism in the device is thermionic emission over the barrier, as in a Schottky diode. The device was first proposed by Shannon, and because of the nature of its band diagram, is known as a *Camel diode*. It is also known as a *modulated barrier photodiode* because of the barrier modulation possibilities. A variation of the device is a *planar-doped barrier* or *triangular barrier diode*, in which the barrier is placed in the i-region of an $n^+ - i - n^+$ structure with a heavily doped p^+ layer. The schematic and band diagram of a triangular doped barrier are shown in Fig. 9.5. The p^+ region is fully depleted and the negative space charge of the ionized acceptors is balanced by the positive space charge of the i-regions. From a solution of the Poisson equation with the appropriate charge densities, it can be shown (Problem 9.2) that the bias-dependent barrier heights to electron flow from left to right and vice versa are given, respectively, by

$$\phi_{b1} = \phi_{b0} + \frac{d_1}{d_1 + d_2} V$$

and

$$\phi_{b2} = \phi_{b0} - \frac{d_2}{d_1 + d_2} V \qquad (9.7)$$

where

$$\phi_{b0} = \frac{q N_A^S d_1 d_2}{\epsilon_s \epsilon_0 (d_1 + d_2)} \qquad \text{(volts)} \qquad (9.8)$$

is the barrier height at zero bias, V is the applied bias, and N_A^S is the sheet acceptor density in the p^+ layer in cm^{-2}. The thicknesses d_1 and d_2 are indicated in Fig. 9.5.

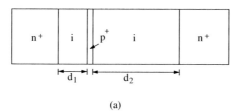

(a)

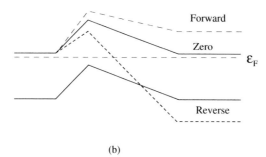

(b)

Figure 9.5 (a) Layer schematic and (b) energy band diagram of triangular barrier photodiode.

If the barrier potential is in excess of several $k_B T/q$, then the carrier transport over the resultant triangular potential barrier is by thermionic emission and the current density in the dark can be expressed as

$$J_{ed} = A^{**} T^2 \exp\left(-\frac{q\phi_{b0}}{k_B T}\right)\left[\exp\left(\frac{q\alpha_2 V}{k_B T}\right) - \exp\left(-\frac{q\alpha_1 V}{k_B T}\right)\right] \qquad (9.9)$$

where $\alpha_1 = \frac{d_1}{d_1 + d_2}, \alpha_2 = \frac{d_2}{d_1 + d_2}$ and A^{**} is the effective Richardson constant. It can be seen that there are some unique features of the planar-doped barrier. The zero-bias barrier height, ϕ_{b0}, can be controlled by varying N_A^S, d_1 and d_2, or by using heterostructures, as we shall see. The capacitance of the diode is given by

$$C \cong \frac{\epsilon_s \epsilon_0 A}{d_1 + d_2} \qquad (9.10)$$

which is independent of bias. Here A is the area of the diode. Finally, it is important to note that in this device carrier injection occurs in both directions over the barrier. The parameters α_1 and α_2 are like diode ideality factors. It is easily seen that if $d_2 \gg d_1$, Eq. 9.9 reduces to that for a Schottky diode.

Modulated barrier photodiodes were originally designed and developed for recti-fying applications. However, it was soon realized that the device could also be used as a very efficient photodiode with high optical gain. As suggested by Eqs. 9.7 and 9.8, the barrier height of a modulated barrier photodiode at any value of applied bias is directly related to the ionized acceptor density in the thin p^+ region. If the diode is illuminated with light of appropriate wavelength, then it can be absorbed in the barrier region, creating electron-hole pairs. Reference is made to Fig. 9.4. The electrons are swept away by the field, while the holes are drawn to the potential minimum at the p^+ layer. The positively charged holes accumulate in the barrier region and neutralize some of the negative space charge, thus reducing the barrier height. As a result, there is a large increase of current due to thermionic emission over the barrier, remember-ing that the current is related exponentially to the barrier height. The increase in the current implies that an optical detector with a large internal gain can be realized. In the device shown in Fig. 9.6, an AlGaAs layer is used on the side on which the diode is illuminated. This layer serves two purposes. First, it acts as a window layer for the photoexcitation, ensuring that most of the incident radiation is absorbed in the barrier region. Second, it forms a more efficient well region for the holes to be trapped, thereby enhancing the barrier-lowering effect. In the following we will analyze the photoresponse characteristics of this device.

Consider the heterojunction modulated barrier photodiode shown in Fig. 9.6. For convenience of discussion, we label the different regions as source, barrier, and drain. Photoexcitation incident on the device from the left, of wavelength corresponding to the bandgap of GaAs, will be mainly absorbed in the drain GaAs layer, since the thickness of the barrier GaAs layer is very small ($\leq 100\text{Å}$). The absorbed photons will generate electron-hole pairs. The electrons will drift down the potential hill and will be collected. The holes will also drift toward the potential minimum (for holes), beyond which it has to surmount a potential barrier of height ($\phi_{b0} - \Delta\phi_b$), where

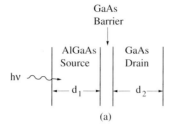

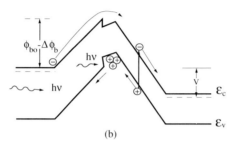

Figure 9.6 (a) Layer schematic and (b) energy band diagram under bias of a heterojunction n-p$^+$-n GaAs/AlGaAs modulated barrier photodiode.

$\Delta\phi_b$ is the barrier lowering due to the accumulation of photogenerated holes at the potential minimum. The trapped holes can be removed by a variety of mechanisms, each of which has a small probability. They can be removed by recombination with electrons, but the probability of this mechanism is small, since electrons spend a very small amount of time in the thin barrier region. Holes can escape by surmounting the potential barrier by thermionic emission, or by diffusion across the source region. In the steady state, the rate of holes moving into the barrier region must be equal to the rate of holes moving out of the potential minimum, or

$$\frac{J_{hd}}{q} + \frac{P_{inc}\eta}{Ah\nu} = \frac{\Delta p}{\tau} \tag{9.11}$$

where J_{hd} is the hole dark current, A is the area of the diode, Δp is the two-dimensional density of holes trapped in the potential minimum, and τ is the effective lifetime of holes, determined by all the escape processes described above. As mentioned above, thermionic emission and diffusion are the most likely escape processes. It is assumed that all the incident power is absorbed in the device. Δp is related to the barrier lowering, and can therefore be expressed by the equation

$$\Delta p = \frac{\tau J_{hd}}{q} \exp\left(\frac{q\Delta\phi_b}{n_f k_B T}\right) \tag{9.12}$$

assuming that the accumulated holes form a Fermi gas. The prefactor on the rhs of Eq. 9.12 is the two-dimensional equilibrium hole density in the potential minimum in the dark. According to this equation, $\Delta\phi_b$ will increase as Δp increases. However, as the barrier is lowered, the hole-trapping process becomes less efficient and there is a saturation of the barrier-lowering effect. This saturation effect is accounted for by the ideality factor n_f. At low values of incident power, $n_f \cong 1$. However, at higher powers, when saturation sets in, n_f has a value less than unity and approaches zero. Also, from Eq. 9.12 it is evident that for $\Delta\phi_b \ll k_BT/q$, Δp is linearly proportional to $\Delta\phi_b$. Substitution of Eq. 9.12 into Eq. 9.11 leads to

$$\Delta\phi_b = \frac{n_f k_B T}{q} \ln\left[1 + \frac{P_{inc}\eta q}{A J_{hd}h\nu}\right] \qquad (9.13)$$

which expresses the barrier lowering in terms of the incident power and other device and material parameters. Some relevant observations can be made from Eq. 9.13. For low values of P_{inc} such that $\frac{P_{inc}\eta q}{A J_{hd}h\nu} \ll 1$, $\Delta\phi_b$ is linearly proportional to P_{inc}. On the other hand, for large values of P_{inc}, such that $\frac{P_{inc}\eta q}{A J_{hd}h\nu} \gg 1$, $\Delta\phi_b \propto \ln P_{inc}$. This illustrates the saturation in the barrier lowering, which occurs due to a reduction in the trapping of holes in the barrier region.

Another important parameter of the device is its responsivity, which is defined as

$$\mathcal{R} = \frac{A(J_e - J_{ed})}{P_{inc}} \qquad (9.14)$$

where J_e and J_{ed} are the electron current densities under photoexcitation and in the dark, respectively, and are given by

$$J_e = A^{**}T^2 \exp\left[-q\left(\frac{\phi_{b0} - \Delta\phi_b}{k_B T}\right)\right]\left[\exp\left(\frac{q\alpha_1 V}{k_B T}\right)\right] \qquad (9.15)$$

and

$$J_{ed} = A^{**}T^2\left[\exp\left(-\frac{q\phi_{b0}}{k_B T}\right)\right]\left[\exp\left(\frac{q\alpha_1 V}{k_B T}\right)\right] \qquad (9.16)$$

where it is assumed that the currents are determined purely by thermionic emission and the diode in Fig. 9.5 or 9.6 is reverse biased (i.e., the left-hand side is biased negatively). Substitution of Eqs. 9.13, 9.15, and 9.16 into Eq. 9.14 leads to

$$\mathcal{R} = \frac{A J_{ed}}{P_{inc}}\left[\left(1 + \frac{P_{inc}\eta q}{A J_{hd}h\nu}\right)^{n_f} - 1\right] \qquad (9.17)$$

where, it may be remembered that $n_f \cong 1$ for low incident power levels and $n_f < 1$ for high incident power levels. It is seen from Eq. 9.17 that at very low incident power levels $\mathcal{R}$ is nearly constant. Also, the maximum value of $\mathcal{R}$ at these low power levels is

$$\mathcal{R}_{max} \cong \frac{\eta q}{h\nu}\frac{J_{ed}}{J_{hd}} \qquad (9.18)$$

while, at high power levels, the responsivity decreases monotonically with incident power according to $(P_{inc})^{n_f - 1}$, remembering that $n_f < 1$ at these power levels. The modulated barrier photodiode is thus a unique photodetection device, in which the responsivity, or optical gain, increases as the incident optical power decreases. This is contrary to the characteristics of a phototransistor, whose responsivity decreases with increasing incident optical power. Optical gains upto 100 have been measured in these devices at low power levels. It should be noted that a modulated barrier photodiode in which the majority carriers are holes, and the barrier is made of n^+ material and traps electrons, will have similar characteristics. The current-voltage characteristics and the optical gain of such a device are shown in Fig. 9.7. Most noticeable is the decreasing optical gain with increasing incident optical power.

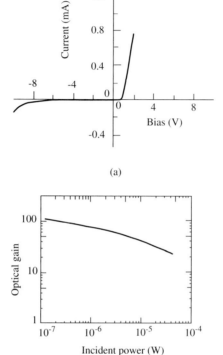

(a)

(b)

Figure 9.7 (a) Current-voltage characteristics and (b) optical gain as function of incident power of 50 μm diameter GaAs/Al$_{0.3}$Ga$_{0.7}$As p^--n^+-p^- modulated barrier photodiode (W. Q. Li, *Some Optoelectronic Devices for Systems Applications*, Ph.D. Thesis, University of Michigan, 1989).

Of obvious interest are the temporal response characteristics and bandwidth of the photodiode. These are determined by the intrinsic carrier transport mechanism and extrinsic circuit parameters. The temporal response of the device to a very short ($\sim$ few picoseconds) pulse excitation is characterized by a very fast rise time (of the order of the excitation pulse) and a slower fall time ($\sim$ few hundred picoseconds). The latter is caused by the trapping of carriers in the barrier region. If a lower gain is acceptable, then the device speed can be enhanced.

In conclusion, the modulated barrier photodiode is a novel device because of several important features. It has the potential of having a larger bandwidth than the conventional phototransistor. Unlike the avalanche photodiode, this device does not have excess noise that is associated with the avalanche process. With its high gain at low incident power, this detector can play an important role in low noise, low signal level detection systems.

EXAMPLE 9.1
Objective.   To express the sensitivity in terms of the barrier lowering.

From Eqs. 9.14, 9.15, and 9.16,

$$\mathcal{R} = \frac{A J_{ed}}{P_{inc}} \left[\exp\left(\frac{q \Delta \phi_b}{k_B T} \right) - 1 \right]$$

Also, the cross-over power, between low and high incident powers, is given by

$$P_{inc}^C = \frac{A J_{hd} h \nu}{\eta q}$$

As an example for $A = 10^{-4}$ cm^2, $J_{hd} = 2 \times 10^{-5}$ A/cm^2, $\eta = 0.5$, and $h\nu = 1.5$ eV, $P_{inc}^C = 6$ nW.

9.4 METAL-SEMICONDUCTOR (SCHOTTKY BARRIER) PHOTODIODE

The Schottky barrier diode that was described in Chapter 4 can be used as a very efficient photodiode. Since it is a majority carrier device, minority carrier storage and removal problems do not exist and therefore higher bandwidths can be expected. The diode is usually operated under a reverse bias. In its simplest configuration, the basic device can be mesa-shaped with an n^+ substrate layer, an n^- ($\sim 10^{15}$ cm^{-3}) absorbing layer, and a thin layer of semitransparent metal (100–300 Å) deposited on it. The area of the diode is made to be 10^{-3} to 10^{-5} cm^2. A thin dielectric antireflection coating is deposited on the metal, and larger top contact pads are deposited through openings in the same dielectric, as schematically shown in Fig. 9.8. This figure shows a high-speed photodiode made on semi-insulating substrate. The equivalent circuit of the Schottky photodiode is similar to that of a p-i-n diode shown in Fig. 8.12, having the same circuit elements. The temporal response speed and frequency bandwidth of the device are determined by the transit time of carriers through the absorption region and the external circuit parameters. In high-speed diodes the absorption region

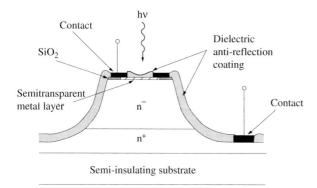

Figure 9.8   Schematic cross section of a Schottky photodiode fabricated on semi-insulating substrate.

is between 0.2 and 0.5 μm thick. This ensures full depletion of this region even at low values of reverse bias and both electrons and holes travel at their respective saturation velocities. Under these conditions the transit time is $\sim$ 1 ps. Similarly, with optimized design of the diode and contact geometry, the time constant of the circuit can be made almost of the same order. Schottky diodes can attain bandwidths of 100 GHz.

The photoresponse characteristics of the Schottky photodiode are rather interesting and can extend over a large spectral range. With respect to the band diagrams shown in Fig. 4.21, at smaller values of photon energy, such that $q\phi_b < h\nu < \mathcal{E}_g$, electrons photoexcited in the metal can surmount the barrier by thermionic emission, transit across the semiconductor, and be collected at the ohmic contact. This process extends the spectral response of the diode to photon energies lower than the bandgap. However, it is not a very desirable mode of operation from the point of view of device speed, since the thermionic emission process can be slow. For $h\nu > \mathcal{E}_g$, photons are absorbed in the semiconductor provided that the metal layer is semi transparent. The electron-hole pairs that are created move in opposite directions with their respective saturation velocities and are collected. This is the most efficient mode of operation of the Schottky diode and is similar to that of a high-speed p-i-n diode. However, compared to the latter, the Schottky diode is a simpler device to operate. Note that a device very similar to a metal-semiconductor diode can be realized by replacing the metal with a low-bandgap semiconductor. For example, a InAs/GaAs heterojunction, whose band diagram is shown in Fig. 9.9, can operate under the same principles. The advantage is that the entire diode structure can be realized by one-step epitaxy, and making contact to the low-bandgap n-type semiconductor is rather simple.

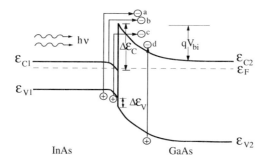

Figure 9.9 Equilibrium energy band diagram of (n)InAs/(n)GaAs heterojunction. a–d indicate the various transitions related to absorption of light of different energies.

9.5 METAL-SEMICONDUCTOR-METAL (MSM) PHOTODIODE

Optoelectronic integrated circuits involve the integration of devices that are compatible from the point of view of fabrication and processing. In photoreceivers, which will be described in Chapter 12, a photodetector is monolithically integrated with a preamplifier. The latter is usually a field-effect transistor (FET) or a heterojunction bipolar transistor (HBT). Depending on the application, one of the several kinds of detectors can be used and different kinds of FETs may also be used. The metal-semiconductor

FET (MESFET) is very simple in configuration, with which the required gain and bandwidth for most applications can easily be obtained. It consists of an n-type doped active region, or channel, on an undoped or high-resistivity buffer layer. A planar integration scheme of a photoreceiver will necessitate the design of a photodetector with either the active layer or the buffer layer with the minimum number of processing steps. A photoconductive detector on the buffer layer is a possibility, but as we have seen, such detectors are plagued with a higher-than-acceptable dark conductivity. Also, if sizable gain is required, then the detector is prone to have a slow temporal response. A metal-semiconductor-metal (MSM) photodiode is a device that is very suitable for such applications. We will briefly study the properties of this device.

The MSM photodiode is made by forming two Schottky contacts on an undoped semiconductor layer. It can be designed such that the region in between is almost completely depleted. In a planar configuration the two Schottky contacts are made of Au, Au/Ti, or tungsten silicide (WSi_x). These can be single contacts, or interdigitated, as shown in Fig. 9.10, with a contact spacing of 1-5 μm. The band diagram of the MSM photodiode with an applied bias V is shown in Fig. 9.11. The flatband voltage V_{FB} can be expressed as

$$V_{FB} = \frac{qN_D L^2}{2\epsilon_s \epsilon_0} \tag{9.19}$$

where two-dimensional effects are ignored. Here L is the electrode spacing and N_D is the donor density in the n-type layer. The dark current of the device is principally determined by thermionic emission over the barrier. With respect to Fig. 9.11, at low biases, electron injection at the reverse-biased contact is the dominant conduction mechanism. As the bias increases, hole injection at the forward-biased contact also becomes a dominant factor. Hole injection tends to dominate after the reach-through condition (the two depletion edges coincide) is reached. The total current density

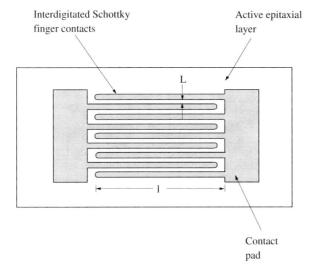

Interdigitated Schottky finger contacts

Active epitaxial layer

L

l

Contact pad

Figure 9.10 Schematic illustration of an interdigitated metal-semiconductor-metal photodiode formed on an epitaxial layer.

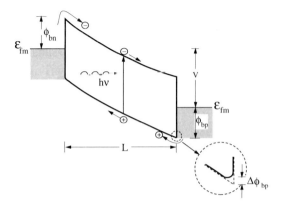

Figure 9.11   Energy band diagram of an MSM photodiode under bias. The inset shows the image force lowering effect. The same effect occurs at the other contact also.

under these conditions is therefore given by

$$J = A_n^{**} T^2 e^{-q(\phi_{bn} - \Delta\phi_{bn})/k_B T} + A_p^{**} T^2 e^{-q(\phi_{bp} - \Delta\phi_{bp})/k_B T} \tag{9.20}$$

where the A^*'s are the respective Richardson constants and the $\Delta\phi$'s are the respective barrier height lowerings due to the image force effect, shown in the inset of Fig. 9.11. The current-voltage characteristics of a GaAs MSM device are shown in Fig. 9.12(a).

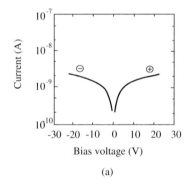

(a)

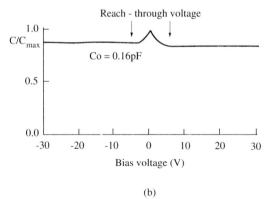

(b)

Figure 9.12   Measured characteristics of GaAs MSM photodiode with tungsten silicide Schottky contacts: (a) dark current-voltage characteristics; (b) dark capacitance-voltage characteristics; and (c) spectral sensitivity as a function of bias (from M. Ito and O. Wada, *IEEE J. Quantum Electronics*, **QE-22**, 1073, ©1986 IEEE).

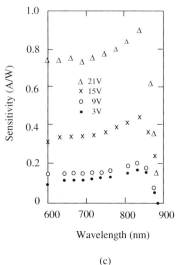

(c)   **Figure 9.12** (*continued*)

The dark current is of the order of 1 nA or lower, which is comparable to a p-i-n photodiode.

The ideal device capacitance for a source-drain configuration with contact (or finger) length l and finger spacing L is given by

$$C = C_0 l \tag{9.21}$$

where C_0 is the gap capacitance of the contacts per unit length. If, as in Fig. 9.10, the device is of interdigitated configuration, having N fingers with width W and spacing L, then

$$C = (N-1)C_0 l \tag{9.22}$$

Here C_0 is given by

$$C_0 = \epsilon_0 (1 + \epsilon_s) \frac{K}{K'} \tag{9.23}$$

where K and K' are elliptic integrals[†] and are functions of L and W. The capacitance-voltage characteristics of a GaAs device with 3 μm contact spacing are shown in Fig. 9.12(b). For both bias polarities the capacitance decreases with bias upto the reach-through voltage, after which it remains nearly unchanged. It is important to note that the capacitance of a MSM diode is smaller than that of a PIN photodiode of equal light-sensitive area by almost a factor of 4. This helps in enhancing the high-speed response of the device.

The frequency response and bandwidth of the MSM photodiode are determined primarily by the transit time of the photogenerated carriers and the charge-up time of the diode. The latter is determined by the extrinsic parasitics. The equivalent circuit

---

[†] See M. Ito and O. Wada, *IEEE J. of Quantum Electronics*, **QE-22**, 1073, 1986.

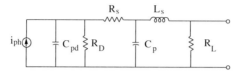

Figure 9.13 Equivalent circuit of MSM photodiode under illumination.

of the MSM photodiode with the parasitic circuit elements are as shown in Fig. 9.13. With careful device design both the parasitic capacitance C_p and the diode capacitance C_{pd} can be made less than 0.1 pF. Typical value of R_S, R_D, R_L, and L_S are 50 Ω, 100 MΩ, 50 Ω, and 0.1 nH, respectively. At the same time, the finger spacing can be made to be less than 1 μm. With these device and circuit parameters, bandwidths of 20–50 GHz can be obtained.

Finally, it is important to study the photoresponse characteristics of the device. The measured spectral response of GaAs devices at different bias values is shown in Fig. 9.12(c). The increase of the responsivity with bias indicates that the MSM diode may have internal gain, even at low bias values. The latter precludes the avalanche multiplication effect. The gain at low biases could result from several mechanisms. A photoconductive gain mechanism can be operative due to traps or surface defects having a long lifetime. Accumulated electrons at the conduction band minimum point can lower the barrier for hole transit. These factors are slightly speculative, and more research is necessary to fully understand the device operation.

In conclusion, the MSM photodiode is a versatile device that has all the desirable attributes of a practical photodetector and is amenable to simple and planar integration schemes. It exhibits gain, has a low dark current, and has a large bandwidth. For devices to operate in the 1.3–1.6 μm range, $In_{0.53}Ga_{0.47}As$ can be used, in which a thin $In_{0.52}Al_{0.48}As$ surface layer can be used to enhance the Schottky barrier height.

9.6 DETECTORS FOR LONG-WAVELENGTH OPERATION

A variety of very important applications require detection capability in the far-infrared (5–20 μm). Two approaches have proved to be successful in long wavelength detection. The first is to use a junction diode and band-to-band transitions in small bandgap materials. Examples are the antimony bearing III–V compounds, which have the lowest bandgaps and the II–VI alloys such as HgCdTe. Use of the antimonides almost automatically necessitates the use of strained semiconductors due to lack of suitable lattice-matched semiconductors. It can be seen from Fig. 1.2 that $InAs_{0.5}Sb_{0.5}$ has the lowest bandgap among the III–V compounds synthesized so far. However, the same material properties that yield small bandgaps also produce a "soft" material, which is difficult to grow and process. Because of the small bandgaps, the dark current in the devices are very large, which necessitates their operation at cryogenic temperatures. In spite of these difficulties, impressive progress has been made in the design and fabrication of long-wavelength detectors and detectivities $D^* \sim 10^{10}$ (cm.Hz$^{1/2}$. W^{-1}) have been measured in these devices.

An interesting scheme for designing detectors for long wavelength applications is shown in Fig. 9.14(a). Use is made of transitions between the bound states in a

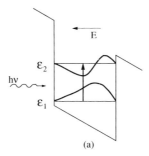

(a)

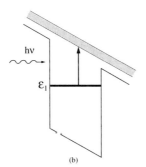

(b)

Figure 9.14 Absorption of long-wavelength light in a quantum well due to (a) intersubband transitions in a wide well and (b) transition from a quasi-bound state to the continuum in a narrow well.

quantum well. The absorption of photons having energy equal to the intersubband separation leads to transition of carriers from the ground state to the first excited state. For III–V quantum wells of width $L_z = 100\text{Å}$, the intersubband energy separation is in the range of 100–200 meV. For example, the absorbance and intersubband transition energies measured in pseudomorphic $In_xGa_{1-x}As/Al_{0.4}Ga_{0.6}As$ multiquantum wells are shown in Figs. 9.15(a) and (b). We will not discuss the detailed theoretical background of intersubband transitions, but two important points need to be made. First, the oscillator strength of a transition from m^{th} to $(m + 1)^{th}$ subband increases with the value of m. Second, the selection rules only allow absorption of electromagnetic radiation when the incident polarization is parallel to the growth (confinement) direction. This can cause difficulties in detecting a 2-D image, since radiation has to be absorbed in a waveguide geometry. However, illumination geometries such as those shown in Fig. 9.16(a) can circumvent the problem. Also, it has been shown that in a quasi-type II superlattice such as InAs/AlGaSb, shown in Fig. 9.16(b), the selection rules allow absorption of radiation with polarization parallel to the growth direction.

Detection devices utilizing intersubband absorption operate in the photoconductive mode. In other words, the photogenerated carriers tunnel out of the wells and create a photocurrent. However, in the process of making this tunneling efficient, the tunneling of residual carriers, and hence the dark current, is also enhanced. Careful designs of the barrier can reduce such tunneling. Significant improvement in the performance of

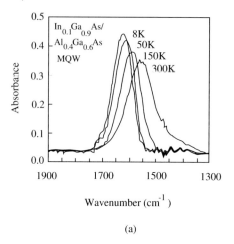

(a)

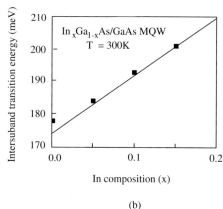

(b)

Figure 9.15 (a) Absorbance due to intersubband transitions measured by Fourier transform infrared spectroscopy and (b) intersubband transition energy as a function of In composition in pseudomorphic $In_xGa_{1-x}As/GaAs$ MQW. The data were obtained by the author and co-workers (X. Zhou et al., *Applied Physics Letters*, **54**, 855, 1989).

these devices is obtained by utilizing photon absorption and carrier transitions between a bound state and the continuum *outside* the well (usually 20 meV above the well). The scheme is illustrated in Fig. 9.14(b). The absorption peak is extended to longer wavelengths and by using small barrier heights, the absorption can be extended to even longer wavelengths. In addition, since the photoexcited carriers are raised to the continuum, the bandwidth of the absorption, or the detector responsivity curve is extended. The collection of photocarriers in a detector based on this principle does not require the tunneling process and, at the same time, the device dark current can be reduced by using thick barriers.

9.7 WAVELENGTH SELECTIVE DETECTION

One of the prime motivations of using optical fibers for communication, replacing conducting cables, is to be able to utilize their enormous bandwidth, which translates to an enhancement in the number of communication channels. For example,

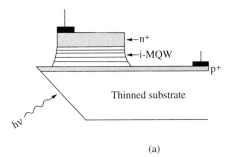

(a)

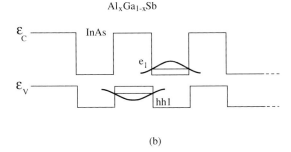

(b)

Figure 9.16 (a) Illumination geometry for intersubband light absorption in a p-i(MQW)-n photodiode and (b) equilibrium band diagram of quasi-type II InAs/Al$_x$Ga$_{1-x}$Sb superlattice for $x \geq 0.3$.

the increase in bandwidth is from several hundred MHz in conductors to several hundred GHz in optical fibers. It is necessary to be able to selectively detect the information coming in at one frequency (or wavelength) when there is information coming in simultaneously at other wavelengths. None of the detectors discussed so far have such wavelength selectivity in their spectral response characteristics. Techniques to overcome this difficulty include use of tuned filters and the use of coherent detection schemes. While these approaches are important and feasible, systems design will ultimately benefit from the use of detectors which display wavelength selectivity.

Use can be made of the excitonic absorption spectrum of a quantum well, which was described in Sec. 3.4, to develop tunable detectors. The spectrum is characterized by sharp resonances resulting from transitions between heavy-hole to electron (e-hh) and light-hole to electron (e-lh) subbands. For a 100 Å quantum well, the separation between these resonances is usually $\sim$ 10 meV. If such a multiquantum well (MQW) is incorporated in the i-region of a p-i-n diode, and a reverse bias is applied, then the exciton peaks move to lower energies, as shown in Fig. 3.16. This phenomenon was described as the *quantum confined Stark effect* (QCSE) in Sec. 3.4, and arises mainly due to a drastic modification of the electron and hole wavefunctions in the well region. The photocurrent-voltage characteristics of the diode, with illumination of energy slightly lower than that of the HH peak in the absorption spectrum, are schematically shown in Fig. 9.17. At very low values of bias the photocurrent is mainly due to the recombination of photogenerated electrons and holes in the wells. As the bias is increased, the current is due to tunneling of photogenerated carriers out

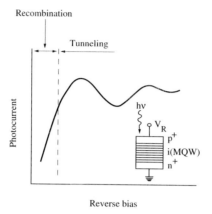

Figure 9.17 Photocurrent-voltage characteristics of p-i(MQW)-n photodiode when spectral energy of illumination is smaller than heavy-hole exciton energy.

of the wells and their transit across the MQW region and eventual collection. Due to the *red shift* of the absorption spectrum with bias, which is the QCSE, the photocurrent-voltage characteristics exhibits peaks similar to that of the absorption spectrum. The operation of the p-i(MQW)-n diode described so far is similar to that of a p-i-n photodiode, except that the photocurrent exhibits a negative differential resistance (NDR). The QCSE and NDR can be exploited to develop efficient modulation and switching devices, which we will learn about in Chapter 11.

Due to the QCSE, the photocurrent-voltage characteristics exhibit very strong wavelength dependence, and this can be exploited to develop tunable detectors. Shown in Fig. 9.18 are the room-temperature reverse-biased photocurrents of a p-i(MQW)-n

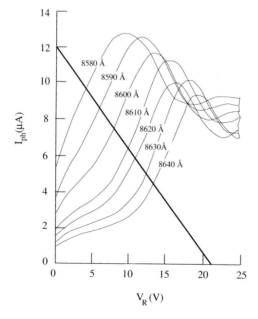

Figure 9.18 Photocurrent-voltage characteristics of a p-i(MQW)-n photodiode for various input wavelengths measured by the author and co-workers. The GaAs(100 Å)/$Al_{0.3}Ga_{0.7}As$ (80 Å) MQW has 40 periods (S. Goswami et al., *IEEE Journal of Quantum Electronics*, **27**(4), 875, ©1991 IEEE).

diode for a series of wavelengths of equal intensity generated by a tunable laser. As can be clearly seen, even at room temperature, at a given bias value the photocurrent is quite distinct for different wavelengths. With an appropriate external load resistance (load line shown in Fig. 9.18) we see that the voltage across the p-i-n diode has a strong wavelength dependence. A change in input wavelength of 100 Å produces a change in the output reference voltage of ~ 10 V. If the input information is coming serially, the output reference voltage in the circuit provides a very selective determination of the state (on or off) of the chosen wavelength. This scheme will enable the detection of up to 50 wavelength channels serially, allowing a reference voltage across the resistor to vary by ~ 0.2 V.

While the serial detection of a given wavelength may be useful for many applications, in communication one would like the detectors to selectively detect information in a channel at wavelength λ_i regardless of the state of the channel at wavelength $\lambda_j (j \neq i)$. The photocurrents shown in Fig. 9.18 are not capable of doing this as can be easily seen by summing various λ_i curves. However, if two p-i-n structures are biased at voltages V_i and $V_i + \Delta V_i$, a simple subtraction of the currents provides the value of $\Delta I_{ph}/\Delta V$. This scheme of operation is illustrated in Fig. 9.19(a). The "parallel" selectivity of this scheme is shown in Fig. 9.19(b) where the values of $\wedge I_{ph}/\Delta V$ at 200°K are plotted as a function of voltage, when information is coming in through four channels at wavelengths 8350, 8370, 8410, and 8430 Å, The light curves for a selected $\lambda_i (= 8350\text{Å})$, represent the results of $\sum_{j \neq i} \Delta I_{ph}(\lambda_j)/\Delta V$ for all possible $(2^3 - 1)$ combination states (i.e., on or off) of the other $(\lambda_j \neq \lambda_i)$ channels.

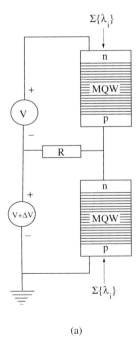

Figure 9.19 (a) Schematic for photocurrent subtraction using two diodes for wavelength selective detection. The potential developed across the resistance R is proportional to the differential photocurrent; (b) values of $\Delta I_{ph}/\Delta V$ for different states of four incoming channels λ_j. The bold curve represents the ON state for $\lambda_i = 8350$ Å only. These data were obtained by the author and co-workers (S. Goswami et al., *IEEE Journal of Quantum Electronics*, **27**(4), 875, ©1991 IEEE).

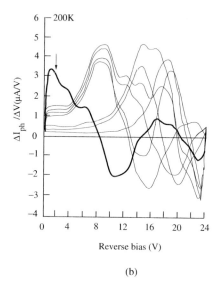

(b)

Figure 9.19 (*continued*)

The bold line represents the value of $\Delta I_{ph}(\lambda_i)/\Delta V$ for the case where only channel λ_i is on.

It is apparent that if the difference current obtained from two p-i(MQW)-n diodes of Fig. 9.19(a) biased at values 1 V apart is compared with a constant current source of appropriate value the state of the channel λ_i can be detected regardless of the state of the other channels, if the devices are biased in the region indicated by the arrow in Fig. 9.19(b). For example, if the diodes are biased at 2.0 $\pm$ 0.5 V, the difference current is more than 2 μA if the λ_i bit is on and less than 2 μA if it is off regardless of the states of the other channels. Operation at low temperature improves the selectivity and therefore can increase the number of parallel channels that can be simultaneously decoded. It may also be noted that an important source of nonselectivity in this scheme comes from the negative resistance region of the photocurrent-voltage characteristics. If this region were not present, the selectivity can be increased to many more channels.

9.8 COHERENT DETECTION

In all the different types of detectors described so far, the optical signal is converted into an electrical signal in the device. This form of detection is known as *direct* detection. Although such direct detection is theoretically very sensitive, in practical receiver systems, the actual sensitivity is nowhere close to the theoretically predicted values. *Heterodyning* is a coherent receiving technique that is commonly used in radio communication. More recently, this technique is rapidly gaining importance in long-distance optical communication systems. In coherent optical detection, a weak received signal is mixed with a strong local oscillator (LO) wave at a close-enough frequency, resulting in an effective signal gain through coherent phase interference.

The gain is proportional to the local oscillator amplitude. A coherent detection system is schematically illustrated in Fig. 9.20. When the carrier wave, modulated with the information signal, is mixed with the local oscillator wave, a photodiode detects the composite wave and produces an electrical signal that is identical to the modulating information (data) signal. This electrical output is at a frequency of 10^9–10^{10} Hz, compared to 10^{16} Hz for the carrier signal. The electrical output at the lower frequency can be picked up by sharp electronic filters operating at the beat frequency and amplified by high-performance electronic circuitry. This gives superior wavelength selectivity, which enables a higher channel density in the communication system.

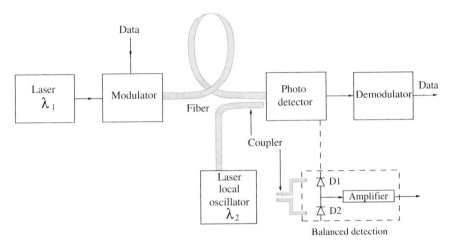

Figure 9.20 Scheme for coherent detection of optical signals. The dashed box represents the scheme with a balanced detection system.

There are, however, some inherent problems in this scheme of detection. The optical gain obtained in a coherent system is very valuable, since it increases the sensitivity. Unfortunately, as the local oscillator signal power is increased, the measured shot noise in the detector and the noise from the laser may outweigh the signal gain. There is, therefore, an optimum local oscillator power that gives the highest signal-to-noise ratio and maximum receiver sensitivity. A balanced detector configuration, as shown by the dashed box in Fig. 9.20, and in Fig. 9.21, in which two identical detectors are connected together, can enhance the beat signal, while suppressing the local oscillator noise. The signal-to-noise ratio of a balanced detection system can be expressed as

$$\frac{S}{N} = \frac{\frac{1}{\hbar\omega}\mathcal{T}(1-\mathcal{T})(\eta_1+\eta_2)^2 E_S^2}{B[\mathcal{T}\eta_1 + (1-\mathcal{T})\eta_2 + \frac{q\gamma_N}{\hbar\omega}E_L^2(\eta_1\mathcal{T}-\eta_2(1-\mathcal{T}))^2]} \tag{9.24}$$

This relation is derived in Appendix 9. In Eq. 9.24 $\mathcal{T}$ and $(1-\mathcal{T})$ are the power transmission and reflection coefficients of the coupler, respectively, η_1 and η_2 are the quantum efficiencies of the two detectors, E_S and E_L are the signal and local

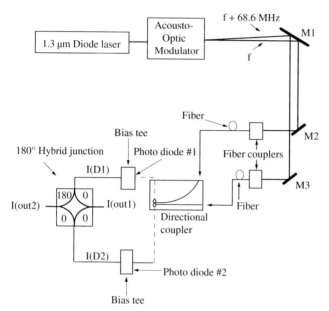

Figure 9.21 Schematic of a balanced detection test set-up in the laboratory.

oscillator field amplitudes, respectively, B is the bandwidth of the receiver, and γ_N is a parameter related to the intensity fluctuations of the local oscillator. γ_N has units of inverse current and typical values range from 10^2 to $10^9 A^{-1}$. As a special case, if we assume a coherent or heterodyne system consisting of two detectors with equal quantum efficiency η and if $T = 0.5$,

$$\frac{S}{N} = \frac{\mathcal{R} E_S^2}{q B}$$

(9.25)

where $\mathcal{R}$ is the responsivity of the detector.

The detectors themselves can be MSM or p-i-n diodes. The advantages of balanced detection in a coherent receiver system and its noise-suppression characteristics can be demonstrated even in a laboratory experiment. The setup is shown in Fig. 9.21. Light from a laser of suitable wavelength at a frequency f is directed through an acousto-optic (AO) modulator, which produces a second weaker beam at $(f + \Delta f)$. The two beams are physically separated and end-fired into the two arms of a directional coupler and phase shifter. Light from the directional coupler is focused onto the two balanced detectors. The photoresponse from the integrated detectors is combined through a 0–180° microwave hybrid coupler. The hybrid coupler provides the sum and difference of the input signals, with the sum output terminated internally. There is a 180° phase shift between the output photocurrent of the two photodiodes. Therefore, at the difference port of the hybrid coupler the signal components of the photocurrent add constructively, whereas the noise-components tend to cancel. It may be noted that in this arrangement the AO modulator is not necessary to study the noise-suppression characteristics of the detector and light from the laser can be directly split by a beam splitter. Figure 9.22 shows the noise-suppression characteristics of balanced InGaAs

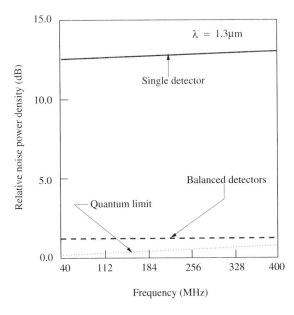

Figure 9.22 Relative intensity noise (RIN) suppression by InGaAs MSM photodiodes measured by the author and co-workers (J. Pamulapati et al., *IEEE Photonics Technology Letters*, **3**(4), 357, ©1991 IEEE).

MSM photodiodes. The bottom curve shows the quantum shot noise level of a single detector, obtained by illuminating the detector with incoherent white light. The top curve represents the noise level produced in the detector due to noise in the laser. The middle curve represents the relative intensity noise (RIN) suppression obtained with balanced detectors, and is approximately 11 dB. Noise suppression up to 30 dB has been measured in such balanced detection systems.

Coherent receiver systems with balanced detection are undoubtedly going to play a major role in long-distance optical communication due to their high sensitivity and selectivity. The disadvantage of the technique, as with any heterodyning technique, is the requirement of stable local oscillators and the number of components, hybrid or monolithically integrated, that are required. Obviously, if a single detector could provide the same selectivity and sensitivity, it would be more useful and economical.

9.9 MICROCAVITY PHOTODIODES

It is evident that the progression toward higher transmission rates in optical communication requires high-speed photodetectors. We have learned earlier that there is a fundamental tradeoff between the quantum efficiency (responsivity) and bandwidth of a photodiode. A large value of the former parameter necessitates the use of a thick ($\sim 2\mu$m) absorption region, while a large bandwidth dictates the need for a thin ($< 0.5\mu$m) absorption region. Such a tradeoff can be avoided by using microcavity photodiodes,[†] such as the one shown in Fig. 9.23. In this diode configuration, a thin

---

[†]R. Kuchibhotla et al., *IEEE Photonics Tech. Lett.*, **3**, 354, 1991.

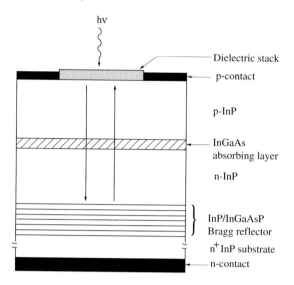

Figure 9.23   Schematic cross section of a microcavity photodiode.

low-bandgap absorption region is placed in the middle of a resonant cavity formed by heavily doped wider-bandgap regions and reflecting mirrors. The bottom reflecting mirror is formed by a $\lambda/4$-superlattice distributed Bragg reflector (DBR), while the top mirror can be a dielectric film or stack. The quarter-wave superlattice reflector is made up of alternate layers of high and low refractive index, and each thickness is equal to $\lambda/4$. The light reflected at each interface interacts constructively at the surface. The reflection coefficient of the entire stack depends on the difference in the refractive indices, and the number of periods and can be made quite large. The design wavelength λ_0 and the superlattice parameters are related by

$$\lambda_0 = 2(n_{rH}d_H + n_{rL}d_L) \tag{9.26}$$

where d represents layer thickness and the subscripts H and L represent layers with high and low refractive indices, respectively. The bandwidth and reflectance of the superlattice stack are given by

$$\Delta\lambda = \frac{4}{\pi} \frac{n_{rH} - n_{rL}}{n_{rH} + n_{rL}} \lambda_0 \tag{9.27}$$

and

$$\Theta_R = \left(\frac{1 - n_r^*}{1 + n_r^*}\right)^2 \tag{9.28}$$

respectively, where

$$n_r^* = \left(\frac{n_{rL}}{n_{rH}}\right)^{2p+1} \tag{9.29}$$

Here p is the number of periods. Equation 9.29 is valid for the structure shown in Fig. 9.23 in which there are $(p + 1)$ layers of InGaAsP and p-layers of InP, each $\lambda_0/4$ thick. These equations are valid for normal incidence. Reflectances close to unity can be achieved with GaAs/AlGaAs and InP/InGaAsP superlattices. With reference to Fig. 9.23, a resonance is built up in the InP cavity at those frequency components of the incoming light for which the round-trip phase shift is a multiple of 2π. Therefore, at a resonance frequency the incoming light is reflected at the two mirrors and the round-trip path length is greatly increased. The absorption and the quantum efficiency are therefore enhanced. However, the transit length remains small as the photogenerated carriers only have to traverse the thickness of the absorption region to be collected. Microcavity photodiodes therefore make the simultaneous achievement of large bandwidth and high-quantum efficiency possible.

PROBLEMS

9.1 A heterojunction n-p-n phototransistor is characterized by $\Delta \mathcal{E}_V = \Delta \mathcal{E}_g$ and the following material and device parameters: $W_b = 0.4\mu m$, $L_e = 8\mu m$, $L_h = 2\mu m$, $D_e = 200$ cm²/s, $D_h = 10$ cm²/s, $N_D = 10^{16}$ cm⁻³, $N_A = 10^{18}$ cm⁻³, $m_{eB}^* = 0.06/m_0$, $m_{hB}^* = 0.5m_0$, $m_{eE}^* = 0.1m_0$, $m_{hE}^* = 0.65m_0$ Calculate and plot the variation of the optical gain Γ_G versus $\Delta \mathcal{E}_V$ when $\Delta \mathcal{E}_V$ changes from 0.05 to 0.5. Assume the value of η to be 0.65. Note that as $\Delta \mathcal{E}_V (= \Delta \mathcal{E}_g)$ changes, the effective masses in the emitter region will change. This change is ignored. Discuss the nature of the plot and comment on its validity in the entire range of $\Delta \mathcal{E}_V$.

9.2 With reference to Fig. 9.5, derive Eq. 9.7.

9.3 From Eq. 9.18, derive the maximum responsivity of a PIN diode. Does this expression agree with that derived in Chapter 8?

9.4 From a measurement of responsivity versus incident power of a modulated barrier photodiode of area 10^{-4} cm², a value of $n_f = 0.5$ is estimated. For a measured $\Delta\phi_b = 3mV$, $P_{inc} = 1.5$ nW, illumination energy $h\nu = 1.5$ eV, and $\eta = 0.5$, calculate the value of J_{hd} at room temperature. Assuming $\tau \cong$ 1ns, estimate the number of holes accumulated at the potential minimum. If the maximum responsivity of the device is 1300 A/W, calculate the electron current under photoexcitation.

9.5 Draw the energy band diagram of a $p^- - n^+ - p^-$ heterojunction modulated barrier photodiode under biased condition and explain its principle of operation.

9.6 For the InAs/GaAs heterojunction photodiode shown in Fig. 9.9, illustrate the expected spectral response. Indicate the different peaks and explain their origins.

9.7 What scheme of detection would you use in an application where the principle requirements are low noise and large bandwidth?

READING LIST

BOARD, K. "New Unorthodox Semiconductor Devices." *Rep. Prog. Phys.*, **48**, 1595–1635, 1985.

CAMPBELL, J. C. "Phototransistors for Lightwave Communications." In *Semiconductors and Semimetals*, Vol. 22, Part D, ed. R. K. Willardson and A. C. Beer, Academic Press, Orlando, FL, 1985.

LINKE, R. A., and HENRY, P. S. "Coherent Optical Detection: A Thousand Calls on One Circuit." *IEEE Spectrum*, **24**, 52–57, 1987.

10

Solar Cells

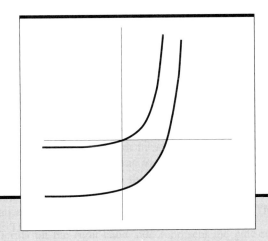

Chapter Contents

10.1 INTRODUCTION

Solar cells constitute a critical technology for overcoming global environmental and energy problems. The invention of the p-n junction in 1949 formed the basis of the discovery of the crystalline Si solar cell by Pearson in 1954. Since then, solar cells have been developed and produced with polysilicon, CdTe, and GaAs. Needless to mention, remarkable progress has been made in the past four decades. Megawatt solar power generating plants have been built, solar cells are being combined with building materials, and very recently the first solar-cell-powered plane demonstrated a transcontinental flight across the United States. Application of solar cells are now an important and integral part of our daily lives, ranging from calculators and wristwatches to solar powered irrigation systems. Over 95% of solar cells in production are silicon based.

The energy output from the sun is primarily electromagnetic radiation, which covers the spectral range of 0.2 to 3.0 μm. The radiation reaching Earth is scattered and absorbed in the atmosphere and the intensity is dependent on the angle of incidence. Depending on this angle, the intensity can vary between 500 and 1000 W/m^2. The power level of the solar spectrum in outer space, where there is no absorption of the radiation, is about 140 mW/cm^2. This is commonly termed the air-mass-zero (AM0) spectrum. On Earth at sea level, with the sun at zenith, the power level is reduced to nearly 100 mW/cm^2. This is the AM1 spectrum. At an angle of incidence that results in twice the path length through the atmosphere, the power level drops to approximately 80 mW/cm^2 and the corresponding spectrum is termed AM2. AM0 and AM2 solar spectra are shown in Fig. 10.1. The conversion of radiation energy into electrical energy is, in general, the *photovoltaic* effect. The most important photovoltaic device is the solar cell. The primary requirement for a material to be applicable to solar cells is a bandgap matching the solar spectrum and high mobilities and lifetimes of charge carriers. These conditions exist in GaAs and many other III–V compounds. In this chapter we learn about single crystal junction solar cells, which is basically a p-n junction with a large surface area, converting solar radiation directly into electrical energy with efficiency greater than 10%. Light, or photons, impinging on an unbiased junction create electron-hole pairs that diffuse toward the junction. As

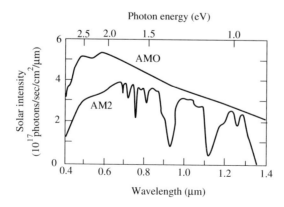

Figure 10.1 AM0 and AM2 solar spectra (from J. L. Shay et al., Conference Record of the *Eleventh Photovoltaic Specialists Conference*, New York, 504, ©1975 IEEE).

will be evident in the next section, electrical power develops across the junction as in a battery, and this power is delivered to an external load. A solar cell can deliver powers of the order of 1 kW/m^2. The schematic of the typical junction solar cell, with its top finger contacts, is shown in Fig. 10.2(a). The reason for this configuration of the contacts will be explained later. The photovoltaic energy conversion process may best be expressed by the equivalent circuit shown in Fig. 10.2(b). An ideal diode is connected in parallel with a constant current (or voltage) source, which represents the photovoltaic energy generated, and with a load resistor.

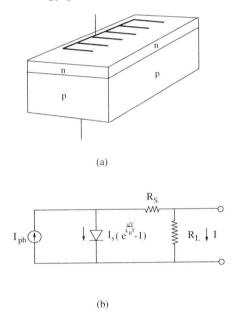

(a)

(b)

I_{ph} $I_s(e^{\frac{qV}{k_BT}}-1)$ R_S R_L I

Figure 10.2   (a) Schematics of a shallow junction solar cell with top "finger" ohmic contacts and (b) idealized equivalent circuit of solar cell. R_S is the cell series resistance and R_L is the load resistance.

10.2 BASIC PRINCIPLES: CURRENT-VOLTAGE CHARACTERISTICS

The solar cell shown in Fig. 10.2(a) consists of a shallow p-n junction formed by diffusion or epitaxy. Figure 10.2(b) represents the simplest equivalent circuit of the cell and contains the constant current source I_{ph}, the load current I, and the reverse saturation current of the diode, I_S. The current-voltage characteristics of such a diode under illumination are shown in Fig. 10.3. It is seen that the curve passes through the fourth quadrant and hence the device can deliver power. The characteristics in the fourth quadrant are represented more clearly in Fig. 10.4, obtained with a rotation of the current axis by 180° around the voltage axis. V_{oc} is the maximum voltage obtainable at the load under open-circuit conditions of the diode, and I_{sc} is the maximum current through the load under short-circuit conditions. The power delivered by the device can be maximized by maximizing the area under the curve in Fig. 10.4, that is, maximizing the product ($I_{sc} \times V_{oc}$). By properly choosing the load resistor, the output power can be as high as 0.8($I_{sc} \times V_{oc}$).

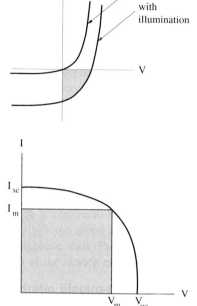

Figure 10.3   Current-voltage characteristics of a junction diode in the dark and under illumination.

Figure 10.4   I–V characteristics of an illuminated solar cell in the fourth quadrant.

Next, consider the band diagrams of a p-n junction in equilibrium in the dark and under illumination, as shown in Fig. 10.5. In the dark, drift of thermally generated minority carriers across the junction constitutes the reverse saturation current. At zero bias this is exactly balanced by a small flow of majority carriers in the opposite direction, resulting in zero net current. Under forward bias, the net direction of current is from the p-side to the n-side. If the junction is illuminated by photons with $h\nu \geq \mathcal{E}_g$, additional electron-hole pairs are created with a generation rate $G(\text{cm}^{-3}.\text{sec}^{-1})$. Then,

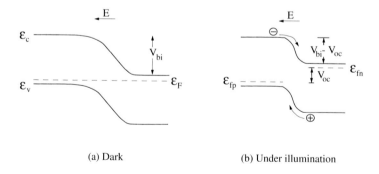

(a) Dark   (b) Under illumination

Figure 10.5   Band diagrams of an unbiased p-n junction solar cell (a) in the dark and (b) under illumination. In the open-circuited diode in (b) the direction of motion of the photogenerated carriers determine the direction of current flow within the cell.

the number of holes created per second within a diffusion length on the n-side is AL_hG, where A is the area of the diode. Similarly, the number of electrons created per second within a diffusion length on the p-side is AL_eG. Thus, the total photogenerated current, I_{ph}, due to the drift of these carriers across the junction is given by

$$I_{ph} = qAG(L_h + L_e) \qquad (10.1)$$

This current is directed from the n-side to the p-side and is opposed to the main diode current from the p-side to the n-side. Therefore, for the illuminated diode,

$$I = qA\left(\frac{L_h}{\tau_h}p_{NO} + \frac{L_e}{\tau_e}n_{PO}\right)\left(e^{qV/k_BT} - 1\right) - qAG(L_h + L_e) \qquad (10.2)$$

Thus, the current is lowered by an amount proportional to the generation rate, as shown in Fig. 10.3. Note that generation and recombination effects within the depletion region are neglected.

Let us now consider the two cases of the short-circuited and open-circuited diode. For the short-circuited diode, $V = 0$, and

$$I_{sc} = I_{ph} = -qAG(L_h + L_e) \qquad (10.3)$$

For the open-circuited diode, $I = 0$, and

$$V = V_{oc} = \frac{k_BT}{q}\ln\left[\frac{L_h + L_e}{\left(\frac{L_h}{\tau_h}\right)p_{NO} + \left(\frac{L_e}{\tau_e}\right)n_{PO}}G + 1\right] \qquad (10.4)$$

which is equivalent to

$$V_{oc} = \frac{k_BT}{q}\ln\left[\frac{I_{ph}}{I_S} + 1\right]$$

$$\cong \frac{k_BT}{q}\ln\left[\frac{I_{ph}}{I_S}\right] \qquad (10.5)$$

Hence, for a fixed I_{ph}, V_{oc} increases logarithmically with decreasing saturation current I_S. An important point to note in Fig. 10.5(b) is that although the band-bending and energy position of the quasi-Fermi levels are similar to a forward-biased diode, the carrier flow is in the *opposite* direction. It is this carrier flow that creates the potential V_{oc} across the device. Therefore, under illumination the junction acts as a power source. The output power is given by

$$P = IV = I_SV\left(e^{qV/k_BT} - 1\right) - I_{ph}V \qquad (10.6)$$

We define the quantities I_m and V_m, shown in Fig. 10.4, as the current and voltage, respectively, for maximum power output $P_m(= I_mV_m)$. The condition for maximum power is obtained by setting $\frac{dP}{dV} = 0$, which from Eq. 10.6 is

$$\frac{d}{dV}\left[I_SV\left(e^{qV/k_BT} - 1\right) - I_{ph}V\right] = 0 \qquad (10.7)$$

The differentiation of Eq. 10.7 leads to the equation

$$I_S \left(e^{qV/k_B T} - 1 \right) + I_S \frac{qV}{k_B T} e^{qV/k_B T} = I_{ph} \tag{10.8}$$

from which

$$V = \frac{k_B T}{q} ln \left(\frac{1 + I_{ph}/I_S}{1 + qV/k_B T} \right) \tag{10.9}$$

Therefore,

$$V_m = \frac{k_B T}{q} ln \left(\frac{1 + I_{ph}/I_S}{1 + q V_m/k_B T} \right)$$

$$= V_{oc} - \frac{k_B T}{q} ln \left(1 + \frac{q V_m}{k_B T} \right) \tag{10.10}$$

Also, from Eqs. 10.6, 10.8, and 10.10,

$$I_m = I_S \frac{q V_m}{k_B T} e^{q V_m/k_B T}$$

$$\cong I_{ph} \left(1 - \frac{1}{q V_m/k_B T} \right) \tag{10.11}$$

Thus,

$$P_m = I_m V_m \cong I_{ph} \left[V_{oc} - \frac{k_B T}{q} ln \left(1 + \frac{q V_m}{k_B T} \right) - \frac{k_B T}{q} \right]$$

$$= I_{ph} \left(\frac{\mathcal{E}_m}{q} \right) \tag{10.12}$$

assuming $V_m \cong V_{oc}$. Here

$$\mathcal{E}_m = q \left[V_{oc} - \frac{k_B T}{q} ln \left(1 + \frac{q V_m}{k_B T} \right) - \frac{k_B T}{q} \right] \tag{10.13}$$

and is defined as the energy per photon delivered to the load at the maximum power point.

The conversion efficiency of a solar cell, η_C, is defined by

$$\eta_C = \frac{P_m}{P_{inc}} = \frac{V_m I_m}{P_{inc}} = \frac{V_m^2 I_S (\frac{q}{k_B T}) e^{q V_m/k_B T}}{P_{inc}} \tag{10.14}$$

where P_{inc} is the incident power.

Next assume a symmetrical junction in which $p_{NO} = n_{PO}$ and $\tau_h = \tau_e$. Under thermal equilibrium conditions the recombination rate is equal to the thermal generation rate, G_T. In other words,

$$\frac{p_{NO}}{\tau_h} = \frac{n_{PO}}{\tau_e} = G_T \tag{10.15}$$

Thus,

$$V_{oc} \cong \frac{k_B T}{q} \ln \left[\frac{G}{G_T} \right] \tag{10.16}$$

This equation apparently suggests that V_{oc} can increase indefinitely with G. However, as G increases, the extra electron-hole pairs increase, which decreases $\tau_{e,h}$. Since p_{NO} and n_{PO} are fixed for a particular material at a given temperature T, G_T also increases. In fact, the maximum attainable value of $V_{oc} = V_{bi}$. Equation 10.16 describes the photovoltaic effect. The variations of I_{sc} and V_{oc} with light intensity are shown in Fig. 10.6.

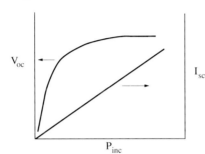

Figure 10.6 Schematic illustration of the variation of V_{oc} and I_{sc} with incident light intensity.

To maximize the output power, both I_{sc} and V_{oc} must be made as large as possible. From Eq. 10.5 it is evident that I_S should be made small, that is, $n_{NO}(N_D)$ and $p_{PO}(N_A)$ must be made as large as possible. In other words the p- and n-sides of the junction must be doped heavily. The term *fill factor* is used to define the power extraction efficiency and is expressed as

$$\text{Fill Factor} = \frac{I_m V_m}{I_{sc} V_{oc}} \tag{10.17}$$

The fill factor is an important figure of merit in solar cell design.

EXAMPLE 10.1

Objective.   A 5 cm^2 Ge solar cell with a dark reverse saturation current of 2 nA has AM1 radiation incident upon it, producing 4×10^{17} electron-hole pairs per second. The electron and hole diffusion lengths may be assumed to be 5 μm. To calculate the short-circuit current and open-circuit voltage of the cell.

From Eq. 10.3,

$$I_{ph} = I_{sc} = 1.6 \times 10^{-19} \times 5 \times 4 \times 10^{17} \times 10 \times 10^{-4}$$

$$= 0.32 mA$$

The open-circuit voltage is obtained from Eq. 10.5 as

$$V_{oc} = 0.026 \ln \left[\frac{3.2 \times 10^{-4}}{2 \times 10^{-9}} \right]$$

$$= 0.31 V$$

10.3 SPECTRAL RESPONSE

The spectral response of a solar cell is given by the variation of the short-circuit current as a function of wavelength of incident light. To derive an expression for the spectral response, we refer to the simple one dimensional geometry of a junction cell shown in Fig. 10.7. Here d is the junction depth from the surface at $x = 0$. It is being assumed that the depletion width of the junction is much smaller than the diffusion lengths L_e or L_h. As an example, for the dopings used in junction solar cells the depletion width is ~ 0.1 μm, whereas L_e and L_h are ~ 1–10 μm. Incident photons with energies in excess of the energy bandgap are absorbed and the photon density in the material varies as

$$\phi = \phi_o \, e^{-\alpha(\lambda)x} \ (sec^{-1}. \ cm^{-2}) \tag{10.18}$$

where it should be remembered that the absorption coefficient α is dependent on λ, the wavelength of incident photons. The electron-hole generation rate is then given by

$$G(x) = \phi_o\alpha \, e^{-\alpha x} \tag{10.19}$$

In the n-side of the junction, the minority carriers (holes) created at a distance x will have a fraction proportional to $\exp[-(d - x)/L_h]$ diffuse to the junction. Therefore the total number of holes on the n-side reaching the junction

$$p_N^{op} = \int_o^d \phi_o \, \alpha \, e^{-\alpha x} \, e^{-(d-x)/L_h} \, dx$$

$$= \frac{\phi_o\alpha}{\alpha - 1/L_h} \left[e^{-d/L_h} - e^{-\alpha d} \right] \tag{10.20}$$

Similarly, the number of minority carriers (electrons) reaching the junction as a result of optical generation in the p-side of the junction,

$$n_P^{op} = \int_d^\infty \phi_o\alpha \, e^{-\alpha x} e^{-(x-d)/L_e} \, dx$$

$$= \frac{\phi_o\alpha}{\alpha + 1/L_e} \, e^{-\alpha d} \tag{10.21}$$

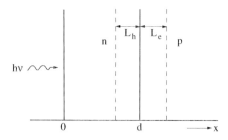

Figure 10.7 One-dimensional geometry of a junction solar cell.

Thus, the total number of photogenerated carriers crossing the p-n junction is given by

$$p_N^{op} + n_P^{op} = \frac{\phi_o \alpha}{\alpha - 1/L_h} \left[e^{-d/L_h} - e^{-\alpha d} \right] + \frac{\phi_o \alpha}{\alpha + 1/L_e} e^{-\alpha d} \qquad (10.22)$$

The generated current is determined by $\left(p_N^{op} + n_P^{op}\right)$ and is therefore proportional to the wavelength. Equation 10.22 represents the spectral response of the solar cell. Depending on the application, it may be advantageous to get larger response in some selected part of the solar spectrum. For example, if better response is required at longer wavelengths ($h\nu \cong \mathcal{E}_g$) where α is small, it is clear from Eq. 10.22 that L_e must be large. The latter is achieved by making the doping on the p-side, N_A, small. However, it was seen earlier from Eq. 10.5 that, to make V_{oc} large, N_A must be made large. Thus, a compromise is necessary and usually in practical cells $N_A \sim 10^{16}$ cm^{-3}. Similarly, if larger response is desired for short wavelengths ($h\nu \gg \mathcal{E}_g$) where α is large and most of the radiation is absorbed near the junction, d must be made small (i.e., a shallow junction is needed).

As discussed in Chapter 4, the current-voltage characteristics of a real p-n junction are given by

$$I = I_S \left[e^{qV/n_f k_B T} - 1 \right] \qquad (10.23)$$

where n_f is usually greater than unity and accounts for recombination within the depletion region. This will reduce the efficiency of the solar cell in comparison with an ideal diode. Again, when d is reduced to enhance the short wavelength response of the cell, the series resistance, R_S, of the top n-type layer increases. The effect on the current-voltage characteristics due to this series resistance is depicted in Fig. 10.8, where it is clear that the output is also reduced. To reduce R_S the top layer, formed by diffusion or epitaxy, is heavily doped. In addition, a distributed fingerlike contact configuration shown in Fig. 10.2(a) is formed, which allows light to be absorbed almost uniformly over the entire top surface. Finally, surface recombination must also be taken into account in the design. In most III–V semiconductors the surface recombination velocity is $\sim 10^5$ cm/s. Thus, if d is made too small to enhance the short wavelength response, surface recombination might dominate over bulk recombination and the efficiency of the cell is reduced.

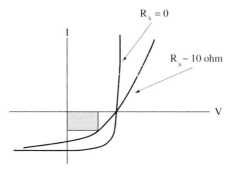

Figure 10.8 Effect of series resistance on the output of a junction solar cell.

The "finger" or stripe top contact described above has two major effects on solar cell performance. First, the short-circuit current is reduced because of the blocking effects of the contacts on light. This reduction is approximately equal to the ratio of the active (nonmetallized) area to the total area. In typical devices, the contact stripes cover 5%–10% percent of the total surface area. The second factor is the introduction of series resistance into the terminal characteristics because of the flow of current transverse to the surface through the sheet resistance of the surface layer. Because of this transverse current flow, the entire p-n junction is not at the same electric potential. Under short-circuit conditions, for example, regions of the cell away from the contact stripes remain under forward bias. A typical potential variation between two stripes is illustrated in Fig. 10.9. It has been shown that the peak value of the potential, ΔV, is proportional to d_f^2, where d_f is the stripe-to-stripe contact spacing.[†] Therefore, the average spacing between the fingers has to be carefully designed. Transparent contacts are also used in solar cell fabrication.

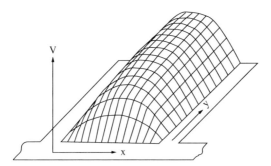

Figure 10.9 Schematic illustration of potential variation between contact stripes (from J. R. Hauser and P. M. Dunbar, *IEEE Trans. Elec. Dev.*, **ED-24**, 305, ©1977 IEEE).

10.4 HETEROJUNCTION AND CASCADED SOLAR CELLS

The homojunction solar cell design considered until now has several practical limitations. It was seen that to enhance the short wavelength response a shallow junction is desirable. Again, to reduce the series resistance of the top layer, it is desirable to dope it heavily. This, unfortunately will enhance the absorption in this layer and reduce the diffusion length. There is the overriding factor of recombination loss of the incident radiation on the surface, characterized by a surface recombination velocity, s_R. This recombination takes place at the surface states, which are present on the free surface of a semiconductor. Certain chemical treatments are known to reduce the density of surface states by almost an order of magnitude. However, with III–V materials and technology there is another attractive alternative.

[†]See J. R. Hauser and P. M. Dunbar, *IEEE Trans. on Electron Devices*, **ED-24**, 305–321, 1977.

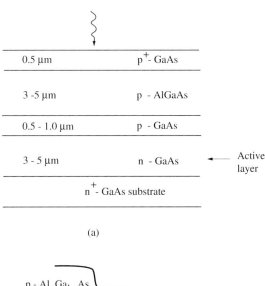

(a)

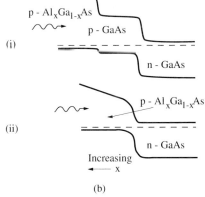

(b)

Figure 10.10 (a) Schematic of heterostructure and (b) band diagrams of GaAs/AlGaAs heterojunction solar cells. Both graded and fixed composition AlGaAs of direct and indirect bandgap can be used as the window layer. The top GaAs contact layer is not represented in (b).

Figure 10.10 shows the schematic and band diagrams of a $Al_xGa_{1-x}As$/GaAs heterostructure solar cell. There are numerous advantages in using such a heterostructure. The AlGaAs layer with a bandgap larger than that of GaAs acts as a transparent window for photons with energies below the AlGaAs bandgap. These photons are absorbed close to or within the depletion region of the p-n junction in GaAs. Thus, the short wavelength response can be enhanced, since the density of interface states at the GaAs/AlGaAs heterointerface is very low. The p-GaAs/p-AlGaAs heterojunction confines the electrons generated in the GaAs layer, giving a low surface recombination velocity, which greatly increases the cell efficiency. The higher bandgap AlGaAs can be made relatively thick and doped heavily without affecting its light transmission characteristics. Since the AlGaAs sheet resistance is in parallel with the GaAs sheet resistances the overall series resistance of the cell is reduced. Finally, a thick large-bandgap window layer improves the radiation tolerance of the solar cell. The current-voltage characteristics of a heterojunction cell are very similar to that of a homojunction one.

It is easier to estimate and visualize the advantages and disadvantages of a heterojunction solar cell by considering a structure in which the p-n junction and heterojunction coincide. Such a structure is shown in Fig. 10.11. In this N-p structure the total built-in potential is

$$q V_{bi} = q(V_{bi,n} + V_{bi,p})$$
$$= \mathcal{E}_{g2} + \Delta\mathcal{E}_C - (\mathcal{E}_C - \mathcal{E}_F) - (\mathcal{E}_F - \mathcal{E}_V) \qquad (10.24)$$

Therefore, if $\Delta\mathcal{E}_C$ is positive, the built-in voltage of a heterojunction can be larger than that of a homojunction. This will reduce the reverse saturation current I_S, and if the photocurrent I_{ph} remains the same, then according to Eq. 10.5 the output V_{oc} will be enhanced. However, in reality, the "notch" at the heterojunction hinders carrier transport across it and the photocurrent can be greatly reduced. Therefore, unless special techniques such as the use of compositional grading is used, there is no enhancement in the output of a heterojunction and the only advantages are the ones mentioned in the previous paragraph.

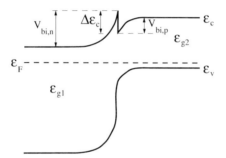

Figure 10.11 Energy band diagram of a N-p heterojunction.

To further increase the conversion efficiency, two or more p-n junctions of different bandgaps can be used. Lower energy photons not absorbed by one cell are absorbed by another cell of lower bandgap. There are two approaches to combining cells of different bandgaps. In the first, the junctions of different materials (with varying bandgaps) are cascaded or stacked together in series by intermediate tunnel junctions. The schematic and I–V characteristics of a tunnel junction are shown in Figs. 10.12(a) and (b). Radiation is first incident upon the cell with the largest bandgap. For example, with a two-cell device, to obtain maximum overall efficiency the bandgap values should be 1.75–1.80 eV for the top cell material and 1.0–1.1 eV for the bottom cell material. It is difficult to find a combination of III–V binary, ternary or quaternary lattice-matched compounds that have these or intermediate bandgaps. Lattice-mismatched or pseudomorphic materials can be employed. With a fairly mature epitaxial technology for GaAs (and AlGaAs) on Si, it is feasible to have Si as the material for the bottom cell and AlGaAs as the material for the top cell. Such cascaded solar cells have been made by the epitaxial techniques discussed in Chapter 1. The second approach to combining cells of different gaps is beam splitting. In this technique spectral splitting with a multilayered dielectric filter allows each cell to be spatially separated and selectively

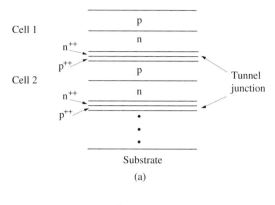

(a)

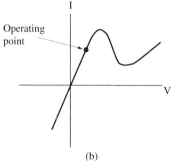

(b)

Figure 10.12 (a) Schematic of a multijunction solar cell interconnected by tunnel junctions. The n^{++} and p^{++} layers of the latter are typically of the order of 1000 Å and 200 Å, respectively; and (b) current-voltage characteristics of a typical tunnel junction.

excited. The main advantage is that the substrates on which the cells are fabricated do not have to be optically transparent to the solar spectrum and current matching—as in the cascaded configuration—is not necessary for system interconnection.

10.5 SCHOTTKY BARRIER CELLS

The schematic energy band diagram of a metal-n-semiconductor junction under illumination is shown in Fig. 10.13. The metal is deposited as a semitransparent film through which most of the light can pass through. There are three photocurrent

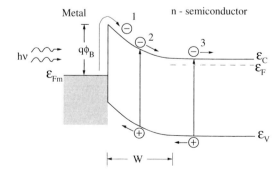

Figure 10.13 Energy band diagram of Schottky barrier solar cell under illumination. The three carrier generation processes are described in the text.

components. Light with energy $h\nu \geq q\phi_b$ is absorbed in the metal and excites electrons over the metal into the semiconductor. Light transmitted through the metal into the semiconductor is mainly absorbed in the depletion region, creating electron-hole pairs. Longer wavelength light is absorbed in the neutral regions and the holes diffuse to the depletion edge to be collected. The first component of photocurrent is negligibly small.

The current-voltage characteristics of the Schottky barrier under illumination is similar to the p-n junction and is given by

$$I = I_S \left(e^{qV/k_B T} - 1 \right) - I_{ph} \qquad (10.25)$$

where

$$I_S = A^{**} T^2 \exp\left(-q\phi_b / k_B T\right) \qquad (10.26)$$

The main advantages of a Schottky barrier cell over a junction diode cell are the possibility of low-temperature processing, since no diffusion steps are involved, and the use of polycrystalline materials. Also, the presence of the depletion region very near the surface reduces the effects of high-surface recombination velocity and improves the spectral response characteristics.

10.6 MATERIALS AND DESIGN CONSIDERATIONS

10.6.1 Materials Requirements

The first requirement of a material to be used in a solar photovoltaic energy conversion device is a bandgap matching the solar spectrum and also having high mobilities and lifetimes of charge carriers. These conditions are met by many II–VI and III–V compounds and Si. Examples are GaAs, CdSe, CdTe, etc. In spite of the high costs of synthesis and fabrication of these semiconductors, they have been used very successfully in space applications, for which cost has not been an important factor. For terrestrial applications, where cost reduction is an important factor, the objective is to find cheaper and perhaps, amorphous materials and establish a simple and less expensive manufacturing process. The cost of a compound depends roughly on the relative abundance of its components and on their melting points. The latter determines energy and material requirements for refining the elements. The melting point and the reactivity of the elements at this temperature determine the energy and materials requirements for crystal growth. Lower melting temperatures also mean low diffusion temperatures and less reactive manufacturing conditions. Thus, high vacuum or inert atmospheres may not be needed for their synthesis.

In general, compounds with high melting points and bandgaps are hard and have lower mobilities and carrier lifetimes. Compounds that are soft and have smaller bandgaps also generally have higher mobilities and carrier lifetimes. The criterion that is usually applied to find such compounds is that for given melting temperatures of the elements low-melting-temperature compounds are those of tetrahedral coordination or chain or layered compounds made of elements from the right of the periodic table. From the point of view of junction devices it is also important to keep in mind that a potential barrier V_{bi} of a minimum height can be built into the material. Similarly, for Schottky barrier devices, which are simpler to manufacture, it is necessary to have a reasonable metal-semiconductor barrier height. These requirements put a constraint on the permissible bandgaps of the semiconductor.

Solar cells made of large bandgap semiconductors are more effective in absorbing photons of higher spectral energy. However, unless cascaded cells are used, photons with energy less than the bandgap are not absorbed in a single cell. Furthermore, the efficiency of a solar cell is more in the AM1 spectrum than in the AM0 spectrum because the ultraviolet component of the latter is largely unabsorbed. From Eqs. 10.3 and 10.5 it is evident that the higher the bandgap of the semiconductor used, the higher is the open circuit voltage and the lower is the short-circuit current. Calculated short-circuit current densities as a function of semiconductor bandgap, assuming unity spectral response, are shown in Fig. 10.14.

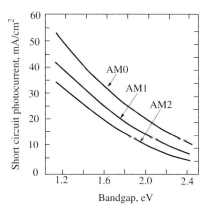

Figure 10.14 Calculated short-circuit current versus semiconductor bandgap for unity spectral response (from H. J. Hovel, *Semiconductor and Semimetals; Vol. 11. Solar Cells*, 38, Academic Press, New York, 1975).

It should be noted, in passing, that in this chapter we have mainly discussed the properties of junction solar cells made of single crystal semiconductors. Solar cells are also made of polycrystalline and amorphous semiconductors, and although the efficiency of such cells may be lower than the single crystal ones, cost-effectiveness sometimes justifies their use in large-scale applications.

10.6.2 Solar Cell Design

Considering heterojunction solar cells made with GaAs/AlGaAs, the important features for high efficiency design are:

1. An ultra-thin (500–1000 Å) AlGaAs surface layer to minimize GaAs surface recombination and optical absorption in this layer.
2. A p-type GaAs active layer with large electron diffusion length (5–10 μm).
3. Broadband antireflection coating on top to minimize reflection losses. It is extremely important to keep the cell series resistance to a low value. To achieve this, the top finger stripes have to be properly designed and the thickness of the p-GaAs layer should be chosen such that the spreading resistance is low. The design of the top contact grid has been discussed earlier. A radial contact grid pattern is also used sometimes. It consists of tapered grid lines of different lengths for carrying current from the irradiated part of the cell to an outer ring contact.

4. Use of solar concentrator systems for obtaining more power per solar cell. Concentrator solar cells of high concentration ratio must also track the sun. Such cells must also be provided with adequate heat dissipation and cooling facilities. Use of concentration factors of $20\times$ to $200\times$ in practical applications is quite common.

The III–V homo- or heterojunction solar cell is usually grown by one of the epitaxial techniques. Si solar cells are realized by diffusion or epitaxy. Processing is done by photolithography and selective etching to delineate the active area, ohmic contact formation to the front and back accompanied by silver plating of the front contact grid, and deposition of a wideband antireflection coating on the front surface consisting of a double layer of Ta_2O_5 and SiO_2. The individual cells are bonded to an aluminum heat sink with high conductivity epoxy and contact to the top fingers are made by ultrasonic bonding of gold wires.

10.6.3 p^+-n-n^+ versus n^+-p-p^+ Cells

The more commonly used type of cell is of the n^+-p-p^+ configuration. Early Si solar cells were made by diffusing an n-type impurity into a p-type Si substrate to form the junction. However, with present epitaxial techniques one can easily realize p^+-n-n^+ cells. In III–V compounds such as GaAs and InP and ternary and quaternary compounds lattice-matched to these, the electron diffusion length is usually larger than the hole diffusion length. This fact would dictate the use of n^+-p-p^+ cells. One also has to consider bandgap-narrowing effects in the top layer due to heavy doping, which depends on the effective mass of carriers in the material. Bandgap narrowing will affect the short wavelength response. Also, it has been observed that V_{oc} can be different for a p^+-n-n^+ and a n^+-p-p^+ cell with the same materials. This arises because of nonuniform generation of electrons and holes in the base or absorption region and unequal diffusion coefficients for electron and holes. As a result an extra voltage called the *Dember potential* appears in the base region, which aids carrier collection for a p-type base, but opposes carrier collection for an n-type base region. The factors mentioned above need to be taken into consideration in designing practical cells.

10.6.4 Dependence of Cell Performance on External Factors

The two important external parameters that affect solar cell performance most are temperature and radiation effects. Remembering that $L = \sqrt{D\tau}$, with increase of temperature, D remains fairly constant or increases slightly, and τ increases. Therefore, the diffusion length L increases with temperature, causing an increase in I_{ph} in accordance with Eq. 10.1. However, since the saturation current increases exponentially with temperature, V_{oc} will decrease rapidly, in accordance with Eq. 10.5. As a consequence, the I–V curve becomes more rounded, degrading the fill factor. The overall effect is a reduction of the cell efficiency with increasing temperature.

For extraterrestrial applications in particular, it is important to consider radiation effects. Radiation in outer space with high energy (MeV) particles creates defects and trapping centers in the material, which decrease the minority carrier lifetimes.

This causes the diffusion lengths to decrease. The decrease in diffusion length is almost inversely proportional to the increase in radiation flux density, as illustrated in Fig. 10.15. This results in a reduction of solar cell power output. It has been shown that certain elements, such as Li, when incorporated in solar cell materials can combine with radiation-induced defects and neutralize their effects. In spite of

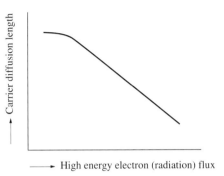

Figure 10.15 Schematic illustration of the dependence of carrier diffusion length on radiation flux.

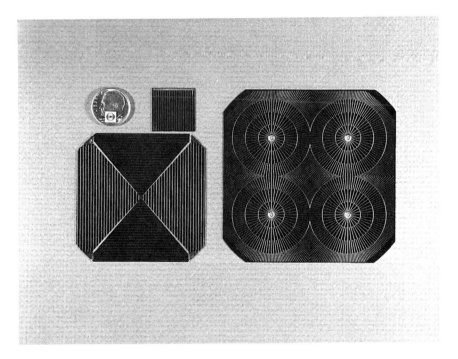

Figure 10.16 Practical Si and GaAs solar cells for space applications. The three larger Si cells range in size from 8×8 cm^2 to 2×2 cm^2 and have an efficiency of 14.5% at AM0. These cells have both contacts on the side. The smallest cell shown is made of GaAs and is 5 mm in diameter. Its efficiency is $\sim 22\%$ at $100\times$ concentration and at $80°$C (courtesy of I. Weinberg, NASA Lewis Research Center, Cleveland).

these limitations, solar cells currently fabricated with III–V materials have exhibited quantum efficiencies of approximately 30%. Practical solar cells and there applications in satellites are illustrated in Figs. 10.16 and 10.17, respectively.

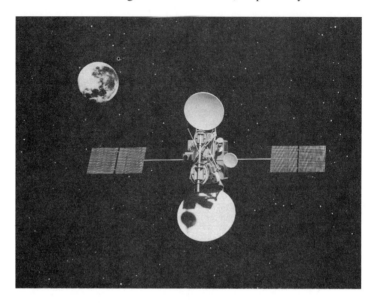

Figure 10.17 The Advanced Communication Technology satellite with arrays of Si solar cells which can generate almost 2KW of power (courtesy of I. Weinberg, NASA Lewis Research Center, Cleveland).

PROBLEMS

10.1 With the help of suitable diagrams (I–V curves) explain the operation of a junction photodiode and the junction solar cell. Can the cell be operated in any other quadrant besides the fourth quadrant?

10.2 Define the fill factor of a solar cell and discuss the factors that influence it.

10.3 With the help of appropriate equations, discuss the temperature dependence of the open-circuit voltage, output power, and the conversion efficiency of a solar cell. Calculate the changes in these parameters for a GaAs solar cell when the temperature increases from 300° to 350°K.

10.4 The reverse saturation current of a Si p-n junction solar cell is $5 \times 10^{-11} A$ and under illumination it produces a short-circuit current of 50 mA. Calculate the open-circuit voltage of the cell under these conditions. If the cell is of area 1 cm^2, estimate the pair generation rate for the specified short-circuit current.

10.5 Discuss the spectral response characteristics of a solar cell. What factors must be considered to make the long wavelength response more efficient?

10.6 A solar cell is to be designed to absorb in the spectral energy range of 0.8 eV to 1.8 eV. Show schematically the design of a cascaded cell with the different heterostructures.

10.7 On a bright day, when the sun is at its zenith, the short-circuit current density of a solar cell is approximately 30 mA/cm^2. What material is the cell made of?

10.8 A Si solar cell has an area of 5 cm^2. Under AM1 illumination it develops $V_{oc} = 0.7V$. Under these conditions $V_m/V_{oc} = 0.85$ and the fill factor is 0.75. Estimate the value of I_m. If the incident power is 50 mW/cm^2, calculate the cell efficiency.

10.9 Under illumination a GaAs solar cell with a dark saturation current of 0.5 nA produces a short-circuit current of 100 mA. Estimate the value of the maximum voltage V_m.

10.10 With reference to the epitaxial techniques described in Chapter 1, describe the technique you would use for producing cascaded cells, with reasons.

R E A D I N G   L I S T

HOVEL, H. J. *Semiconductors and Semimetals: Vol. 11. Solar Cells* (ed. R. K. Willardson and A. C. Beer), Academic Press, New York, 1975.

MILNES, A. G. *Semiconductor Devices and Integrated Electronics*, Van Nostrand Reinhold, New York, 1980.

SZE, S. M. *Physics of Semiconductor Devices*, 2nd ed. Wiley Interscience, New York, 1981.

11

Optoelectronic Modulation and Switching Devices

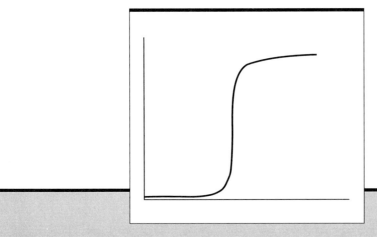

Chapter Contents

11.1 INTRODUCTION

In any system designed for the transfer of information over short or long distances, the signal carrying the information is superimposed on a high-frequency electromagnetic wave called the carrier and transmitted through a suitable medium. At the receiving end, the carrier and information are detected, and the information, almost exactly resembling the transmitted information, is extracted by suitable means. The two processes are known as *modulation*, and *demodulation*, respectively. Modulation is necessary for a variety of practical reasons. The size of the radiating elements is inversely related to transmission frequency, and therefore a high-frequency carrier is useful. Other reasons are reduction of noise, possibility of multiplexing, and the possibility of overcoming equipment limitations in a certain frequency range. The attenuation and information-carrying capacity of the high-frequency carrier signal are related to its frequency. In general, the higher the carrier frequency, the larger is the available transmission bandwidth. The relative frequencies and wavelengths in the electromagnetic spectrum were listed in Table 1.4. It is clear that there is the potential of an increased usable bandwidth by a factor of 10^4–10^5 in using a carrier at optical frequencies. A related advantage is the reduction in component size. Thus, while microwave and millimeter wave transmission systems have been very successful, their modulation bandwidth is of the order of a GHz. In order to utilize the enormous bandwidth of an optical carrier, transmission of signals modulated at several gigahertz is necessary. This will increase the information carrying capacity of the optical fiber or other suitable transmitting media.

Early attempts to building practical optical communication links were slowed down due to two main reasons: lack of coherent sources and lack of suitable guiding media. Transmission of light from incoherent sources through the atmosphere was severely affected by atmospheric conditions, which, to make matters worse, change with time. Discoveries of the laser and the optical fiber changed everything. They provided, respectively, a coherent source and a dielectric guiding medium, removing the necessity of line-of-sight transmission through the atmosphere. In addition to the enormous bandwidth, optical fibers provide the additional advantages of low transmission loss, ruggedness, flexibility, ease of maintenance, better electrical isolation, and reduction of crosstalk. The schematics of a practical optical fiber communication system is shown in Fig. 11.1. The role of the modulator is to impress the information, in analog or digital form, onto the continuous wave signal from the laser or LED source. The same arguments are also true for short-distance chip-to-chip communication links and in local area networks (LANs).

In an optical transmission system, there are also needs for optical switching and logic devices. The former is important for switching and routing signals, while the latter can possibly replace electronic devices for digital computing. For example, switching devices can be useful at the nodes of an optical communication system, such as the one shown in Fig. 11.1, or in a wavelength-division switching system shown in Fig. 11.2. In the latter, the electronic input modulates the sources of different

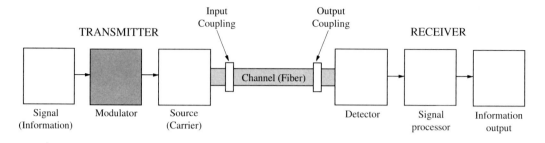

Figure 11.1 Schematics of a typical optical-fiber communication system.

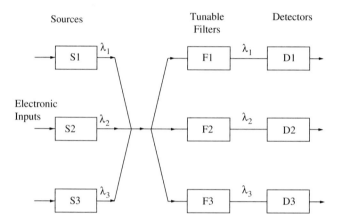

Figure 11.2 Schematics of a wavelength division-switching system.

wavelengths. The different channels are combined and split again to be detected separately by a wavelength-selective detection system.

Optically controlled hi-lo or on-off switching devices can also perform optical logic functions and can function as memory devices. An optical device with an input-output characteristic shown in Fig. 11.3 demonstrates two-level logic and can be a very important component of an optical computing system. Other electronic functions that could be realized by optical devices are amplification and memory.

It is important to distinguish two classes of switching and logic devices at this point. The first are all-optical and are termed "photonic" switching or logic devices.

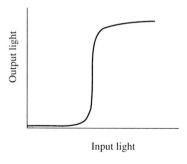

Figure 11.3 Schematic illustration of thresholding input-output characteristics.

Their operation is usually based on nonlinear optical phenomena and require large amounts of input optical power. Optoelectronic switching and modulation devices usually operate at lower power levels and can provide all the advantages. In addition, they can provide the possibility of an electrical input, in addition to the optical input, which can be very useful in cascaded circuits.

11.2 ANALOG AND DIGITAL MODULATION

Modulation is the process by which the waveform of a high-frequency carrier wave is modified suitably to transmit information. Usually, the information signal, at a much lower frequency, is impressed on the high-frequency carrier wave. As originally classified, two basic types of modulation were identified, according to the final shape of the carrier waveform: *continuous wave* (cw), in which the carrier is usually a sinusoidal waveform, and *pulse modulation*, in which the carrier is a periodic stream of pulses. Modulation is best described as a frequency translation process in which the information signal is shifted to higher frequencies. More recently, the modulation process has been classified as *analog* or *digital*. In analog modulation the information signal or wave varies the light from the source, or the high-frequency signal, in a continuous manner. Thus, both could be sinusoidal, as shown in Fig. 11.4(a). There is always a onc-to-one correspondence between the information signal and the magnitude of the modulated carrier. In digital modulation, discrete changes (on-off) in the intensity of the carrier are caused by the information signal. Information is then transmitted by the high-frequency signal as a series of discrete pulses (0 and 1), as shown in Fig. 11.4(b).

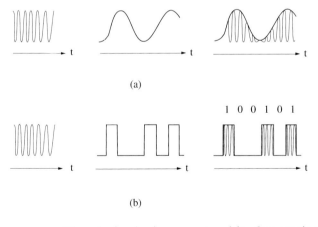

Figure 11.4   (a) Analog and (b) digital modulation of a high-frequency carrier signal.

Though simpler in concept and implementation, analog modulation suffers from a few practical drawbacks. It requires a higher signal-to-noise ratio at the receiver or detector. For large bandwidth applications the laser is driven at high-current levels, at which the light-current characteristics are very nonlinear. Analog modulation may be more suited for low modulation frequencies. Digital modulation is more suited for large bandwidth optical transmission and reception.

Having described the basic forms of modulation, it is important to mention at this point that the devices to be described in this chapter are basically amplitude and phase modulators, which can be used as external modulators in both digital and analog modulation schemes. In this scheme of modulation the cw light from the source is incident on the modulator in which the signal is impressed through the bias circuit. The output of the device is modulated light.

11.3 FRANZ-KELDYSH AND STARK EFFECT MODULATORS

The Franz-Keldysh and Stark effects were described in Chapter 3. Both effects result in absorption of photons with energies smaller than the bandgap with the application of an electric field. The phenomenon is, in general, described as electroabsorption. Electroabsorption modulators based on these effects can be fabricated by choosing the light wavelength to correspond to an energy smaller than the bandgap. At zero field, the light is fully transmitted while it is attenuated with application of a field. The sequence of the bias pulses can correspond to the signal to be transmitted. However, since the effects are very weak, either very large fields, or long ($\sim$ several millimeters) devices are needed to enhance the electro-absorption effect. Therefore, practical modulation devices cannot be realized based on these rather weak electro-absorption effects.

11.4 QUANTUM WELL ELECTRO-ABSORPTION MODULATORS

We have seen in Chapter 3 that when a transverse electric field is applied to a multiquantum well (MQW) the transition energy near the absorption edge, given by Eq. 3.65, is drastically modified. Modification of the electron and hole envelope functions due to the field results in a modification of the subband energies, and as a result a pronounced red shift of the absorption edge is observed. This shift, known as the quantum confined Stark effect (QCSE), is much larger than that produced by the Stark effect or Franz-Keldysh effect in bulk semiconductors. Roughly, large changes in the absorption are expected when the potential drops across a single well becomes comparable to the value of one of the energy terms in Eq. 3.65. Therefore, for a field of $(1 - 10) \times 10^4$ V/cm, the shift in the absorption edge and the exciton resonances in a 100 Å GaAs/AlGaAs MQW is of the order of 10–50 meV.

The conventional technique of applying a transverse bias across a MQW is to apply a reverse bias across a p-i(MQW)-n diode as shown in Fig. 11.5. Bias-dependent absorption measured in such a device having 40 periods of GaAs(90 Å)/$Al_{0.3}Ga_{0.7}As$ (80 Å) at room temperature is shown in Fig. 3.16. The red shift in the absorption spectrum is clearly evident. It may also be noticed that the shift is accompanied by a quenching of the peak height of the excitonic resonances. This is due to the transformation of the electron and hole bound states in the quantum well into quasi-bound states and the corresponding decrease in the overlap of the electron and hole wavefunctions.

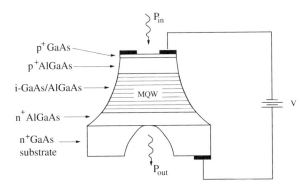

Figure 11.5 Mesa-etched p-i(MQW)-n electroabsorption modulator with a vertical illumination geometry.

Figure 11.5 shows the schematics of a mesa-etched modulator in which light is transmitted vertically through the device, normal to the plane of the quantum well layers. Since the GaAs substrate is usually not transparent to the light that is transmitted, it is usually etched selectively under the active region of the device. The total thickness of the MQW region is typically $\sim 1\mu$m, the diameter of the mesa diode is between 50 and 100 μm and the optical window on top of the device is typically half the diode area. The entire device is made by photolithography, selective wet-chemical or dry etching, and ohmic contact formation. In InP-based devices with InGaAs/InAlAs or InGaAs/InP MQW, etching of the InP substrate is not necessary, since it is transparent to the incident light at the operating wavelength. The device can also be realized in guided wave form for integrated optical applications. Here the light to be modulated is incident on one of the facets of the waveguide, which typically has a dimension of 2 μm $\times$ (4 − 10) μm for single-mode transmission. More importantly, since the device can be a couple of hundred microns in length, very few quantum wells are needed for electro-absorption and QCSE. Such a device is also called a *traveling wave modulator*. The typical layout for a guided wave modulator is shown in Fig. 11.6.

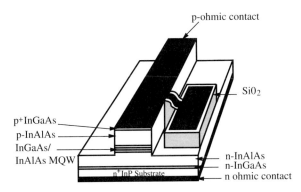

Figure 11.6 Schematic illustration of InP-based guided-wave MQW electroabsorption modulator.

Important operational parameters of the QCSE modulator are the modulation bandwidth, modulation depth or index, and operating bias. The latter becomes important if compatibility with VLSI operation is desired. The modulation bandwidth is determined by intrinsic and extrinsic factors. The intrinsic factors are related to carrier dynamics. In particular, the parameters of interest are exciton formation and dissociation times and the time taken to remove electrons and holes from the respective wells by thermionic emission and tunneling. The rate equation for carriers can be expressed, from Eq. 3.97 as

$$\frac{dn}{dt} = G - \frac{n}{\tau} - \frac{n}{\tau_{es}} \tag{11.1}$$

where τ is the recombination lifetime and τ_{es} is the escape time. Tunneling times depend on carrier masses, the applied electric field and the barrier height and thickness. With proper tailoring of these parameters, tunneling times $\sim$ 10–100 ps can be achieved. Therefore, if the carriers can be removed fast enough from the quantum wells, intrinsic modulation bandwidths $\sim$ 50–100 GHz are achievable. The external limitation to modulation speed of the p-i-n diode will be the RC time constant, as in any junction diode. The total capacitance is the sum of the intrinsic layer capacitance $C_i = \epsilon_s A/d_i$ and the bonding pad capacitance, $C_{pad} = \epsilon_{ox} A_{pad}/d_{ox}$, where d_i is the intrinsic layer thickness and d_{ox} is the thickness of the oxide layer between the contact pad and the semi-insulating substrate. This oxide layer also serves as a surface passivation layer for the diode. The resistance of the device is primarily limited by the contact resistance of the ohmic contacts, and to a small extent by the resistance of the neutral regions of the diode. Modulation bandwidths greater than 20 GHz have been measured in guided wave devices having a total capacitance of 0.2 pF at zero bias. Obviously, for high-speed devices, the circuit elements including bond-wire inductances become very important.

The modulation index, on-off ratio, or extinction ratio of the device is the ratio of the high and low intensities of the transmitted light. The ratio ultimately depends on the value of the field-dependent absorption coefficient at the operating voltage. For a large modulation index it is important to have a strong excitonic absorption, and a large change in the absorption per unit applied bias. The bias-dependent transmission characteristics of a device at room temperature are shown in Fig. 11.7. The excitonic absorption peaks are enhanced with lowering of temperature and the modulation index

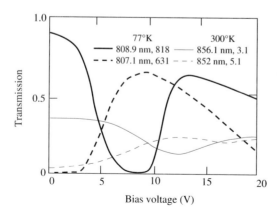

Figure 11.7 Operating characteristics of a p-i-n type GaAs/AlGaAs MQW modulator with 100 wells. The inset lists the operating wavelength and the maximum modulation ratio for each characteristic (courtesy of R. Sahai, Rockwell International, Thousand Oaks, CA).

is expected to improve, too. The modulation index can be increased by operating a vertically illuminated device in the reflection mode, instead of the transmission mode. This is achieved by either depositing a metal reflector at the substrate end, or using a quarter-wave multilayer dielectric mirror (see Sec. 9.9). Such techniques eliminate the need for substrate etching if the photon energy is larger than the substrate material bandgap. Furthermore, since the incident light makes at least two passes through the MQW region, the absorption is enhanced, and the contrast ratio is increased by at least a factor of two. The operating bias can be reduced by almost a factor of two by using coupled quantum wells, in which the overlap of the electron and hole wavefunctions is decreased very rapidly with application of electric field. Hence, the QCSE in such structures is very strong.

11.5 ELECTRO-OPTIC MODULATORS

11.5.1 Birefringence and the Electro-Optic Effect: Application to Phase Modulation

The constitutive relations in material media are

$$\mathbf{D} = \epsilon_0 \mathbf{E} + \mathbf{P}$$

$$\mathbf{B} = \mu_0 \mathbf{H} + \mathbf{M} \tag{11.2}$$

where $\mathbf{P}$ and $\mathbf{M}$ are the electric and magnetic polarization vectors, respectively. In *anisotropic* media, the induced polarization due to the change in the distribution of electrons within it will depend, in magnitude and direction, on the direction of the applied field. Thus,

$$\mathbf{P}_i = \epsilon_0 \sum_j \chi_{ij}^e \mathbf{E_j} \tag{11.3}$$

where χ_{ij}^e is known as the susceptibility tensor and $\mathbf{P}$ and $\mathbf{E}$ are usually complex amplitudes of harmonic, time-varying quantities. The choice of axes with respect to the crystal structures becomes very important in defining χ_{ij}^e. For example, in a system of rectangular cartesian coordinates:

$$\begin{bmatrix} P_x \\ P_y \\ P_z \end{bmatrix} = \epsilon_0 \begin{bmatrix} \chi_{11}^e & \chi_{12}^e & \chi_{13}^e \\ \chi_{21}^e & \chi_{22}^e & \chi_{23}^e \\ \chi_{31}^e & \chi_{32}^e & \chi_{33}^e \end{bmatrix} \begin{bmatrix} E_x \\ E_y \\ E_z \end{bmatrix} \tag{11.4}$$

It is also useful to keep in mind the corresponding relation:

$$\epsilon_{ij} = \epsilon_0 \left(1 + \chi_{ij}^e \right) \tag{11.5}$$

If the coordinate axes are chosen in such a way that the off-diagonal elements of the susceptibility tensor vanish, then these directions define the *principal dielectric axes* of the crystal. For example, along the x principal dielectric axis

$$P_x = \epsilon_0 \chi_{11}^e E_x$$

$$\epsilon_{11} = \epsilon_0 (1 + \chi_{11}^e) \tag{11.6}$$

and similar equations are true in the y and z directions.

There are some important consequences of the dielectric anisotropy that we have just learned about. It is immediately obvious, that for a linearly polarized plane wave propagating in the z-direction, its polarization vector will depend on the direction of the electric field according to Eq. 11.6. Furthermore, by virtue of Eq. 11.5, the phase velocity of the propagating beam will depend on the direction of polarization of the electric field. This phenomenon is known as *birefringence,* and is not observed in isotropic media. Imagine now that a linearly polarized wave, with equal electric field components in the *x* and *y* directions is launched in the *z*-direction. As a result of birefringence, it can be shown from Maxwell's equations (Problem 11.1) that the propagation constants are different in the *x*- and *y*-directions. The *x*- and *y*-components of the wave get out of phase and the wave gets elliptically polarized. This is the basis of *electro-optic modulation* and the phenomenon can be used to make efficient field-induced phase modulators with appropriate crystals.

The phenomenon of birefringence and the two resulting polarizations are very conveniently described by what is called the *index ellipsoid,* expressed as

$$\frac{x^2}{n_{rx}^2} + \frac{y^2}{n_{ry}^2} + \frac{z^2}{n_{rz}^2} = 1 \tag{11.7}$$

where $n_{rx}, n_{ry},$ and n_{rz} are the indices in the directions of the major axes of the ellipsoid. Here, wave propagation is in the z-direction. The index ellipsoid is also conceptually useful in finding the polarization directions for any arbitrary propagation direction in the crystal. The intersection of the plane normal to this direction passing through the origin and intersecting with the ellipsoid gives an ellipse. The directions of the major and minor axes of this ellipse are the two polarization directions. The waves propagating in the two directions are sometimes called *ordinary* and *extraordinary* rays.

GaAs and InP and compounds based on them are isotropic. However, they are also noncentrosymmetric crystals.[†] In other words, they lack inversion symmetry. The simplest manifestation of this is that the application of an electric field in a certain direction in the crystal is not equivalent to an electric field applied in the opposite direction. This induced birefringence is the *electro-optic effect,* which is the change in the refractive index of the crystal in the directions of the ordinary and extraordinary rays due to the application of an electric field. The field-dependent change in the refractive index can be expressed by the equation

$$\Delta\left(\frac{1}{n_r^2}\right) = r^l \mathbf{E} + s^q \mathbf{E}^2 \tag{11.8}$$

There are higher order (in **E**) terms, but these are not very significant. Here r^l is called the *linear electro-optic coefficient* and s^q is called the *quadratic electro-optic coefficient.* If the value of r^l is very large, the corresponding electro-optic effect is called the *Pockels effect.* If s^q is large and makes the quadratic term dominant, then

---

[†] These are cubic crystals of the $\bar{4}$3m symmetry group which have fourfold axes of symmetry along the cube edges [(100) direction] and threefold axes of symmetry along the cube diagonals [(111) directions].

the corresponding electro-optic effect is called the *Kerr effect*. The Pockels effect is accurately defined by the equation[†]

$$\Delta \left(\frac{1}{n_r^2} \right)_i = \sum_{j=1}^{3} r_{ij}^l E_j \tag{11.9}$$

where $i (= 1 - 6)$ are the six terms of the index ellipsoid under an applied field and $j (= 1 - 3)$ correspond to the three cartesian coordinates. The coefficient r_{ij}^l is the electro-optic tensor described by a (6×3) matrix. In crystals that have inversion symmetry, $r_{ij}^l = 0$. Symmetry considerations determine the zero or nonzero values of the different r_{ij}^l coefficients. For example, in materials such as GaAs, which crystallize in the zincblende structure, the only three and equal nonzero linear coefficients are $r_{41}^l = r_{52}^l = r_{63}^l$. The value of the linear electro-optic coefficient for GaAs and two other nonsemiconducting electro-optic crystals are listed in Table 11.1. It is immediately obvious that the electro-optic effect is weak in GaAs. This is also true for all III–V compounds.

TABLE 11.1 COMPARISON OF THE LINEAR ELECTRO-OPTIC COEFFICIENT OF GaAs WITH OTHER CRYSTALS.

Material	r_{ij}^l $(10^{-12}$ m/V$)$
GaAs	1.6
KH$_2$PO$_4$ (KDP)	10.6[1]
LiNbO$_3$ (lithium niobate)	30.8[2]

[1] r_{63}^l,
[2] r_{33}^l

We will first discuss the operation of the linear electro-optic modulator. A possible experimental configuration is shown in Fig. 11.8(a). The incident light of energy much smaller than the bandgap energy propagates through the crystal in the z-direction. The x and y axes are the principal dielectric axes and the wave is incident with its plane of polarization inclined at 45° to each of these axes. If the wave incident at z = 0 is described by $E = A_0 e^{j\omega t}$, then the polarized components along the two principal axes, x and y are each

$$E_x = E_y = \frac{A_0 e^{j\omega t}}{\sqrt{2}} \tag{11.10}$$

When an electric field is applied in the z-direction, the Pockels effect, in the limit of

---

[†]See A. Yariv, *Introduction to Optical Electronics*, 2nd ed., Holt, Rinehart and Winston, New York, 1976.

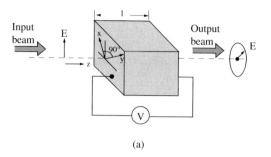

(a)

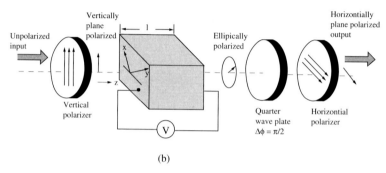

(b)

Figure 11.8 (a) Electro-optic phase modulation and (b) electro-optic amplitude modulation.

a small refractive index change, can be expressed as

$$\Delta\left(\frac{1}{n_r^2}\right) = -\frac{2\Delta n_r}{n_r^3} = r_{ij}^l E_z \qquad (11.11)$$

where E_z is the applied electric field in the z-direction. Equation 11.11 expresses the refractive index change along the principal axes and can be rewritten as

$$n_{rx} = n_{r0} + \frac{n_{r0}^3}{2}r_{ij}^l E_z$$

$$n_{ry} = n_{r0} - \frac{n_{r0}^3}{2}r_{ij}^l E_z$$

$$n_{rz} = n_{rE} \qquad (11.12)$$

where n_{rE} is the index in the z-direction. The two different refractive indices along the privileged directions (x and y) cause the two waves to travel with different propagation constants, which are described by

$$E_x = \frac{A_0}{\sqrt{2}}e^{j[\omega t - \frac{2\pi}{\lambda}n_{rx}z]}$$

$$E_y = \frac{A_0}{\sqrt{2}}e^{j[\omega t - \frac{2\pi}{\lambda}n_{ry}z]} \qquad (11.13)$$

Equation 11.12 also implies that the field in the z-direction creates an index ellipsoid with unequal axes in the x and y directions. In a sample of length l, the phase difference at the output plane between the two components is given by

$$\Delta\phi(l) = \frac{2\pi}{\lambda}(n_{rx} - n_{ry})l$$

$$= \frac{2\pi}{\lambda}r_{ij}^{l}n_{r0}^{3}V \tag{11.14}$$

where $V = E_z/l$ is the applied bias. The two waves, which are launched with their polarization (E-fields) orthogonal to each other, could be TE and TM modes. Therefore, the phase shift is also sometimes expressed as

$$\Delta\phi_{TE-TM} = \frac{2\pi}{\lambda}r_{ij}^{l}n_{r0}^{3}V \tag{11.15}$$

which is independent of l. The phase difference is also known as *electro-optic retardation*.

EXAMPLE 11.1
Objective.   To calculate the change in refractive index in GaAs for an applied electric field of 2×10^5 V/cm.

From Eqs. 11.11 and 11.12 and Table 11.1,

$$\Delta n_r = \pm\frac{1}{2} \times 1.6 \times 10^{-12} \times (3.6)^3 \times 2 \times 10^7$$

$$\text{or } |\Delta n_r| = 4.7 \times 10^{-4}$$

11.5.2 Electro-Optic Amplitude Modulation

It is clear from Eq. 11.13 that if the phase shift between the two components at the output is $\pi/2$, then the input linearly polarized wave is changed to a circularly polarized wave at the output. In this case, the two components at the output are

$$E_x = \frac{A_0}{\sqrt{2}}\cos\omega t$$

$$E_y = \frac{A_0}{\sqrt{2}}\cos\left(\omega t - \frac{\pi}{2}\right) = \frac{A_0}{\sqrt{2}}\sin\omega t \tag{11.16}$$

which describe a circularly polarized wave. Similarly, if the phase change is $\Delta\phi = \pi$, then the components at the output are

$$E_x = \frac{A_0}{\sqrt{2}}\cos\omega t$$

$$E_y = -\frac{A_0}{\sqrt{2}}\cos\omega t \tag{11.17}$$

which describes a linearly polarized wave, with its polarization vector oriented at 90°
to the input polarization vector. Now imagine the configuration shown in Fig. 11.8(b),
where the input polarizer launches a wave with a vertical polarization. Also, assume
that an output polarizer oriented at 90° (horizontal polarization) is placed in the path
of the output beam. If the voltage V is adjusted such that $\Delta\phi = \pi$, then when
$V = 0$, the output beam is blocked off, while when $V = V_\pi = \lambda/(2r_{ij}^l n_{r0}^3)$, the wave
is fully transmitted. Therefore, the arrangement serves as an electro-optic amplitude
modulator. It can be easily shown (Problem 11.3) that the ratio of the input and output
intensities, or the modulation index, is given by

$$\frac{I_{out}}{I_{in}} = \sin^2 \frac{\Delta\phi}{2}$$

$$= \sin^2 \left[\frac{\pi}{2} \frac{V}{V_\pi} \right] \tag{11.18}$$

This equation forms the basis of the Pockels electro-optic amplitude modulator. The
variation of transmission with voltage is shown in Fig. 11.9. It is clear that the
variation of transmission is nonlinear with voltage and for small V, from Eq. 11.18,
the transmission is proportional to V^2.

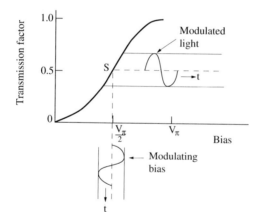

Figure 11.9   Principle of small-signal amplitude modulation with electro-optic modulator.

What we have just discussed is, in essence, large-signal or on-off modulation.
For small-signal modulation, the modulator is usually biased at some point, say S,
and a small-signal modulating voltage is superimposed. An added advantage is that
about the point S the transmission characteristics are fairly linear. The bias point
S is usually fixed by inserting a birefringent crystal in the path of the beam, which
introduces a $\pi/2$ phase shift between the components, at the input or output of the
modulator, as shown in Fig. 11.8(b). Such a component is called a *quarter-wave
plate*. The total phase difference between the components at the output of the output
polarizer is

$$\Delta\phi_{total} = \frac{\pi}{2} + \Delta\phi \tag{11.19}$$

From Eq. 11.18, the transmission in this case is given by

$$\frac{I_{out}}{I_{in}} = \sin^2\left[\frac{\pi}{4} + \frac{\pi}{2}\frac{V}{V_\pi}\right]$$

$$= \frac{1}{2}\left[1 + \sin\frac{\pi V}{V_\pi}\right] \tag{11.20}$$

For small V, in comparison to V_π, the change in transmission is linear with V. Thus, if a modulating voltage $V_0 \sin \omega t$ is applied to a modulator around the bias point $V_\pi/2$ determined by a quarter-wave plate, the transmitted intensity is given by

$$\frac{I_{out}}{I_{in}} = \frac{1}{2} + \frac{\pi V_0}{2V_\pi}\sin \omega t \tag{11.21}$$

where $\pi V_0/V_\pi$ should be $\ll 1$. The output intensity in this case is also linearly modulated at frequency ω. If the condition $\pi V_0/V_\pi \ll 1$ is not fulfilled, the output will contain higher-order harmonics of ω.

The electro-optic modulators that we have considered until now are called *longitudinal electro-optic modulators*, since the electric field is applied in the same direction as the optical beam. This scheme suffers from two disadvantages. First, the electrodes or ohmic contacts at the ends, which are difficult to form in this geometry, must transmit light. The contacts must therefore either be semitransparent, or have apertures in them. These can result in nonuniform transmission and losses. Practical modulators are therefore operated as *transverse electro-optic modulators* in which the electric field is applied normal to the direction of propagation. This is illustrated in Fig. 11.10(a). The beam propagates along the y privileged direction and its polarization is at an angle of 45° to the x privileged direction. The field is applied in the z-direction. The phase difference is given by

$$\Delta\phi = \phi_x - \phi_z \tag{11.22}$$

or

$$\Delta\phi = \frac{2\pi l}{\lambda}\left[(n_{r0} - n_{rE}) - \frac{n_{r0}^3}{2}r_{ij}^l\frac{V}{d}\right] \tag{11.23}$$

where d is the thickness of the crystal. Note that in this case $\Delta\phi$ contains a term which does not depend on the applied voltage. This is the built-in birefringence and can be disadvantageous if this term is much larger than the voltage-dependent term. In isotropic crystals such as GaAs $n_{r0} = n_{rE}$ and

$$\left|\Delta\phi\right| = \frac{\pi l}{\lambda d}n_{r0}^3 r_{41}^l V \tag{11.24}$$

Note that since the y direction is the direction of propagation, the phase shift given by Eq. 11.24 is smaller than that given by Eq. 11.14. A typical GaAs electro-optic

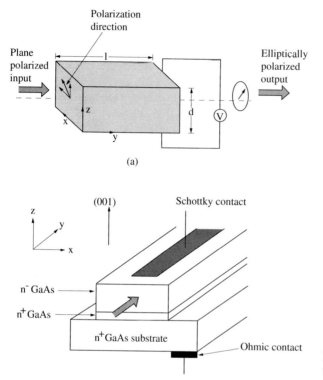

Figure 11.10 (a) Geometry of transverse electro-optic modulator and (b) schematic of a GaAs waveguide modulator.

modulator is shown in Fig. 11.10(b). Such ridge waveguide structures are fabricated by lithography, wet and dry etching, and appropriate contact formation after first cleaving the sample along the (011) directions. The bias is applied by means of a reverse-biased p-n junction or Schottky diode along the (001) (growth) direction. The optical beam is coupled into the guide with its polarization at 45° to the edges of the guide, so as to launch both TE and TM modes. The measured phase shift between these modes in a InGaAlAs/InP device is shown in Fig. 11.11. It is important to note that the electro-optic effect observed in zincblende crystals such as GaAs is much smaller than that observed in lithium niobate. It may also be noted that in a transverse modulator $\Delta\phi$ or Δn_r can be enhanced by increasing l and decreasing d. In other words, a long and thin structure is desirable. However, increasing the length of the device will increase the diode capacitance, which will degrade the high-frequency modulation capability, and increase the losses. If the thickness of the guide is reduced, the guide will again become lossy due to lack of mode confinement. In passing, it is important to realize that the phase shift or electro-optic retardation in a crystal such as GaAs is dependent on the crystallographic direction of the applied field. It is also useful to note that the electro-optic effect is very small in Si and Ge.

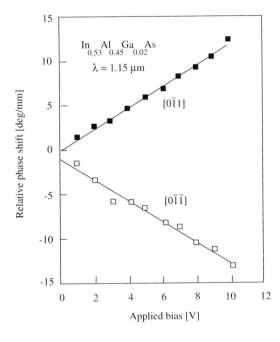

Figure 11.11 TE-TM phase shift for the orthogonal $[0\bar{1}1]$ and $[0\bar{1}\bar{1}]$ propagation directions in a $In_{0.53}Al_{0.45}Ga_{0.02}As$ waveguide electro-optic modulator measured by the author and co-workers (J. Pamulapati and P. Bhattacharya, *Applied Physics Letters*, **56**, 103, 1990).

EXAMPLE 11.2

Objective.   To calculate the value of V_π for a GaAs amplitude modulator when the wavelength of the incoming light is 1.1 μm. The waveguide is 1 μm thick and 1.5 mm long.

From Eq. 11.24

$$V_\pi = \frac{\lambda d}{l r_{ij}^l n_{r0}^3}$$

$$= \frac{1.1 \times 10^{-6} \times 10^{-6}}{1.5 \times 10^{-3} \times 1.6 \times 10^{-12} \times (3.6)^3}$$

$$= 11.8V$$

As in electro-absorption modulators, the ultimate modulation bandwidth of the electro-optic modulator is determined by internal and external parameters. The internal limit to modulation frequency is the plasma oscillation frequency, which in III–V compounds is ~ 100 GHz. The external limitations are imposed by the capacitance and series resistance of the device. Therefore, the device area has to be small and the series resistance resulting from the contacts has to be minimized. Sometimes the internal resistance of the modulating source itself is important. In other words, if f_0 is an average modulation frequency, then most of the applied voltage will drop across the source resistance R_s if $R_s > 1/2\pi f_0 C$. This problem is alleviated by using a

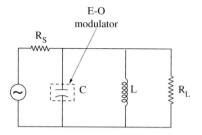

Figure 11.12   Biasing circuit for high-frequency modulation of electro-optic device.

circuit as shown in Fig. 11.12. The value of the inductance L is chosen to satisfy the condition

$$\omega^2 \;=\; 4\pi^2 f_0^2 \;=\; \frac{1}{LC} \tag{11.25}$$

such that at resonance ($f = f_0$) the impedance of the circuit is equal to the load resistance R_L, which is chosen to be much larger than R_s. This arrangement ensures that the modulating voltage drops across the modulator. The power needed to obtain a maximum or peak phase shift (Problem 11.6) is also an important consideration and gives an indication of the device efficiency.

11.5.3 The Quadratic Electro-Optic Effect: Quantum Well Modulators

In quantum wells, the electro-optic effect is different from that in bulk semiconductors, where the linear term in Eq. 11.8 is dominant. In a quantum well heterostructure, there will be a strong interaction of the electrical field with the optical wave. Close to the subband transition energies the absorption is dominated by excitonic effects and the electroabsorption is greatly enhanced. This was described as the quantum confined Stark effect (QCSE), which is a quadratic effect with respect to electric field. In other words, the transition energies vary quadratically with electric field. As a result, the absorption coefficient and refractive indices will vary quadratically with electric field. Thus, it is expected that the quadratic electro-optic effect will be pronounced in quantum wells. It is important to remember that the QCSE is an excitonic effect and therefore the energy of the light to be modulated should be close to the e_1–hh_1 transition energy. The quadratic electro-optic effect may also be observed in single-layer devices if the photon energy is very close to the bandgap energy.

By analogy with Eq. 11.12, the refractive indices in the directions of the ordinary and extraordinary rays can be written as

$$n_{rx} = n_{r0} - \frac{1}{2}n_{r0}^3 s_{12}^q E_z^2$$

$$n_{ry} = n_{r0} - \frac{1}{2}n_{r0}^3 s_{11}^q E_z^2 \tag{11.26}$$

where s_{11}^q and s_{12}^q are the quadratic electro-optic coefficients. Accordingly, the phase shift is given by

$$\Delta\phi \;=\; \frac{\pi}{\lambda}n_{r0}^3(s_{12}^q - s_{11}^q)E_z^2 l \tag{11.27}$$

If the linear electro-optic effect is also present, the total phase change is given by

$$\Delta\phi = \frac{\pi l}{\lambda}n_{r0}^3\left[r_{41}^l E_z + (s_{12}^q - s_{11}^q)E_z^2\right] \tag{11.28}$$

Here s_{11}^q and s_{12}^q are the electro-optic coefficients describing the index ellipsoid for the quadratic effect.

There is another important difference between bulk and quantum well materials. In bulk semiconductors there is no anisotropy and therefore there is no built-in birefringence. The layering of multiple materials with alternating indices of refraction will induce a birefringence with an extraordinary index of refraction, which is perpendicular to the layer plane. The mean refractive indices,

$$n_{r\parallel}^2 = \frac{d_1 n_{r1}^2 + d_2 n_{r2}^2}{d_1 + d_2} \tag{11.29}$$

and

$$n_{r\perp}^2 = \frac{n_{r1}^2 n_{r2}^2 (d_1 + d_2)}{d_1 n_{r2}^2 + d_2 n_{r1}^2}, \tag{11.30}$$

can be obtained by examining the boundary value problem for plane waves, which relate to the polarization of the optical electric field vector parallel or perpendicular to the quantum well. In Eqs. 11.29 and 11.30, n_{r1} and d_1 refer to refractive index of the well material and the thickness of the well, respectively. Likewise, n_{r2} and d_2 refer to the index of refraction of the barrier material and the barrier thickness, respectively. For a multiple quantum well, d_1 is replaced by $N_z d_1$, where N_z is the number of wells. Similarly d_2 is replaced by $N_b d_2$, where N_b is the number of barriers. The built-in birefringence in the MQW can be expressed as

$$\Delta\phi_0 = (\beta_{TE} - \beta_{TM})l = \Delta n_r l\left(\frac{2\pi}{\lambda}\right), \tag{11.31}$$

where

$$\Delta n_r = n_{r\parallel} - n_{r\perp}. \tag{11.32}$$

The total phase shift in a MQW can therefore be expressed as

$$\Delta\phi = \Delta n_r l\left(\frac{2\pi}{\lambda}\right) + \frac{\pi l}{\lambda}n_{r0}^3\left[\mp r_{41}^l E_z + (s_{12}^q - s_{11}^q)E_z^2\right] \tag{11.33}$$

Further changes in the electro-optic properties are expected if the quantum wells are biaxially strained. As discussed in Chapter 1, such strain changes the energy separation and coupling between the heavy-hole and light-hole excitonic transitions. In the case of biaxial compressive strain, the heavy-hole absorption dominates the absorption edge, while in the case of tensile strain the light-hole absorption dominates. Since the TE mode couples to the heavy hole, while the TM mode couples to the light hole, the field-dependent HH-LH splitting is expected to play an important role in determining the electro-optic properties. Experimental results obtained from both effects, quantum confinement and strain, are depicted in Fig. 11.13. Variation of the In composition in the well allows the quantum wells to be under biaxial compressive or tensile strain. The waveguide configuration allows a long interaction length, whereby the multiple quantum wells provide adequate modulation of the electro-optic coefficients to observe significant changes. When the photon energy is much smaller than the band-edge absorption energy, the device behaves as a linear electro-optic modulator, as with bulk semiconductors. However, as the photon energy approaches the bandedge absorption energy, excitonic effects dominate, and the quadratic electro-optic effect becomes important. This is seen in Fig. 11.13, where the photon energy is 15 meV below the absorption edge. Another striking and important result displayed in this figure is that the electro-optic effect is enhanced in compressively strained quantum wells and suppressed in tensilely strained wells, compared to lattice-matched ones. This is shown to be true from the electroabsorption behavior of these materials in Appendix 10. For III–V semiconductors oriented in the (111) direction, application of biaxial strain induces a piezoelectric field in the same direction, the effect of which is to change the electro-optic properties.

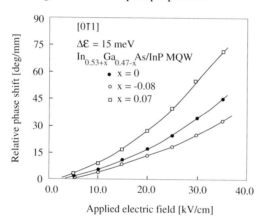

Figure 11.13 Phase shift as a function of applied bias for lattice-matched and biaxially strained InGaAs/InP MQW electro-optic modulators measured by the author and co-workers (J. Pamulapati et al., *Journal of Applied Physics,* **69,** 4071, 1991).

Some differences between electro-absorption and electro-optic modulators may now be obvious. Electro-absorption modulators draw a photocurrent, while electro-optic devices do not. The photocurrent can represent a heat dissipation problem. Temperature effects have to be considered, since changes in temperature will also alter the refractive index of a material. Also, since electro-absorption modulators operate near the semiconductor bandedge, absorption losses are higher and the insertion loss of guided-wave devices tends to be high. It may also be noted that the quadratic electro-optic effect due to electro-absorption leads to chirping effects in modulation.

11.5.4 Modulation by Carrier Injection

In addition to the modulation principles outlined above, p-n junction carrier-injection modulators have also been demonstrated. In these devices the injection of free carriers changes the refractive index. This effect can be quite pronounced if the injection current is of the order of a few tens of millamperes. Carrier injection can also be used in a quantum well to quench the band-edge absorption. This is done in a device called the Barrier Reservoir and Quantum Well Electron Transfer (BRAQWET) modulator.[†] Carrier injection into the quantum well is usually done by transfer- or modulation-doping from an adjacent large bandgap doped semiconductor with the help of an applied bias across a n-i-n structure. The injected carriers give rise to band-filling effects and results in a blue shift of the absorption edge. The resulting change in refractive index and phase modulation can be quite large even for small applied voltages. It is important to note that while in the QCSE modulator the oscillator strength is only moved spectrally, in this device the existing oscillator strength is switched on and off completely.

11.6 OPTICAL SWITCHING AND LOGIC DEVICES

11.6.1 Introduction
Interest in "optical computing" has been around for a few decades and has gone through crests and troughs. Optical computing using lenses and Fourier transforms have been demonstrated. However, versatile optical computers that can replace, or even effectively compete with electronic computers remain elusive. Some demonstrations of optic logic, utilizing the nonlinear effects in multi-quantum wells have been made. However, these schemes run into problems, since enormous amounts of switching energy are required and heat dissipation becomes an important issue when arrays of such devices on a single wafer are considered for computing applications. There is a need for low-power photonic switching devices, which can be used effectively for computing and logic applications. In this respect, devices based on the QCSE seem to hold promise. This effect allows one to tailor the device response suitably for the demonstration of simple, optically controlled switching and logic devices, which are next described.

11.6.2 Self-Electro-Optic Device
In addition to QCSE, it is important to understand the photocurrent behavior of the p-i(MQW)-n diode described in Sec. 11.4. Since the QCSE involves a quadratic Stark effect, large electric fields are necessary for any useful shift of the absorption edge (e.g., $E \sim 70$ kV/cm is needed for a 15 meV shift of the heavy-hole peak in a 1 μm GaAs/AlGaAs MQW with 100 Å wells). At the correspondingly large bias values, the MQW diode also behaves as an optical detector. Figure 11.14 shows a schematic of the photocurrent response of a p-i(MQW)-n structure as a function of applied bias. Optical detection in a MQW involves (a) absorption of the incident photons, (b) recombination of some of the photoexcited electron-hole pairs, and (c) tunneling of electrons and holes through the barriers in opposite directions and their collection at the contact regions to generate an external photocurrent. At low fields, the tunneling rate is small and the recombination process dominates, resulting in a very small

---

[†]M. Wegner et al., *Appl. Phys. Lett.*, **55**, 583–585, 1989.

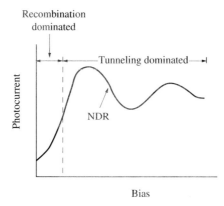

Figure 11.14 Photocurrent-voltage characteristics of p-i(MQW)-n diode.

photocurrent. At higher fields, the tunneling rate is enhanced and the photocurrent follows the absorption spectra of the MQW. The internal quantum efficiency of a MQW structure has been shown to be approximately unity at fields above 10 kV/cm at room temperature. The simultaneous role of an optical detector and a modulator is a unique electro-optic property of an MQW structure. At appropriate wavelengths, a strong negative differential resistance (NDR) region is observed in the photocurrent versus bias voltage relationship of a p-i(MQW)-n structure. This arises because the photocurrent results from the change in absorption coefficient of the MQW due to QCSE. The NDR occurs where the heavy-hole (HH) and the light-hole (LH) peaks cross the photon energy of the input light. As long as the electron and hole densities in the quantum wells are less than 10^{11} cm^{-2}, the absorption process is essentially linear.

The potential of the negative resistance region of the photocurrent-voltage characteristics has been exploited to develop a number of photonic switching and logic devices. The first and most important of these is called the Self-Electro-optic Effect Device (SEED).[†] This device exhibits photonic switching, bistability, and optically induced oscillations due to the negative differential resistance in the photocurrent. The basic SEED circuit with a series resistor is shown in Fig. 11.15(a). The switching action is demonstrated in Fig. 11.15(b). When the light intensity changes from $\Im_2$ to $\Im_1$, the voltage across the device shifts from V_1 to V_2, causing a transmittance change from T_1 to T_2. Note that this large voltage change, with respect to the optical power change, cannot be achieved without using MQW excitonic transitions. Another important point to note is that the transmission through the device is also changed by almost a factor of two at the same time. Therefore, the device provides an integrating and thresholding capability.

The general principle of the SEED is that the photocurrent flowing through the circuit, including the series resistor, changes the voltage across the modulator, which in turn influences its absorption and transmission. As a consequence the photocurrent is changed. The photonic switching operation is understood in terms of the character-

---

[†]D. A. B. Miller et al., *IEEE J. Quantum Electronics*, **QE-21**, 1462, 1985.

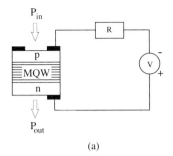

(a)

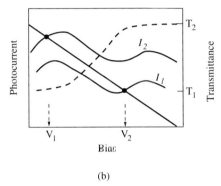

Bias

(b)

Figure 11.15   (a) SEED circuit with feedback resistor and (b) its excitonic switching property.

istics shown in Fig. 11.16. Light of energy lower than the HH resonance in the absorption spectrum is incident on the device. Since the diode is typically reverse-biased, the HH peak is red-shifted to lower energies, and most of the light can be transmitted if its energy coincides with the low absorption region between the HH and LH resonances. Therefore, for low-input power, most of the light is transmitted and the output power increases in proportion to the input power. As the light intensity increases, the

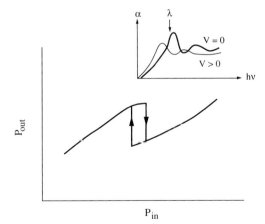

Figure 11.16   Schematic illustration of photonic switching characteristics. The inset shows the light energy with respect to the heavy-hole exciton peak.

photocurrent increases and the voltage drop across the series resistor will increase. Since the bias voltage remains constant, the reverse bias across the diode decreases, which shifts the HH absorption peak to higher energies and the transmission drops. As the input power increases further, the output power will increase again. Thus, the state of the device is altered solely by light intensity. Such photonic switching can also be illustrated with two beams, one for transmission and one for control. The hysteresis observed in the characteristics of Fig. 11.16 is due to the asymmetric shape of the absorption resonances. The feedback due to the resistor is truly opto-electronic.

11.6.3 The Bipolar Controller-Modulator

In the SEED, the path and effects of the signal and control beam are the same and are therefore indistinguishable in their internal effects. Furthermore, in order to make the SEED more compatible with the optical power levels available in optoelectronic integrated circuit (OEIC) technology, it is important to have gain in the circuit. Gain is also essential for larger tolerance in the devices as well as large fan-out and cascadability. One scheme that can be implemented to achieve gain is to connect a bipolar phototransistor in series with the SEED. The control or switching beam is incident on the transistor, while the signal is incident upon and transmitted through the SEED, which functions as the modulator. However, due to the requirement of current continuity the modulator current needs to be high, since the transistor current is usually quite high. This can only be achieved by using the p-i(MQW)-n modulator in waveguide form.

Gain can also be realized by using a heterojunction bipolar transistor (HBT) with a MQW in the base-collector region. This device provides a number of advantages. Since a transistor operates vertically, a large uniform transverse electric field can be applied to the collector-base junction to cause the QCSE. The base terminal provides extra controllability for efficient optical and electronic coupling. Incorporation of the MQW in the collector region effectively allows the realization of a p^+-i(MQW)-n^+ modulator by selective etching of the emitter and so the control signal on the HBT and the information signal (to be modulated) on the modulator can be physically separated. More importantly, the entire structure of the n-p^+-i(MQW)-n^+ MQW-HBT and the p^+-i(MQW)-n^+ modulator can be realized by single-step epitaxy. The schematic of the integrated MQW-HBT along with its equivalent circuit are shown in Fig. 11.17. The transistor amplifies the photocurrent generated in the MQW and provides a voltage feedback to the MQW-HBT by changing its collector-base voltage. The modulator and controller are connected in parallel, and the load is connected in series with the controller and the modulator. The signals detected in both the controller and the modulator give a feedback to themselves via the load, and the absorption of light in both the controller and the modulator is altered. The parallel connection of the controller and the modulator allows the sum of the input signals in these devices to control the modulation of light. It is conceivable that the p-i-n structure could also form a laser, which would become important in the design of OEICs to be discussed in the next chapter. The structure is therefore very compatible with OEIC applications. An important point to realize is that amplification of the photocurrent by transistor action allows low-power photonic switching.

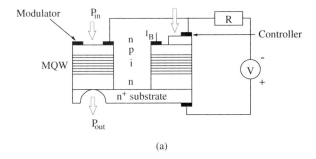

(a)

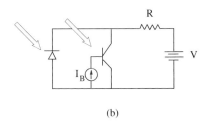

(b)

Figure 11.17 (a) Schematic illustration of integrated controller-modulator and (b) its equivalent circuit.

Figures 11.18(a) and (b) show the measured output characteristics for two different switching conditions. In the figures the load lines are also shown. Figure 11.18(a) shows the collector current for different values of optical power at zero base current, while Fig. 11.18(b) shows the results for different base currents at fixed optical power. Thus, efficient switching can be carried out by optical or electronic signals. The important point to realize is the presence of amplification of photocurrent, which provides for higher sensitivity to light and reduced load resistance. Another point is that the I–V curves can be shifted either by optical power or base bias. Also, the nonlinear gain characteristic of the HBT makes the negative differential resistance stronger than in the simple p-i-n structure.

It is possible to conceive of two classes of applications for the integrated controller-modulator device. In one application the change in light input in the controller would alter the light passing through the modulator. The modulated light would then be an input to the next controller stage and the process could be continued. This application will require careful optical alignment of one modulator to the next stage controller and would be very useful for special optical computing architectures. Another class of application would involve an on-off optical signal to simply change the optical state of one or more modulators connected to it. The choice of the photon energy with respect to the HH exciton peak energy is usually different in the two cases. Some special functions that can be realized by the controller-modulator circuit are now briefly described.

11.6.3.1 Optoelectronic Amplification.

In the use of the controller-modulator (C-M) circuit for amplification of an optical signal, a photon energy approximately 15 meV below the exciton peak at zero bias is usually chosen. The transmittance-

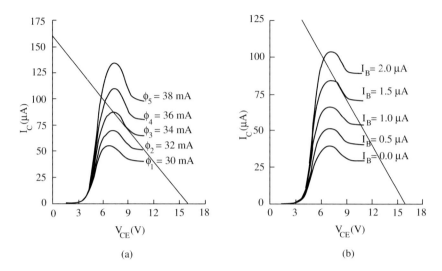

Figure 11.18   Collector current versus collector-emitter characteristic of GaAs/AlGaAs MQW-HBT; (a) at $I_B = 0 \mu A$, for different illumination levels. The illumination (ϕ) is indicated by the diode laser drive current in mA; and (b) for different base bias at constant illumination ($\phi = 27$ mA). The measurements have been made by the author and co-workers (S. Goswami, *A Quantum Well Phototransistor for Switching Applications*, Ph.D. thesis, University of Michigan, 1992).

voltage curve for this choice is shown in Fig. 11.19 along with the circuit configuration. The large gain of the controller will allow a small change in the controller input intensity to produce a large change in the modulator output. Figure 11.19 also shows the modulated light P_m through the modulator as a function of incident power P_i through the controller. The results are shown for two values of the base current. As can be seen, the improved gain resulting from the larger base current allows a larger amplification. Such optoelectronic amplification can find important applications in optical systems.

11.6.3.2 Programmable Memory Device.   The MQW-HBT can be operated as a simple flip-flop programmable memory element. The controller-modulator, constituting a single memory cell, can be conceived to form an element of a larger two-dimensional array. For its operation, the individual cell or entire array needs to be illuminated with a constant and uniform low-level photon flux with a wavelength that gives the desirable photocurrent-voltage characteristics. The operating principles of the device are schematically shown in Fig. 11.20(a). At a photon flux ϕ_{op}, when the base current is I_B^0, the photocurrent-voltage curve provides two stable points for the load line. The high-voltage point V_H also corresponds to high transmittance through the MQW region, and the low-voltage stable point at V_L corresponds to a low transmittance. If the base current is made near-zero, the load line has only one stable point at A and when the base current is restored to I_B^0, the stable point at bias V_H is

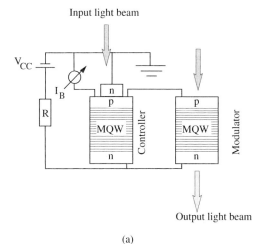

(a)

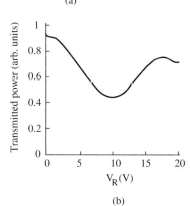

(b)

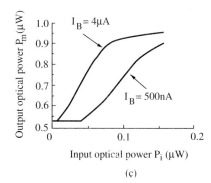

(c)

Figure 11.19 Demonstration of optoelectronic amplification with GaAs/AlGaAs MQW-HBT: (a) circuit configuration, (b) transmittance-voltage curve for the wavelength of light used, and (c) measured amplification of input light. The experiments were done by the author and co-workers (S. Goswami et al., *IEEE Journal of Quantum Electronics*, **27**, 760, ©1991 IEEE).

set. If the base current is made higher (I_B^1), there is again only one stable operating point at B. Now, when the holding base current I_B^0 is restored the low-voltage point V_L is chosen. The state of the device can thus be efficiently altered and maintained by the base current. This device is fully compatible with HBT digital technology

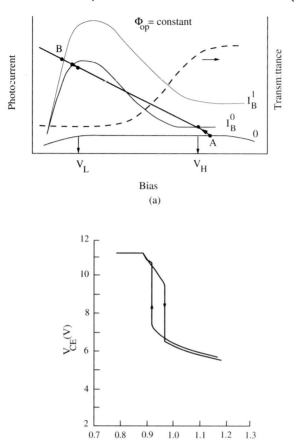

Figure 11.20   (a) Principle of a MQW-HBT flip-flop programmable memory element and (b) experimental demonstration of holding with a GaAs/AlGaAs MQW device (W. Q. Li, et al., *Electronics Letters*, **27**, 31, 1991).

and only requires a constant uniform optical illumination. Also, it is very simple in that it requires only one transistor. It can also be read either electronically through the voltage levels or optically by the transmittance through the MQW region. The bistable flip-flop characteristic, demonstrating the switching and holding behavior of a GaAs/AlGaAs MQW device is shown in Fig. 11.20(b). For the sake of comparison it may be noted that an all-electronic bistable flip-flop circuit requires at least four transistors.

11.6.3.3 Tunable Threshold Logic Gate.

Figure 11.21(a) shows the circuit diagram of an MQW-HBT threshold gate. For operation of the gate the wavelength of the light should be above the excitonic peak, so that the complete heavy-hole response appears in the photocurrent spectrum of the MQW-HBT. The current in the

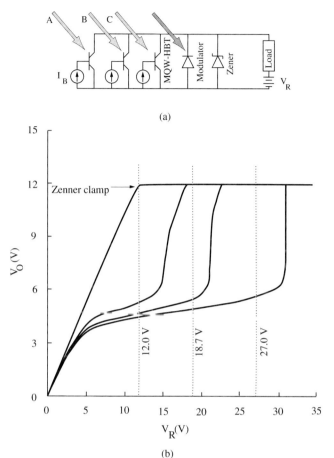

(a)

(b)

Figure 11.21 (a) Circuit diagram of a thresholding gate with three controllers and a modulator and (b) output characteristics of the gate (S. Goswami et al., *IEEE Journal of Quantum Electronics*, **28**, 1636, ©1992 IEEE).

load resistor is a superposition of the currents in the MQW-HBTs. The p-i(MQW)-n diodes or modulators in parallel with the MQW-HBTs are used to modulate light of the same wavelength. The photocurrent of the modulators also passes through the load, but the electronic gain of MQW-HBTs ensures that this photocurrent is much smaller than that of the MQW-HBT. Figure 11.21(b) shows the experimental output characteristics for a 3-input threshold gate. A suitable load resistance with a variable voltage supply is used. The Zener diode connected in parallel with the MQW-HBTs prevents damage from accidental high voltages. A tunable laser is used as the light source. When none of the MQW-HBTs are illuminated the output voltage increases linearly with the supply voltage, as there is no current in the circuit and hence no potential drop across the load. From Fig. 11.21(b) it is seen that for a 12 V supply the output is 1 when none of the inputs are 1 (illuminated), but becomes 0 when any of them is 1. This is the NOR function (Table 11.2). For a supply voltage of 18.7 V, the circuit is an INVERSE CARRY gate as the output is 0 when two or more of the inputs are 1, and otherwise the output is 1. When the

TABLE 11.2 THE TRUTH TABLE FOR THE MQW-HBT SWITCHING GATE AT DIFFERENT VOLTAGES.

State #	Inputs			Logic Functions		
	A	**B**	**C**	**NAND (V3)**	**ICARRY (V2)**	**NOR (V1)**
0	0	0	0	1	1	1
1	0	0	1	1	1	0
2	0	1	0	1	1	0
3	1	0	0	1	1	0
4	0	1	1	1	0	0
5	1	0	1	1	0	0
6	1	1	0	1	0	0
7	1	1	1	0	0	0

supply voltage is about 27 V the gate output is 0 whenever all of the inputs are 1. Hence, the gate now performs the NAND operation. Thus, the function of the gate can be altered simply by changing the supply voltage. In addition, the access to the base terminal of the MQW-HBT allows fine-tuning of the photocurrents in the MQW-HBTs so that a small optical misalignment can be compensated by electrical means.

11.6.4 Switching Speed and Energy

Optics can provide, among other advantages, massive parallelism and large connectivity. Optical processing holds the promise of reconfigurable interconnects. Optical interconnects also show a great potential for being programmable. In other words, light beams can be directed from individual sources to individual detectors on the basis of a specific command, depending on the task. This can be realized by the generalized crossbar switch shown in Fig. 11.22. In this architecture all possible inputs can be connected to all possible outputs via the two-dimensional switch array. Each laser from the one-dimensional source array would illuminate an entire

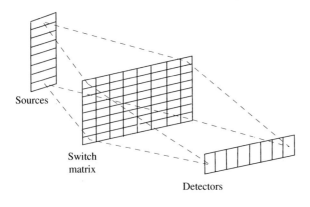

Sources

Switch matrix

Detectors

Figure 11.22 Optical crossbar switch with one-dimensional arrays of sources and detectors and a two-dimensional switch array.

column of the two-dimensional array. Each detector from the one-dimensional detector array would receive from an entire row of the two-dimensional array. Each
element of the optical-switching array could be an optical or opto-electronic switch,
such as the SEED or C-M device discussed in the previous section. The transmission of each individual switch could be programmed in real time, based on information coming in sequentially or in parallel. An array of SEED devices is shown in
Fig. 11.23.

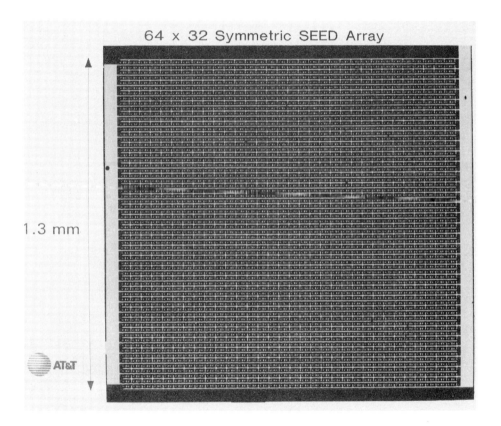

Figure 11.23   A 64 × 32 array of SEED devices (courtesy of D. A. B. Miller,
AT&T Laboratories, Holmdel, NJ).

In a switching device the important operational parameters are switching time
(or speed) and switching energy. The switching power is obtained by dividing the
switching energy by switching time. The basic operating speed of a bipolar device
is limited by the gain-bandwidth product. The gain required in these bipolar devices
is not very large. Current gains of 10–20 are deemed sufficient. The bandwidth is
then principally limited by the device capacitance, carrier transit times, and the base
spreading resistance. In a simple transistor design with optimized parameters, the

temporal response time, limited only by the RC time constant, is 100–500 ps, which is probably adequate for most logic applications, particularly in parallel architectures. The switching energy of the bipolar controller-modulator circuit is estimated to be 100 fJ. This optical switching energy can be reduced, at the expense of electrical energy, by increasing the base current. The switching energy can also be reduced by improved device design.

Only one type of modulation and switching device, based on the QCSE in quantum wells, has been described here. Many other kinds of all-optical and opto-electronic switching devices can demonstrate similar properties. Examples are bistable optical devices (BODs), also based on MQW phenomena, LEDs, junction devices, and many others. Key parameters for the success of any particular technology are switching energy and switching time. In many applications, where large arrays of such devices

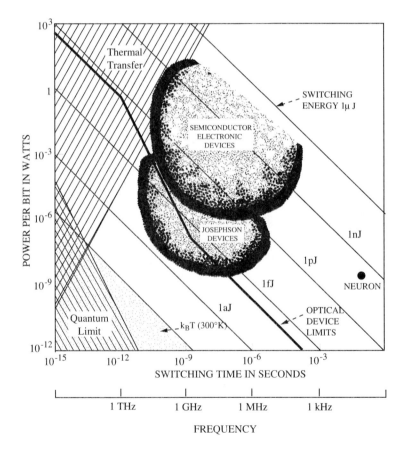

Figure 11.24 Switching powers and times for different technologies (from P. W. Smith, *Bell Syst. Tech. Journal*, **61**, 1975, 1982. Copyright ©1982 AT&T. All rights reserved. Reprinted with permission).

are required, power dissipation problems may eventually preclude a particular technology. As a concluding note, the switching powers of electronic and optical devices are shown in Fig. 11.24. It is clear that Josephson Junction devices can still outperform most technologies. It is also of interest to see how these devices compare with the *neuron*—a biological switching device.

PROBLEMS

11.1   A wave propagating in the z-direction in an anisotropic crystal has field components E_x and H_y only. Starting from Maxwell's equations, show that the propagation constants are different in the x- and y-directions.

11.2   Describe, with equations if necessary, the linear and quadrate electro-optic effects in a compound semiconductor.

11.3   Derive Eq. 11.18, where the symbols have their usual meanings.

11.4   Discuss means by which the built-in birefringence may be enhanced in an electro-optic modulator. In (111)-oriented crystals such as GaAs, the applications of biaxial strain, say by alloying with In, induces a built-in piezoelectric field in the (111)-direction. Design a modulator based on this principle and explain the expected results.

11.5   An electro-optic modulator is made with a GaAs/AlGaAs multiquantum well (MQW). The MQW forms the i-region of a p^+-i-n^+ diode grown on (100) GaAs. Describe and explain the expected $\Delta\phi$–V characteristics as the incident light wavelength is changed from 1.1 μm to 0.85 μm.

11.6   The maximum modulation bandwidth of an electro-optic modulator is given by $\Delta f = (2\pi R_L C)^{-1}$ where R_L is the load resistance of the circuit and C is the device capacitance. Show that the power needed to obtain a peak retardation $|\Delta\phi|_m$ is given by

$$P = \frac{|\Delta\phi|_m^2 \lambda^2 \epsilon A \Delta f}{4\pi l n_{r0}^6 (r_{ij}^l)^2}$$

where $n_{r0}l$ is the length of the optical path in the crystal, A is the cross-sectional area of the crystal normal to l and ϵ is the dielectric constant at the modulation frequency.

11.7   Two p-i(MQW)-n modulators are aligned such that the light incident on and transmitted by the first modulator is also transmitted through the second modulator. The first modulator has a tunable voltage V_1 applied to it. The second modulator has a bias V_2 applied to it and has a feedback resistor connected in series to the voltage supply and the modulator. If a fixed-intensity light beam is incident on the first modulator, plot the voltage across the second modulator, V_0, as a function of V_1. Discuss some possible applications of the characteristics that you obtain.

READING LIST

HINTON, H. S., and MIDWINTER, J. E. eds. "Photonic Switching," in *Progress in Lasers and Electro-Optics* (series ed. P. W. E. Smith), IEEE Press, New York, 1990.

WILSON, J., and HAWKES, J. F. B. *Optoelectronics: An Introduction*, Prentice Hall International, Englewood Cliffs, NJ, 1983.

YARIV, A. *Introduction to Optical Electronics*, 3rd ed., Holt, Rinehart and Winston, New York, 1985.

12

Optoelectronic Integrated Circuits

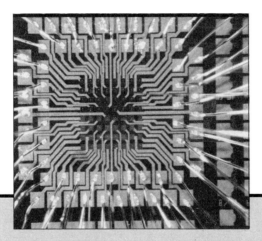

Chapter Contents

12.1 INTRODUCTION

Optoelectronics and optoelectronic devices have made tremendous progress in the past two decades. In the preceding chapters, we read about the materials that are used to make these devices and their properties. We also learned about the operating principles of the discrete devices themselves. Over the years, the role of these devices has changed from display devices and sensors to active components in information handling systems. The latter includes optical communication systems and computing systems. Before proceeding further, it is perhaps constructive to examine the role of photonics and electronics. Optics provides the inherent advantages of large bandwidth, parallelism, and reconfigurable configurations. However, optics does not provide input-output isolation, as electronic devices do, and it can be difficult to focus multiple beams in a parallel system. It is therefore logical to *couple* electronic and photonic devices, resulting in *optoelectronic integration.*

An important aspect of both optical communication systems and computing systems is the interconnect medium. The performance of conventional electrical interconnects is adversely affected by increases in reactance and reflections due to impedance mismatch at the higher frequencies. Multilevel board technology is being developed, where chip-to-chip communication is achieved through via-holes in the wafers. Even these fail to perform as operating frequencies approach the gigahertz range. An attractive alternative, therefore, is an optical interconnect medium, which can take the form of free space, integrated optical waveguides, or optical fibers. Optical interconnects and transmission media provide, in addition to large bandwidth and high-speed data transmission, immunity to mutual interference and crosstalk and freedom from capacitive loading effects. The large bandwidth translates eventually to system size reduction, reduced system power, and increased fanout capabilities.

Optoelectronic integrated circuits (OEIC) therefore involve the integration of electronic and optical components and optical interconnects (Fig. 12.1). It is expected that the monolithic integration of electronic and optical devices on the same chip will lead to high-speed, high-sensitivity, compactness and reliability, all at a low cost. Successful integration architectures pose many challenges. First the component devices, which may have different layer structures, unlike electronic integrated circuits, must be of very high quality. Second, there is the question of compatibility and impedance matching between the devices and the devices with the interconnect media. Last, but not the least, there is the question of cost, which becomes an important factor in systems design and realization.

12.2 NEED FOR INTEGRATION: HYBRID AND MONOLITHIC INTEGRATION

An optoelectronic device is a good example of the complementary and collaborative role of electrons and photons to perform a single function—emission or detection.

Photograph on page 465 is of an individually addressable 8×8 vertical cavity surface-emitting laser (VCSEL) array (courtesy of S. Swirhun, Bandgap Technology Corporation, Broomfield, CO).

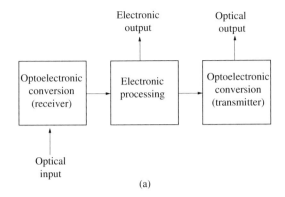

(a)

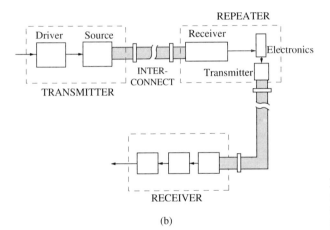

(b)

Figure 12.1 (a) Block diagram showing the essential elements of a OEIC and (b) layout of an optical-fiber link.

Electrons and photons interact effectively in a direct bandgap semiconductor to produce optoelectronic conversion. As mentioned in the previous section, photons and electrons have their own strengths and weaknesses. As in any optoelectronic device where electrons and photons are involved in producing the device characteristics, it is possible to envision, in a larger sense, an optoelectronic system in which multiple functions are separately performed by electronic and optoelectronic devices. Such a system, by analogy with the integrated circuit (IC), can be called an OEIC. Electronic functions such as switching or amplification can be combined with detection and also light transmission in such an integrated chip. Before proceeding further, it is of interest to entertain the nontrivial question: Is such integration necessary? The answer is not simple and is very application dependent. The need for integration arises from a variety of needs; speed and bandwidth, functionality and multifunction capabilities, compactness, low parasitics, to name a few. It should be cautioned, however, that such systems may never achieve the density of very large scale integration (VLSI) systems used in Si-based electronics. The need for such large-scale integration may not arise.

Nonetheless, it is clear that the OEIC will play a pivotal role in the development of future optoelectronic systems.

There are two principal forms of optoelectronic integration—hybrid and monolithic. In hybrid integration, as the name suggests, discrete devices on separate functional blocks or chips are connected using electronic (leads) or optical (fiber) interconnects. An example is the integration of a junction laser with its driver circuit consisting of a bipolar transistor to form a transmitter. An advantage of such hybrid integration is the possibility of using high-performance discrete devices as components. The disadvantages are lack of compactness and enhanced parasitic effects in terms of interconnects, bonding and lead wires. Parasitics are considerably reduced in "flip-chip" bonding, in which two chips containing component devices and circuits are interconnected by indium bumps. In monolithic integration all active and passive components are fabricated on the same chip. However, unlike silicon ICs in which almost all the parts are made with the same materials and same processing steps, the heterostructures and processing steps of the different components of a OEIC can be different. This makes the realization of high-performance monolithic OEICs a real challenge. One stands to gain in such monolithic integration in terms of size reduction, reduction of parasitics and the consequent achievement of higher-circuit speed and bandwidth.

Monolithic integration can be achieved in either a vertical or horizontal configuration. In the vertical scheme, electronic and optical device structures are epitaxially grown sequentially with an isolation layer in between. This is schematically shown in Fig. 12.2(a). An obvious disadvantage of this scheme is the lack of planarity. The horizontal scheme of monolithic integration can be classified into the planar compatible type, shown in Fig. 12.2(b), where both devices are made from the same heterostructure, or planar regrown shown in Fig. 12.2(c), where one of the devices is selectively regrown after growth of the first device. Although this technique provides a large freedom in the choice of device heterostructures, the regrown interface can have a large density of traps and other electrically active defects that can affect the performance of the regrown device.

12.3 APPLICATIONS OF OPTOELECTRONIC INTEGRATED CIRCUITS

It is quite possible that a majority of signal-processing functions in communication and computing applications will be performed by electronic circuits. There are important advantages in doing this. Silicon-based microelectronics technology provides the advantages of low-cost, large-scale integration and ruggedness. However, photonic devices and circuits can serve unique functions that are complementary to those provided by the electronic counterparts. Also, new functional capabilities can emerge by integrating electronic and photonic devices and circuits.

The first requirement for successful integration is the development of high-performance discrete devices with minimized parasitic effects to ensure the desired high-frequency performance, if required. Such devices, both electronic and optoelectronic, such as transistors, lasers, detectors, etc., are available. These ensure a good de-

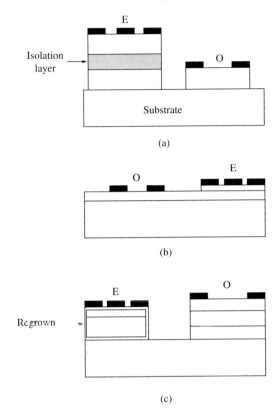

Figure 12.2 Schemes of monolithic integration: (a) vertical integration, (b) planar-compatible scheme, and (c) planar scheme using regrowth.

gree of success for hybrid integration. Monolithic integration however, presents a unique set of challenges. Due to dissimilar material requirements, the individual devices may have to be fabricated with nonoptimal materials and lattice-mismatched heteroepitaxy or be realized by epitaxial regrowth. Interfacial and bulk defect densities can therefore become important considerations. Interconnection and interfacing of the electronic and optoelectronic devices are important issues, which depend closely on the integration architecture used—vertical or planar. Some devices may even have to be fabricated differently. For example, the facets of normal, edge-emitting lasers are created by cleaving. This technique cannot obviously be used in integrated circuits, where the laser is coupled, *on-chip*, with a passive or active component. In-situ dry-etching techniques have to be developed to create the laser facets. OEICs can suffer from several types of cross-talk effects—electrical, thermal, and optical. The last one becomes important if the OEIC chip contains a light source as a component device. Finally, the OEIC chip has to be interfaced with other chips or the transmission media. This involves interconnection with electrical circuitry and with optical fibers. Each of these areas requires an intense research and development effort, much of which is already under way. If most of these problems can be solved, the applications listed below stand to gain from the compactness, low cost, ruggedness, and superior performance of optoelectronic integrated circuits.

One of the primary areas where photonic devices and OEICs will make an impact is telecommunications, driven by the large bandwidth and light weight of optical fibers. The object is to bring fiber systems to the home and individual subscribers in the form of telephone links and broadcast cable TV. This implies that optoelectronic technologies have to be extended to the subscriber loop. These systems will necessitate the development of lasers with precise frequency control and tunability and wavelength-selective detectors and receivers. It is envisaged that data transmission rates of several tens of gigabits/sec will be attained in these circuits and systems. Similar requirements also hold for optical computing systems. However, since links over shorter distances are needed for these applications (Fig. 12.3), it is possible to use GaAs-based technology without incurring too much of losses in the optical fibers. Since such systems are essentially digital systems, the bit error rates (BER) must be very low ($< 10^{-15}$) even in a noisy environment and they should have high reliability and redundancy in critical paths with transmission error correcting capability. Once again, architecture and interfacing of active and passive components are critical issues.

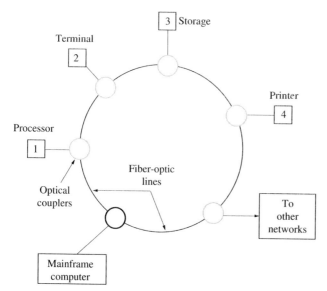

Figure 12.3 Elements of a local-area network (LAN) using optical interconnects.

Optoelectronic integration is also important for radar applications. The schematic of a microwave or millimeter wave phased-array radar system is shown in Fig. 12.4. In principle, a phase-shifted and modulated (high-frequency) optical signal is injection locked to a free-running microwave, or millimeter-wave oscillator, which forms one element of a phased-array radar. Beam steering in such radars is achieved by incorporating a progressive phase shift $\Delta\phi$ between successive elements. Instead of using phase shifters at microwave frequencies, which are bulky and cumbersome, it

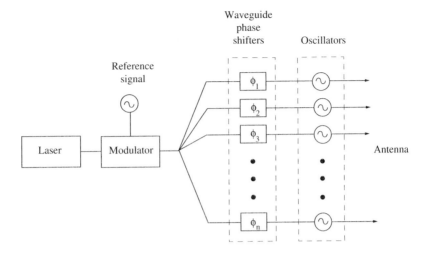

Figure 12.4   Schematic block diagram of an optoelectronic phased array antenna system.

is more efficient to use optical phase shifters. Typically, these are electro-optic phase modulators that were described in Chapter 11. An optical signal is modulated at the operation frequency by current (internal) modulation of the laser or by an external modulator, such as the quantum well electroabsorption modulator described in Chapter 11. The modulated and phase-shifted signal is coupled by injection locking to a microwave oscillator, which forms a single element of the phased array. Each element of the phased array therefore consists of a source, a modulator, a phase-shifter, a waveguide, and an oscillator. These devices can be combined by hybrid integration, but for the sake of compactness (since the phased array will contain many radiating elements) and ruggedness, it is more desirable to realize the array by monolithic integration. The resulting OEIC chip, even for a single element, is extremely complex. Such radars will be used on the ground, on aircrafts ("smart skins"), and in satellites.

12.4 MATERIALS AND PROCESSING FOR OEICS

The choice of materials and the epitaxial techniques used to synthesize them for OEICs are dictated by the same requirements as those for high-performance discrete devices. These are operating wavelength, lattice-matching considerations, and additionally, the choice of the component devices. The technology associated with GaAs and related materials is more matured because of the developments in IC technology, and it is envisaged that local-area networks, computer interconnects, and optical information processing will heavily rely on GaAs-based materials. InP-based materials will undoubtedly be more applicable to OEICs for long-distance fiber communication. The second approach to the use of materials for OEICs is heteroepitaxy, or the use of mismatched materials that include III–V or II–VI compounds on similar semiconductors, or GaAs and InP-based compounds on Si. There are potential advantages and disadvantages in this type of heteroepitaxy. The biggest problem in these mismatched materials is the large density of misfit dislocations that are created at the heterointerface and which propagate upward as threading dislocations into the active regions of the device. In spite of this shortcoming, and in view of the low cost and large density of Si electronic devices and circuits, a plausible configuration for high-performance OEICs is to have "pockets" of III–V optoelectronic devices in a "sea" of Si VLSI. Similar reasoning has justified the heteroepitaxy of GaAs on InP substrates and encouraging results are also being obtained from devices made with these materials.

In the area of processing for OEICs, the present day capabilities of lithography and nanofabrication for microelectronics are more than adequate. In fact, it is not clear if feature sizes of individual devices will need to become smaller than what has already been achieved in microelectronics. However, it is agreed upon without doubt that special processing techniques will have to be invoked to meet the special requirements of OEICs. These are briefly described in the following.

The fabrication of lower-dimensional quantum confinement structures, such as quantum wires and quantum boxes, will involve epitaxy followed by nanolithography. The desirable feature sizes are ~ 100–400 Å, which can only be achieved by electron-beam lithography with very advanced electron optics and masked ion beam lithography. Growth on patterned substrates and the incorporation of lithographic tools in the growth environment may also be required. Fabrication of optical devices on chips will require advanced dry-etching capabilities, particularly for the etching of mirrors and integrated optical components. Due to incompatibility of the structures of the different devices on an OEIC chip and because of the desire to fabricate planar circuits, regrowth is a processing option. Conventionally, such regrowth involves growing the first device structure, taking the wafer out of the growth environment, and selectively etching portions of its surface for growth of the next structure. The wafer is next reintroduced into the growth system, and the structure for another device is grown. Conventional regrowth produces an interface with a large density of defects (Fig. 12.5), which can prove to be detrimental to the operation of the regrown device or the whole circuit. An attractive alternate possibility is in-situ etching, patterning, and processing, followed by growth. For example, with a focused ion beam (FIB) system, which can be incorporated in or close to the growth environment, it is possible to combine ion beam writing and dry etching with molecular beam epitaxy. Low-dose Ga ions are used to produce localized damage for subsequent selective etching and epitaxy. A schematic of the apparatus and the process sequence are shown in Fig. 12.6. Integrated optical components may also be produced by techniques such as diffusion-induced disordering of quantum well materials, which is discussed in Sec. 12.6.

In conclusion, the advent and requirements of OEICs have placed special demands on epitaxy and materials processing. The success in these areas will dictate, to a large extent, the design and implementation of novel OEIC architectures.

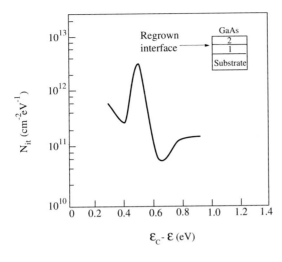

Figure 12.5 Interface state distribution at a regrown MBE GaAs interface measured by the author and co-workers. Regrowth (2) was done after etching the as-grown (1) layer surface with $3NH_4OH:1H_2O_2:50H_2O$ and annealing in an arsenic ambient for 20 min (D. Biswas et al., *Jour. of Electronic Materials*, **18**, 137, ©1989 IEEE).

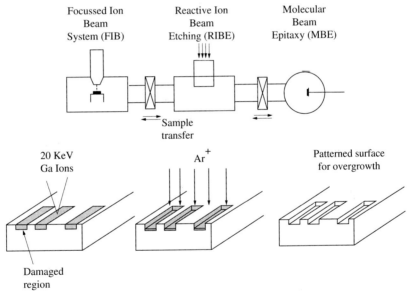

Figure 12.6 In-situ processing and growth using FIB, RIBE and MBE techniques.

12.5 INTEGRATED TRANSMITTERS AND RECEIVERS

The overall objective of optoelectronic integration is to create three-dimensional functionality. Vertical integration has shortcomings, both electrically and in terms of processing, where electrical interconnects will have to be made between different circuit components at different levels. The problems of capacitive coupling and interconnection are reduced in a planar architecture. However, growth, regrowth and processing become more complex and it can become difficult to realize such a complex chip while maintaining optimized performance of the component devices. The impact of these issues on actual circuit performance will become clear in the following sections.

12.5.1 Front-End Photoreceivers

In the design of an optical-fiber communication system, whether for use in long-distance communication or for bussing data over short distances, and operating at low or high data rates, a key element is the receiver. The basic purpose of the receiver is to detect the incident light and convert it into an electrical signal containing the information impressed on the light at the transmitting end. The important performance characteristics of a photoreceiver are operating bandwidth and sensitivity. Sensitivity plays a vital role in deciding the number of repeaters needed in a long-haul communication system. The receiver sensitivity is defined as the minimum amount of optical power level needed at the receiver input so that the signal-to-noise ratio is greater than a given value. In digital communications systems, this translates to having a bit error rate less than a given value. Other features such as dynamic range become important in specific applications.

The block diagram of a photoreceiver circuit is shown in Fig. 12.7. The first two blocks consisting of the photodetector and the low-noise preamplifier are considered the front end of the photoreceiver, while the rest of the circuit perform equalization, pulse shaping, and gain control functions. Also, in most photoreceiver circuits multiple stages of amplification are included to increase the gain of the signal. The overall performance of the circuit is mostly dictated by the front end, which consists of a photodiode and a single-stage amplifier. The desirable features of the photodetector are high quantum efficiency, low capacitance, small response time, and low dark current. Avalanche photodiodes, PIN photodiodes, and metal-semiconductor-metal (MSM) photodetectors are among the most commonly used photodetectors for this application. For the preamplifier circuit, low noise with high output gain are very important because it is the first stage of amplification. The input capacitance and parasitic resistances as well as the leakage currents of the amplifier should be minimized. Both field-effect transistors (FETs) and bipolar transistors have been used in the preamplifier circuit.

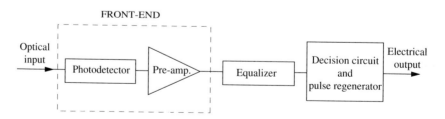

Figure 12.7 Block diagram of a photoreceiver circuit.

The selection of the integration scheme for the photodetector and the low-noise pre-amplifier circuit is also very important in determining the overall photoreceiver performance.   Three types of integration are commonly used and are shown in Fig. 12.8; the low-input impedance design, the high-input impedance design, and the transimpedance design. The latter two are most commonly used. The high-input impedance design requires the addition of an equalization circuit to extend the value of the bandwidth due to the large RC time constant. The transimpedance design is probably the most popular because no equalization is usually required. A field-effect transistor can also be used as an active feedback element, replacing the passive resistor and increasing the impedance value to as high as 80 KΩ. This significantly reduces the overall thermal noise.

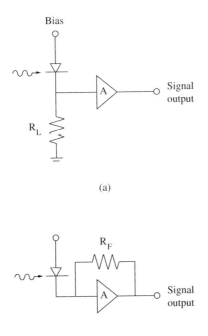

(a)

(b)

Figure 12.8   Photoreceiver circuit diagrams for (a) low, high, and (b) transimpedance designs.

Figure 12.9 schematically illustrates the monolithic integration of an $In_{0.53}Ga_{0.47}As$ photodiode with a $In_{0.53}Ga_{0.47}As/In_{0.52}Al_{0.48}As$ modulation doped FET (MODFET) by regrowth on InP. The MODFET consists of a layer of undoped low-bandgap material forming a heterojunction with a highly doped high-bandgap material. Due to the difference in the electron affinities of the two layers, electrons are transferred from the high-bandgap material to the low-bandgap material to form a quasi two-dimensional electron gas (2DEG). The main advantage of such a structure is that the electrons are separated from their parent donors and the coulombic scattering is greatly reduced. This results in higher carrier mobility and drift velocity. Typically, a high-bandgap

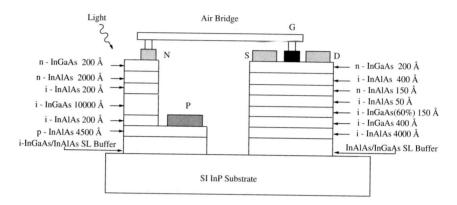

Figure 12.9   InP-based front-end photoreceiver utilizing MBE regrowth (from R. Lai et al., *Electronics Letters*, **27**, 364, 1991).

undoped spacer layer is added between the highly doped high-bandgap layer and the undoped low-bandgap layer to further separate the electrons from their parent donors. The MODFET structure has an advantage in the proximity of the channel to the gate (typically less than 500 Å), which becomes important as the gate length is shrunk to submicron dimensions. Furthermore, the noise figures and noise temperatures exhibited by MODFETs are lower than other FETs. The InGaAs/InAlAs/InP heterostructure system is a superior candidate compared to the GaAs/AlGaAs system for the MODFET because of the large conduction band discontinuity ($\Delta \mathcal{E}_c = 0.5$ eV), which allows for increased confinement of carriers in the quasi two-dimensional electron gas, and higher mobility ($> 10,000$ cm^2/V.s) and peak velocity in the In-GaAs layer. Further improvement in device performance can be made by increasing the In composition in the In$_x$Ga$_{1-x}$As channel layer. This results in even higher mobilities and peak velocities (approximately a 15% increase for x = 0.65). Pseudomorphic In$_x$Ga$_{1-x}$As/In$_{0.52}$Al$_{0.48}$As/InP MODFETs have demonstrated extrapolated cutoff frequencies greater than 200 GHz and maximum oscillation frequencies greater than 400 GHz. Also, noise figures as low as 1.4 dB with an associated gain of 6.6 dB at W-band have been demonstrated. In the integration scheme shown in Fig. 12.9, the PIN diode is grown in the first epitaxial step and the MODFET is regrown.

12.5.1.1 Photoreceiver Noise Considerations.

Noise and bandwidth are important parameters of a photoreceiver circuit. In order to optimize performance in terms of these parameters, materials and device design must be carefully tailored. For simplicity, we will consider the integration of a FET with a PIN diode. A similar analysis can be made for other combinations of amplification and detection devices.

The front-end noise current of the FET can be expressed as[†]

$$\overline{i_N^2} = 2q I_{ph} I_{P1} B + \frac{4k_B T}{R_L} I_{P2} B + 2q(I_g + I_D) I_{P2} B + \frac{4k_B T \theta_m}{g_m} (2\pi C_T)^2 f_o I_f B^2$$

$$+ \frac{4k_B T \theta_m}{g_m} (2\pi C_T)^2 I_{P3} B^3 + \frac{4k_B T \theta_m}{g_m} (2\pi C_T)^2 \left(\sum_{r=1}^{n} \frac{I_r A_T}{\tau_T} \right) B^2 \qquad (12.1)$$

The first term is the signal noise associated with the photocurrent I_{ph}, the second term is the thermal noise due to the load resistance in the high- and low-input impedance designs or the feedback resistance in the transimpedance design, and the third term represents shot noise due to the leakage current in the gate, I_g and the dark current in the photodiode, I_D. The fourth term represents the 1/f noise in the FET device, where f_0 is the noise corner frequency. The fifth term represents the noise associated with the channel conductance of the FET and the sixth term is due to noise resulting from traps in the channel or buffer regions. I_{P1}, I_{P2}, and I_{P3} are known as the Personick integrals and I_f and I_r are the 1/f noise integral and trap integral, respectively. θ_m is a materials related parameter, g_m is the FET transconductance, τ_T is the trap emission time constant and A_T is a constant dependent on trap density and transistor parameters. To minimize the noise, it is evident that low gate leakage current and low PIN photodiode dark current, low total input capacitance and high transconductance are needed. At higher frequencies and bit rates the noise is dominated by the noise associated with the channel conductance. This term is proportional to the cube of the bandwidth of the circuit and the square of the total capacitance, C_T, given by

$$C_T = C_{gs} + C_j + C_p \qquad (12.2)$$

where the terms on the right-hand side represent the gate-source capacitance in the FET, the PIN diode capacitance, and parasitic capacitance. It has been shown that the condition for minimum photoreceiver noise yields[‡]

$$C_{gs} = C_j + C_p \qquad (12.3)$$

It can be readily seen that increasing the cutoff frequency of the FET is very important in minimizing the overall photoreceiver noise. The cut-off frequency f_T of the transistor is given by

$$f_T = \frac{g_m}{2\pi C_{gs}} \qquad (12.4)$$

From Eq. 12.1, an expression for the receiver sensitivity, ηP, can be derived for

[†]R. G. Smith and S. D. Personick, "Receiver Design for Optical Fiber Communication." in *Semiconductor Devices for Optical Communication* (ed. H. Kressel), 88, Springer-Verlag, Heidelberg, 1987.
[‡]Y. Archambault et al., *Jour. Lightwave Technology*, **LT-5**, 355, 1987.

digital applications, as [†]

$$\eta P = \left(\frac{1+r}{1-r}\right)\left(\frac{h\nu}{q}\right) Q\sqrt{\overline{i_N^2}} \tag{12.5}$$

where $(1 + r)/(1 - r)$ is the factor due to nonzero extinction, r is the extinction ratio, which is defined as the ratio of power recived in the 0 state to that in the 1 state, Q is a parameter dependent upon the bit error rate, $h\nu/q$ is the photon energy, and η is the quantum efficiency of the detector. For a bit error rate of 10^{-9}, Q is 6.[‡] A corresponding equation can be derived for the signal to noise ratio in an analog photoreceiver. If the input light power is expressed as

$$P(t) = P_o(1 + m\sin(\omega t)) \tag{12.6}$$

where m represents the modulation index and P_o is the cw light input power, then the signal to noise ratio is

$$\frac{S}{N} = \frac{m^2 I_o}{2\sqrt{\overline{i_N^2}}} \tag{12.7}$$

where I_o represents the dc photocurrent level from the cw light signal, P_o. The equations given above can be used to evaluate the noise performance of practical photoreceiver circuits. For example, the calculated input noise spectral density versus bit rate for a InP-based photoreceiver consisting of a 0.25 μm gate MODFET and high impedance (20 KΩ) front end is shown in Fig. 12.10. The parameters in the

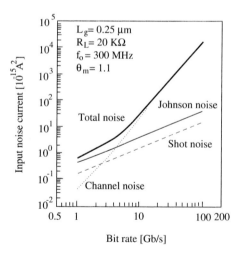

Figure 12.10 Calculated noise spectral density versus bit rate for an InP-based photoreceiver with a 20 K ohm load resistance and a 0.25 μm gate length MODFET (from R. Lai, *Monolithic Integration of InGaAs/InAlAs/InP Electronic and Optoelectronic Devices and Circuits*, Ph.D. thesis, University of Michigan, 1991).

[†]R. G. Smith and S. D. Personick in "Receiver Design for Optical Fiber Communication Systems," *Semiconductor Devices for Optical Communication*, (ed. H. Kressel), 88, Springer-Verlag, Heidelberg, 1987.

[‡]The relationship is BER $= \frac{1}{2}erfc\left(\frac{Q}{\sqrt{2}}\right)$

different noise terms of Eq. 12.1 are extracted from the equivalent circuit models for actual photoreceiver circuits. Trap noise and 1/f noise are assumed to have negligible contributions. It is clear from Fig. 12.10 that for lower bit rates the thermal noise due to the load resistance is most significant, while at higher bit rates, the channel noise in the MODFET becomes more dominant. The corresponding calculated receiver sensitivities are shown in Fig. 12.11 for the high-impedance ($R_L = 20K\,\Omega$) design. FET gate lengths of 1.0 and 0.25 μm have been considered. The improvement in the sensitivity of up to 4 dB by using the 0.25 μm gate length MODFET device is primarily due to the decrease in the channel noise of the MODFET. However, in the transimpedance and low-impedance designs, the thermal noise due to the input load resistor is dominant and the benefit of using the 0.25 μm gate length MODFET is negligible. It is evident from these calculations that the resistance values must be kept at a very high level to take full advantage of the benefits of submicron gate length FETs and to achieve high receiver sensitivities.

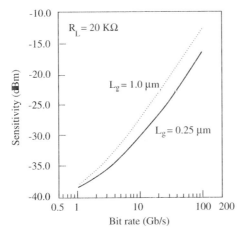

Figure 12.11   Calculated sensitivities versus bit rate for an InP-based photoreceiver with a 20 K ohm input load resistance and a 1.0 μm (dotted) and a 0.25 μm (solid) gate length MODFET (from R. Lai, *Monolithic Integration of InGaAs/InAlAs/InP Electronic and Optoelectronic Devices and Circuits*, Ph.D. thesis, University of Michigan, 1991).

12.5.1.2 Photoreceiver Bandwidth Considerations.

The most important parameters that determine the bandwidth of the receiver are the transit time of the generated carriers in the diode and the RC time constant of the circuit. The frequency response of the PIN photodiode, $J(\omega)$, limited by transit time consideration was given by Eq. 8.75. The electrical frequency response, $H(\omega)$, of the overall circuit includes the diode capacitance and resistances. This can be calculated using a microwave simulation program such as SPICE. The frequency response for the complete receiver can then be expressed as

$$J_0(\omega) \; = \; J(\omega)H(\omega) \tag{12.8}$$

The thickness of the absorbtion and transit layer of the photodiode should be optimally designed. For thicknesses of 1 μm or less of this layer, the electrical circuit frequency response typically determines the overall frequency response of the photoreceiver.

However, there is a bandwidth-noise tradeoff in the selection of the value for the load resistor. In many designs, the noise is the main concern because equalization can be used in later stages of the photoreceiver, thereby increasing the effective bandwidth of the photoreceiver. The calculated and measured electrical frequency response of a PIN MODFET front-end photoreceiver circuit, shown in Fig. 12.12(a), are given in Fig. 12.12(b). Higher frequencies can be achieved by optimizing device and circuit parameters. The photomicrograph of a InP-based photoreceiver made by two-step MBE is shown in Fig. 12.13.

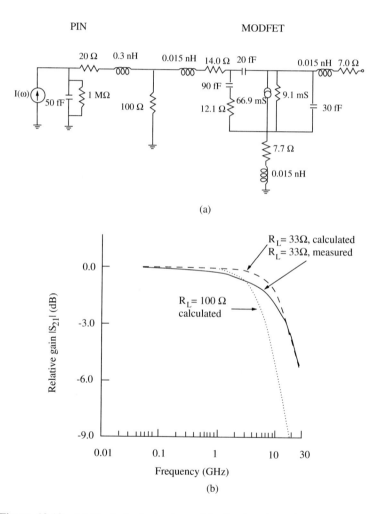

Figure 12.12 (a) Equivalent circuit model of a low input impedance PIN-MODFET (0.25 μm gate length) front-end photoreceiver determined from S-parameter measurements and (b) calculated and measured frequency response of the circuit (from R. Lai et al., *Electronics Letters*, **27**, 364, 1991).

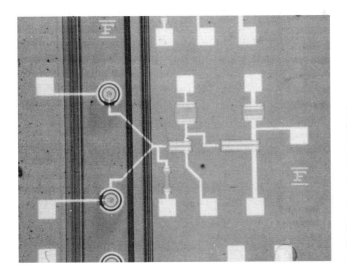

Figure 12.13 Photomicrograph of $In_{0.53}Ga_{0.47}As/In_{0.52}Al_{0.48}As$ PIN-MODFET (1 μm gate) photoreceiver with dual detectors for balanced detection. The detectors have an active area of 15 μm diameter. The two devices were grown by MBE in two steps, with the MODFET being regrown in the second step (courtesy of O. Wada, Fujitsu Limited, Atsugi, Japan).

12.5.2 OEIC Transmitters

A transmitter circuit includes a light source such as a high-power LED or a laser. Integration of the laser with the associated electronics—particularly the driver circuitry in the form of a transistor—is more complicated than the fabrication of a photoreceiver. This is because the laser has more stringent materials and processing requirements than a photodetector. First, the laser structure is nearly 4 μm high, which makes the processing steps for integration with an incompatible heterostructure for the electronic device very difficult. Second, the optical cavity in an edge-emitting laser needs to be defined by two end mirrors. Third, electrical and optical confinement needs to be achieved in the lateral dimension. Finally, the operation of the laser necessitates efficient heat sinking of the whole chip. In spite of these disadvantages it is worthwhile to integrate the component devices to ensure a higher modulation bandwidth. The latter is ensured by the close proximity of the devices and the associated reduction of parasitics from the devices and interconnects. An OEIC transmitter using the same heterostructure for the laser and the driver transistor is shown in Fig. 12.14. However, for the optimization of the performance of both devices, separate optimized structures are more desirable.

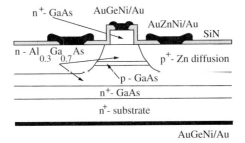

Figure 12.14 Schematic cross-sectional diagram of a double-heterojunction bipolar transistor which also functions as a transverse-injection laser (from Y. Hasumi et al., *IEEE Electron. Dev. Lett.*, **EDL-8**, 10, © 1987 IEEE).

Early transmitter circuits consisted of a single edge-emitting laser, whose facet was created by cleaving, integrated with a single transistor. The cross section of a laser-MESFET and its equivalent circuit are shown in Fig. 12.15. More recently, laser facets are formed *on-chip* by microcleaving or by using dry-etching techniques such as ion beam etching. Figure 12.16 shows the cross section of a simple transmitter circuit fabricated by using these techniques. The circuit also contains a photodiode for monitoring the laser output power. The laser is a SQW GRIN-SCH device. Figure 12.17 shows the corresponding equivalent circuit, which includes three FETs and a 50 ohm input load resistance. The identical FETs labeled Q_1 and Q_2 form a differential amplifier or current source that provides the advantages of common mode rejection and noise reduction. The inputs at the respective gates are V_{G1} and V_{G2}. The FET labeled Q_3 acts as a constant current source and provides the drive current. The current source I_B is a dc source to bias the laser at threshold, remembering that the inputs at V_{G1} and V_{G2} are small-signal modulated signals. The photodiode PD acts as a monitoring device. If the (equal) transconductance of the two FETs is g_m

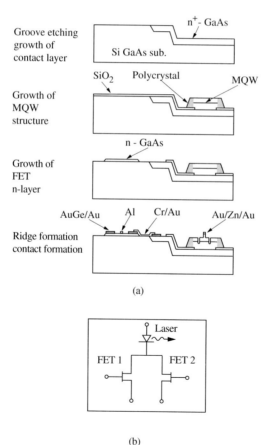

Figure 12.15 (a) Fabrication process of GaAs-based laser-MESFET OEIC with 3 MBE steps and (b) equivalent circuit with 2 FETs (courtesy of O. Wada, Fujitsu Limited, Atsugi, Japan).

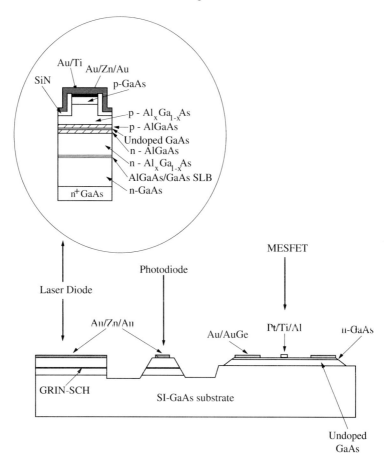

Figure 12.16   Cross section of integrated transmitter circuit with monitoring photodiode formed by 2-step MBE, ion beam etching, and microcleaving of laser facets. The laser structure was grown in the first epitaxial step. The n-ohmic contacts to the laser and photodiode are not shown (from M. Kuno et al., *Appl. Phys. Lett.*, **49**, 1575, 1986).

and the slope in the lasing portion of the light-current characteristic is S, the output power, P_{out}, of the laser can be simply expressed as[†]

$$P_{out} = \frac{g_m S}{2}(V_{G1} - V_{G2}) \qquad (12.9)$$

It may be noted that the detector will be almost transparent to the laser output, since the emission of a laser occurs at energies slightly below the absorption edge of the material.

[†]O. Wada, "Recent Progress in OEICs," SPIE Proceeding No. 797, 224–235, 1987.

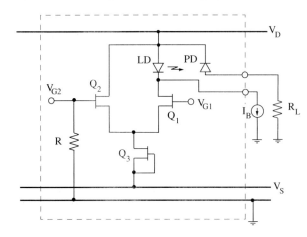

Figure 12.17 Circuit diagram of single channel laser-MESFET transmitter shown in Fig. 12.16 (from M. Kuno et al., *Appl. Phys. Lett.*, **49**, 1575, 1986).

Simple GaAs-based circuits have been described here to illustrate the techniques of fabrication and operating principles. More complex circuits using both GaAs- and InP-based materials are also being made.

EXAMPLE 12.1
Objective.   To calculate the input bias difference ($V_{G1} - V_{G2}$) for a transmitter circuit in which the laser has a slope efficiency of 0.4 mW/mA/facet and produces an output power of 4 mW. The transconductance of the two FETs are 20 mS.

From Eq. 12.9,

$$V_{G1} - V_{G2} = \frac{4 \times 2}{0.4 \times 20} = 1V.$$

12.5.2.1 Equivalent Circuit of Integrated Transmitter.   The internal limit to the modulation bandwidth of a laser is set by the relaxation oscillation frequency, f_r, given by Eq. 7.138. The modulation bandwidth can be increased by a large photon density in the cavity and a short photon lifetime. The former, in turn, is achieved by driving the laser well above threshold, which can become a critical issue, and more so in OEICs, due to the problem of heat dissipation. The power dissipation in a laser can be expressed as[†]

$$P_d = P_{in} - P_{out}$$

$$= I_B \left(\frac{h\nu}{q} + I_B R_s \right) - \frac{h\nu}{q} (I_B - I_{th}) \eta_P \qquad (12.10)$$

Here η_P is the power efficiency, I_B is the drive (or bias) current and R_s is the series resistance in the circuit. It is clear that the power dissipation can be kept small, even

---

[†]S. Margalit and A. Yariv, "Integrated Electronic and Photonic Devices," in *Semiconductor and Semimetals, Vol. 22*, Part E (Vol. ed. W. T. Tsang), eds. R. K. Willardson and A. C. Beer, Academic Press, Orlando, FL, 1985.

for a large drive current, if the series resistance and threshold current are small and η_P is large.

The intrinsic model of a laser diode can be represented by an RLC circuit that has a resonance frequency $f = 1/2\pi\sqrt{LC}$. Noise sources, due to the shot noise of electrons and photons, are also incorporated in this equivalent circuit. However, in practice, the combined impedance of this RLC circuit is smaller than those of parasitic circuit elements. The extrinsic limit to the internal modulation frequency is set by these parasitic elements. The most important parasitic elements are the series resistance of the diode, R_s, the bondwire induction L_b, and the parasitic capacitance C_p between the bonding pad and ground plane. This capacitance is drastically reduced by fabricating the laser diode on a semi-insulating substrate. The equivalent circuit of a laser monolithically integrated with a FET is shown in Fig. 12.18. An advantage of having the FET in the circuit is that of impedance matching. The modulating high-frequency input is fed to the gate of the FET, which is terminated with a 50 ohm resistance and therefore the laser is matched to the 50 ohm input through the FET. This enhances the modulation bandwidth. Usually two inputs are fed to the laser: a dc current for biasing and the ac modulating input. A high-frequency bias-T circuit is used to separate the two inputs.

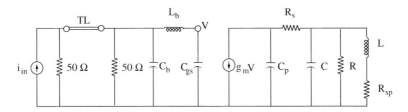

Figure 12.18   Equivalent circuit of a junction laser monolithically integrated with a FET driver matched to a 50 ohm transmission line. C_b is the bondwire capacitance, C_{gs} and g_m are the gate-source capacitance and transconductance of the FET, and R_{sp} originates from the shot noise associated with the spontaneous emission of photons. The other circuit elements are explained in the text.

12.5.3 Complex Circuits and Arrays

What has been described upto this point are circuits that perform single functions such as reception or transmission and are made of very few active and passive components. For example, the front-end photoreceiver can be realized by integrating a detector with a FET or HBT and a few passive elements. However, it is necessary and desirable to perform more complex functions on a single chip. As high data rate local-area networks are being installed in an increasing number of locations due to the merging of a high density of fibers in each of these locations, it becomes necessary to use multiple optoelectronic devices, or arrays, for reducing component cost and weight. In the simplest sense an array consists of an assemblage of identical devices monolithically integrated on the same chip. Arrays can serve a variety of different needs. One- or two-dimensional arrays of identical sources provide larger-output power and, with proper design, near-single longitudinal mode characteristics. If their respective phases are progressively tuned, then such a phased array of sources can be used for electronic beam steering. Wavelength tunability is also possible in a two-dimensional array of sources. Similar advantages can be had from arrays of detectors. Arrays of optoelectronic devices are becoming increasingly important in the transmission, routing, and reception of signals in optical-fiber systems. For example, in star networks, information from several locations are brought in at the node of the star and are rerouted again. Arrays are also useful in parallel architectures, such as in the interconnection of computers. Parallel data transmission between the computers becomes more efficient than serial data transmission, whereby the downloading time of data is considerably reduced.

The simplest configuration of four-channel transmitter and receiver arrays are shown in Figs. 12.19(a) and (b). In the transmitter array the interval between lasers

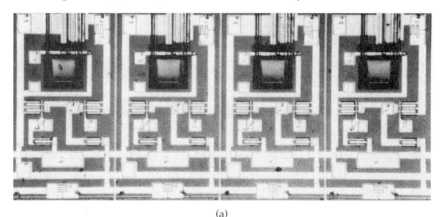

(a)

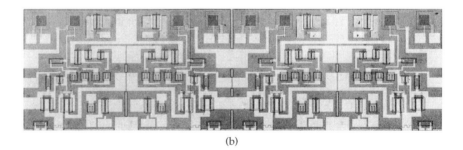

(b)

Figure 12.19 Photomicrographs of simple OEIC arrays: (a) four-channel transmitter array consisting of GRINSCH-SQW GaAs/AlGaAs laser integrated with 2 μm gate GaAs MESFET. The outer facets of the lasers are cleaved while the inner facets are microcleaved (the large holes are to facilitate this). Monitoring photodiodes are placed behind the lasers; (b) four-channel transimpedance receiver array made with MOVPE-grown GaAs MSM photodiodes and 2 μm gate MESFETs (courtesy of O. Wada, Fujitsu Limited, Atsugi, Japan).

is 1 mm and the chip size is 4×2 mm^2. The lasers are edge-emitting devices. The circuit operates up to a bit rate of 2 gigabits/sec. The crosstalk between channels is less than -20 dB at a frequency of 600 MHz. Similarly, the receiver array consists of MSM photodiodes and MESFETs made with GaAs. The spacing between photodiodes is 1 mm and the chip size is 1.2×4 mm^2.

As a concluding note it is important to mention that the development of arrays, and OEICs in general, is in its infancy. In addition to the immense technological roadblocks, a lot of understanding of impedance-matching problems, electrical and optical interference, and materials growth and compatibility needs to be developed.

12.6 GUIDED WAVE DEVICES

It should be evident by now that in complex OEICs it would be advantageous to have optical waveguides and other passive or quasi-passive optical components monolithically integrated with the active components: sources, detectors, modulators, and electronic devices. Conceivably one could use optical fibers, but the application of these would be limited both from the point of view of integration, and that of functionality. Lithium niobate (LiNbO$_3$) and similar dielectric materials that have a large electro-optic coefficient are very suitable for most of these applications, but it is impossible to monolithically integrate such components with active semiconductor devices. Guided wave components are required for routing optical signals on a chip and also for the functions of directional coupling, filtering, and modulation. The study of these devices made with suitable dielectric materials falls in the realm of integrated optics, which has been discussed in detail in several texts. In this section, the properties of some simple and essential guided wave components, that are important for OEICs, are briefly described.

12.6.1 Waveguides and Couplers

As seen in Chapter 6 a waveguide is a region of dielectric through which light is propagated, surrounded by dielectric regions or air having a smaller dielectric constant. Therefore, to form a guide, it is essential to employ techniques that will effectively and selectively create regions of varying refractive index.

The simplest technique of delineating a guiding region is by introducing free carriers. This is because in a semiconductor material with a large density of free carriers the refractive index is lowered—from that in pure material—due to the negative contribution of the free-carrier plasma to the dielectric constant. The lowering of the refractive index due to free carriers is expressed by[†]

$$\Delta n_r = -\frac{n\lambda_0^2 q^2}{8\pi^2\epsilon_0 n_r m^* c^2} \tag{12.11}$$

where n_r is the refractive index of the undoped semiconductor at a free-space wavelength λ_0. For example, in GaAs $\Delta n_r = -0.02$ when $n = 1 \times 10^{19}$ cm^{-3} for

---

[†]E. Garmire, "Semiconductor Components for Monolithic Applications," in *Topics in Applied Physics*, Vol. 7, 2nd ed. (ed. T. Tamir), Springer-Verlag, Heidelberg, 1985.

$\lambda_0 = 1\mu$m. This change in refractive index is large enough for light confinement. Based on this principle waveguides can be produced either by growth of an undoped epitaxial layer on a highly doped substrate, or by implantation damage. In the second scheme, the waveguide-cladding layer interface is not well defined, since the implantation profile is not rectangular, but Gaussian. In the epitaxial technique the interface is more abrupt. At the top of the guide, optical confinement is provided by the index change at the semiconductor-air interface. Guiding is also achieved by compositional variation in the vertical direction, such as that by GaAs-AlGaAs heterojunctions, and the index difference and optical confinement in this case are made possible by the different bandgaps. Examples were seen in the design of edge-emitting LEDs and lasers.

For obtaining single-mode guiding and propagation, it is necessary to delineate the guiding region in the lateral direction also, by causing an index change. This is achieved in a ridge waveguide, a buried channel guide, or a strip-loaded guide. In the first, the lateral waveguide dimensions are delineated by wet- or dry-etching, or a combination of both. A dry-etching process such as ion-milling or reactive ion etching, which provides control, is followed by a wet-etching process, which smoothes the surface. Buried channel waveguides can be created by a variety of techniques. The simplest one conceptually is by regrowth where, for example, a GaAs waveguide is grown and delineated and a higher-index AlGaAs confining layer is regrown by LPE or MOCVD. Selective diffusion or ion implantation can also be used. A more novel technique is diffusion induced disordering. A multiquantum well guiding layer is first grown epitaxially. It is then masked selectively and the regions adjacent to the guiding region are doped by implantation and annealing or by diffusion. Either process converts the ordered MQW or superlattice structure into a disordered random alloy, usually with a lower refractive index, providing lateral optical confinement. The various processes are illustrated in Fig. 12.20. The last type of waveguide structure is the strip-loaded guide. The formation of a dielectric or metal stripe on the guiding layer alters the refractive index of the semiconductor underneath it and confines light.

For any application it is necessary to ensure that the guides provide low propagation loss. If the guides are made of high-quality, defect-free epitaxial layers, then the major sources of loss are surface scattering and absorption. Therefore, etching and formation techniques become critical in the fabrication of low-loss waveguides. An important figure of merit for a waveguide is its loss coefficient γ, which we have seen in Chapters 6 and 7. The value of γ, which determines the insertion loss of a waveguide, is mainly determined by free-carrier absorption and scattering at bulk and surface imperfections. Therefore, material quality and processing become important in determining γ. Transmission of optical power in the guide is given by

$$P(z) = P(0)e^{-\gamma z} \qquad (12.12)$$

from which the guide loss is

$$\mathcal{L} = 4.3\gamma \quad (dB/cm) \qquad (12.13)$$

Low-loss waveguides have values of $\mathcal{L}$ ranging from 0.1 to 1.0 dB/cm.

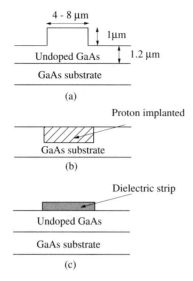

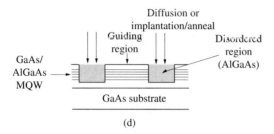

Figure 12.20 Techniques for fabricating waveguides: (a) ridge, (b) buried channel by implantation, (c) strip loaded, and (d) disordered MQW. Typical dimensions of a single mode guide are indicated in (a).

The simplest guided wave integrated optical circuit element is a directional coupler, which is useful for transferring optical energy from one waveguide to another and therefore from one region of an OEIC chip to another. The directional coupler consists of two parallel waveguides between which the transfer of optical energy occurs due to the overlapping of waveguide modes. This energy exchange requires that the light propagating in both guides have nearly the same velocity and propagation vector. If these parameters in the two channels are exactly identical, then the power propagating in the two guides is given by (Appendix 11)

$$P_1(z) = \cos^2(\kappa z)e^{-\gamma z}$$
$$P_2(z) = \sin^2(\kappa z)e^{-\gamma z} \tag{12.14}$$

Here z is the direction of propagation and κ is the coupling constant given by

$$\kappa = \frac{2\beta_y^2 b e^{-bd}}{\beta_z W(\beta_y^2 + \beta_z^2)} \tag{12.15}$$

where b is called the extinction coefficient, d is the separation between the guides, W is the width of each guide, and β_z and β_y are the mode propagation constants in the propagation and transverse directions, respectively. From Eq. 12.14, the coupling length of a directional coupler, l_C, defined as the length at which total transfer of power takes place, is given by

$$l_C = \left(m + \frac{1}{2}\right)\frac{\pi}{\kappa}, \quad m = 0, 1, 2, \cdots \tag{12.16}$$

It should be remembered that in real couplers the two guides may not be identical, and the propagation constants may differ by a small amount $\Delta\beta_z$. The coupling constant is then given by

$$\kappa_r^2 = \kappa^2 + \left(\frac{\Delta\beta_z}{2}\right)^2 \tag{12.17}$$

and the power-flow equations in the two guides are modified.

The photomicrograph of a straight ridge waveguide dual-channel coupler is shown in Fig. 12.21. Curved sections of guides are important elements of integrated optics. Sharp bends constitute a large radiation loss. For the bends to be nearly loss-free, the radius of curvature at the bend must be much larger than λ, the wavelength of the light propagating in the guide. Finally, there are various forms of leaky guides and couplers where the index of the coupling region is so adjusted that most of the radiation leaks out into an optical detector or other such device. Other passive components that are required in integrated optics are lenses, mirrors, and gratings, for the purposes of beam focusing, reflection, and filtering.

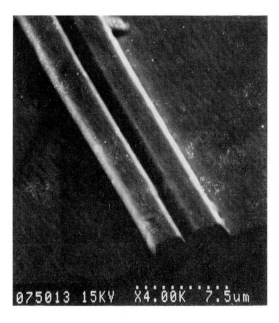
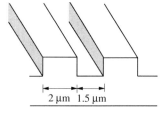

2 μm 1.5 μm

Figure 12.21 Photomicrograph of a GaAs dual-channel single mode ridge waveguide coupler with $W = 2\mu$m and d = 1.5 μm.

12.6.2 Active Guided Wave Devices

The integrated optical components described in the previous section are essentially passive, and are used for routing optical signals. There are also quasi-passive or active guided wave components that can be integrated in OEICs with active optoelectronic devices. Examples of active guided wave devices that we have already learned about are the laser and the electro-optic modulator. There are other active guided wave devices that are used as modulators, interferometers, and filters. These have been traditionally designed with materials with a large electro-optic coefficient, such as lithium niobate. For use in OEICs, these devices are being made with semiconductors.

A simple guided-wave modulation/switching device based on the electro-optic effect is the Mach-Zehnder interferometer, shown in Fig. 12.22. The incoming optical beam is split equally between the two branches of the Y and then recombined at the Y at the other end. Care must be taken in materials growth and processing such that the tapers are very gradual to reduce bend losses, and that there is spatial uniformity. With no applied bias, the phase shift in the two arms are equal and at the output Y the two waves interfere constructively and all the power (minus the insertion loss of the device) appears at the output. If the net phase shift difference between the two arms is π, which can be achieved by the application of a suitable bias, the modes in the two branches are 180° out of phase and no power emerges at the output. Thus, the output power can be controlled in a tunable manner by varying the voltage. The phase difference $\Delta\phi$ between the two arms is given by Eq. 11.24. For $W = 2 - 3 \ \mu\text{m}$ and l of the order of a few millimeters, the voltage required for full extinction ($\Delta\phi = \pi$) is of the order of a few volts in compound semiconductor devices. Eventually, very complex integrated optical guided wave active devices will have to be successively realized with GaAs- or InP-based semiconductors.

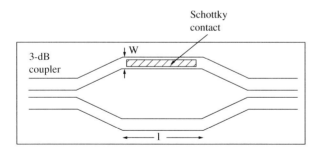

Figure 12.22 Schematic of a guided-wave Mach-Zehnder interferometer.

In passing it may be mentioned that depending on the type of devices integrated on a chip, the latter has also been described as a photonic integrated circuit (PIC) instead of OEIC.

12.7 PROSPECTS FOR OPTICAL INTERCONNECTS

Long-distance optical-fiber communication has received a significant boost with the advent and development of low-loss and low-dispersion silica fibers for operation in the spectral range of 1.1–1.6 μm. Therefore, in the natural course of development, most existing fiber-optic links are for high-data rate, point-to-point, long-distance communication. There has been increasing awareness of the fact that optical fibers could also be used in short-distance, high-connectivity applications. For example, the ultimate speed of VLSI chips are not limited by the speed of the individual devices themselves, but by the interconnect delays. The time taken to transmit a signal across a conducting wire is determined by an RC time constant that is proportional to the square of the wire length. There has therefore been growing interest in replacing such interconnects with optical interconnects, through which data can travel with the speed of light.

Optical interconnects can be used in an integrated optoelectronic chip, between two chips, or between printed circuit boards, each of which can support one or more complex circuits. Interconnects are also needed between a group of devices that can perform calculations, share data and perform input-output operations. These are usually short-distance, high-connectivity applications, and are termed local-area networks (LANs). Different network topologies are used in the connectivity scheme and two examples, known as star and ring networks are shown in Fig. 12.23. For future computing systems, massively parallel network schemes are envisaged and this requires rapid interconnection between integrated circuits. Ultimately, high connectivity and parallelism, high-speed, low-loss, and minimal interference and crosstalk make optical interconnects very attractive. Needless to say, in addition to long-distance communication, optical interconnects are important for optical computing, LANs, and even for interfacing with individual subscribers and homes.

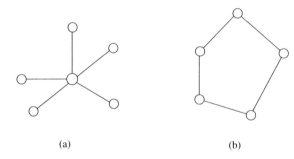

(a)   (b)

Figure 12.23   Local-area network topologies: (a) star network and (b) ring network.

Three types of optical interconnects can basically be used for inter- and intra-chip and interboard connections. These are optical waveguides, which we have already seen, free-space communication, and optical fibers. Free-space communication includes holographic techniques and are used in optical cross-bars for imaging and pattern recognition.

In high-speed LANs, optical interconnects will principally link functional modules, as shown in Fig. 12.3. Coupling is accomplished by the use of optical-fiber backplanes and backplane couplers. This kind of high-speed, high data-rate interconnection usually involves distances from a few hundred meters upto a kilometer. For chip-to-chip connections on a single board, very short length (~ 0.5 m) interconnects, sometimes in massively parallel architecture, are usually required. Therefore losses and dispersion do not pose as critical parameters.

A common fiber, as shown in Fig. 12.24, is an optical guide characterized by an inner high-index core and outer low-index cladding. Typical core and cladding diameters of single-mode fibers are 5 and 125 μm, respectively. Fiber properties are largely determined by core and cladding diameters and refractive indices. Important parameters in the use of fibers are fiber insertion and coupling losses and intermodal distortion. This distortion is approximately given by $3.3\Delta n_r (ns/m)$ where Δn_r is the refractive index difference between core and cladding. Distortion also increases as the fiber core diameter and numerical aperture (Eq. 5.13) are increased to make coupling into it easier.

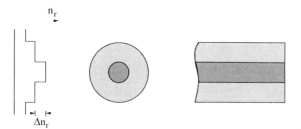

Figure 12.24 Schematic illustration of an optical fiber and its index profile.

There are difficult technological hurdles which have yet to be overcome in fiber interconnect technology. The first is interfacing or coupling with active or passive devices such as detectors, lasers, and waveguides. A possible scheme is illustrated in Fig. 12.25. The second aspect arises from the fact that bidirectional transmission, or the fabrication of fiber couplers and power splitters/combiners, may require splitters and isolators. These components need to be fabricated to provide low coupling and insertion losses. The efficiency of optical-fiber links is determined by what are known as power budget and rise-time budget calculations. The important common parameter in both is the length of the fiber. The first calculation is related to the sensitivity of the system, while the second is related to the bandwidth. A brief mention should be made of the choice of semiconductors, which form the active devices, and hence decide the operating wavelength, in any application. For long-distance communication

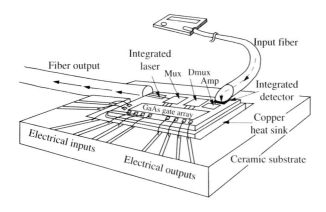

Figure 12.25 Optical fiber interconnects to a OEIC transceiver chip (from L. Hutcheson, *Jour. Electronic Materials*, **18**, 259, ©1989 IEEE).

in the 1.1 to 1.6 μm spectral range extending over several kilometers, fiber losses and dispersion become very critical and InP-based alloys and heterostructures such as InGaAs/InGaAsP/InP are used. For short-distance communication and LANs, losses and dispersion are not critical factors and GaAs-based devices operating in the 0.8–0.9 μm range are perfectly suitable.

Finally, before concluding this section, a comment will be made on the use of optical fibers as active elements. In other words, we seek the answer to the question, "Can fibers perform functions other than passive guiding?" Attempts have been made in this respect to fabricate useful active devices on fibers to couple with the signal being transmitted. Thus, a fraction of the signal may be made to leak out in predetermined sections and excite active devices. Acousto-optic modulators and photodetectors have been placed on fibers. The photocurrent produced in the detector can be used as a monitoring or routing signal. Success has also been achieved in making traveling wave fiber amplifiers, or lasers, by rare-earth doping of conventional silica-glass fibers. For example, doping the fiber material with Er atoms produces a three-level lasing medium. More importantly, the lasing transition of Er^{3+} in silica glass is at 1.5 μm, which makes it perfect for use with InP-based communication systems. Such lasers are optically pumped at one end and the lasing, or amplified coherent signal, is made to exit at the other end. The section of fiber therefore constitutes an active optical amplifier, which will eliminate or reduce the need for repeater stations in long-haul communication links.

In conclusion, the prospects for optical-fiber interconnects in selectively replacing electrical interconnects are extremely bright, and work is ongoing around the world to make it a real success. It is perhaps appropriate to conclude this chapter and the

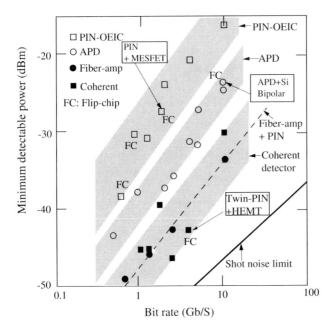

Figure 12.26 Performance of optical receivers using different devices and technologies (courtesy of O. Wada, Fujitsu Limited, Atsugi, Japan).

text with a comparison of optical receiver performance using different devices and technologies. This is illustrated in Fig. 12.26. This figure, in a way, summarizes the development of optoelectronic devices.

PROBLEMS

12.1 Distinguish between hybrid and monolithic integration and discuss their relative merits and demerits.

12.2 Describe, with figures, the materials, devices, and OEIC designs of transmitters, receivers and repeaters for a transatlantic optical-fiber communication link.

READING LIST

BUCKMAN, A. B. *Guided-Wave Photonics.* Saunders College Publishing, Harcourt Brace Jovanovich College Publisher, Orlando, FL, 1992.

DAGENAIS, M., et al. "Applications and Challenges of OEIC Technology: A Report on the 1989 Hilton Head Workshop," *Jour. Lightwave Technology*, **8**, 846, 1990.

PALAIS, J. C. *Fiber-Optic Communications*, 2nd ed. Prentice Hall, Englewood Cliffs, NJ, 1988.

SMITH, R. G., and PERSONICK, S. D. "Receiver Design for Optical Fiber Communication Systems." In *Semiconductor Devices for Optical Communication*, 2nd ed. (ed. H. Kressel), Topics in Applied Physics, **39**, 88–160, Springer-Verlag, Heidelberg, 1987.

WILLARDSON, R. K., and BEER, A. C. eds. *Semiconductors and Semimetals: Vol. 22. Lightwave Communications Technology* (Volume ed; W. T. Tsang), *Part E: Integrated Optoelectronics.* Academic Press, Orlando, FL, 1985.

List of Symbols

$\mathbf{A}$	magnetic vector potential
A_n	numerical aperture of optical fiber
A_{21}	Einstein coefficient
a, a_o	lattice constants
a_I	impurity concentration gradient in graded junction
B	frequency bandwidth
$\mathbf{B}$	magnetic flux density
B_r	band-to-band recombination coefficient
B_T	base transport factor of transistor
B_{21}	Einstein coefficient
b	extinction coefficient of waveguide
C_d	diffusion capacitance of junction
C_j	junction capacitance
c	velocity of light
c_{ij}	elastic constant
$\mathbf{D}$	electric flux density
$D_{e(h)}$	diffusion constant for electrons (holes)
d_m	distance to which mode extends in laser cavity
E	electric field
$\mathcal{E}$	energy
$\mathcal{E}_{C(V)}$	energy of conduction (valence) band edge
$\Delta\mathcal{E}_{C(V)}$	conduction (valence) band offset
$\mathcal{E}_{D(A)}$	ionization energy of donor (acceptor) level
$\mathcal{E}_{ex}, \mathcal{E}_{ex}^l$	exciton binding energies
$\mathcal{E}_F$	Fermi level in semiconductor
$\mathcal{E}_{Fi}$	intrinsic Fermi level (or intrinsic level)
$\mathcal{E}_{fn(p)}$	electron (hole) quasi-Fermi levels
$\Delta\mathcal{E}_f$	energy separation between quasi-Fermi levels $\mathcal{E}_{fn}$ and $\mathcal{E}_{fp}$
$\mathcal{E}_g$	bandgap of semiconductor
$\mathcal{E}_{ie(h)}$	threshold energy for impact ionization of electrons (holes)

496

$\mathcal{E}_m$	energy per photon delivered to load in solar cell circuit at maximum power point
$\mathcal{E}_p$	phonon energy
$\mathcal{E}_{ph}$	photon energy
$\mathcal{E}_{ST}$	strain energy
F	force (of attraction or repulsion)
F_i	flux of species i in MBE
F_T	transmission function at semiconductor-air interface
$F_{1/2}(\eta)$	Fermi integral where $\eta = \frac{\mathcal{E}_F - \mathcal{E}_C}{k_B T}$
f	lattice mismatch or *misfit*
f_{BE}	Bose-Einstein distribution function
f_{CV}, f'_{CV}	oscillator strengths for direct allowed and forbidden transitions
$f(\mathcal{E})$	Fermi-Dirac distribution function
f_o	noise corner frequency
f_S	force constant or stiffness factor
G	generation rate of carriers
G_C	channel conductance in photoconductor $(= 1/R_C)$
G_T	thermal generation rate
g, g_{th}	gain coefficient, threshold gain of laser
g_d, G_d	reverse-, forward-biased conductance of p-n diode
g_m	transconductance of amplifier
H	magnetic field
$H_{k''k'}$	matrix element for optical transitions
$h, \hbar$	Planck's constant and reduced Planck's constant
h_c	critical thickness of strained semiconductor
I	current
I_f	$1/f$ noise integral
$I_{P(1-3)}$	Personick integrals
I_s	reverse saturation current of diode
I_{SC}	short-circuit current of solar cell
$\mathfrak{I}$	light intensity
i_B	photocurrent due to background radiation
i_D, I_D	dark current in photodetector
i_{GR}	generation-recombination noise current
i_J	Johnson noise current
i_N	total noise current
i_{ph}, I_{ph}	photocurrent
i_S	signal photocurrent
J, J_{dr}, J_{diff}	current density, drift current density, diffusion current density
J_s	reverse saturation current density of diode
J_{th}	threshold current density of laser
J_{th}^o	transparency current density of laser
$\kappa, \kappa_{AB}, \kappa_r$	coupling constant for coupled modes
$\mathbf{k}, \mathbf{k_o}$	wavevector
$\mathbf{k_a}$	complex part of refractive index
k_B	Boltzmann's constant

k_r	electron-to-hole ionization coefficient ratio
$\mathcal{L}$	waveguide loss coefficient
L_D	Debye length
$L_{e(h)}$	diffusion length for electron (hole)
L_z	thickness of quantum well
l	laser cavity length
l_C	coupling length of directional coupler
$\mathbf{M}$	magnetic polarization
M_c	number of equivalent minima in conduction band
$M_{e(h)}$	multiplication coefficient for electrons (holes)
m	modulation index of ac signal
$m_{ce(h)}$	conductivity effective mass of electron (hole)
$m_{de(h)}$	density of states effective mass of electron (hole)
$m^*_{e(h)}$	electron (hole) effective mass
$m_{l(t)}$	longitudinal (transverse) effective mass
m_o	rest mass of electron
m^*_r	reduced effective mass
$N_{C(V)}$	effective density of states in conduction (valence) band
$N_{D(A)}$	donor (acceptor) density
$N(\mathcal{E})$	density of states function
$N_J(\mathcal{E})$	joint density of states
N_p	photon density per unit energy interval
N_{ST}	surface state density
N_T	trap density
n, n_o	density of electrons, density of electrons in thermal equilibrium in semiconductor
$\Delta n(p)$	density of excess carriers
$n_\mathcal{E}, n_v$	photon mode density
n_f	diode ideality factor
n_i	intrinsic carrier concentration in semiconductor
n_{nom}	injected carrier density that makes lasing medium transparent
n_r	refractive index of semiconductor
n_{th}	threshold carrier concentrating in laser
P	transition probability, optical power
$\mathbf{P}$	electric polarization
$P_{em(abs)}$	emission (absorption) probability
P_i	equilibrium vapor pressure of effusion species i in MBE
P_{inc}	incident optical power
P^C_{inc}	cross-over power of modulated barrier photodiode
P_o	power output of laser or LED
p, p_o	density of holes, density of holes in thermal equilibrium in semiconductor
p_{CV}	momentum matrix element for optical transition
Q	quality factor of Fabry-Perot cavity
q	electronic charge
R, R_S	volume recombination rate, surface recombination rate
R_{abs}	absorption rate

R_{sp}	total spontaneous emission rate
R_{st}	total stimulated emission rate
R_1, R_2	laser mirror reflectance
$\mathcal{R}$	responsivity of detector or light source
r_d	Bohr radius
r^l	linear electro-optic coefficient
$r_{sp}(\mathcal{E})$	spontaneous emission rate
$r_{st}(\mathcal{E})$	stimulated emission rate
$r(\omega)$	frequency response of LED
$\mathbf{S}$	Poynting vector
S_{ij}	stress
$S(v)$	lineshape function of spontaneous emission spectrum
$s_{e(h)}$	capture cross section of traps to electrons (holes)
s^q	quadratic electro-optic coefficient
s_R	surface recombination velocity
$\mathcal{T}$	transmission coefficient of coupler in coherent receiver
t	time
t_d	delay in laser response to large-signal modulation
t_{tr}	carrier transit time
U	alloy scattering potential
$u_\mathcal{E}, u_v$	number of photons per mode, or photon occupation
$V_a, V_{r(f)}$	applied bias, reverse (forward) bias
V_{bi}	built-in potential
V_{BR}	breakdown voltage of diode
V_d	diffusion potential
V_{FB}	flatband voltage of diode
V_{OC}	open circuit voltage of solar cell
V_{pt}	punchthrough voltage
V_π	bias voltage required to produce a phase shift of π radians in electro-optic modulator
$\vartheta_{e(h)}, \vartheta_{th}, \vartheta_D, \vartheta_{diff}$	electron (hole) velocity, thermal velocity of carriers, drift velocity, diffusion velocity
W	depletion layer width in a junction
Z	atomic number
α	absorption coefficient
$a_{e(h)}$	impact ionization coefficient for electrons (holes)
α_{fc}	free-carrier absorption coefficient
α_i^s	sticking coefficient
α_{np}	nonparabolicity parameter of band
α_T	current transfer ratio of transistor
β	mode propagation constant
β_r	compensation ratio
β_T	transistor current gain
$\beta_{TE(TM)}$	propagation constant of TE (TM) mode
Γ, Γ_m	confinement factor for propagating mode

Γ_G	optical gain
γ	loss coefficient in laser cavity or waveguide
γ_E	emitter injection efficiency in transistor
γ_N	parameter related to intensity fluctuations of local oscillator in a coherent receiver
$\epsilon_o, \epsilon_r, \epsilon_s, \epsilon_\infty$	permittivity of free space, relative dielectric constant, static dielectric constant, high-frequency dielectric constant
$\zeta(t)$	electron-hole recombination probability
η_c	coupling efficiency of optical fiber
η_d	differential quantum efficiency of laser
η_i, η_{ext}	internal, external quantum efficiency
$\eta_{in}, \eta_e, \eta_{o(P)}$	injection efficiency, extraction efficiency and overall power conversation efficiency of light-emitting diode
η_L	luminous efficiency of a light source
ηP	photoreceiver sensitivity
η_r	radiative efficiency
Θ_R	reflectivity
θ_a	acceptance angle of optical fiber
θ_c	critical angle
λ	wavelength of light
$\mu_{e(h)}$	mobility of electrons (holes)
ν	frequency
ν_{PR}	Poisson's ratio
ρ	resistivity of semiconductor
ρ_C	specific contact resistance
ρ_s	sheet resistance of ohmic contact
σ	conductivity of semiconductor
σ_i	impact ionization cross-section
σ_s	scattering cross section
σ_{ST}	strain-related material constant
τ, τ_r, τ_{nr}	lifetime, radiative lifetime, nonradiative lifetime
τ_C	relaxation time of carriers
$\tau_{e(h)}$	electron (hole) recombination lifetime
τ_{es}	escape time of carriers
τ_o	value of τ_r for large carrier density
Φ_o	photon flux
ϕ_b	potential barrier height
$\phi_{m(s)}$	metal (semiconductor) work function
$\varphi(\mathcal{E}), \varphi(\nu)$	radiation density or total photon density
$\bar{\varphi}$	average value of total photon density
χ^e	electric susceptibility
χ_s, χ_1, χ_2	semiconductor electron affinity
ω	angular frequency
ω_r	relaxation oscillation angular frequency of laser

Appendix 1

Important Properties of Common Semiconductors

Properties	Si	GaAs	$In_{0.53}Ga_{0.47}As$
Crystal structure	diamond	zinc blende	zinc blende
Atomic number	14		
Atoms/cm^3	5.02×10^{22}		
Electronic shell configuration	$1s^2 2s^2 2p^6 3s^2 3p^2$		
Atomic weight	28.09		
Lattice constant a (Å)	5.4309 (298.2°K)	5.6533 (300°K)	5.8687 (300°K)
Nearest neighbor distance $a\sqrt{3/4}$ (Å)	2.34	2.44	2.54
Melting point (°C)	1412	1237	
Breakdown field (V/cm)	$\sim 3.0 \times 10^5$	$\sim 4.0 \times 10^5$	
Density (g/cm^3)	2.329 (298°K)	5.3176 (298°K)	
Dielectric constant ($\epsilon_r = \epsilon_s$)	11.7	13.2 (300°K)	
Diffusion constant (cm^2/s)	37.5 (electrons) (300°K) 13 (holes) (300°K)	207 (electrons, at 300°K) 10 (holes, at 300°K)	
Effective density of states in the conduction band (cm^{-3})	2.9×10^{19} (300°K)	4.4×10^{17} (300°K)	
Effective density of states in the valence band (cm^{-3})	1.1×10^{19} (300°K)	8.2×10^{18} (300°K)	
Electron effective mass	density of states $1.18m_o$ (300°K)	$0.067m_o$(0°K) $0.063m_o$(300°K)	$0.045m_o$ (300°K)
Hole effective mass	hh $0.49m_o$(300° K) lh $0.16m_o$(300°K) density of states $0.81m_o$ (300°K)	density of states $0.53m_o$ (also see Table 2.1)	hh $0.465m_o$ (along [001]) $0.56m_o$ (along [110]) $0.60m_o$ (along [111]) lh $0.0503m_o$
Electron affinity (V)	4.01	4.07	
Energy gap (eV)	1.12 (300°K)	1.424 (300°K) 1.507 (77°K) 1.519 (0°K)	0.74 (300°K)
Index of refraction	3.42	3.66	
Intrinsic carrier concentration (cm^{-3})	1.02×10^{10} cm^{-3} (300°K)	5.0×10^6 (300°K)	
Intrinsic Debye length (μm)	24	2250 (300°K)	
Intrinsic resistivity (ohm-cm)	3.16×10^5 (300°K)	10^8 (300°K)	
Mobility (cm^2/V.s)	1450 (electron, at 300°K) 500 (hole, at 300°K)	8500 (electron, at 300°K) 400 (hole, at 300°K)	14,000 (electron, at 300°K), 100,000 (electron, at 77°K) 400 (hole, at 300°K)
Optical phonon energy (eV)	0.063	0.036	

Appendix 2

Dispersion Relation of a Diatomic Lattice

We will derive simple equations that predict the phonon energies as a function of their oscillation frequency. For simplicity, we will consider a one-dimensional lattice and assume that the vibrations of atoms along this lattice chain is coupled. This implies that the forced displacement of one atom spreads along the lattice as a plane wave. The lattice vibration (phonon) or random motion of the atoms can be looked upon as a superposition of many such plane waves:

$$U = \sum_n u = \sum_n A \, e^{j(\omega t + ka)} \tag{A2.1}$$

where a is the interatomic spacing, or lattice constant, and A is a constant. We are interested in the dispersion relation, ω versus k, for the lattice vibrations. Instead of analyzing the case of the monatomic chain, consisting of only one kind of atom, we will consider the more relevant diatomic chain, which contains atoms of two different chemical species, as in compound semiconductors, or ionic crystals. In terms of the crystalline structure, such materials have two (or more) atoms per primitive cell. Because of crystal symmetry, only the values of the wavevector in the first Brillouin zone need to be considered. The diatomic chain is illustrated in Fig. A2.1, where the two kinds of atoms are spaced distance a apart. Atoms of mass m are located at the odd-numbered lattice-points $2n - 1, 2n + 1 \cdots$, and atoms of mass M are located at the even-numbered lattice points $2n, 2n + 2 \cdots$. Here n is an integer and $M > m$. The behavior of elastic waves on a line of similar atoms is not different from that of elastic waves on a homogeneous line, provided the wavelength is large in comparison with the lattice constant. It is of interest to explore the motion of the atomic chain, if the wavelength is made shorter. Two simplifying assumptions are made: the displacement of the nth atom, x_n, from its equilibrium position is small, and Hooke's law is obeyed. The second condition implies that the restoring force is proportional to the displacement. The equation of motion for an atom of mass M, under the assumption of nearest neighbor interactions, is

$$M \ddot{x}_{2n} = -f_S(x_{2n} - x_{2n+1}) - f_S(x_{2n} - x_{2n-1})$$

The term on the left-hand side is the force acting on the $(2n)^{th}$ atom. The first term on the right is the increase in length of the bond between atoms $2n$ and $(2n + 1)$ and the second term on the right is the increase in length of the bond between atoms $2n$ and $(2n - 1)$. If both bonds increase in length, the two forces on atom $2n$ will be oppositely directed. f_S is the *force constant*

2n-1   2n 2n+1 2n+2 |←a→|←a→|

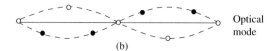

m   M   m   M   m   M   m   M

(a)

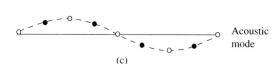

Optical
mode

(b)

Acoustic
mode

(c)

Figure A2.1 (a) A linear diatomic chain representing a solid possessing two atoms per primitive cell, (b) the transverse optical mode of vibration, and (c) the transverse acoustic mode of vibration.

or *stiffness factor* and is the restoring force per unit displacement. Combining the terms together in Eq. A2.2, we get

$$M\ddot{x}_{2n} = f_S(x_{2n+1} + x_{2n-1} - 2x_{2n}) \tag{A2.2}$$

Similarly, the equation of motion of the atoms of mass m is

$$m\ddot{x}_{2n+1} = f_S(x_{2n+2} + x_{2n} - 2x_{2n+1}) \tag{A2.3}$$

The plane-wave solutions to Eqs. A2.3 and A2.4 are of the form

$$x_{2n} = Ae^{j(\omega t + 2nka)} \tag{A2.4}$$

and

$$x_{2n+1} = Be^{j[\omega t + (2n+1)ka]} \tag{A2.5}$$

which lead, on substitution into Eqs. A2.3 and A2.4, to

$$-\omega^2 M A = f_S B(e^{jka} + e^{-jka}) - 2f_S A$$
$$-\omega^2 m B = f_S A(e^{jka} + e^{-jka}) - 2f_S B \tag{A2.6}$$

where B is another constant. A nontrivial solution exists for this set of homogeneous equations only if the determinants of the coefficients of the unknown quantities, A and B, vanish. In other words,

$$\begin{vmatrix} 2f_S - M\omega^2 & -2f_S\cos ka \\ -2f_S\cos ka & 2f_S - m\omega^2 \end{vmatrix} = 0 \tag{A2.7}$$

which leads to

$$\omega^2 = f_S\left(\frac{1}{m} + \frac{1}{M}\right) \pm f_S\left[\left(\frac{1}{m} + \frac{1}{M}\right)^2 - \frac{4\sin^2 ka}{Mm}\right]^{1/2} \tag{A2.8}$$

At $k = 0$, the two roots are

$$\omega^2 = 2f_S\left(\frac{1}{m} + \frac{1}{M}\right)$$ (A2.9)

and

$$\omega^2 = 0.$$ (A2.10)

For small k, the roots are

$$\omega^2 \cong 2f_S\left(\frac{1}{m} + \frac{1}{M}\right)$$ (A2.11)

and

$$\omega^2 \cong \frac{2f_S}{m+M}k^2a^2$$ (A2.12)

At $k = \pi/2a$, the roots are

$$\omega^2 = 2f_S/m$$ (A2.13)

and

$$\omega^2 = 2f_S/M.$$ (A2.14)

The desired dispersion relation, or the variation of ω with k, as expressed by Eq. A2.9, is shown in Fig. A2.2 for the case when $M > m$. There are two branches and these are called the *optical branch* and the *acoustic branch*. These can be understood more clearly by considering the motion of the two types of atoms shown in Fig. A2.1. The ratio of the amplitudes of the transverse waves associated with the two types of atoms is $\frac{A}{B}$. In the optical branch, from Eqs. A2.7 and A2.10, we get at $k = 0$

$$\frac{A}{B} = -\frac{m}{M}$$ (A2.15)

which implies that the two types of atoms vibrate against each other in such a way that the center of mass of the primitive cell remains fixed. Such motion of the atoms, if they are oppositely charged, can be excited by the electric field associated with an optical wave. The name of the branch is derived from this. The photon wavevector has a value $2\pi/\lambda$, which in the wavelength range of interest is much smaller than the limit $\pi/2a$. It may also be noted in Fig. A2.2 that wavelike solutions do not exist for energies between $\hbar(2f_S/M)^{1/2}$ and $\hbar(2f_S/m)^{1/2}$. The energy difference corresponds to the forbidden energy gap at the boundary of the first Brillouin zone, where Bragg reflection occurs. The value of k corresponding to frequencies in the forbidden gap are complex, indicating the existence of damped waves. The other solution at small values of k, from Eq. A2.11 gives

$$\frac{A}{B} = 1$$ (A2.16)

which implies that the two different atoms and their mass centers move together. This is the case in acoustic vibrations and therefore the corresponding dispersion relation is termed the acoustic branch.

Some important and relevant points may be noted here. The optical branch only exists in a diatomic lattice, such as GaAs, InP, and their derivative compounds. Indeed, if $M = m$, the

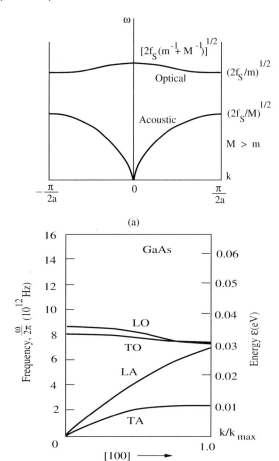

(a)

(b)

Figure A2.2 (a) Dispersion relation for the optical and acoustic modes of lattice vibration and (b) phonon spectra in GaAs (from J. L. T. Waugh and G. Dolling, *Physical Review*, **132**, 210, 1963).

optical branch coincides with the acoustic branch. Note that in this case the lattice constant becomes equal to $a/2$ and therefore the dimension of the first Brillouin zone increases by a factor of 2. In a polar crystal such as NaCl, there are ionic charges on the atoms and the lattice gets polarized in antiphase vibrations. A local space charge is created and we get polar vibrations in a polar crystal. It may be remembered that GaAs is a slightly ionic crystal and polar vibrations exist in the lattice. As the atomic vibration, or disturbance passes through the solid, the local periodicity of the lattice changes and the propagation of electrons through the lattice is affected. In fact, the local disturbance is incredibly powerful for scattering electrons and holes. The vibration passing through the crystal can be regarded as a particle that is called a *phonon*.

In the analysis just outlined we have considered a linear atomic chain. The vibrational spectrum and phonon dispersion curves in a three-dimensional real crystal are more complex. There are multiple nondegenerate and degenerate dispersion curves corresponding to transverse and longitudinal acoustic and optical waves. Also, the dispersion relations vary in the different crystal directions.

Appendix 3

The Fermi Integral and Carrier Concentration in Degenerate Semiconductors

The Fermi integral is given by Eq. 2.74

$$F_{1/2}(\eta) = \frac{2}{\sqrt{\pi}} \int_o^\infty \frac{x^{1/2} dx}{1 + \exp(x - \eta)}$$

where

$$x = \frac{\mathcal{E} - \mathcal{E}_C}{k_B T}, \quad \eta = \frac{\mathcal{E}_\Gamma - \mathcal{E}_C}{k_B T}$$

When $\eta \ll -1$,

$$F_{1/2}(\eta) \cong \exp(\eta) \tag{A3.1}$$

and when $\eta \gg 1$,

$$F_{1/2}(\eta) \cong \frac{4\eta^{3/2}}{3\sqrt{\pi}} \quad \text{(Problem 2.20)} \tag{A3.2}$$

For $-10 < \eta < 10$, the Fermi integral is given by[†]

$$F_{1/2}(\eta) = \exp\left(-0.32881 + 0.74041\eta - 0.045417\eta^2 - 8.797 \times 10^{-4}\eta^3 + 1.5117 \times 10^{-4}\eta^4\right) \tag{A3.3}$$

We need to formulate some relations to obtain the Fermi level from the carrier concentration. The Fermi distribution can be expanded into a power series

$$f(y) = y(1 - y + y^2 - y^3 \cdots) \tag{A3.4}$$

where $y = \exp[-(\mathcal{E} - \mathcal{E}_F)/k_B T]$. Therefore, for $\eta = (\mathcal{E}_F - \mathcal{E}_C)/k_B T < 0$, the electron concentration is

$$n = n_o\left[1 - \frac{e^\eta}{2\sqrt{2}} + \frac{e^{2\eta}}{3\sqrt{3}} - \frac{e^{3\eta}}{4\sqrt{4}} + \cdots\right] \tag{A3.5}$$

[†]M. Shur, *Physics of Semiconductor Devices*, Prentice Hall, Englewood Cliffs, NJ, 1990.

where, according to Eqs. 2.73 and 2.78,

$$n_o = Ae^{\eta} \tag{A3.6}$$

and

$$A = 2.5 \times 10^{19} \left(\frac{m_{de}}{m_o} \frac{T}{300} \right)^{3/2} \quad (cm^{-3}) \tag{A3.7}$$

For a highly degenerate semiconductor with $\eta \geq 5$, it can be shown that[†]

$$n = n'_o \left(1 + \frac{\pi^2}{12\eta^2} \right) \tag{A3.8}$$

where

$$n'_o = B\eta^{3/2} \tag{A3.9}$$

and from Eq. 2.110,

$$B = 1.9 \times 10^{19} \left(\frac{m_{de}}{m_o} \frac{T}{300} \right)^{3/2} \quad (cm^{-3}) \tag{A3.10}$$

The problem remains with the transition region $(-1 < \eta < 5)$. Usually, for $\eta \leq 0$, Eq. A3.5 is used with the first two or three terms. For $\eta \geq 1$ Eq. A3.8 is used. An equation that gives the carrier concentration as a function of the Fermi energy reasonably well in the range $-1 \leq \eta \leq 6$ follows from Eq. A3.5 and is

$$\eta = \ln \left(\frac{n}{N_C} \right) + 2^{-3/2} \left(\frac{n}{N_C} \right) \tag{A3.11}$$

This is also known as the Joyce-Dixon approximation.[‡]

[†]S. Wang, *Fundamentals of Semiconductor Theory and Device Physics*, Prentice Hall, Englewood Cliffs, NJ, 1989.
[‡]W. B. Joyce and R. W. Dixon, *Applied Physics Letters*, **31**, 354, 1977.

Appendix 4

Radiation Density and Photon Density

We will first derive an expression for the mode density in a frequency interval between ν and $\nu + d\nu$. We will calculate the number of modes in a cube of refractive index n_r and side L, assuming that L is much larger than the wavelength of the mode under consideration. As is usual, a propagating wave can be described by $\exp(j\mathbf{k} \cdot \mathbf{r})$ where

$$\mathbf{r} = x\mathbf{x} + y\mathbf{y} + z\mathbf{z} \tag{A4.1}$$

and

$$\mathbf{k}_x = \left(\frac{2\pi}{L}\right)l\mathbf{x}, \mathbf{k}_y = \left(\frac{2\pi}{L}\right)s\mathbf{y}, \mathbf{k}_z = \left(\frac{2\pi}{L}\right)m\mathbf{z} \tag{A4.2}$$

with

$$k^2 = k_x^2 + k_y^2 + k_z^2 = \left(\frac{2\pi}{\lambda}\right)^2 = \left(\frac{2\pi \nu n_r}{c}\right)^2 \tag{A4.3}$$

A combination of l, s, and m describes a mode. From Eq. A4.2 the volume associated with each mode in k-space is $(2\pi/L)^3$. Therefore the number of modes per unit volume of k-space in the frequency range 0 to ν (or having wavevectors in the interval 0 to k) is obtained by dividing the volume of a sphere with radius ν (or k) by the volume per mode $(2\pi/L)^3$. In other words,

$$N_\nu = \frac{8\pi^3 \nu^3 n_r^3 L^3}{3c^3 \pi^2}$$

$$= \frac{k^3 L^3}{3\pi^2} \tag{A4.4}$$

Note that there are two possible polarizations (TE and TM) per mode and therefore a factor of 2 is included in Eq. A4.4. Since the chosen volume of the enclosure is $V_\nu = L^3$, the number of modes per unit volume in the frequency range 0 to ν is

$$\frac{N_\nu}{V_\nu} = \frac{8\pi \nu^3 n_r^3}{3c^3} \qquad (cm^{-3}) \tag{A4.5}$$

Our initial assumption was that the enclosure size L is much larger than the wavelength of the enclosed mode. This implies a large mode density closely spaced in frequency. We can therefore obtain the mode density per unit frequency interval per unit volume between v and $v + dv$ by taking the derivative of Eq. A4.5 with respect to v. In other words

$$n_v = \frac{1}{V_v}\frac{dN_v}{dv} = \frac{8\pi v^2 n_r^3}{c^3} \qquad (s.cm^{-3}) \tag{A4.6}$$

Similarly, the mode density per unit energy interval is

$$n_{\mathcal{E}} = \frac{8\pi n_r^3 \mathcal{E}^2}{c^3 h^3} \qquad ((eV)^{-1}.cm^{-3}) \tag{A4.7}$$

This equation gives the number of electromagnetic modes with different l, s, m, per unit volume, having energy in the interval between $\mathcal{E}$ and $\mathcal{E} + d\mathcal{E}$.

The radiation density of the modes in a frequency interval dv is given by

$$\varphi(v)dv = vn_v u_v dv \tag{A4.8a}$$

and in an energy interval $d\mathcal{E}$ by

$$\varphi(\mathcal{E})d\mathcal{E} = hv n_{\mathcal{E}} u_{\mathcal{E}} d\mathcal{E} \tag{A4.8b}$$

The first term on the right-hand side of Eq. A4.8(b) is the energy per photon, the second term is the mode density, and the third term is the average number of photons per mode in the energy interval between $\mathcal{E}$ and $\mathcal{E} + d\mathcal{E}$. It is given by the Bose-Einstein distribution

$$u_{\mathcal{E}}d\mathcal{E} = \frac{1}{\exp(hv/k_B T) - 1}d\mathcal{E} \tag{A4.9}$$

Using Eqs. A4.6 and A4.9 the ideal blackbody radiation density, or the spectral radiation density at temperature T, is given by

$$\varphi(v) = \frac{8\pi n_r^3 v^3}{c^3}\frac{1}{\exp(hv/k_B T) - 1} \qquad (cm^{-3}) \tag{A4.10}$$

Similarly, in terms of energy

$$\varphi(\mathcal{E}) = \frac{8\pi n_r^3 \mathcal{E}^3}{c^3 h^3}\frac{1}{\exp(\mathcal{E}/k_B T) - 1} \qquad (cm^{-3}) \tag{A4.11}$$

Equations A4.10 or A4.11 are the mathematical statements of Planck's radiation law. The *photon density* per unit energy interval $N_p = n_{\mathcal{E}} u_{\mathcal{E}}$. In other words,

$$N_p = \frac{8\pi n_r^3 \mathcal{E}^2}{c^3 h^3}\frac{1}{\exp(\mathcal{E}/k_B T) - 1} \qquad ((eV)^{-1}.cm^{-3}) \tag{A4.12}$$

Appendix 5

Parameters for Uniformly Doped Abrupt GaAs Junction at 300°K[†]

————

†From G. L. Miller, D. V. Lang and L. C. Kimmerling,"Capacitance Transient Spectroscopy" in *Annual Reviews of Materials Science,* **7**, 387, 1977.

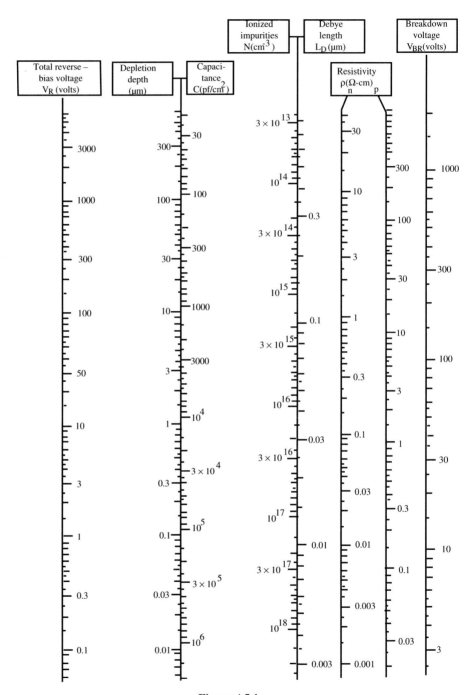

Figure A5.1

Appendix 6

Frequency Response of Light-Emitting Diode

The intrinsic limit to the frequency response of an LED is caused by charge storage and diffusion capacitance of the p-n junction. These phenomena are ultimately linked to the minority carrier lifetime. For our analysis we assume a one-dimensional case (Fig. A6.1) in which electrons are injected into a p-type active region of width d. Therefore, the p-n junction is at $x = 0$ (which might also be a heterojunction) and excess carriers are assumed to recombine with a lifetime τ throughout thickness d. The narrow depletion layer width in a forward-biased junction is neglected. The excess electron density $n(x)$ in the active region is lost by spontaneous recombination and diffusion. Therefore, when injection ceases, the rate of change of the excess carriers is given by

$$\frac{\partial n(x)}{\partial t} = -R_{sp} + D_e \frac{\partial^2 n}{\partial x^2} \tag{A6.1}$$

where the current due to drift is neglected. If we assume that $R_{sp} = \frac{n(x)}{\tau}$ (where $\tau \sim \tau_o \cong 0.5\text{ns}$), then

$$\frac{\partial n(x)}{\partial t} = -\frac{n(x)}{\tau} + D_e \frac{\partial^2 n}{\partial x^2} \tag{A6.2}$$

In the case of small-signal modulation of the diode a time-dependent current is superimposed on a dc bias. Therefore, the excess carriers will have a steady-state and a frequency-dependent

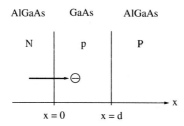

Figure A6.1   One-dimensional geometry of heterojunction LED.

513

component. In other words,

$$n(x, t) \; = \; n_o(x) + n_1(x)e^{j\omega t} \tag{A6.3}$$

Substituting Eq. A6.3 into Eq. A6.1 and separating dc and frequency-dependent parts, we can write

$$D_e \frac{\partial^2 n_o}{\partial x^2} \; - \; \frac{n_o}{\tau} \; = \; 0 \tag{A6.4a}$$

$$D_e \frac{\partial^2 n_1}{\partial x^2} \; - \; \frac{n_1(1 + j\omega\tau)}{\tau} \; = \; 0 \tag{A6.4b}$$

By analogy with the definition $L_e = (D_e\tau)^{1/2}$ we define $L_e(\omega) = [D_e\tau/(1 + j\omega\tau)]^{1/2}$. Therefore, Eq. A6.4(b) can be rewritten as

$$\frac{\partial^2 n_1(x)}{\partial x^2} \; - \; \frac{n_1(x)}{L_e^2(\omega)} \; = \; 0 \tag{A6.5}$$

The frequency response of the LED, $r(\omega)$ is defined as

$$r(\omega) \; = \; \frac{\phi_1(\omega)}{J_1(\omega)/q} \tag{A6.6}$$

where the numerator is the ac photon flux and the denominator is the electron flux. J_1 is the ac current density. At the p-n junction, where it is assumed that only electrons cross the boundary, $n_1(x) = n_{10}$. It is also assumed that $d \gg L_e$, so that $n_1(d) = 0$. Under these boundary conditions the solution of Eq. A6.5 is given by

$$n_1(x) \; = \; n_{10}e^{-x/L_e(\omega)} \tag{A6.7}$$

Then

$$\phi_1(\omega) = \frac{1}{\tau} \int_o^d n_1(x)dx$$

$$\cong \frac{n_{10}L_e(\omega)}{\tau} \tag{A6.8}$$

and

$$J_1(\omega) = qD_e \frac{\partial n_1(x)}{\partial x}$$

$$= -qD_e \frac{n_{10}}{L_e(\omega)} \tag{A6.9}$$

Substituting Eqs. A6.8 and A6.9 in Eq. A6.6 we get

$$r(\omega) \; = \; \frac{|L_e(\omega)|^2}{\tau D_e} \; = \; \frac{1}{[1 + \omega^2\tau^2]^{1/2}} \tag{A6.10}$$

The same result can be derived when d is finite, but there is a heterojunction barrier at $x = d$.

Appendix 7

Propagation Modes in a Symmetric Planar Slab Waveguide

This problem has been solved by many authors.[†] The three-layer waveguide structure is shown in Fig. 6.3. The confining layers, with indices of refraction n_{r1} and $n_{r3}(= n_{r1})$ are assumed to extend to infinity in the $+x$ and $-x$ directions, respectively. For the case of TE plane waves traveling in the z-direction with propagation constant β the solutions can be expressed as

$$E_y'(x, z, t) = E_y(x)e^{j(\omega t - \beta z)} \tag{A7.1}$$

where the transverse function $E_y(x)$ has the general form

$$E_y(x) = \begin{cases} Ae^{-sx} & 0 \leq x \leq \infty \\ B\cos(hx) + C\sin(hx) & -t_g \leq x \leq 0 \\ D\,e^{p(x+t_g)} & -\infty \leq x \leq -t_g \end{cases} \tag{A7.2}$$

$A, B, C, D, s, h,$ and p are all constants that can be determined by matching the boundary conditions, which require the continuity of E_y and $H_z = (j/\omega\mu)\partial E_y/\partial x$. Since the permeability μ and frequency ω are assumed to be constant, $A, B, C,$ and D can be determined by making E_y and $\partial E_y/\partial x$ continuous at the boundary between region 1 and region 2 ($x = 0$), and E_y continuous at $x = -t_g$. Note that in Fig. 6.3 the guide thickness $t_g = 2d$. The solution for E_y can be expressed in terms of a single constant C'

$$E_y(x) = \begin{cases} C'e^{(-sx)}, & 0 \leq x \leq \infty \\ C'[\cos(hx) - (s/h)\sin(hx)], & -t_g \leq x \leq 0 \\ C'[\cos(ht_g) + (s/h)\sin(ht_g)]e^{p(x+t_g)} & -\infty \leq x \leq -t_g \end{cases} \tag{A7.3}$$

$s, h,$ and p can be obtained using Eqs. A7.1 and A7.3 and the Maxwell's equation for the three regions giving

$$s = (\beta^2 - n_{r1}^2 k^2)^{1/2},$$

$$h = (n_{r2}^2 k^2 - \beta^2)^{1/2},$$

$$p = (\beta^2 - n_{r3}^2 k^2)^{1/2},$$

$$k = \omega/c = 2\pi/\lambda \tag{A7.4}$$

[†]See, for example, H. F. Taylor and A. Yariv, *Proc. IEEE*, **62**, 1044, 1974.

Taking $\partial E_y / \partial x$ from Eq. A7.3 and making it continuous at $x = -t_g$ yields the condition

$$\tan(ht_g) = \frac{p + s}{h(1 - ps/h^2)} \tag{A7.5}$$

which is a transcendental equation. In case of the three-layer planar asymmetric waveguide, $n_{r3} \gg n_{r1}$. Of course, n_{r2} must still be greater than n_{r3} if waveguiding is to occur. For the symmetric waveguide case $n_{r1} = n_{r3}$. Therefore, at cut-off, when the fields become oscillatory in regions 1 and 3,

$$\beta = k n_{r1} = k n_{r3} \tag{A7.6}$$

Substituting Eq. A7.6 into A7.4 we find that $p = s = 0$ and

$$h = k(n_{r2}^2 - n_{r1}^2)^{1/2} = k(n_{r2}^2 - n_{r3}^2)^{1/2} \tag{A7.7}$$

and substituting Eq. A7.7 into Eq. A7.5 we get the condition

$$\tan(ht_g) = 0 \tag{A7.8}$$

or

$$ht_g = m\pi, m = 0, 1, 2, 3, \ldots \tag{A7.9}$$

Thus, for waveguiding of a given mode to occur, we must have

$$\Delta n_r = (n_{r2} - n_{r1}) > \frac{m^2 \lambda^2}{16(n_{r1} + n_{r2})d^2}, m = 0, 1, 2, \ldots \tag{A7.10}$$

where $t_g = 2d$ has been substituted. The three lowest order TE modes of the symmetric guide are shown in Fig. 6.3.

Appendix 8

Modes in an Asymmetric Slab Waveguide

Reference is made to the discussion in Appendix 7. If one calculates and compares the modes of a symmetric waveguide ($n_{r1} = n_{r3}$) with an asymmetric waveguide of half the thickness, then it is seen that for well-confined modes the lower half of the ($m = 1$) mode of the symmetric guide corresponds to the $m = 0$ mode of the asymmetric guide. This fact is used to calculate the mode cut-off condition for the asymmetric guide. For a symmetric waveguide of thickness $2t_g = 4d$, the condition is, from Eq. 7.10,

$$\Delta n_r = n_{r2} - n_{r3} > \frac{m^2 \lambda^2}{4(n_{r2} + n_{r3})(2t_g)^2}, \quad m = 0, 1, 2, 3, \cdots \tag{A8.1}$$

However, the asymmetric waveguide supports only those modes that correspond to the odd modes of a symmetric guide of twice its thickness. Therefore, for an asymmetric guide of thickness $t_g = 2d$

$$\Delta n_r = \frac{m'^2 \lambda^2}{64(n_{r2} + n_{r3})d^2} \tag{A8.2}$$

where m' can only take odd values. Substituting

$$m' = (2m + 1), \quad m = 0, 1, 2, \cdots \tag{A8.3}$$

the cut-off condition is

$$\Delta n_r = n_{r2} - n_{r3} > \frac{(2m + 1)^2 \lambda^2}{64(n_{r2} + n_{r3})d^2} \tag{A8.4}$$

Appendix 9

Signal-to-Noise Ratio in Balanced Detection System

In a dual-detector balanced detection scheme,[†] a weak received signal is combined with a stronger local oscillator signal. The two beams are normally aligned in polarization so that the fields add. The inputs to the beam-splitter, shown in Fig. A9.1 are

$$E_S = \sqrt{2}\hat{E}_S e^{j(\omega_1 t + \phi_1)}$$

$$E_L = \sqrt{2}\hat{E}_L e^{j(\omega_2 t + \phi_2)} \tag{A9.1}$$

where the average signal and local oscillator powers are E_S^2 and E_L^2, respectively. E_S and E_L are assumed to be real. If the beam-splitter is lossless, linear, and symmetric, then the output fields $\hat{E}_1$ and $\hat{E}_2$ are given by

$$\begin{bmatrix} \hat{E}_1 \\ \hat{E}_2 \end{bmatrix} = e^{j\phi_r} \begin{bmatrix} \sqrt{1-\mathcal{T}} & \sqrt{\mathcal{T}}e^{j\pi/2} \\ \sqrt{\mathcal{T}} & \sqrt{1-\mathcal{T}} \end{bmatrix} \begin{bmatrix} \hat{E}_S \\ \hat{E}_L \end{bmatrix} \tag{A9.2}$$

where $\mathcal{T}$ is the transmission coefficient of the beam splitter and ϕ_r is the phase shift of the reflected field. The photocurrent from a detector can be expressed by

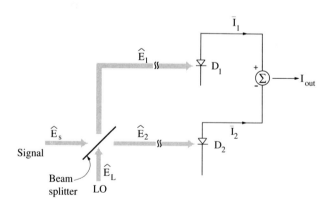

Figure A9.1 Scheme for balanced detection.

[†]See G. L. Abbas et al., *Journal of Lightwave Technology*, **LT-3**, 1110, 1985.

$$\bar{I} = \frac{\eta q}{\hbar \omega} \cdot \frac{1}{2} |\hat{E}|^2$$

$$= \mathcal{R} \cdot \frac{1}{2} |\hat{E}|^2 \tag{A9.3}$$

where $\mathcal{R}$ is the responsivity of the detector. Therefore, for detector D_1

$$\bar{I}_1 = \frac{\eta_1 q}{\hbar \omega} \left[(1 - T) E_S^2 + T E_L^2 + 2\sqrt{T(1-T)} E_S E_L \cos \left\{ (\omega_1 - \omega_2)t + \phi_1 - \phi_2 - \frac{\pi}{2} \right\} \right] \tag{A9.4}$$

and for detector D_2

$$\bar{I}_2 = \frac{\eta_2 q}{\hbar \omega} \left[T E_S^2 + (1 - T) E_L^2 + 2\sqrt{T(1-T)} E_S E_L \cos \left\{ (\omega_1 - \omega_2)t + \phi_1 - \phi_2 + \frac{\pi}{2} \right\} \right] \tag{A9.5}$$

The dc components of the photocurrents are

$$I_{1,dc} = \frac{\eta_1 q}{\hbar \omega} T E_L^2 \left[\frac{1-T}{T} \left(\frac{E_S}{E_L} \right)^2 + 1 \right] \tag{A9.6}$$

and

$$I_{2,dc} = \frac{\eta_2 q}{\hbar \omega} (1 - T) E_L^2 \left[\frac{T}{1-T} \left(\frac{E_S}{E_L} \right)^2 + 1 \right] \tag{A9.7}$$

If $E_L \gg E_S$, as is usually the case,

$$I_{1,dc} = \frac{\eta_1 q}{\hbar \omega} T E_L^2 \tag{A9.8}$$

$$I_{2,dc} = \frac{\eta_2 q}{\hbar \omega} (1 - T) E_L^2 \tag{A9.9}$$

The total noise power can be expressed as

$$N = 2q B (I_{1,dc} + I_{2,dc}) + 2q \gamma_N B (I_{1,dc} - I_{2,dc})^2 \tag{A9.10}$$

where the first term on the right is the shot noise and the second term is the excess noise. γ_N is a parameter related to the intensity fluctuations of the local oscillator. Substituting Eqs. A9.8 and A9.9 into A9.10,

$$N = \frac{2q^2 E_L^2 B}{\hbar \omega} \left[\eta_1 T + \eta_2 (1 - T) + \frac{q \gamma_N}{\hbar \omega} E_L^2 \{ \eta_1 T + \eta_2 (1 - T) \}^2 \right] \tag{A9.11}$$

The ac photocurrents at the difference frequency are

$$I_1 = \frac{2\eta_1 q}{\hbar \omega} \sqrt{T(1-T)} E_S E_L \cos \left\{ (\omega_1 - \omega_2)t + \phi_1 - \phi_2 - \frac{\pi}{2} \right\} \tag{A9.12}$$

$$I_2 = \frac{2\eta_2 q}{\hbar \omega} \sqrt{T(1-T)} E_S E_L \cos \left\{ (\omega_1 - \omega_2)t + \phi_1 - \phi_2 + \frac{\pi}{2} \right\} \tag{A9.13}$$

The total photocurrent is

$$I = I_1 - I_2$$

$$= \frac{2q}{\hbar \omega} \sqrt{T(1-T)} E_S E_L (\eta_1 + \eta_2) \cos \left\{ (\omega_1 - \omega_2)t + \phi_1 - \phi_2 - \frac{\pi}{2} \right\} \tag{A9.14}$$

The signal-to-noise ratio is

$$\frac{S}{N} = \frac{\frac{1}{2}|I|^2}{N} \tag{A9.15}$$

Substituting Eqs. A9.11 and A9.14 into A9.15 we get

$$\frac{S}{N} = \frac{\frac{1}{2}\frac{4q^2}{(\hbar\omega)^2}T(1-T)E_S^2 E_L^2(\eta_1+\eta_2)^2}{\frac{2q^2 E_L^2 B}{\hbar\omega}\left[\eta_1 T + \eta_2(1-T) + \frac{q\gamma_N}{\hbar\omega}E_L^2\{\eta_1 T + \eta_2(1-T)\}^2\right]} \tag{A9.16}$$

which simplifies to

$$\frac{S}{N} = \frac{\frac{1}{\hbar\omega}T(1-T)(\eta_1+\eta_2)^2 E_S^2}{B\left[T\eta_1 + (1-T)\eta_2 + \frac{q\gamma_N}{\hbar\omega}E_L^2(\eta_1 T + \eta_2(1-T))^2\right]} \tag{A9.17}$$

As a special case, if $T = 0.5$ and the quantum efficiencies of the two detectors are equal, $\eta_1 = \eta_2 = \eta$, then

$$\frac{S}{N} = \frac{\mathcal{R}E_S^2}{qB} \tag{A9.18}$$

which is independent of the local oscillator power.

Appendix 10

Electroabsorption in Biaxially Strained Quantum Wells

Shown below in Fig. A10.1 are the calculated transition energies of the $e_1 - hh_1$ and $e_1 - lh_1$ transitions in lattice-matched and biaxially strained 90 Å $In_{0.53 \pm x}Ga_{0.47 \mp x}As/InP$ quantum wells. It is clear that, compared to the lattice-matched case, the difference between the transition energies at any applied transverse electric field *increases* in the case of compressive strain and decreases in the case of tensile strain. In fact, by approximating the heavy-hole and light-hole excitonic resonances as δ-functions, it can be shown that $\Delta n_r \sim |\mathcal{E}_{lh} - \mathcal{E}_{hh}|$ (Problem 3.9). The latter separation has a quadratic dependence on electric field. Therefore Δn_r increases quadratically with field and the difference between lattice-matched, compressively strained, and tensilely strained cases is obvious.

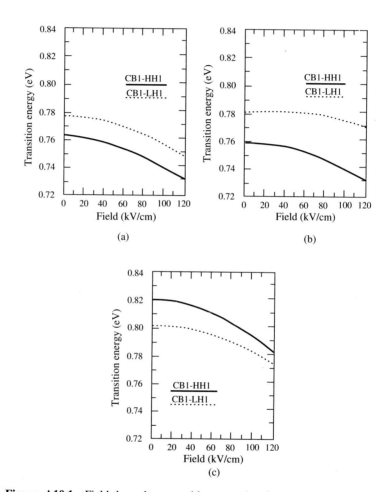

Figure A10.1 Field-dependent transition energies for (a) lattice-matched, (b) compressively strained, and (c) tensilely strained $In_{0.53\pm x}Ga_{0.47\mp x}As/InP$ quantum wells. The well width is 90 Å (courtesy of J. Singh and J. P. Loehr, University of Michigan).

Appendix 11

Power Flow in Dual-Channel Directional Coupler

In terms of the complex amplitudes of the two plane waves flowing in a linear directional coupler shown in Fig. A11.1, the coupled mode equations (Eq. 7.56) can be written in terms of the complex amplitudes a_1 and a_2 as

$$\frac{da_1}{dz} = -j\beta a_1(z) - jka_2(z)$$

$$\frac{da_2}{dz} = -j\beta a_2(z) - jka_1(z) \qquad \text{(A11.1)}$$

Here it is assumed that the propagation constants in the two guides are equal (i.e., $\beta_1 = \beta_2 = \beta = \beta_r - j\frac{\gamma}{2}$). β_r is the real part of β and γ is the loss coefficient. If it is assumed that light is coupled into guide 1 at $z = 0$, the boundary conditions are

$$a_1(0) = 1$$

$$a_2(0) = 0 \qquad \text{(A11.2)}$$

Then the solutions to Eq. A11.1 are described by

$$a_1(z) = \cos(kz)e^{j\beta z}$$

$$a_2(z) = -j\sin(kz)e^{j\beta z} \qquad \text{(A11.3)}$$

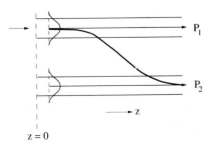

Figure A11.1 Schematic of mode propagation in a dual-channel coupler.

The power flow in the two guides are then given by

$$P_1(z) = a_1 a_1^* = \cos^2(kz)e^{-\gamma z}$$

$$P_2(z) = a_2 a_2^* = \sin^2(kz)e^{-\gamma z} \tag{A11.4}$$

The power flow in the two guides is schematically shown in Fig. A11.2.

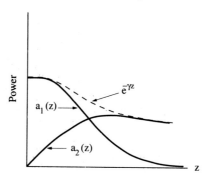

Figure A11.2 Schematic illustration of power flow in a dual channel coupler with equal propagation constants.

Index

Q

R

S

U

T

V